Fourier, Hadamard, and Hilbert Transforms in Chemistry

Fourier, Hadamard, and Hilbert Transforms in Chemistry

Edited by
Alan G. Marshall
The Ohio State University
Columbus, Ohio

Plenum Press • New York and London

Library of Congress Cataloging in Publication Data

Main entry under title:

Fourier, Hadamard, and Hilbert transforms in chemistry.

Bibliography: p.
 Includes index.
 1. Fourier transform spectroscopy. I. Marshall, Alan G., 1944-
QD96.F68F68   543'.0858   81-20984
ISBN 0-306-40904-6   AACR2

© 1982 Plenum Press, New York
A Division of Plenum Publishing Corporation
233 Spring Street, New York, N.Y. 10013

All rights reserved

No part of this book may be reproduced, stored in a retrieval system, or transmitted
in any form or by any means, electronic, mechanical, photocopying, microfilming,
recording, or otherwise, without written permission from the Publisher

Printed in the United States of America

Dedicated
to
the Memory of

Willis H. Flygare

PREFACE

In virtually all types of experiments in which a response is analyzed as a function of frequency (e.g., a spectrum), transform techniques can significantly improve data acquisition and/or data reduction. Research-level nuclear magnetic resonance and infra-red spectra are already obtained almost exclusively by Fourier transform methods, because Fourier transform NMR and IR spectrometers have been commercially available since the late 1960's. Similar transform techniques are equally valuable (but less well-known) for a wide range of other chemical applications for which commercial instruments are only now becoming available: for example, the first commercial Fourier transform mass spectrometer was introduced this year (1981) by Nicolet Instrument Corporation. The purpose of this volume is to acquaint practicing chemists with the basis, advantages, and applications of Fourier, Hadamard, and Hilbert transforms in chemistry. For almost all chapters, the author is the investigator who was the first to apply such methods in that field.

The basis and advantages of transform techniques are described in Chapter 1. Many of these aspects were understood and first applied by infrared astronomers in the 1950's, in order to improve the otherwise unacceptably poor signal-to-noise ratio of their spectra. However, the computations required to reduce the data were painfully slow, and required a large computer. In 1965, Cooley and Tukey published a fast Fourier transform algorithm that reduced the computation time by a factor of $N/\log(N)$, making it possible to transform data sets of useful size (say, 8192 points) in an acceptably short time (about 10 sec for 8192 points, compared to about 6 hours with the conventional algorithm). On-line FT computations could thus be done by a minicomputer. Moreover, analog-to-digital converters with suitable speed (50,000 words/sec) and accuracy (12-bit per word) were available, and commercial stand-alone FT-IR and FT-NMR spectrometers were produced within 3 years.

Application of transform methods to other experiments required additional theoretical and/or technical developments, and thus occurred later. For example, FT-rotational spectrometry (first demonstrated in 1974) used the same pulse excitation as FT-NMR, but required a much wider spectral range and thus faster analog-to-digital converters (e.g., 100 MHz at 1 bit per word). Different excitation waveforms were necessary for FT ion cyclotron resonance mass spectrometry (frequency-sweep, 1973), FT faradaic admittance (pseudo-random noise, 1977), and FT dielectric relaxation (voltage step, 1975). Two-dimensional FT NMR (1975) and the conceptually similar FT ENDOR experiment (1972) required significant theoretical groundwork. FT mu spin resonance analysis (the mu atom is a very light isotope of the hydrogen atom) followed very soon after wide use of the μSR technique itself (ca. 1975). Hadamard transform methods were developed independently (again, first for IR applications) about 1968. Hilbert transform techniques were first applied to spectroscopy (in this case for NMR) in 1978.

Chapter 1 (Marshall) presents a general basis for describing the advantages of Fourier, Hadamard, and Hilbert transform techniques in acquisition of data, enhancement of desired spectral features (e.g., signal-to-noise or resolution), and suppression of instrumental artifacts (deconvolution of imperfect excitation). The next three chapters introduce general aspects of these three transform types. Chapter 2 (Sloane) describes the construction and properties of Hadamard "codes", which are perhaps the conceptually simplest "multiplex" spectrometric method. Chapter 3 (Dumoulin & Levy) contains practical details in applying the fast Fourier transform algorithm to large data sets using a computer with limited memory. Chapter 4 (Marshall) provides the first review of applications for the "DISPA" (dispersion versus absorption) data reduction based on the Hilbert transform, as a means for identifying and distinguishing between spectral line-broadening mechanisms.

The succeeding chapters are loosely organized in order of increasing frequency of the spectrum of interest. Beginning at low frequency (0-2 MHz), Chapter 5 (Comisarow) describes the FT mass spectrometry technique, and gives some very recent applications showing the advantages of the technique for unraveling complex mass spectra at high ionic masses. At somewhat higher radiofrequency, Chapter 6 (Klainer et al.) reviews the state of the art in development and applications of FT nuclear quadrupole resonance spectrometers, and suggests a number of areas for immediate future work. Chapter 7 (Cole & Winsor) gives a history of advances in time-domain reflectometry as a source of dielectric relaxation data, including a review of all detection methods in current use. Chapter 8 (Flygare) is a comprehensive treatment of the theory and experimental complications of FT rotational spectroscopy--this technique appears especially promising for spectra of weakly associated van der Waals molecules.

Chapter 9 (Morris) critically compares the advantages of the principal two-dimensional FT-NMR experiments, and includes a comprehensive literature survey through mid-1980. Chapter 10 (Mims) gives a brief review of the somewhat parallel development of electron spin echo FT spectroscopy. Chapter 11 (Dalal) critically compares the

PREFACE

advantages of the "double-resonance" approach of Chapter 10 and direct FT-NMR for radicals containing quadrupolar nuclei. Chapter 12 introduces FT methods in mu spin resonance spectrometry, an area new to many chemists, but with implications important in the study of isotope effects. Chapter 13 (de Haseth) reviews the continuing development of the FT-IR technique, including the new areas of photo-acoustic spectroscopy, gas- and liquid-chromatography/FT-IR, and silicon impurity determinations. Chapter 14 (Nordstrom) extends interferometric detection to the optical frequency range, with proposed applications in atomic absorption spectroscopy.

The remaining two chapters offer somewhat different approaches. Chapter 15 (Smith) details various possible pseudorandom sequences as spectral sources over the frequency range used for a.c. polarography, and gives several applications of this quite new technique. The final Chapter 16 (McCreery & Rossi) describes a new FT technique which was first published (by those authors) this year (1981), in which Fourier transformation of the pattern of light diffracted near the edge of an electrode is used to discover the concentration profile of electroactive species very near to the electrode.

This volume is a successor to "Transform Techniques in Chemistry", edited by Peter R. Griffiths three years ago (Plenum, 1978). Since virtually all the work described in the present volume has been produced in just those intervening three years, the accelerating increase in use of transform techniques by chemists is obvious. Future directions are expected to include more uses of hard-wired fast Fourier transform processors (a 1024-point FFT can now be performed in about 15 millisec), array processors (especially for two-dimensional Fourier transforms), and use of other excitation waveforms (e.g., pseudo-random excitation in FT mass spectrometry). The market share of FT-spectrometers can be expected to increase in all areas in which commercial FT-instruments are available.

The editor wishes to thank all various contributors for providing authoritative, up-to-date, critical summaries of the state of the art in various transform techniques. The manuscripts were converted to their present form by Alan Marshall and Dixie Fisher. Finally, this volume is dedicated in recognition of the late Professor W. H. Flygare, who contributed his own comprehensive chapter at a time when he was already very ill.

CONTENTS

Advantages of Transform Methods in Chemistry.........................1
   A.G. Marshall

Hadamard and Other Discrete Transforms in Spectroscopy.............45
   N.J.A. Sloane

Processing Software for Fourier Transform Spectroscopies...........69
   C.L. Dumoulin and G.C. Levy

Dispersion versus Absorption (DISPA): Hilbert Transforms
     in Spectral Line Shape Analysis..............................99
   A.G. Marshall

Fourier Transform Ion Cyclotron Resonance Spectroscopy............125
   M.B. Comisarow

Fourier Transform Nuclear Quadrupole Resonance Spectroscopy.......147
   S.M. Klainer, T.B. Hirschfeld, and R.A. Marino

Fourier Transform Dielectric Spectroscopy.........................183
   R.H. Cole and P. Winsor, IV

Pulsed Fourier Transform Microwave Spectroscopy...................207
   W.H. Flygare

Two-Dimensional Fourier Transform NMR Spectroscopy................271
   G.A. Morris

Endor Spectroscopy by Fourier Transformation of the
     Electron Spin Echo Envelope.................................307
   W.B. Mims

Advances in FT-NMR Methodology for Paramagnetic Solutions:
     Detection of Quadrupolar Nuclei in Complex Free
     Radicals and Biological Samples.............................323
   N.S. Dalal

Fourier Transform μSR..345
    J.H. Brewer, D.G. Fleming, and P.W. Percival

Fourier Transform Infrared Spectrometry...........................387
    J.A. de Haseth

Aspects of Fourier Transform Visible/UV Spectroscopy..............421
    R.J. Nordstrom

Fourier Transform Faradaic Admittance Measurements (FT-FAM):
    A Description and Some Applications..........................453
    D.E. Smith

Optical Diffraction by Electrodes: Use of Fourier
    Transforms in Spectroelectrochemistry........................527
    R.L. McCreery and P. Rossi

List of Contributors..549

Index...551

ADVANTAGES OF TRANSFORM METHODS IN CHEMISTRY

Alan G. Marshall

Departments of Chemistry and Biochemistry
The Ohio State University
140 W. 18th Avenue
Columbus, OH   43210

INTRODUCTION

Transform techniques offer three main advantages for chemists. First, transform techniques provide a variety of simple procedures for manipulating digitized data: smoothing or filtering to enhance signal-to-noise ratio; resolution enhancement (via either narrower line width or more points per line width); changing spectral line shapes (as from Lorentzian to Gaussian); generation of a dispersion spectrum from an absorption spectrum; generation of integrals or derivatives; and clipping to reduce data storage requirements. Second, Fourier methods can be used to remove any known irregularities in the excitation waveform, so that the corrected ("deconvoluted") response reflects only the properties of the sample, and not the effect of the measuring instrument. Third, "coded" or "multiplex" detection, followed by Fourier or Hadamard "decoding" can offer a multiplex or Fellgett advantage of up to $\sqrt{N}$ in signal-to-noise ratio (or 1/N in time) compared to a scanning instrument, where N is the number of data points in the frequency spectrum.

Absorption and dispersion spectra: steady-state frequency-response

It is useful to begin by reviewing the origin, form, and detection of absorption and dispersion spectra. All the necessary aspects can be demonstrated from the simple mechanical analog (Figure 1) of a weight of mass m, suspended from a spring of force constant k, subject to frictional resistance f, and driven by an external force F oscillating at angular frequency ω:[1]

$$m \frac{d^2x}{dt^2} = -kx - f \frac{dx}{dt} + F_0 \cos(\omega t) \qquad [1]$$

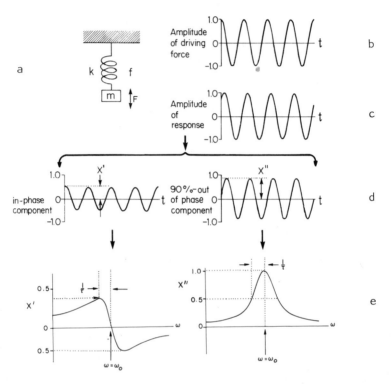

Figure 1. Motion of a driven, damped weight on a spring. (a) Mass m suspended from spring of force constant k and frictional coefficient f, driven by a sinusoidal force $F_0 \cos(\omega t)$. (b) Driving force amplitude versus time. (c) Steady-state displacement of driven mass versus time--note that displacement is in general not exactly in step with driver. (d) Displacement analyzed into components that are exactly in-phase or 90°-out-of-phase with driver. (e) Amplitudes (i.e., maximum displacements) of the components in (d), for various choices of driving frequency, ω.

Advantages of Transform Methods in Chemistry

The steady-state displacement <u>response</u> (Figure 1c) to a continuous sinusoidally time-varying driver <u>excitation</u> (Figure 1b) always oscillates at the same <u>frequency</u> as the driver, but not usually with the same <u>phase</u>; i.e., the curve in Figure 1c is somewhat displaced in time from that of Figure 1b. The total displacement (Figure 1c) can always be analyzed into components of amplitude x' and x" that are exactly in-phase or 90°-out-of-phase with the driver (Figure 1d), in much the same spirit that a vector in a plane is analyzed into its x- and y-components. If the steps shown in Figure 1b-d are repeated for various choices of driving frequency, ω, then the component amplitudes x' and x" vary with driving frequency as shown in Figure 1e.

The spectroscopic generality of the analysis in Figure 1 is that for relatively <u>small</u> displacement, x, virtually <u>any</u> driving force,

$$F(x) = a_0 + a_1 x + a_2 x^2 + \cdots \qquad [2]$$

can be represented by just the first two terms of Equation 2. a_0 can be eliminated by choosing a suitable reference frame, leaving a force of the form,

$$F(x) = a_1 x = -kx \qquad [3]$$

which is analogous to the restoring force of a mechanical spring. For example, although an electron may be bound to an atom by a Coulomb attraction, the displacement of the electron by the oscillating electric field of a light wave can be represented by the model of Figure 1, providing that the electron displacement is sufficiently small.

The reason for analyzing the displacement into in-phase and 90°-out-of-phase components is that their amplitude spectra represent the variation of refractive index and power absorption with frequency of the incident radiation. The "dispersion" spectrum is so named because it is the variation of refractive index with frequency that leads to the spreading out ("dispersion") of white light by a prism.

Figure 1e clearly shows that the same information is available from either the dispersion or absorption spectrum. The "natural" or "resonant" frequency

$$\omega_0 = \left(\frac{k}{m}\right)^{1/2} \qquad [4]$$

is obtained from the <u>midpoint</u>, and the frictional resistance is resistance is available from the <u>width</u> of either spectrum:

$$\frac{1}{\tau} = \frac{f}{2m} \qquad [5]$$

The spectral line position thus gives information about the system (i.e., the spring strength and mass), while the line width defines the strength of interaction of the spring with its surroundings (via the frictional coefficient).

It is worth noting that the dispersion-mode frequency spectrum is not simply the derivative,

$$\frac{d\,A(\omega)}{d\omega} = \text{absorption-mode derivative spectrum.} \qquad [6]$$

of the absorption-mode spectrum. Although the dispersion and the absorption-mode derivative spectra have qualitatively similar appearance, their line widths and line shapes are distinctly different. For the Lorentzian line shape of Figure 1e, for example, Figure 2 shows that the peak-to-peak separation for the absorption derivative is smaller than for the dispersion by a factor of $1/\sqrt{3}$.[1] Experimentally, the absorption-mode derivative spectrum is the usual display mode in steady-state ESR spectroscopy (see Chapter 4), and is becoming more popular in steady-state optical spectroscopy.[2]

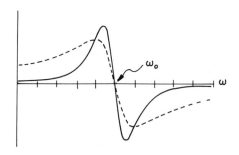

Figure 2. Dispersion (-----) spectrum and corresponding absorption-mode derivative (———) spectrum. Both spectra have been computed from the Lorentzian line shapes of Figure 1e. The line shape differences are clearly evident.

The model based on Equation 1 leads to the fundamental absorption and dispersion line shapes (Figure 1e) of spectroscopy. In addition, various chemically important relaxation phenomena can be modeled by setting m = 0 in Equation 1 (driven, damped, massless weight-on-a-spring):

$$f \frac{dx}{dt} + kx = F_0 \cos(\omega t) \qquad [7]$$

The steady-state displacement may again be analyzed into components in-phase and 90°-out-of-phase with the driver as in Figure 1, to give the plots shown in Figure 3. The mathematical line shapes are very similar to those obtained in Figure 1e, except that the curves are now centered at zero frequency, and the width is now given by

$$\frac{1}{\tau} = \frac{f}{k} \qquad [8]$$

Experimentally, these line shapes appear in steady-state plots of dielectric or ultrasonic susceptibility versus frequency. The in-phase and 90°-out-of-phase dielectric spectral amplitudes can be combined in a method based on Hilbert transforms (see Chapter 4).

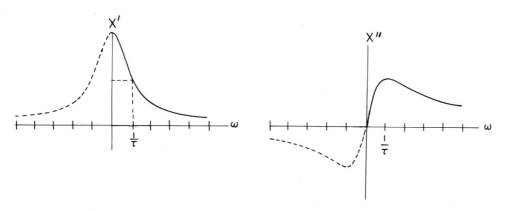

Figure 3. Amplitudes, x' and x", of the in-phase and 90°-out-of-phase components of the steady-state displacement of a frictionally damped, massless spring driven by a sinusoidal force, $F_0 \cos(\omega t)$.

Real and complex notation

Physically, x' and x" are properly identified as the in-phase and 90°-out-of-phase steady-state amplitudes defined in Figure 1d. In other words, for a mathematically <u>real</u> driving force,

$$F(t) = F_0 \cos(\omega t), \qquad [9]$$

the mathematically <u>real</u> solution to the <u>real</u> Equation [1] or [7] can be expressed,

$$x(t) = x' \cos(\omega t) + x'' \sin(\omega t) \qquad [10]$$

However, in solving Equations [1] or [7], it is <u>mathematically</u> convenient to add an <u>imaginary</u> term to the driving force to give

$$F(t) = F_0 \cos(\omega t) + i F_0 \sin(\omega t)$$

$$= F_0 \exp[i\omega t] \qquad [11]$$

and then solve the resulting <u>complex</u> Equation [1] or [7] to obtain the <u>complex</u> displacement,

$$X = \mathcal{X} \exp[i\omega t] \qquad [12]$$

It is then straightforward to show that the <u>complex</u> amplitude can be written

$$\mathcal{X} = x' - ix'' \qquad [13]$$

The final result of these manipulations is that the <u>real part</u> of (complex) X is simply

$$Re(X) = Re[\mathcal{X} \exp[i\omega t)]$$

$$= Re[(x' - ix'')(\cos(\omega t) + i \sin(\omega t)]$$

$$= x' \cos(\omega t) + x'' \sin(\omega t) \qquad [14]$$

Therefore, the <u>real</u> part of the <u>complex</u> solution to the <u>complex</u> form of Equation [1] or [7] is the same as the <u>real</u> solution of the <u>real</u> form of Equation [1] or [7]. The two main advantages of complex notation are (a) simpler algebra in solving Equations [1] or [7], and (b) automatic separation of the in-phase and 90°-out-of-phase components (as the real and imaginary parts of a complex amplitude).

It is therefore common to refer to x' and x" as the mathematically "real" and "imaginary" parts of a "complex" quantity, $\mathcal{X}$, even though x' and x" clearly represent physically (and mathematically) <u>real</u> in-phase and 90°-out-of-phase amplitudes of a <u>real</u> displacement.

Advantages of Transform Methods in Chemistry

Transient time-domain response to impulse excitation

Historically, most of the spectral responses discussed in the succeeding chapters first came into general use in the form of a steady-state response (usually absorption-mode rather than dispersion-mode) to a "continuous-wave" oscillating driving force. More recently, the same information has come to be extracted from the time-domain response of the same system to a sudden impulse.

Consider the same weight-on-a-spring systems of Equations [1] or [7], but this time in the absence of any driving force:

$$m \frac{d^2x}{dt^2} + f \frac{dx}{dt} + kx = 0 \quad\quad [1a]$$

or
$$f \frac{dx}{dt} + kx = 0 \quad\quad [7a]$$

If the spring is initially at rest (i.e., $x = 0$), nothing happens. But if the spring is stretched initially to $x = x_0$ by a sudden pull (impulse excitation), then the spring displacement, x, will keep changing until friction eventually damps its motion back to zero:

$$x = x_0 \exp[-t/\tau] \cos(\omega_0 t), \quad \frac{1}{\tau} \equiv \frac{f}{2m} \quad\quad [15]$$

or $\quad x = x_0 \exp[-t/\tau], \quad \frac{1}{\tau} \equiv \frac{f}{k} \quad\quad [16]$

as shown in Figure 4.

For the mass-on-a-spring (Figure 4a), we can discover the "natural" spring frequency simply by counting the number of spring oscillations per second. Moreover, we can extract the same damping constant, $1/\tau$, from Figure 4a or 4b as from steady-state experiments on the same systems in Figures 1 or 3. In other words, we can discover the natural frequency of a tuning fork, either by humming at it until we find the resonant pitch, or by striking it and listening to its natural oscillation.

Whenever the same parameters are available from two different curves (e.g., ω_0 and τ from Figure 1 or Figure 4a), there is some mathematical relation between the curves. For the "linear" system we have considered (i.e., displacement is proportional to driving amplitude F_0) the time-domain and frequency-domain responses are connected by a Fourier transform. Similarly, absorption and dispersion spectra both yield the same information, and are related by a Hilbert transform (see Chapter 4). In this Chapter, we will next develop some simple Fourier transform properties for continuous curves such as Figures 1-4, and then show the advantages of applying similar relations to discrete data sets consisting of actual physical responses sampled at equally-spaced intervals.

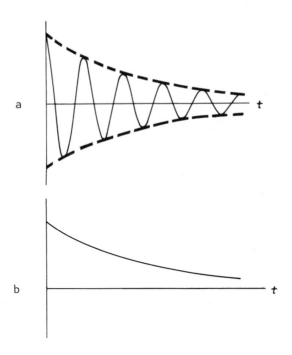

Figure 4. Transient displacement of a mass m suspended from a frictionally damped spring driven by a sinusoidally time-varying force, $F_0 \cos(\omega t)$. (a) $m \neq 0$; (b) $m = 0$.

FOURIER TRANSFORM PAIRS: A PICTORIAL LIBRARY

In the previous section, we established a correspondence between the <u>transient time-domain</u> response (exponentially damped cosine wave) to a sudden "<u>impulse</u>" excitation and the <u>steady-state frequency-domain</u> response (Lorentzian absorption and dispersion spectra) to a <u>continuous</u> excitation. The Fourier transform may be thought of as the mathematical <u>recipe</u> for going from the time-domain to the frequency-domain. In this section, we shall introduce the mathematical forms of the transforms, along with pictorial examples of several of the most important signal shapes.

In physical applications, Fourier transforms are commonly used to connect a <u>single</u> time-domain signal, $f(t)$, to <u>five</u> different kinds of spectra, according to the following mathematical recipes.[3]

$$A(\omega) = \frac{1}{\pi} \int_{-\infty}^{\infty} f(t) \cos(\omega t)\, dt = \text{ABSORPTION SPECTRUM} \qquad [17]$$

$$D(\omega) = \frac{1}{\pi} \int_{-\infty}^{\infty} f(t) \sin(\omega t)\, dt = \text{DISPERSION SPECTRUM} \qquad [18]$$

$$F(\omega) = \frac{1}{2\pi} \int_{-\infty}^{\infty} f(t)\, e^{-i\omega t}\, dt = \text{COMPLEX SPECTRUM} \qquad [19]$$

$$M(\omega) = \left\{ [A(\omega)]^2 + [D(\omega)]^2 \right\}^{1/2} = \text{MAGNITUDE SPECTRUM} \qquad [20]$$

$$2|F(\omega)|^2 = \tfrac{1}{2}\left\{ [A(\omega)]^2 + [D(\omega)]^2 \right\} = P(\omega)$$
$$= \text{POWER SPECTRUM} \qquad [21]$$

The "inverse" transforms for connecting the frequency-domain amplitudes to the time-domain signal are very similar:

$$f(t) = \int_{-\infty}^{\infty} [A(\omega) \cos(\omega t) + D(\omega) \sin(\omega)t]\, d\omega \qquad [22]$$

$$= \int_{-\infty}^{\infty} F(\omega)\, e^{+i\omega t}\, d\omega \qquad [23]$$

A major property from Equations 17-23 is that if a forward Fourier transform [e.g., conversion of $f(t)$ to $A(\omega)$ and $D(\omega)$] is followed by an inverse transform, the successive integrals must be multiplied by a net factor of $(1/2\pi)$ in order to give back the original $f(t)$. We have chosen to introduce the factor of $(1/2\pi)$ in Equation 19; another convention is to use a factor of $(1/2\pi)^{1/2}$ for each of the forward and inverse transforms. Both conventions (and others) are in common use, as discussed in detail in Reference 3.

Absorption, dispersion, magnitude ("absolute-value"), and power spectra for several physically important waveforms are shown in Figure 5. For lower frequencies, for which phase-sensitive detection is available (e.g. FT-NMR, FT-NQR, FT-microwave), absorption-mode is usually the preferred display, because it is narrowest and most symmetrical. At higher frequencies or in cases where the time-domain waveform consists of random or pseudo-random noise (see below), only the magnitude or power spectrum is available. It is possible to generate many other useful waveforms from this library and use of the "convolution" theorem (see next Section).

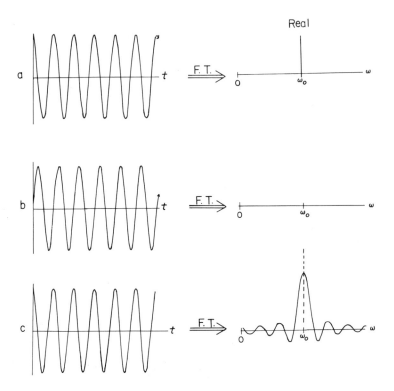

Figure 5. Pictorial library of Fourier transform pairs. The left-most curves represent time-domain signals. The frequency-domain spectra corresponding to each time-domain signal are shown at right. The time-domain curves are: (a) infinitely long cosine wave; (b) infinitely long sine wave; (c) cosine square wave of duration, T.

Advantages of Transform Methods in Chemistry

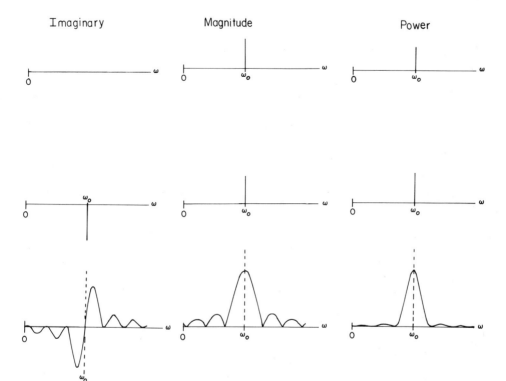

Sinusoid ⇌ Delta-function

Figures 5a and 5b show that the Fourier transform of an infinitely persisting time-domain cosine or sine wave is a spike in the freqeuncy domain. Because the time-domain sine wave is 90"-out-of-phase with respect to the time-domain cosine, their Fourier transforms appear as "real" or "imaginary" spikes, respectively, in accord with our previous claim that complex notation serves to keep the in-phase and 90°-out-of-phase components separated as "real" and "imaginary" parts of a complex quantity.

Square wave ⇌ Sinc function

In Figure 5c, we observe the same time-domain cosine wave as in Figure 5a, but for only a finite period, T sec. The result is that the frequency spectrum is now broadened from an infinitely sharp spike to a signal whose frequency width is of the order of (1/T) Hz. This result is an example of a classical "uncertainty principle": the product of the time-domain width (T) and the frequency-domain width (1/T) is constant. In other words, the only way to determine the frequency of a time-domain signal with perfect accuracy (i.e., infinite frequency "resolution") is to observe it for an infinite length of time.

In fact, it is quite generally true that the narrower the signal in one domain, the wider will be the signal in the Fourier transform domain. The simple picture of Figure 5c thus shows that a sufficiently narrow square pulse can serve as a radiation source over a broad frequency bandwidth (i.e., the central region of the "real" spectrum of Figure 4c). For example, a 10 μsec square pulse is equivalent to irradiation with essentially flat amplitude over about $0.1 \times (1/10^{-5}) = 10$ kHz. Thus, in a typical proton FT-NMR experiment, a simple square pulse can be used to excite NMR signals over the usual range of ^{1}H NMR chemical (frequency) shifts.

A further important conclusion follows from Figure 5d. Since all experimental signals eventually die away, attempts to improve experimental resolution by increasing the observation period must ultimately end in acquiring only noise for most of the (latter) part of the experiment. Thus, if the noise is unrelated to the signal (see below), the classical uncertainty principle translates into the experimental result that the product of signal-to-noise ratio and resolution is fixed. One of the great advantages of Fourier transforms in spectroscopy is this capability to increase resolution at the expense of signal-to-noise ratio without any mechanical adjustments to the spectrometer, simply by increasing the length of the observation period. Fourier transform spectrometers thus effectively operate with continuously variable exit slit width--to a limit approaching zero, so that resolution is unaffected by the measuring instrument.

Advantages of Transform Methods in Chemistry

Exponential ⇌ Lorentzian

Figure 5d confirms that the Fourier transform of an infinitely decreasing exponential time-domain signal gives the familiar Lorentzian absorption and dispersion frequency-domain line shapes derived in the previous Section. Again, the faster the exponential decay (i.e., the narrower the time-domain signal), the broader is the frequency-domain line width.

The time-domain traces of Figures 5c and 5d represent two extremes for experimental measurement. In Figure 5c, the signal is observed for such a short period that the signal does not decay at all during the observation, while in Figure 5d, the signal decays completely during the observation period. Actual signals will thus display frequency-domain line shapes intermediate between the "sinc" function of Figure 5c and the Lorentzian of Figure 5d (see below).

Gaussian ⇌ Gaussian

Figure 5e shows the remarkable result that the Fourier transform of a Gaussian (time-domain signal) is also a Gaussian (frequency-domain signal). This property can be especially useful in manipulating spectral line shapes (see Apodization, below).

Frequency-sweep ⇌ Bandwidth function

Figure 5f shows the Fourier transform of the time-domain signal corresponding to a conventional slow scan through a range of frequencies. In the limit that the scan rate is infinitely slow, the frequency-domain magnitude or power spectrum is simply a constant amplitude over the scanned frequency range, with zero amplitude elsewhere. In the illustrated example, the scan rate is finite, so that the frequency-domain power spectrum is not perfectly flat, but shows some variation in amplitude over the nominal scan range. [Since the phase angle accumulates quadratically as the frequency of the time-domain signal is linearly increased during the sweep, the absorption- and dispersion-mode spectra exhibit wild variations in amplitude with frequency.[4]]

Random noise ⇌ "White spectrum"

Figure 5g shows an example of time-domain random noise. Since the phase of the signal can be taken as random at any given instant, the result of averaging many such traces will be to give absorption and dispersion spectra that are zero at all frequencies. Intuitively, a trace giving positive absorption at a given frequency will, on the average, cancel another trace giving equal and opposite (negative) absorption; a more formal treatment confirms this result (Ref. 1, Chapter 21). However, the magnitude spectrum represents a root-mean-square average noise, and is non-zero with constant amplitude up

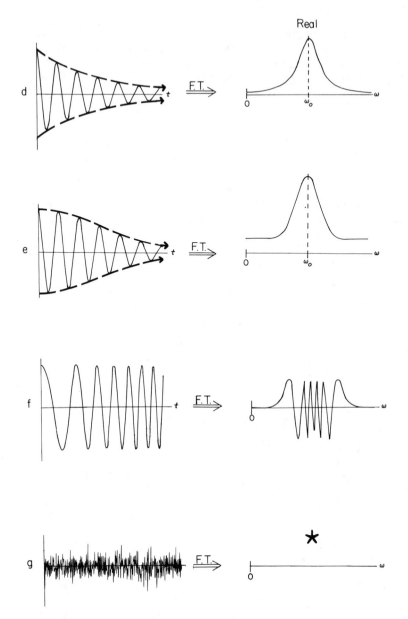

Figure 5. Pictorial library of Fourier transform pairs, continued. (d) infinitely decreasing exponential cosine; (e) infinitely decreasing Gaussian cosine; (f) frequency-sweep; (g) random noise. * indicates spectra averaged over many experiments.

Advantages of Transform Methods in Chemistry

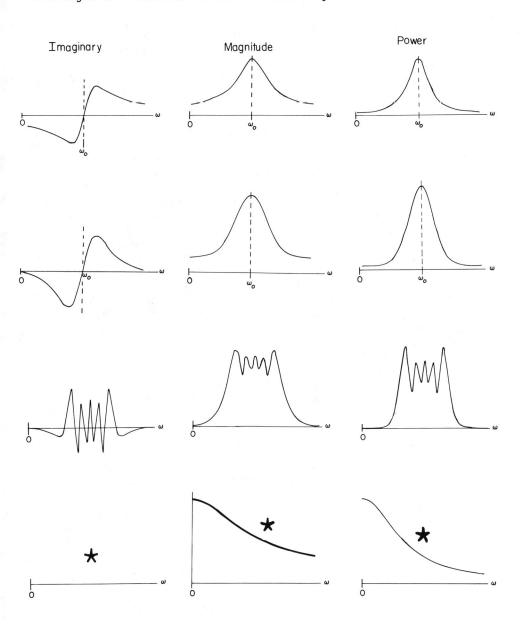

to a characteristic frequency, ω_c, that is inversely related to the time constant, τ_c, corresponding to the average time period between random changes in the time-domain signal amplitude or phase:

$$\omega_c = \frac{1}{\tau_c} \qquad [24]$$

τ_c is the "correlation time" for the random process.

Figure 5g thus demonstrates that random noise can serve as a spectral radiation source, provided that the time-domain fluctuations in electric or magnetic field are sufficiently rapid to span the frequency range of interest. Random noise resulting from molecular motion or chemical reactions leads to so-called "induced" transitions in spectroscopy (e.g., T_1 and T_2 in magnetic resonance spectroscopy), in the absence of any externally applied fields. Alternatively, a spectrum can be excited by externally applied electric or magnetic fields whose amplitude or phase varies randomly with time (see Smith Chapter), in so-called "stochastic" excitation.[5] Finally, an excitation consisting consisting of a series of pulses of pseudo-random spacing, phase, or amplitude can be used to construct an excitation spectrum of arbitrary shape, so that 1 or 2 or more spectral "windows" can be excited without irradiating the remainder of the spectrum.[6]

Causal functions: origin of dispersion spectrum

All physical time-domain waveforms are causal; that is, the function is defined starting at time zero. It is this causal aspect that results in a dispersion as well as an absorption spectrum, as will now be explained.

A causal function can always be analyzed into even and odd components,

$$f_{even}(-t) = f_{even}(t) \qquad [25a]$$

$$f_{odd}(-t) = - f_{odd}(t) \qquad [25b]$$

as shown in Figure 6 for the function,

$$f(t) = \exp(-t), \qquad 0 \leq t < \infty \qquad [26]$$

Next, it should be obvious that

$$\int_{-\infty}^{\infty} f_{even}(t)\, dt = 2 \int_{0}^{\infty} f_{even}(t)\, dt \qquad [27a]$$

and $\quad \int_{-\infty}^{\infty} f_{odd}(t)\, dt = 0 \qquad [27b]$

Advantages of Transform Methods in Chemistry

Finally, since $\cos(\omega t)$ is even and $\sin(\omega t)$ is odd, and since[1]

Even·Even	= Even	[28a]
Even·Odd	= Odd	[28b]
Odd·Odd	= Even	[28c]

it is clear that the dispersion spectrum is introduced by the need to include an <u>odd</u> component in representing any <u>causal</u> time-domain function.

For the interferometry experiment (see de Haseth and Nordstrom Chapters), it is in fact possible to detect both halves of the "interferogram" to produce a theoretically even function, whose Fourier transform has no dispersion component. However, it is usual to detect only about half the interferogram (see below), so that a dispersion component is introduced after all.

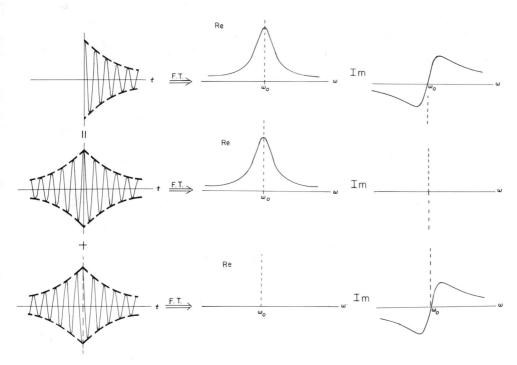

Figure 6. Fourier transform (top right) of a causal function (top left), and its even (middle plots) and odd (lowermost plots) components, showing how the dispersion signal arises from the odd component (see text).

Circularly polarized vs. linearly polarized signals

Up to now, we have considered only the positive-frequency half of the various spectra available from Fourier transformation of a time-domain signal. Although physical measurements are indeed conducted at positive frequencies, there can nevertheless be contributions to the positive-frequency spectrum from the negative-frequency region. The most common effects from the negative-frequency region appear when linearly polarized radiation is used to excite motion that is naturally circular. Time-domain signals arising from circular rather than linear motion are found in ion cyclotron resonance (Chapter 5), nuclear quadrupole resonance (Chapter 6), pure rotational spectroscopy (Chapter 8), nuclear magnetic resonance (Chapters 9, 11), electron spin resonance (Chapter 10) and mu spin resonance (Chapter 12).

The exponentially damped cosine again provides a simple illustration of these effects. Figure 7c shows the full cosine Fourier transform (absorption-mode spectrum) at both negative and positive frequencies for a damped cosine representing a linearly polarized electric or magnetic field. Figures 7b and 7c follow our previous convention of representing in-phase and 90°-out-of-phase components as real and imaginary parts of a complex number, to give an absorption spectrum corresponding to a field that is left- or right-circularly polarized. Figures 7a to 7c also show that a linearly-polarized signal may be analyzed into a sum of left- and right-circularly polarized components.

In Figures 7a to 7c, the natural frequency ω_0 of the circular motion is large compared to the width of the absorption signal:

$$\omega_0 \gg \frac{1}{\tau} \qquad [29]$$

in which τ again denotes the time constant for exponential damping of the time-domain signal. In this limit (i.e., narrow peaks at high natural frequency), the peaks from the negative-and positive-frequency regions of the absorption spectrum of Fig. 7a do not overlap. However, for broader peaks at low natural frequency (Fig. 7d),

$$\omega_0 \lesssim \frac{1}{\tau} \qquad [30]$$

the negative-frequency peak extends into the positive-frequency region and contributes to the observed positive-frequency spectrum, as shown in Figure 7c. Thus, it is common to consider only one of the circularly-polarized components of a linearly-polarized signal in most NMR, NQR, ICR, μSR, and pure rotational experiments, because the limit of Equation 28 is usually satisfied. For magnetic resonance at low magnetic fields,[7] or for excitation waveforms approaching d.c.,[4] it may be necessary to consider the negative-frequency region.

Advantages of Transform Methods in Chemistry

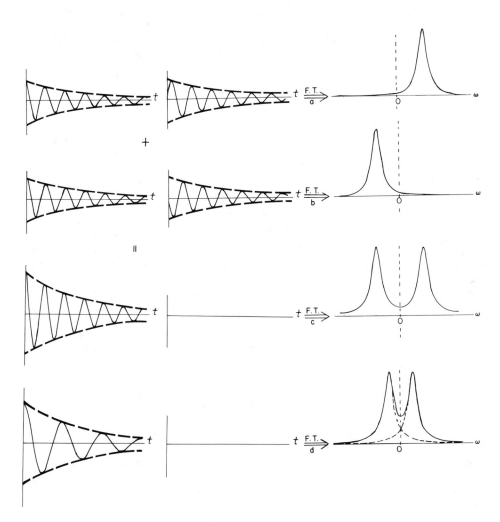

Figure 7. Absorption-mode spectra (right) of linearly and circularly polarized time-domain signals (left). (a) x- and y-components of a field right-circularly polarized about z. (b) Same as (a) for a left-circularly polarized field. (c) Same as (a) for a field linearly polarized along x. (d) Same as (c), for a signal with the same relaxation time (i.e., same frequency-domain line width), but smaller natural frequency--note overlap (see text).

APODIZATION: CHANGING THE SHAPE OF A SPECTRUM

In previous sections, we examined several physically important noise-free signals. [We did briefly consider the effect of noise as a radiation source, but did not consider noise contributions to the observed response to an excitation.] In the absence of noise, a signal of any shape can be analyzed to determine its parameters (e.g., spectral line position, width, area, etc.). However, noise superimposed on a signal can obscure its information content, and it may therefore become desirable to sacrifice one kind of information (e.g., resolution) in order to improve the quality of other information (e.g., signal-to-noise ratio). When an already acquired signal is modified before Fourier transformation, the modification is called apodization (literally, "removal of feet", named after early efforts to smooth FT/IR line shapes--see de Haseth Chapter).

Enhancement of signal-to-noise ratio or resolution

Generalizing from the library examples of Figure 5, we recognize that the longer a time-domain signal is acquired, the narrower is the corresponding frequency-domain spectral line, and the better is the spectral resolution. However, if the signal decreases with time, while the noise level remains constant with time, it follows that the signal-to-noise (S/N) ratio decreases with longer acquisition period. The trade-off between S/N ratio and resolution is therefore simple and direct in Fourier transform spectroscopy: S/N ratio is optimized using short acquisition period, and resolution is optimized using long acquisition period.

Once a given set of (say) time-domain data points has already been acquired, it is still possible to enhance either S/N ratio or resolution. In order to enhance S/N ratio, we need simply weight the initial time-domain data points more than those near the end of the acquisition period. Convenient weight functions include the boxcar truncation of Figure 5c and the exponentially decreasing weight function of Figure 5d. Either weight function will broaden the width of the spectral lines obtained by Fourier transforming the weighted time-domain signal as shown in Figure 5c or 5d.

For example, if the original time-domain signal is an exponentially damped cosine (Figure 8a), then multiplying the time-domain signal by the weight function, $\exp(-t/\tau_0)$, will increase the absorption-mode line width by $2/\tau_0$ s^{-1}, or about $0.6/\tau_0$ Hz, for τ_0 in sec. Figure 8b shows that the effect of this apodization is to enhance the S/N ratio at the expense of degrading the resolution. Conversely, if the original signal is multiplied by $\exp(+t/\tau_0)$, then Figure 8c shows that the resulting frequency-domain absorption line width is now narrower by about $0.6/\tau_0$ Hz (i.e., resolution is improved), but has poorer signal-to-noise ratio.

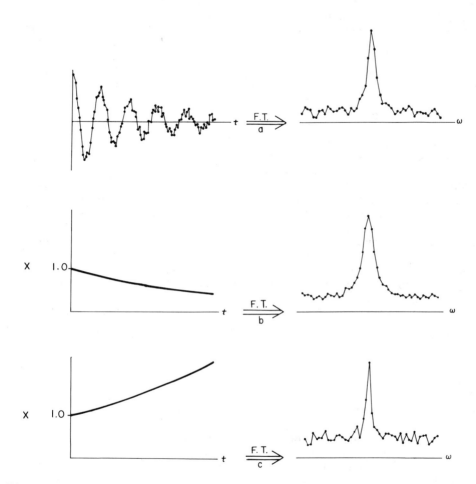

Figure 8. Signal-to-noise or resolution enhancement produced by apodization of a time-domain transient signal. (a) Exponentially damped cosine, $\exp(-t/\tau)\cos(\omega_0 t)$, with added noise. (b) Curve (a), multiplied by a weight factor, $\exp(-t/2\tau)$, before Fourier transformation. (c) curve (a), multiplied by a weight factor, $\exp(+t/2\tau)$, before Fourier transformation.

Change in line shape: conversion of Lorentzian to Gaussian

One problem with the Lorentzian line shape that characterizes many Fourier transform experimental spectra arises from the broad "tailing" of the absorption amplitude (see Figure 5d). Overlap between several neighboring peaks can thus affect the apparent resolution within the summed envelope. A Gaussian peak, on the other hand, is broader at the top, but its amplitude drops off rapidly starting at about 1 line width away from the peak center. Thus, one might hope to improve the apparent resolution in an envelope of many component Lorentzian peaks, by converting each Lorentzian line to a a Gaussian shape.

The Lorentz-to-Gauss conversion can be accomplished with two successive apodizations, as illustrated in Figure 9. The method begins from a time-domain data set consisting of a sum of damped cosines (all with the same damping constant) of the form shown in Figure 5d. This time-domain data is first apodized with an increasing exponential weight function, whose time constant is chosen to be equal to the time constant of the original damping, so that the apodized time-domain signal is effectively converted from the shape of Figure 5d to the box shape of Figure 5c. If the signal were Fourier transformed at this stage (Figure 9c), resolution would be enhanced, but at the expense of signal-to-noise ratio. However, if the time-domain data set is subjected to a second apodization consisting of a Gaussian weight function (Figure 5e), the Fourier transform of the doubly-apodized transient will have Gaussian peak shapes (Figure 9d). The parameters of the two apodizations in Figure 9 were chosen so that the final Gaussian spectrum effectively improved the signal-to-noise ratio without sacrificing resolution (compare Figures 9a and 9d).

Filtering

Filters are most commonly used to reduce high-frequency fluctuations in a slowly varying signal (low-pass filter, as in NMR spectra), or to eliminate slowly varying baseline drift in a spectrum with sharp features (high-pass filter, as in Raman spectroscopy). Electronic filters are used to perform these functions at stages before the signal is recorded; filtering of the already-acquired signal is known as digital filtering. A typical digital filtering procedure is to Fourier transform the original data, then suppress either the low-frequency part (high-pass filter) or high-frequency part (high-pass filter) of the frequency spectrum, and then inverse Fourier transform to give a filtered spectrum. For example, Figure 10 shows digital filtering to eliminate the (slowly varying) baseline from a Raman spectrum. The spectrum is first Fourier transformed, then apodized to eliminate all high-frequency components, so that subsequent inverse Fourier transformation yields just the baseline, which is then subtracted from the original spectrum.

Advantages of Transform Methods in Chemistry

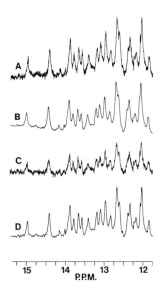

Figure 9. Apodizations of an experimental FT-NMR signal. (A) Fourier transform of the original unweighted free induction decay (F.I.D.) time-domain response to a 90°-pulse excitation. (B) Signal-to-noise enhancement: F.I.D. weighted by the factor, $\exp(-\pi \cdot LB \cdot t)$, with LB = 3.0 Hz, before F.T. (C) Resolution enhancement: F.I.D. weighted as in (B), but with LB = -0.5 Hz. (D) Exponential apodization (LB=-3.5) followed by Gaussian weighting by $\exp(-bt^2)$, with $b = -a/(2 \cdot GB \cdot AQ)$, for GB = 0.05 and AQ = 0.8 sec acquisition period. 400 MHz ^1H FT-NMR spectra acquired by the author (297 K, one-pulse suppression of H_2O, 1000 transients for 1 mM sample of E. coli tRNA^{Val} kindly provided by Prof. Brian Reid.

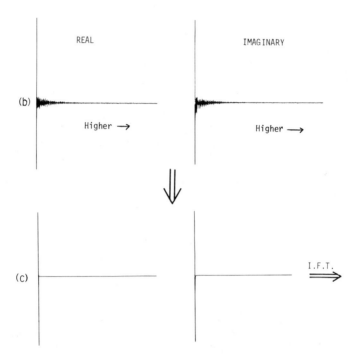

Figure 10. Digital filtering to remove baseline drift from a Raman spectrum. (a) Digitized Raman spectrum of toluene on hydrated zeolite, 200-3800 cm^{-1}. (b) "Frequency" spectrum obtained from Fourier transform of (a). (c) "Frequency spectrum apodized with a weight function (filter) exp[-400I/N], in which N is the number of data points, and I is the index of a given data point. (d) Inverse Fourier transform of (c) to give the baseline alone. (e) (a) minus (d) to give baseline-eliminated display. [Plots provided by courtesy of Bob Julian, Nicolet Instrument Corporation, 5225 Verona Road, Madison, WI.]

Advantages of Transform Methods in Chemistry

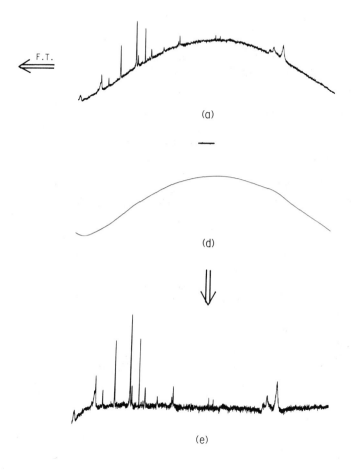

(a)

(d)

(e)

Differentiation and integration using transforms

Differentiation of a signal is commonly used to detect unresolved peak shoulders in ESR (see Chapter 4) and optical[2] spectroscopies, and in many other contexts (see Smith Chapter, for example). However, the noise level is much increased in the derivative display. Ordinary smoothing of the derivative may be undesirable, because the smoothing algorithm may also obscure spectral details.

In such situations, Fourier transforms can provide a means for performing the differentiation, as well as digital filtering or other apodizations for noise reduction (see preceding section). A flow chart showing the interrelations between a function and its integral or derivative is shown in Figure 11.

CONVOLUTION

Convolution arises in the following common experimental situation. Consider a conventional scanning spectrometer in which a slit is scanned across a dispersed spectrum (Figure 12a). Because the slit has finite width, the lines in the detected spectrum (Fig. 12d) will be broader than in the true spectrum (Fig. 12b). In order to obtain the detected spectrum from the actual spectrum, one must sum (integrate) the intensity across the slit width for each slit position, as the slit is scanned across the spectral range (Fig. 12b). The detected response is said to represent the convolution of the true response with the instrumental function. Generally speaking, the response of any linear detector can be described similarly as a convolution of the true response with some detector function.

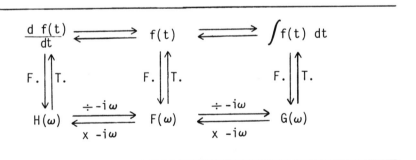

Figure 11. Differentiation and integration using Fourier transforms. Multiplying (dividing) the Fourier transform of a function by $(-i\omega)$, followed by inverse Fourier transformation will produce the derivative (integral) of the original function.

Advantages of Transform Methods in Chemistry

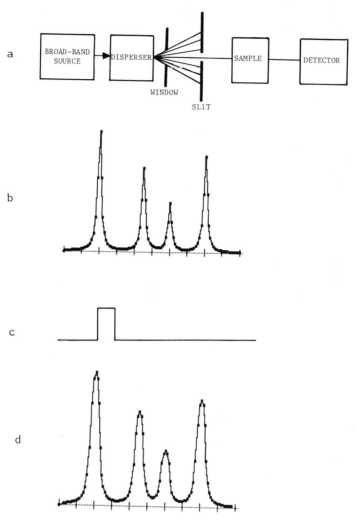

Figure 12. Convolution, as shown by the detection of a sample spectrum from a conventional steady-state scanning spectrometer (a). The true spectrum (b) is scanned through a slit of finite width (c). The detected spectrum (d) is derived by multiplying (b) with (c) and integrating for each slit position as the slit is scanned across the spectrum (i.e., adding all the light that passed through the slit at each slit position). Note the broadening effect of the "convolution" process (compare (d) to (b)).

Mathematically, the convolution of the functions h(t) and e(t) to give f(t) is described by[8]

$$f(t) = h(t) ☆ e(t)$$

$$= \int_{-\infty}^{\infty} h(t') e(t-t') dt' \qquad [31]$$

Graphically, Equation 31 simply requires that one of the two original functions be reversed left-to-right, the two functions then multiplied point-by-point along the abscissa, and the resultant points all added together. The process is repeated for all possible displacements of one of the functions relative to the other. Graphical examples of convolution are shown in Figures 13, 15 and 16.

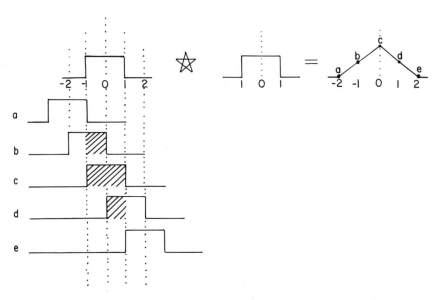

Figure 13. Graphical constructions to show the convolution of two simple square pulses (e(t) and h(t) of Eq. 31) to yield a triangular function (f(t) of Eq. 31). Holding the first square function fixed, the second function is moved from left to right, and the two functions multiplied together, as shown in the shaded segments. Each point, a to e, of the convolution represents the area of the product shown at the lower left (see text).

Advantages of Transform Methods in Chemistry   29

Integration becomes multiplication: the convolution theorem

The relationships between convolution and Fourier transforms are shown in Figure 14. The basic theorem can be stated in several equivalent forms, of which Equation 32 is perhaps the simplest.

If $f(t) = h(t) ☆ e(t)$ = convolution of $e(t)$ with $h(t)$,

then $F(\omega) = H(\omega) \cdot E(\omega)$, [32]

in which $F(\omega)$, $H(\omega)$, and $E(\omega)$ are the Fourier transforms of $f(t)$, $h(t)$, and $e(t)$, respectively.

In other words, the convolution operation (basically, an integration) in one domain becomes a (much simpler) multiplication in the Fourier domain. Thus, just as logarithms convert <u>multiplication</u> into <u>addition</u> [$\log(a \cdot b) = \log(a) + \log(b)$] in the "log <u>domain</u>", convolution converts <u>integration</u> into <u>multiplication</u> in the Fourier domain.

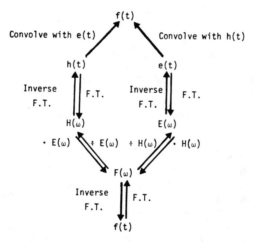

Figure 14. Interrelations between convolution and Fourier transforms Examples are shown in the next two figures.

Using the convolution theorem, we can immediately expand our pictorial library of Fourier transform pairs (Figure 5) to include any others that can be obtained by multiplying any two of the original waveforms together. For example in Figure 15, an exponentially damped sinusoid extending to infinite time is truncated by multiplication with a square weight function persisting only to time T. The Fourier transform of the resulting product could be computed directly from Equations 17-20, but is more easily constructed by graphical convolution of the Fourier transforms of the two functions, as shown in the Figure.

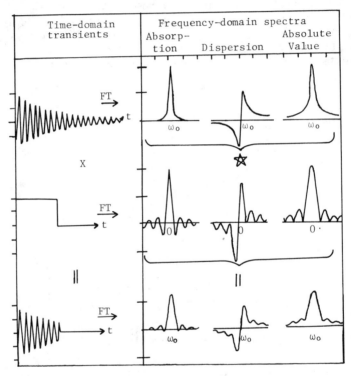

Figure 15. Fourier transform of a truncated damped cosine, obtained by applying the convolution theorem to the pictorial Fourier transform library of Figure 5. (a) Exponentially damped sinusoid persisting to infinite time, and its cosine and sine Fourier transforms. (b) Square wave truncation function and its cosine and sine transforms. (c) Time-domain product of (a) and (b). The Fourier transform of (c) can be obtained either by computation, or from (visual) convolution of the Fourier transforms of (a) and (b)--see text.

Advantages of Transform Methods in Chemistry   31

Deconvolution: Unfolding the response from the excitation

If h(t) is the "ideal" response of a linear system to an impulse (i.e., an infinitely short pulse) excitation, and e(t) is an actual excitation waveform, then the observed response of the system is the convolution of h(t) with e(t) according to Equation 31. In general, the shape of the "ideal" response h(t) to an impulse excitation will not be obvious from the shape of the observed response f(t).

However, since the convolution theorem of Equation 32 can be rewritten as

$$H(\omega) = \frac{F(\omega)}{E(\omega)} \quad , \qquad\qquad [33]$$

the "true" spectrum, $H(\omega)$, can be "unfolded" ("deconvoluted") from the observed spectrum, $F(\omega)$, by dividing $F(\omega)$ by the Fourier transform $E(\omega)$ of the excitation waveform, if the excitation waveform is measurable or theoretically calculable. The impact of this procedure is enormous--it is in principle now possible to eliminate any effect of the measuring instrument upon the desired spectral response.

An example of the value of convolution is shown in the simulated spectra of Figure 16. Figures 16a and 16b are the cosine Fourier transform and magnitude spectra of a linearly increasing frequency sweep excitation waveform, $\cos(\omega_1 t + (1/2)at^2)$. This excitation is commonly used in FT-ICR (see Comisarow Chapter), and has also been used in "correlation" NMR.[9] Because the excitation frequency increases linearly with time, the accumulated phase angle increases quadratically with time to give the wildly oscillating absorption-mode spectrum of Figure 16a. Moreover, even the magnitude spectrum of Figure 16b (from which phase considerations are absent) exhibits non-uniform excitation magnitude over the swept frequency range.

Figure 16c shows the cosine Fourier transform (i.e., absorption-mode spectrum) of the response of a system of 6 damped oscillators to the frequency-sweep excitation. The phase variation of the excitation waveform produces the oscillations in the absorption spectrum of the response. In addition, the apparent peak heights of the magnitude-mode spectrum of the response (Fig. 16d) are non-uniform because of the non-uniform magnitude of the excitation itself (Figure 16b).

However, dividing the (cosine) Fourier transform of the response (Fig. 16c) by the (cosine) Fourier transform of the excitation (Fig. 16a) gives the "deconvoluted" or "true" absorption-mode spectrum of Figure 16e, with correct relative peak amplitudes. The "deconvolved" spectrum (Fig. 16e) thus exhibits narrower peaks of more accurate height than does the magnitude spectrum of the directly observed response (Figure 16d).

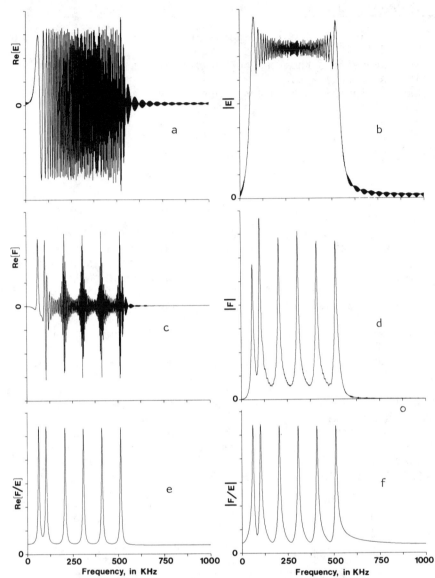

Figure 16. Deconvolution of response to frequency-sweep excitation. (a) Cosine Fourier transform of linearly increasing frequency sweep time-domain waveform. (b) Magnitude spectrum of excitation. (c) Cosine Fourier transform of time-domain response to excitation. (d) Magnitude spectrum of response. (e) = (c)/(a). (f) Magnitude spectrum of spectrum obtained by complex division of Fourier transform of excitation by Fourier transform of response. See text for discussion.

FOURIER AND HADAMARD (MULTIPLEX) CODES: DISCRETE SIGNAL SAMPLING

To this stage, we have considered only <u>continuous</u> waveforms and their various Fourier transform spectra. Experimentally, however, it is now usual to obtain a desired point-by-point <u>discrete</u> spectrum by suitable transformation of the point-by-point <u>sampled</u> output of a output of a detector. In this section, we will examine various recipes for relating a discretely sampled response to the desired absorption or magnitude spectrum.

Suppose that a given spectral range is divided into N segments, or channels, and that a detector measures the overall spectral amplitude or intensity from each channel in turn. [In a dispersive (prism or grating) spectrometer, one could imagine that a single exit slit of width, range/N, is moved a distance range/N at a time until the whole range is covered--see Figure 12b.]

If the N <u>desired</u> spectral elements (amplitudes or intensities at N different slit positions) are designated as $x_1, x_2, \cdots, x_N$, and the N <u>observed</u> amplitudes or intensities are designated as $y_1, y_2, \cdots, y_N$, then the following relations apply to measurements from such a moving-slit spectrometer:

$$y_1 = 1 \cdot x_1 + 0 \cdot x_2 + 0 \cdot x_3 + 0 \cdot x_4 \quad [34a]$$

$$y_2 = 0 \cdot x_1 + 1 \cdot x_2 + 0 \cdot x_3 + 0 \cdot x_4 \quad [34b]$$

$$y_3 = 0 \cdot x_1 + 0 \cdot x_2 + 1 \cdot x_3 + 0 \cdot x_4 \quad [34c]$$

$$y_4 = 0 \cdot x_1 + 0 \cdot x_2 + 0 \cdot x_3 + 1 \cdot x_4 \quad [34d]$$

Number of times each unknown element, x_i, is measured $= 1 \quad 1 \quad 1 \quad 1$

It is useful to think of the experiment as a "code" that connects the <u>observed</u> y_i-values to the <u>unknown</u> desired spectral elements, x_i. The obvious advantage of one-slit-at-a-time scanning experiments is then seen to be the simplicity of the "code":

$$\underbrace{\begin{pmatrix} y_1 \\ y_2 \\ y_3 \\ y_4 \end{pmatrix}}_{\text{Observations}} = \underbrace{\begin{pmatrix} 1 & 0 & 0 & 0 \\ 0 & 1 & 0 & 0 \\ 0 & 0 & 1 & 0 \\ 0 & 0 & 0 & 1 \end{pmatrix}}_{\text{CODE}} \underbrace{\begin{pmatrix} x_1 \\ x_2 \\ x_3 \\ x_4 \end{pmatrix}}_{\text{Unknowns}} \quad , \quad \underset{\sim}{y} = \underset{\approx}{A}\,\underset{\sim}{x} \quad [35]$$

Because the "code" matrix is just the identity matrix, the subsequent inverse code for recovering the unknowns from the observables is simply:

$$\underset{\approx}{A}^{-1} = \underset{\approx}{A} \qquad [36]$$

$$\underbrace{\begin{pmatrix} x_1 \\ x_2 \\ x_3 \\ x_4 \end{pmatrix}}_{\text{Unknowns}} = \underbrace{\begin{pmatrix} 1 & 0 & 0 & 0 \\ 0 & 1 & 0 & 0 \\ 0 & 0 & 1 & 0 \\ 0 & 0 & 0 & 1 \end{pmatrix}}_{\text{INVERSE CODE}} \underbrace{\begin{pmatrix} y_1 \\ y_2 \\ y_3 \\ y_4 \end{pmatrix}}_{\text{Observations}} \quad ; \quad \begin{array}{l} x_1 = 1 \cdot y_1 \\ x_2 = 1 \cdot y_2 \\ x_3 = 1 \cdot y_3 \\ x_4 = 1 \cdot y_4 \end{array} \qquad [37]$$

Although extraction of the N unknown spectral elements from the N observations (i.e., the inverse code) is clearly trivial for the one-slit-at-a-time scanning experiment, the great disadvantage (see Equations 34) is that each unknown element is detected only <u>once</u> during the N observations. If signal-independent noise is present, the noise amplitude in a given channel may be treated as a random walk about zero (the average noise level), and the root-mean-square distance away from zero after N steps in a random walk is proportional to $N^{1/2}$. The <u>signal</u>, on the other hand, accumulates as N, since the signal is the same in each measurement. Therefore, an obvious problem with the above single-channel detector is that there are too many zeroes in the "code" matrix.

It follows that if we could somehow detect <u>each</u> channel in <u>every</u> measurement, then after N measurements, each unknown element would have been measured N times, and the signal-to-noise ratio for the repeated measurement would be better by a factor of $N/N^{1/2}$, or $N^{1/2}$. This enhancement in signal-to-noise ratio is known as the <u>Fellgett</u> or <u>multichannel</u> advantage, and is discussed in reference 10. Among the large number of possible "codes" designed to gain the Fellgett advantage by increasing the number of non-zero code elements, the <u>Hadamard</u> and <u>Fourier</u> codes are especially simple and adaptable to experiments.

Hadamard code

For an arbitrary code relating the y_i to the x_i, the problem of recovering the desired x_i from the observed y_i can be difficult or worse.[11] The very special feature of both the Hadamard and Fourier codes is that the desired <u>inverse</u> code can again be found trivially from the specified <u>original</u> code. In mathematical terms, these particular code matrices are said to be "well-conditioned".[11] Consider again the 4-channel experiment, but this time with the Hadamard code:

Advantages of Transform Methods in Chemistry 35

$$y_1 = 1 \cdot x_1 + 1 \cdot x_2 + 1 \cdot x_3 + 1 \cdot x_4$$
$$y_2 = 1 \cdot x_1 - 1 \cdot x_2 - 1 \cdot x_3 + 1 \cdot x_4$$
$$y_3 = 1 \cdot x_1 - 1 \cdot x_2 + 1 \cdot x_3 - 1 \cdot x_4$$
$$y_4 = 1 \cdot x_1 + 1 \cdot x_2 - 1 \cdot x_3 - 1 \cdot x_4$$

[38]

Number of times each unknown element, x_j, is measured = 4   4   4   4

$$\underbrace{\begin{pmatrix} y_1 \\ y_2 \\ y_3 \\ y_4 \end{pmatrix}}_{\text{Observables}} = \underbrace{\begin{pmatrix} 1 & 1 & 1 & 1 \\ 1 & -1 & -1 & 1 \\ 1 & -1 & 1 & -1 \\ 1 & 1 & -1 & -1 \end{pmatrix}}_{\text{HADAMARD CODE}} \underbrace{\begin{pmatrix} x_1 \\ x_2 \\ x_3 \\ x_4 \end{pmatrix}}_{\text{Unknowns}}$$

[39]

Equations 39 show that with the Hadamard code, each unknown element x_j is observed N times with the same absolute weight factor; namely, the absolute value of each of the matrix elements in the code is unity.

$$\underset{\sim}{y} = \underset{\approx}{H} \cdot \underset{\sim}{x}$$ [40a]

$$|H_{ij}| = 1$$ [40b]

Equation 40b is the key property from which the full Fellgett advantage can be realized--it is as if all the spectral slits are open at once.

If the first row and column of the Hadamard code of Equation 39 are deleted, it becomes clear that each row of the remaining array differs from the preceding row by cyclic permutation. This property carries two immediate advantages. First, it is no longer necessary to construct a separate code for each measurement--see Sloane Chapter and reference 10 for examples of Hadamard mask construction. Second, construction of the desired inverse transformation is trivial:

$$\underset{\approx}{H^{-1}} = \frac{1}{N} \underset{\approx}{H}$$ [41]

or, for the 4-channel case,

$$\begin{pmatrix} x_1 \\ x_2 \\ x_3 \\ x_4 \end{pmatrix} = \frac{1}{4} \begin{pmatrix} 1 & 1 & 1 & 1 \\ 1 & -1 & -1 & 1 \\ 1 & -1 & 1 & -1 \\ 1 & 1 & -1 & -1 \end{pmatrix} \begin{pmatrix} y_1 \\ y_2 \\ y_3 \\ y_4 \end{pmatrix} \qquad [42]$$

$$\underbrace{\text{Desired unknowns}} \quad \underbrace{\text{HADAMARD INVERSE CODE}} \quad \underbrace{\text{Observations}}$$

For example,

$$\begin{aligned} x_2 &= \tfrac{1}{4}(y_1 - y_2 - y_3 + y_4) \\ &= \tfrac{1}{4}(x_1 + x_2 + x_3 + x_4 \\ &\quad - x_1 + x_2 + x_3 - x_4 \\ &\quad - x_1 + x_2 - x_3 + x_4 \\ &\quad + x_1 + x_2 - x_3 - x_4) \\ &= \tfrac{1}{4}(4x_2) = x_2 \qquad \text{Q.E.D.} \qquad [43] \end{aligned}$$

A Hadamard code may be constructed whenever

$$N = 2^m, \quad m = 2,3,4,\cdots \qquad [44]$$

and in other special cases discussed in the Sloane Chapter on Hadamard transform spectroscopy. In practice, actual Hadamard transform infrared spectrometers use a code that is derived from (but different from) the above examples, because it is experimentally simpler to block half of the slits (i.e., let $H_{mn} = 0$ for half of the elements in any one row of the Hadamard code) than to collect light reflected back from half of the slit positions and subtract that intensity from the transmitted intensity (i.e., $H_{mn} = -1$ for half of the elements of any one row of the code). See the Sloane chapter for details.

Fourier code

The Fourier code is based on the same properties just developed for the Hadamard case:

$$\underset{\sim}{y} = \underset{\approx}{F} \cdot \underset{\sim}{x} \qquad [45a]$$

and $|F_{mn}| = 1.$ [45b]

The general formula for the code is

Advantages of Transform Methods in Chemistry

$$F_{mn} = \exp[2\pi imn/N] = \cos(2\pi mn/N) + i \sin(2\pi mn/N) \qquad [46]$$

or, $$y_m = \sum_{i=0}^{N-1} F_{mn} x_n \qquad [47]$$

or, for the 4-channel case,

Observations FOURIER CODE Unknowns

$$\begin{pmatrix} y_0 \\ y_1 \\ y_2 \\ y_3 \end{pmatrix} = \begin{pmatrix} e^0 & e^0 & e^0 & e^0 \\ e^0 & e^{2\pi i/4} & e^{2\cdot 2\pi i/4} & e^{2\cdot 3\pi i/4} \\ e^0 & e^{2\cdot 2\pi i/4} & e^{2\cdot 2\cdot 2\pi i/4} & e^{2\cdot 2\cdot 3\pi i/4} \\ e^0 & e^{2\cdot 3\pi i/4} & e^{2\cdot 3\cdot 2\pi i/4} & e^{2\cdot 3\cdot 3\pi i/4} \end{pmatrix} \begin{pmatrix} x_0 \\ x_1 \\ x_2 \\ x_3 \end{pmatrix} \qquad [47a]$$

$$= \begin{pmatrix} 1 & 1 & 1 & 1 \\ 1 & e^{i\pi/2} & e^{i\pi} & e^{i3\pi/2} \\ 1 & e^{i\pi} & e^{i2\pi} & e^{i3\pi} \\ 1 & e^{i3\pi/2} & e^{i3\pi} & e^{i9\pi/2} \end{pmatrix} \begin{pmatrix} x_0 \\ x_1 \\ x_2 \\ x_3 \end{pmatrix} \qquad [47b]$$

or, $$\begin{pmatrix} y_0 \\ y_1 \\ y_2 \\ y_3 \end{pmatrix} = \begin{pmatrix} 1 & 1 & 1 & 1 \\ 1 & i & -1 & -i \\ 1 & -1 & 1 & -1 \\ 1 & -i & -1 & i \end{pmatrix} \begin{pmatrix} x_0 \\ x_1 \\ x_2 \\ x_3 \end{pmatrix} \qquad [47c]$$

The great computational simplicity of the Fourier code is that the desired <u>inverse</u> code may again be computed trivially from the original code:

$$F_{mn}^{-1} = \frac{1}{N} F_{mn}^{*} = \frac{1}{N} \exp[-2\pi imn/N] \qquad [48]$$

in which the * denotes complex conjugate. For example, for N = 4,

$$\begin{pmatrix} x_0 \\ x_1 \\ x_2 \\ x_3 \end{pmatrix} = \frac{1}{4} \begin{pmatrix} 1 & 1 & 1 & 1 \\ 1 & -i & -1 & i \\ 1 & -1 & 1 & -1 \\ 1 & i & -1 & -i \end{pmatrix} \begin{pmatrix} y_0 \\ y_1 \\ y_2 \\ y_3 \end{pmatrix} \quad [49]$$

$\underbrace{}_{\text{Desired unknowns}} \quad \underbrace{}_{\text{INVERSE FOURIER CODE}} \quad \underbrace{}_{\text{Observations}}$

so that (for instance),

$$x_3 = \frac{1}{4}(y_0 + iy_1 - y_2 - iy_3)$$

$$= \frac{1}{4}(x_0 + x_1 + x_2 + x_3$$
$$+ ix_0 - x_1 - ix_2 + x_3$$
$$- x_0 + x_1 - x_2 + x_3$$
$$- ix_0 - x_1 + ix_2 + x_3)$$

$$= \frac{1}{4}(4x_3) = x_3 \quad \text{Q.E.D.}$$

For the <u>physical applications</u> described in subsequent Chapters, it is usual to sample a spectrometer (or interferometer) time-domain (or pathlength) signal, $y(t)$, at N different times. Since each time-domain point represents a weighted sum of oscillations at <u>all</u> the frequencies in the detected range (e.g., we hear all the tones at once while an orchestra plays), we have now-familiar situation of N observables (N time-domain data points) related to N desired frequency-domain amplitudes by some sort of code:

$$\begin{aligned} y(t_0) &= F_{00}x(\omega_0) + F_{01}x(\omega_1) + \cdots + F_{0,N-1}x(\omega_{N-1}) \\ y(t_1) &= F_{10}x(\omega_0) + F_{11}x(\omega_1) + \cdots + F_{1,N-1}x(\omega_{N-1}) \\ &\vdots \\ y(t_{N-1}) &= F_{N-1,0}x(\omega_0) + F_{N-1,1}x(\omega_1) + \cdots + F_{N-1,N-1}x(\omega_{N-1}) \end{aligned} \quad [50]$$

The data sampling times, t_i, in Equations 50 can be spaced arbitrarily. The unique feature of the <u>Fourier</u> code is that if we choose <u>equally-spaced</u> time-domain samples,

$$t_n = \frac{nT}{N}, \quad n = 0, 1, 2, \cdots, N-1 \quad [51]$$

then the (Fourier) code for finding the N frequency-domain amplitudes at N equally-spaced frequencies,

$$\omega_m = \frac{2\pi m}{T}, \quad m = 0, 1, 2, \cdots, N-1 \qquad [52]$$

is just
$$F_{nm} = \exp[i\omega_m t_n]$$
$$= \exp[2\pi i nm/N] \qquad [53]$$

whose inverse is
$$F_{nm}^{-1} = \frac{1}{N} \exp[-i\omega_m t_n]$$
$$= \frac{1}{N} \exp[-2\pi i nm/N] \qquad [54]$$

The mathematical representation of the Fourier code elements as <u>complex</u> numbers simply corresponds to specifying the <u>phase</u> of the wave at each component frequency. For example, a purely real F_{nm} value corresponds to a pure cosine wave at frequency, $2\pi m/T$, whereas a purely imaginary F_{nm} value corresponds to a pure sine wave at the same frequency.

The Fourier and Hadamard codes are examples of <u>multiplex</u> codes designed to detect an entire spectrum from each sampled data point, so as to realize the Fellgett signal-to-noise advantage when noise is independent of signal. Of all imaginable codes, these codes have two optimal properties. First, each code element has an absolute value of unity: $|H_{nm}| = |F_{nm}| = 1$, so that each spectral element is observed with unit weight in <u>each</u> measurement (rather than one-at-a-time as with a conventional single-slit scanning instrument). Second, the desired <u>inverse</u> code for recovering the spectral amplitudes or intensities from the samples detector signals is trivially constructed from the original code (Equations 41 and 54). Moreover, when the number of data points, $N = 2^n$, n = integer, a fast algorithm[12] provides rapid Fourier (a similar algorithm provides rapid Hadamard[13]) transforms: e.g., less than 10 sec for 8192 data points using a minicomputer, and less than 1 sec with an array processor.

Nyquist frequency

Although the discrete <u>cosine</u> Fourier transform yields N calculated frequency-domain values for N measured time-domain data points, the second half of the cosine F.T. data is a mirror image of the first half, and thus gives no new information. The other half of the time-domain information is contained in the first N/2 frequency-domain values of the <u>sine</u> Fourier transform. [It is possible to put all the information into the cosine transform by first adding N zeroes to the time-domain signal (known as "zero-filling") before Fourier transformation of the resulting 2N time-domain points. The first N of the 2N cosine Fourier transform data points then contain all available spectral information.][14]

The highest frequency at which non-redundant information is available from a discrete cosine (or sine) Fourier transform is called the Nyquist frequency, $\nu_{Nyquist}$, given by

$$\nu_{Nyquist} = \frac{N}{2T} \qquad [55]$$

in which T is the total time-domain acquisition period. Since N time-domain points are collected in T sec, the sampling frequency is

$$\nu_{sampling} = \frac{N}{T} \qquad [56]$$

so that

$$\nu_{Nyquist} = \frac{1}{2} \nu_{sampling} \qquad [57]$$

For example, Fourier transform representation of a time-domain NMR signal containing frequencies up to 10,000 Hz ($\nu_{Nyquist}$ = 10,000) requires that the time-domain signal be sampled at 20,000 points/sec. If the available computer data storage table is 8,192 data points, then the acquisition period, T = 8,192/20,000 = 0.4096 sec. The frequency domain points will be spaced at intervals of 1/T = 2.44 Hz, defining the digital resolution. It is thus possible to specify any two of the three parameters: spectral range, acquisition period, and number of time-domain data points. Higher digital resolution requires longer acquisition period (and thus results in poorer signal-to-noise ratio), using either smaller spectral range or more time-domain data points.

Foldover (aliasing)

It is logical to wonder what will happen to time-domain signals containing oscillations at frequencies higher than the Nyquist frequency. The problem is closely related to the appearance of a spinning stagecoach wheel in a movie or television picture, in which the image consists of discrete pictures displayed in rapid succession. As the wheel picks up speed, the spokes first appear to rotate faster up to a certain rate, then slow down, then rotate in the opposite direction, and so on. The maximum displayable frequency is the Nyquist frequency. Higher frequencies are "folded over" or "aliased", to a frequency obtained by reflection about the Nyquist frequency as shown in Figure 17.

Aliasing is usually regarded as a problem to be avoided, since we are generally interested in the true frequencies in a signal, rather than their folded-over counterparts. Aliasing may be avoided most simply by not exciting signals outside the range of interest, as in frequency-sweep excitation in NMR and ICR.[4,5] In addition, it is usual to employ a bandpass filter in the detector, to suppress frequencies outside the Nyquist range.[15] Occasionally (see Nordstrom Chapter), aliasing may be used to advantage when computer data storage and/or digitizing rates are not large enough to satisfy the Nyquist criterion over the frequency range of interest--peaks of known frequency may then be observed at their folded-over frequencies, when (as in atomic electronic spectra) it is the <u>intensities</u> rather than the absorption <u>frequencies</u> that are important.

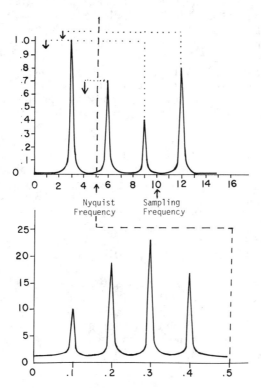

Figure 17. Demonstrate of foldover aliasing. (a) Hypothetical spectrum, with peaks located at their true frequencies. (b) Discrete cosine Fourier transform of the time-domain signal corresponding to (a), with sampling and Nyquist frequencies as shown. The peaks in (b) have correct relative intensities, but are folded-back to lower apparent displayed frequencies.

SUMMARY

The main advantage of Fourier and Hadamard transforms in data acquisition is a multiplex improvement in signal-to-noise ratio, due to detection of the whole spectrum at once rather than just one slit-width at a time. The resulting Fellgett improvement by a factor of up to $\sqrt{N}$, where N is the number of spectral resolution elements (data points) is fully attainable only when the noise is independent of the signal, and thus need not obtain for detectors in the visible-u.v., or for particle detection (e.g., Hadamard ESCA spectroscopy), for which shot noise proportional to the square root of signal strength may cancel all of the Fellgett multichannel improvement (see ref. 1 for a fuller discussion of noise and the Fellgett advantage).

The discussion in this Chapter has been directed toward analysis of time-domain signals. However, the same principles apply toward spatial interferograms obtained with Michelson or other interferometers--the connection is explained in Chapter 20 of ref. 1 and in the de Haseth and Nordstrom Chapters. In addition to the Fellgett advantage (which corresponds to opening the exit slit in a conventional dispersive spectrometer), interferometers and doubly-multiplexed Hadamard transform spectrometers provide a "Jacquinot " or "etendue" ("throughput") advantage that corresponds to opening the entrance slit as well, as detailed in the respective Chapters.

The advantages of Fourier transforms for manipulation of existing data sets have been demonstrated in several examples. The main uses are apodization for digital filtering (e.g., baseline smoothing) and for signal-to-noise or resolution enhancement, deconvolution for removal of spectral distortions introduced by the measurement process, and Hilbert transforms (Chapter 4) to detect hidden signals. In addition, convolution (and the convolution theorem) can be used for simple graphical construction of time- or frequency-domain curves whose analytical evaluation is otherwise tedious.

Practical details of data acquisition and reduction (e.g., phase correction, choice of excitation waveform and response apodization, etc.) have been left to the remaining Chapters. The intent here has been to present an overview from which it should be possible to better understand the motives for the specific transform procedures presented in the remaining Chapters.

ACKNOWLEDGMENTS

Many of the figures in this chapter were produced on an Apple II microcomputer. Tao-Chin Lin and Dixie Fisher helped in preparation of several figures. This work was supported in part by the American Chemical Society (Petroleum Research Fund 11458-AC6), the Ohio State University, and the Alfred P. Sloan Foundation (1976-80).

REFERENCES

1. Marshall, A. G. 1978, Biophysical Chemistry (Wiley, NY), Chapters 13-22.
2. Cahill, J. E. 1979, American Laboratory (November), 79-85.
3. Champeney, D. C. 1973, Fourier Transforms and their Physical Applications (Academic Press, NY), Chapter 2.
4. Marshall, A. G. and Roe, D. C. 1980, J. Chem. Phys. $\underline{73}$, 1581-1590.
5. Ernst, R. R. 1970, J. Magn. Reson. $\underline{3}$, 10-27; Kaiser, R. 1970, J. Magn. Reson. $\underline{3}$, 28-43.
6. Tomlinson, B. L. and Hill, H. D. W. 1973, J. Chem. Phys. $\underline{59}$, 1775-1784.
7. Abragam, A. 1961, The Principles of Nuclear Magnetism (Clarendon Press, Oxford), pp. 53-57.
8. Bracewell, R. 1965, The Fourier Transform and Its Applications, (McGraw-Hill, N.Y.), Chapter 3.
9. Dadok, J. and Sprecher, R. F. 1974, J. Magn. Reson. $\underline{13}$, 243; Gupta, R. K., Ferretti, J. A. and Becker, E. D. 1974, J. Magn. Reson. $\underline{13}$, 275.
10. Marshall, A. G. and Comisarow, M. B. 1978, "Transform Techniques in Chemistry", ed. P. R. Griffiths (Plenum, NY), Chapter 3.
11. Forsythe, G. and Moler, C. B. 1967, Computer Solution of Linear Algebraic Systems (Prentice-Hall, Englewood Cliffs, NJ).
12. Cooley, J. W. and Tukey, J. W. 1965, Math. Comp. $\underline{19}$, 297.
13. Harwit, M. and Sloane, N. J. A. 1979, Hadamard Transform Optics (Academic Press, NY), Appendix A.6.
14. Bartholdi, E. and Ernst, R. R. 1973, J. Magn. Reson. $\underline{11}$, 9.
15. Marshall, A. G., Marcus, T. and Sallos, J. 1979, J. Magn. Reson. $\underline{35}$, 227-230.

HADAMARD AND OTHER DISCRETE TRANSFORMS IN SPECTROSCOPY

N. J. A. Sloane

Mathematics and Statistics Research Center
Bell Laboratories
Murray Hill, NJ 07974

INTRODUCTION

This paper is an introduction to the use of Hadamard and other matrices for increasing the accuracy to which the spectrum of a beam of light can be measured. In the most favorable case if the spectrum has n components the mean squared error in each component is reduced by a factor of n/4. These schemes have the additional merit that the instrumentation required is relatively simple.

The main advantage of discrete transforms in optics lies in their simplicity. This can be illustrated by considering an equivalent problem which is even easier to describe, namely the problem of weighing several small objects. The basic idea, which was perhaps first suggested by Yates[1], is that by weighing the objects in groups rather than one at a time it may be possible to determine the weights more accurately.

For example suppose we want to weigh seven objects, numbered $1, 2, \cdots, 7$. A spring balance, with a single pan, is available to do the weighing. Small random errors are always present in these measurements, but we assume that the balance has been well calibrated. More precisely we assume that the balance gives the correct weight except for a small random error e. The average value of e is zero, but the average of e^2 is say σ^2--this is the variance of e, or the mean squared error in the measurement.

If we simply weigh the objects one at a time on the balance, we make seven measurements $\eta_1, \cdots, \eta_7$. These differ from the true weights $\psi_1, \cdots, \psi_7$ by the (unknown) errors $e_1, \cdots, e_7$:

$$\eta_1 = \psi_1 + e_1$$
$$\eta_2 = \psi_2 + e_2$$
$$\vdots$$
$$\eta_7 = \psi_7 + e_7 \qquad [1]$$

The mean squared error in the i-th measurement is

$$\epsilon_i = \text{average } (\eta_i - \psi_i)^2$$
$$= \text{average } (e_i^2) = \sigma^2$$

On the other hand if we are clever we first put objects 1,2,3 & 5 on the balance and weigh them, then objects 2,3,4 & 6, then 3,4,5 & 7, and so on, thus making the following seven measurements:

$$\eta_1 = \psi_1 + \psi_2 + \psi_3 + \psi_5 + e_1$$
$$\eta_2 = \psi_2 + \psi_3 + \psi_4 + \psi_6 + e_2$$
$$\eta_3 = \psi_3 + \psi_4 + \psi_5 + \psi_7 + e_3$$
$$\eta_4 = \psi_1 + \psi_4 + \psi_5 + \psi_6 + e_4$$
$$\eta_5 = \psi_2 + \psi_5 + \psi_6 + \psi_7 + e_5$$
$$\eta_6 = \psi_1 + \psi_3 + \psi_6 + \psi_7 + e_6$$
$$\eta_7 = \psi_1 + \psi_2 + \psi_4 + \psi_7 + e_7 \qquad [2]$$

Again $e_1, \cdots, e_7$ are the unknown errors in the weighings. To find the weights we "solve" these equations for $\psi_1, \cdots, \psi_7$ pretending that the e_i's are not present. Thus we cover up $e_1, \cdots, e_7$ and solve the equations for $\psi_1, \cdots, \psi_7$. The results, which we call $\hat{\psi}_1, \cdots, \hat{\psi}_7$ since they are not really the true values but only estimates for them, are:

$$\hat{\psi}_1 = (\eta_1 - \eta_2 - \eta_3 + \eta_4 - \eta_5 + \eta_6 + \eta_7)/4$$
$$\hat{\psi}_2 = (\eta_1 + \eta_2 - \eta_3 - \eta_4 + \eta_5 - \eta_6 + \eta_7)/4$$
$$\hat{\psi}_3 = (\eta_1 + \eta_2 + \eta_3 - \eta_4 - \eta_5 + \eta_6 - \eta_7)/4$$
$$\hat{\psi}_4 = (-\eta_1 + \eta_2 + \eta_3 + \eta_4 - \eta_5 - \eta_6 + \eta_7)/4$$
$$\hat{\psi}_5 = (\eta_1 - \eta_2 + \eta_3 + \eta_4 + \eta_5 - \eta_6 - \eta_7)/4$$
$$\hat{\psi}_6 = (-\eta_1 + \eta_2 - \eta_3 + \eta_4 + \eta_5 + \eta_6 - \eta_7)/4$$
$$\hat{\psi}_7 = (-\eta_1 - \eta_2 + \eta_3 - \eta_4 + \eta_5 + \eta_6 + \eta_7)/4 \qquad [3]$$

We will see later how to write down these expressions very easily. The observant reader will have noticed how the coefficient matrix in these expressions is obtained from the transpose of the coefficient matrix in Equations 2.

Substituting Equations 2 into 3 we find that the estimates of ψ_i are related to the true weights by

$$\hat{\psi}_1 = \psi_1 + (e_1 - e_2 - e_3 + e_4 - e_5 + e_6 + e_7)/4$$
$$\vdots$$
$$\hat{\psi}_7 = \psi_7 + (-e_1 - e_2 + e_3 - e_4 + e_5 + e_6 + e_7)/4 \qquad [4]$$

The mean squared errors in these estimates are all equal. For example,

$$\epsilon_1 = \text{average } (\hat{\psi}_1 - \psi_1)^2$$

$$= \text{average } \frac{(e_1 - e_2 - e_3 + e_4 - e_5 + e_6 + e_7)^2}{16}$$

$$= \frac{7\sigma^2}{16}$$

provided we assume that the errors in different weighings are independent of each other.

Thus by weighing the objects in groups we have reduced the mean squared error in the estimates of the weights by a factor of 16/7. This is perhaps not in itself impressive. But when there are n objects to be weighed instead of seven the same method reduces the mean squared error by a factor of

$$\frac{(n+1)^2}{4n} \simeq \frac{n}{4}$$

(Provided that n is a number of the right form. This is not a serious restriction and will be discussed later.) We have therefore obtained a considerable improvement in accuracy. Furthermore when n is large there is an algorithm resembling the fast Fourier transform for solving the Equations 2 to get the estimates $\hat{\psi}_i$ (see Section A.6.2. of reference 2). A rule (like that implied by Equations 2) for specifying which objects are to be placed on the balance in each measurement is called a <u>weighing design</u>. There is an extensive literature on weighing designs--see for example references 2-11.

Precisely the same improvement in mean squared error can be obtained in measuring the spectrum of a beam of light, provided certain conditions are satisfied. Instead of n objects whose weights are to be determined, we have a beam of light divided into n components of different wavelengths, and we wish to find their intensities. We must translate

object	into	light at a particular wavelength or range of wavelengths,
weight	into	intensity, and
spring balance	into	optical detector,

but apart from that the two problems are mathematically the same.

Of course the beam of light must first be separated into its different wavelengths. This is usually done by a grating or prism, shown schematically in Figure 1, which divides the light into say n different components whose intensities $\psi_1, \cdots, \psi_n$ we wish to determine. A graph of $\psi_1, \cdots, \psi_n$ versus wavelength is the <u>spectrum</u> that we are trying to find.

One way to determine these intensities is to measure each in turn with a detector, blocking out the other components with a mask as shown in Figure 1. This is the conventional type of spectrometer, sometimes called a <u>monochromator</u>. But by analogy with the weighing problem just described, it is not surprising that we can often do better by measuring the light at several different wavelengths simultaneously. To do this the light leaving the grating or prism is interrupted by a mask which allows certain components to pass freely and blocks the others (see Figure 2). The light passing through the mask is focused by a lens onto the detector, and the total intensity η of these components is measured. The mask is now changed and the intensity of another set of components is measured. This is repeated until n measurements have been made.

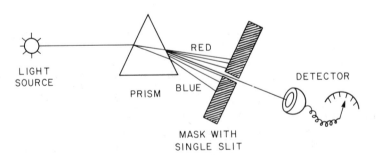

Figure 1. Conventional spectrometer (or monochromator) with a single exit slit.

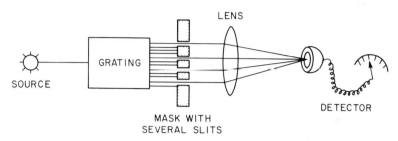

Figure 2. A multiplexing spectrometer with several exit slits, allowing much more light to reach the detector.

In order to give a mathematical description of what happens, let us introduce a matrix $W = (w_{ij})$ to specify the mask. The entry w_{ij} in the i-th row and j-th column is 1 if the j-th slit in the mask is open during the i-th measurement, and 0 if it is closed. Of course the same matrix can be used to specify a weighing design, in which case $w_{ij} = 1$ if the j-th object is placed on the pan during the i-th weighing, and $w_{ij} = 0$ if not. For example the weighing design of Equations 2 is described by the matrix

$$W = \begin{pmatrix} 1 & 1 & 1 & 0 & 1 & 0 & 0 \\ 0 & 1 & 1 & 1 & 0 & 1 & 0 \\ 0 & 0 & 1 & 1 & 1 & 0 & 1 \\ 1 & 0 & 0 & 1 & 1 & 1 & 0 \\ 0 & 1 & 0 & 0 & 1 & 1 & 1 \\ 1 & 0 & 1 & 0 & 0 & 1 & 1 \\ 1 & 1 & 0 & 1 & 0 & 0 & 1 \end{pmatrix} \quad [5]$$

(As we shall see, this is an S-matrix of order 7.) In both the weighing and the optical problem the n measurements tell us the values of

$$\eta_1 = w_{11}\psi_1 + \cdots + w_{1n}\psi_n + e_1$$
$$\eta_2 = w_{21}\psi_1 + \cdots + w_{2n}\psi_n + e_2$$
$$\vdots$$
$$\eta_3 = w_{n1}\psi_1 + \cdots + w_{nn}\psi_n + e_2 \quad [6]$$

If we define the column vectors

$$\eta = \begin{pmatrix} \eta_1 \\ \vdots \\ \eta_n \end{pmatrix} \qquad \psi = \begin{pmatrix} \psi_1 \\ \vdots \\ \psi_n \end{pmatrix} \qquad e = \begin{pmatrix} e_1 \\ \vdots \\ e_n \end{pmatrix}$$

then these equations can be written as a single matrix equation:

$$\eta = W\psi + e \qquad [7]$$

This is the fundamental equation describing the measurements. Just as before we "solve" Equation 7 for ψ by pretending that $e = 0$, obtaining

$$\hat{\psi} = W^{-1}\eta \qquad [8]$$

as our estimate for the unknown spectrum ψ. To find the mean squared error we substitute Equation 7 into Equation 8, obtaining

$$\hat{\psi} = \psi + W^{-1} e$$

We wish to choose the matrix W so as to minimize the mean squared errors in the estimates of the spectrum, i.e., the numbers

$$\epsilon_i = \text{average}(\hat{\psi}_i - \psi_i)^2, \quad i = 1,\cdots,n$$

We shall see that if W is taken to be what is called an S-matrix, then

$$\epsilon_i = \frac{4n}{(n+1)^2}\sigma^2 \cong \frac{\sigma^2}{n/4} \qquad [9]$$

If n is large this is very much smaller than σ^2. Figure 3 shows a spectrum obtained in this way using an instrument with n = 255 slits, compared with the same spectrum when measured by the same instrument operated as a monochromator, i.e., with only a single slit.

This technique, of improving the performance of a spectrometer by measuring light at several wavelengths simultaneously, is called multiplexing, and was perhaps first proposed by Golay in 1949.[12,13] Since then the principle has been used by many people, including Fellgett,[14,15] Ibbett, Aspinall and Grainger,[16] Decker and Harwit,[17] Nelson and Fredman,[18] and starting in 1969 in a series of papers by Martin Harwit and the author.[11,19-26] Equation 9 was first given in ref. 23. The present description is of course only a brief introduction to the subject. For a more complete account the reader is referred to our book.[2]

So far we have said nothing about the conditions needed before this reduction in mean squared error can be attained. The chief requirement is that the error e_i in the i-th measurement (i.e., the i-th reading of the spring balance or optical detector) be independent of the quantity being measured, for $i = 1, 2, \cdots, n$. In the weighing problem this implies that the objects to be weighed should be light in comparison with the mass of the balance, and in the optical problem that the noise in the detector be independent of the

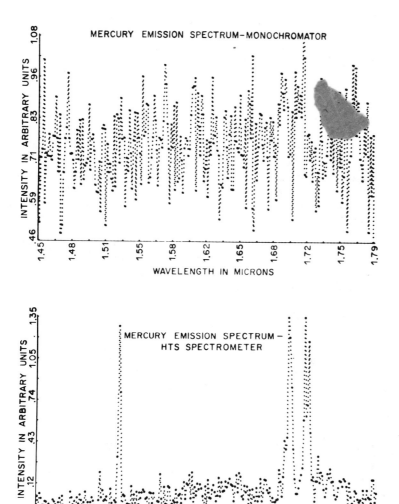

Figure 3. Mercury vapor infra-red emission spectrum from a conventional monochromator (top) and from the same instrument using a mask constructed from a 255 x 255 S-matrix (after Decker[27]).

incoming signal, a condition which for example often applies to infra-red measurements.

There are also some less critical assumptions that have been made to simplify the analysis, namely that

$$\text{average } (e_i) = 0$$
$$\text{average } (e_i^2) = \sigma^2$$
$$\text{average } (e_i e_j) = 0 \qquad [10]$$

for $i,j = 1,\cdots,n$; $i \neq j$. Finally, in order for an S-matrix to exist, the number of measurements n must be a number of the form $4a + 3$ for some integer a, and a Hadamard matrix of order $4a+4$ must exist (see below). If in addition we want a <u>cyclic</u> S-matrix then the value of n is somewhat further restricted. A detailed analysis of when multiplexing is worthwhile may be found in Chapter 4 of ref. 2.

The next section gives an account of the basic theory and states the main results. The following section is devoted to Hadamard and S-matrices (and others), and the chapter concludes with a brief mention of some further topics.

MORE ABOUT WEIGHING DESIGNS

So far we have only mentioned spring balances, with a single pan. Chemical balances, with two pans, also have a place in our analysis. The difference is that in Equations 5 and 6 the entries in the matrix W can now take three values: +1 if the object is placed in the left pan, -1 if it is in the right pan, and 0 if it is omitted from that measurement.

For example, if n=4 objects are to be weighed, we might make the following measurements:

$$\eta_1 = \psi_1 + \psi_2 + \psi_3 + \psi_4 + e_1$$
$$\eta_2 = \psi_1 - \psi_2 + \psi_3 - \psi_4 + e_2$$
$$\eta_3 = \psi_1 + \psi_2 - \psi_3 - \psi_4 + e_3$$
$$\eta_4 = \psi_1 - \psi_2 - \psi_3 + \psi_4 + e_4 \qquad [11]$$

This means that in the first weighing all four objects are placed in the left pan, in the second weighing objects 1 & 3 are in the left pan and 2 & 4 in the right, and so on. The corresponding weighing matrix W is

Hadamard and Other Discrete Transforms in Spectroscopy

$$W = \begin{pmatrix} 1 & 1 & 1 & 1 \\ 1 & -1 & 1 & -1 \\ 1 & 1 & -1 & -1 \\ 1 & -1 & -1 & 1 \end{pmatrix} \qquad [12]$$

(This is an example of a Hadamard matrix of order 4.) The estimates for the weights are

$$\hat{\psi}_1 = \tfrac{1}{4}(\eta_1 + \eta_2 + \eta_3 + \eta_4)$$

$$\hat{\psi}_2 = \tfrac{1}{4}(\eta_1 - \eta_2 + \eta_3 - \eta_4)$$

$$\hat{\psi}_3 = \tfrac{1}{4}(\eta_1 + \eta_2 - \eta_3 - \eta_4)$$

$$\hat{\psi}_4 = \tfrac{1}{4}(\eta_1 - \eta_2 - \eta_3 + \eta_4) \qquad [13]$$

which using Equation 11 implies that

$$\hat{\psi}_1 = \psi_1 + \tfrac{1}{4}(e_1 + e_2 + e_3 + e_4), \text{ etc.}$$

The mean squared errors in the estimates are

$$\epsilon_1 = \text{average}(\hat{\psi}_1 - \psi_1)^2$$

$$= \sigma^2/4, \text{ using Equation 10,}$$

$$= \epsilon_2 = \epsilon_3 = \epsilon_4,$$

an improvement by a factor of 4 over weighing the objects one at a time.

The general weighing design (using either a spring balance or a chemical balance) is still described by an equation of the form

$$\eta = W\psi + e \qquad [7]$$

This equation applies even if we make more measurements than there are unknowns, provided we allow W to be an m x n rectangular matrix, in which m is the number of measurements and n the number of unknowns.

Given the measurements η, the best estimate for ψ is

$$\hat{\psi} = W^{-1}\eta \qquad [14]$$

$$= \psi + W^{-1}\epsilon$$

if W is a square invertible matrix, or

$$\hat{\psi} = W^+\eta \qquad [15]$$

in the general case, where W^+ is the Moore-Penrose generalized inverse of W.[28-34] The most important example of a generalized inverse occurs when the columns of W are linearly independent. Then $W^T W$ is an invertible n x n matrix (the T denotes transpose), and $W^+ = (W^T W)^{-1} W^T$; hence $\hat{\psi} = (W^T W)^{-1} W^T \eta$.

These estimates are best in several senses: they are the best linear unbiased estimates of the unknowns, and are also the best least-squares estimates (see refs. 2, 23, 32, 33, 35 and 36). However it is worth mentioning that there are also good arguments in favor of using a biased estimate, such as the James-Stein estimator -- see refs. 37-40.

Of course this analysis applies equally well to our spectrometers, in which case W represents the masks used and ψ the unknown spectrum. The only restriction is that we cannot design masks with entries that are -1, so the results for chemical balance designs do not apply to the optical case. We could simulate a chemical balance by using two detectors and taking the difference of their readings (this is analyzed in Section 3.5.4. of ref. 2), but the results cannot be compared directly with those obtained for masks of 0's and 1's alone, since using two detectors produces an unfair advantage. We are assuming throughout that only one detector or balance is available.

The next question is to decide how good the estimates of Equations 14 and 15 are. Ideally we would like to minimize all the mean squared errors $\epsilon_1, \cdots, \epsilon_n$. But usually this is impossible and some other criterion must be used. Two of the most common criteria for judging a weighing design or mask W are the following (see refs. 3, 7-11).

W is said to be <u>A-optimal</u> if it minimizes the <u>average mean squared error</u>

$$\epsilon = \frac{\epsilon_1 + \cdots + \epsilon_n}{n}$$

Hadamard and Other Discrete Transforms in Spectroscopy

W is D-optimal if the determinant of W^TW is maximized. If W is square this is equivalent to maximizing the determinant of D itself. A D-optimal design minimizes the volume of the region in which the estimate $\hat{\psi}$ is expected to lie.

These criteria do not always agree. Probably A-optimality is the most important, provided that the individual ϵ_i's are roughly equal.

The average mean squared error ϵ is related to W by the following very simple formula:

$$\epsilon = \frac{1}{n}\sigma^2 \; \text{Trace}(W^TW)^{-1} \qquad [16]$$

where the trace of a matrix is the sum of the entries on its main diagonal (see ref. 2, p. 53, for a proof). An equivalent statement is that ϵ is σ^2/n times the sum of the squares of the entries of W^{-1}. Equation 16 is also valid in the more general case when W is an m x n matrix with linearly independent columns (ref. 2, p. 84). Thus an A-optimal weighing design or mask is one that minimizes

$$\text{Trace}(W^TW)^{-1}$$

We have therefore arrived at the following mathematical question.

Which n x n matrix W with entries 0, 1, and -1 (or entries 0 and 1) has the smallest value of Trace $(W^TW)^{-1}$? [17]

Enough is known about the answer to this question to solve any practical problem in weighing designs or optics. On the other hand a number of important mathematical questions still remain open.

The main results are as follows. In 1944 Hotelling[6] proved that the mean squared error in each unknown is bounded from below by

$$\epsilon_i \geq \frac{\sigma^2}{n} \qquad [18]$$

and furthermore that

$$\epsilon_i = \frac{\sigma^2}{n}, \text{ for } i = 1, 2, \cdots, n, \qquad [19]$$

attaining the minimum value, if and only if W is a Hadamard matrix of order n (these matrices are defined in the next section). A proof of his result is given in Section 3.2.4. of ref. 2. Unfortunately the entries in a Hadamard matrix are +1's and -1's, so this result only applies to the chemical balance weighing problem and not to the optical case. But it does more or less solve the former problem, and enables us to state a fundamental principle:

Hadamard matrices make the best chemical balance weighing designs, i.e., the best matrices with entries +1, 0, and -1. If there are n unknowns and a Hadamard matrix of order n is used, the mean squared error in each unknown is reduced by a factor of n; or in other words the signal-to-noise ratio is increased by a factor of $\sqrt{n}$.

Furthermore it can be shown that Hadamard matrices are also D-optimal (cf. ref. 11, Section 3.2.5 of ref. 2). Equations 11 and 12 describe a weighing design based on a Hadamard matrix of order 4.

When only 0's and 1's can be used the situation is not so clear-cut. Harwit and I conjecture that the best matrices are the S-matrices described in the next section, and that Equations 18 and 19 should therefore be replaced by

$$\epsilon \geq \frac{4n\sigma^2}{(n+1)^2} \qquad [20]$$

with equality if and only if W is an S-matrix. Note that the right-hand side of Equation 20 is roughly

$$\frac{\sigma^2}{n/4},$$

and we have lost a factor of 4 compared with Hadamard matrices. This is the price that must be paid for using (0,1)-matrices instead of (0,±1)-matrices.

We have not quite been able to prove this result. What we have proved however (refs. 2, 11) is that, for large n, S-matrices are within a few percent of being A-optimal, and asymptotically they are A-optimal. In fact we have proved that, for any n x n matrix of 0's and 1's,

$$\epsilon \geq \frac{4\sigma^2}{(n+1)^{(n+1)/n}} \qquad [21]$$

$$\cong \frac{4\sigma^2}{n} \quad \text{for large n.}$$

Recently Joseph Fischer[41] has improved this by showing

$$\epsilon \geq \begin{cases} \dfrac{4(n-1)^2 \sigma^2}{n^3}, & \text{if n is even} \\[2ex] \dfrac{4(n-1)\sigma^2}{n(n+1)}, & \text{if n is odd} \end{cases} \qquad [22]$$

These bounds are very close indeed to the conjecture [20].

Thus for all practical purposes the following principle has been established.

S-matrices make the best spring balance weighing designs and the best (0,1)-masks for multiplexing spectrometers. If there are n unknowns and an S-matrix of order n is used, the mean squared error in each unknown is reduced by a factor of $(n+1)^2/4n \cong n/4$; or in other words the signal-to-noise ratio is increased by a factor of $(n+1)/2\sqrt{n} \cong \sqrt{n}/2$.

It can also be shown that S-matrices are D-optimal (cf. ref. 2, p. 58). Equations 2 and 5 describe a weighing design based on an S-matrix of order 7.

HADAMARD MATRICES, S-MATRICES, AND OTHERS

In this section we describe some of the more important matrices used in weighing designs and spectrometers.

A <u>Hadamard matrix</u> H_n of order n is an n x n matrix of +1's and -1's with the property that the scalar product of any two distinct rows is 0. Thus H_n must satisfy

$$H_n H_n^T = n I_n \qquad [23]$$

where I_n is the n x n identity matrix. These matrices were first introduced by Hadamard in 1893.[42] Examples of Hadamard matrices of orders 1, 2, 4 and 8 are shown in Figure 4. If H_n is a Hadamard matrix, then so is any matrix obtained from H_n by multiplying any of the rows and columns by -1. In this way we can always suppose that H_n is arranged to have all elements of the first row and first column equal to +1. Such a Hadamard matrix is said to be <u>normalized</u>. For example, all the Hadamard matrices in Figure 4 are <u>normalized</u>.

It is easy to prove that if a Hadamard matrix of order n exists, then n must be 1, 2, or a multiple of 4. It is generally believed that Hadamard matrices exist of every order that is a multiple of 4. A large number of different constructions are known (see refs. 43-45), and at present the smallest order that has not been constructed is 268.

Since $H_n^{-1} = n^{-1} H_n^T$, recovering the spectrum from the measurements, as in Equation 14, is particularly simple--compare Equation 13. Furthermore from Equation 16 we have

$$\epsilon = \frac{1}{n}\sigma^2 \text{ Trace } (\frac{1}{n} I_n) = \frac{\sigma^2}{n},$$

in agreement with Equation 19.

To define an S-matrix we begin with a normalized Hadamard matrix H_n of order n. Then an S-matrix of order n-1, S_{n-1}, is the (n-1) x (n-1) matrix of 0's and 1's obtained by omitting the first row and column of H_n and then changing +1's to 0's and -1's to 1's. The S-matrices of orders 1, 3, and 7 obtained from Figure 4 are shown in Figure 5. It is not difficult to show that an S-matrix of order n satisfies

$$S_n S_n^T = \frac{1}{4}(n+1)(I_n+J_n), \qquad [24]$$

$$S_n J_n = J_n S_n = \frac{1}{2}(n+1) J_n \qquad [25]$$

and

$$S_n^{-1} = \frac{2}{n+1}(2S_n^T - J_n) \qquad [26]$$

where J_n is an n x n matrix of 1's. An n x n matrix of 0's and 1's is an S-matrix if and only if Equations 24 and 25 are satisfied. These matrices were first introduced into spectroscopy in ref. 23.

$$H_1 = \begin{bmatrix} 1 \end{bmatrix}, H_2 = \begin{bmatrix} 1 & 1 \\ 1 & - \end{bmatrix}, H_4 = \begin{bmatrix} 1 & 1 & 1 & 1 \\ 1 & - & 1 & - \\ 1 & 1 & - & - \\ 1 & - & - & 1 \end{bmatrix}$$

$$H_8 = \begin{bmatrix} 1 & 1 & 1 & 1 & 1 & 1 & 1 & 1 \\ 1 & - & 1 & - & 1 & - & 1 & - \\ 1 & 1 & - & - & 1 & 1 & - & - \\ 1 & - & - & 1 & 1 & - & - & 1 \\ 1 & 1 & 1 & 1 & - & - & - & - \\ 1 & - & 1 & - & - & 1 & - & 1 \\ 1 & 1 & - & - & - & - & 1 & 1 \\ 1 & - & - & 1 & - & 1 & 1 & - \end{bmatrix}$$

Figure 4. Examples of Hadamard matrices.

Hadamard and Other Discrete Transforms in Spectroscopy 59

Again the simple form of S_n^{-1} makes it easy to recover the spectrum from the measurements--this explains Equation 3 above. When S_n is used as the mask W we find from Equations 16 and 24 that the average mean squared error is

$$\epsilon = \frac{4n\sigma^2}{(n+1)^2},$$

in agreement with the right-hand side of Equation 20.

In order for an n x n S-matrix to exist n must be 1 or a number of the form 4a + 3. Infinitely many examples can be obtained from the known constructions of Hadamard matrices. The most important examples for practical purposes are <u>cyclic</u> S-matrices, having the property that each row is a cyclic shift to the left (or, to the right) of the previous row. Equation 5 is an example.

If S is cyclic then instead of using n different masks in the spectrometer, one for each measurement, we can make do with just one mask of length 2n - 1. This reduces the weight and cost of the instrument. Consider Equation 5 for example. The first row is

1 1 1 0 1 0 0.

$$S_1 = \begin{bmatrix} 1 \end{bmatrix}, \quad S_3 = \begin{bmatrix} 1 & 0 & 1 \\ 0 & 1 & 1 \\ 1 & 1 & 0 \end{bmatrix}$$

$$S_7 = \begin{bmatrix} 1 & 0 & 1 & 0 & 1 & 0 & 1 \\ 0 & 1 & 1 & 0 & 0 & 1 & 1 \\ 1 & 1 & 0 & 0 & 1 & 1 & 0 \\ 0 & 0 & 0 & 1 & 1 & 1 & 1 \\ 1 & 0 & 1 & 1 & 0 & 1 & 0 \\ 0 & 1 & 1 & 1 & 1 & 0 & 0 \\ 1 & 1 & 0 & 1 & 0 & 0 & 1 \end{bmatrix}$$

<u>Figure 5</u>. Examples of S-matrices.

We construct a mask of length 2n - 1 = 13 by placing two copies of the first row side-by-side and omitting the last 0 or 1:

1 1 1 0 1 0 0 1 1 1 0 1 0

(see Figure 6). By moving this mask to the left across a framing mask or window of length n = 7, each row of S appears in turn. Figure 7 shows a much larger mask, of length 2 x 255 - 1 = 509, corresponding to n = 255.

How are such masks constructed? There are three known ways to construct a cyclic S-matrix. They are described in detail in Section A.2 of ref. 2. Here we shall just sketch the most important construction, which is based on <u>maximal length shift-register sequences</u>. This produces cyclic S-matrices of orders $n = 2^m - 1$ for all integers m, i.e., orders n = 1, 3, 7, 15, 31, 63, 127, 255, 511, 1023, $\cdots$. In this construction the first row of the matrix is taken to be one period's worth of the output from a shift-register whose feedback polynomial is a <u>primitive polynomial</u> of degree n. The exact definition of these polynomials need not concern us here, since Table 1 gives examples of degrees 1 through 20, enough to construct S-matrices of order up to $2^{20} - 1 = 1048575$.

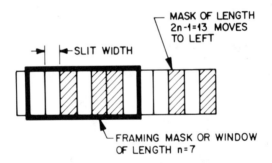

Figure 6. Mask of length 2n - 1 = 13 and framing mask of length n = 7 corresponding to the cyclic S-matrix shown in Equation 5.

Figure 7. Encoding mask of length 2 x 255 - 1 = 509 corresponding to a cyclic S-matrix of size 255 x 255. The first 255 slits are used for the first measurement, slits 2 through 256 for the second measurement, and so on.

Table 1. Primitive polynomials of degree ≤ 20.

Degree	Polynomial
1	$x + 1$
2	$x^2 + x + 1$
3	$x^3 + x + 1$
4	$x^4 + x + 1$
5	$x^5 + x^2 + 1$
6	$x^6 + x + 1$
7	$x^7 + x + 1$
8	$x^8 + x^6 + x^5 + x + 1$
9	$x^9 + x^4 + 1$
10	$x^{10} + x^3 + 1$
11	$x^{11} + x^2 + 1$
12	$x^{12} + x^7 + x^4 + x^3 + 1$
13	$x^{13} + x^4 + x^3 + x + 1$
14	$x^{14} + x^{12} + x^{11} + x + 1$
15	$x^{15} + x + 1$
16	$x^{16} + x^5 + x^3 + x^2 + 1$
17	$x^{17} + x^3 + 1$
18	$x^{18} + x^7 + 1$
19	$x^{19} + x^6 + x^5 + x + 1$
20	$x^{20} + x^3 + 1$

The way the polynomial determines the shift-register is best illustrated by an example. Figure 8 shows the shift-register corresponding to the polynomial

$$x^4 + x + 1.$$

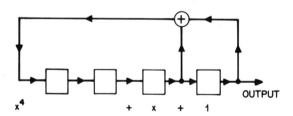

Figure 8. Shift-register corresponding to the primitive polynomial $x^4 + x + 1$.

The output from this shift-register has period $n = 2^m - 1 = 15$, and (if it initially contains 1000) one period's worth of the output is

0 0 0 1 0 0 1 1 0 1 0 1 1 1 1.

Figure 9 shows an S-matrix of order 15 having this sequence as its first row.

Incidentally the output sequence from one of these shift registers is called a <u>pseudo-noise sequence</u> or <u>m-sequence</u>. These have been extensively studied and have many applications--see for example refs. 46-49.

The other two constructions of cyclic S-matrices are (i) the quadratic residue construction, which produces matrices of every order n that is a prime number of the form $4a + 3$, i.e., 3, 7, 11, 19, 23, 31, 43, 47, 59, 67, 71, 79, 83, $\cdots$, and (ii) the twin prime construction, which works whenever $n = p(p + 2)$ and both p and p + 2 are prime numbers--for example n = 15, 35, 143, 323, 899, $\cdots$. For further details see refs. 2, 43, 50-52.

$$S_{15} = \begin{bmatrix} 0 & 0 & 0 & 1 & 0 & 0 & 1 & 1 & 0 & 1 & 0 & 1 & 1 & 1 & 1 \\ 0 & 0 & 1 & 0 & 0 & 1 & 1 & 0 & 1 & 0 & 1 & 1 & 1 & 1 & 0 \\ 0 & 1 & 0 & 0 & 1 & 1 & 0 & 1 & 0 & 1 & 1 & 1 & 1 & 0 & 0 \\ 1 & 0 & 0 & 1 & 1 & 0 & 1 & 0 & 1 & 1 & 1 & 1 & 0 & 0 & 0 \\ 0 & 0 & 1 & 1 & 0 & 1 & 0 & 1 & 1 & 1 & 1 & 0 & 0 & 0 & 1 \\ 0 & 1 & 1 & 0 & 1 & 0 & 1 & 1 & 1 & 1 & 0 & 0 & 0 & 1 & 0 \\ 1 & 1 & 0 & 1 & 0 & 1 & 1 & 1 & 1 & 0 & 0 & 0 & 1 & 0 & 0 \\ 1 & 0 & 1 & 0 & 1 & 1 & 1 & 1 & 0 & 0 & 0 & 1 & 0 & 0 & 1 \\ 0 & 1 & 0 & 1 & 1 & 1 & 1 & 0 & 0 & 0 & 1 & 0 & 0 & 1 & 1 \\ 1 & 0 & 1 & 1 & 1 & 1 & 0 & 0 & 0 & 1 & 0 & 0 & 1 & 1 & 0 \\ 0 & 1 & 1 & 1 & 1 & 0 & 0 & 0 & 1 & 0 & 0 & 1 & 1 & 0 & 1 \\ 1 & 1 & 1 & 1 & 0 & 0 & 0 & 1 & 0 & 0 & 1 & 1 & 0 & 1 & 0 \\ 1 & 1 & 1 & 0 & 0 & 0 & 1 & 0 & 0 & 1 & 1 & 0 & 1 & 0 & 1 \\ 1 & 1 & 0 & 0 & 0 & 1 & 0 & 0 & 1 & 1 & 0 & 1 & 0 & 1 & 1 \\ 1 & 0 & 0 & 0 & 1 & 0 & 0 & 1 & 1 & 0 & 1 & 0 & 1 & 1 & 1 \end{bmatrix}$$

<u>Figure 9</u>. S-matrix of order 15 constructed from the output of the shift-register in Figure 8.

If $\psi = (\psi_1, \cdots, \psi_n)^T$ is any column vector of real numbers, the product

$$\eta = H_n \psi$$

is called the _Hadamard transform_ of ψ. Similarly

$$\eta = S_n \psi$$

may be called an _S-transform_. Our weighing designs and masks thus make it possible to measure the Hadamard transform (or S-transform) of the unknown vector rather than the vector itself--see Equation 7. This explains why these instruments are called Hadamard transform spectrometers. Also one could argue that the reason these schemes reduce the mean squared error is that the Hadamard transform of a vector is less susceptible to noise than the vector itself (but see also ref. 53).

Of course if n is not a multiple of 4 (in the chemical balance problem) or not of the form 4a + 3 (in the mask problem), we cannot use Hadamard or S-matrices. This is not a serious difficulty, since we can usually increase n to the next number of the desired type. Nevertheless it is worth mentioning that for these other values of n there are many good matrices available (although none quite so good as Hadamard and S-matrices).

One interesting class of matrices with entries 0, 1, and -1, for example, are _conference matrices_. These are similar to Hadamard matrices but have a slightly different defining equation. They also give rise to good weighing designs. A conference matrix C_n of order n is an n x n matrix with diagonal entries 0 and other entries +1 or -1 which satisfies

$$C_n C_n^T = (n - 1) I_n \qquad [27]$$

The name arises from the use of such matrices in the design of networks having the same attenuation between every pair of terminals (see refs. 10, 45, 54-58).

n must be even for C_n to exist. But if n is a multiple of 4 these are inferior to Hadamard matrices, so we shall assume n has the form 4a + 2. In this case, by suitably multiplying rows and columns by -1, C_n can be put in the form

$$C_n = \begin{pmatrix} 0 & 1 & 1 & \cdots & 1 \\ 1 & & & & \\ 1 & & B_{n-1} & & \\ \vdots & & & & \\ 1 & & & & \end{pmatrix} \qquad [28]$$

where $B_{n-1} = (b_{ij})$ is a symmetric matrix. Several constructions for conference matrices are known. The most useful for our purpose is Paley's construction. Let $n = 4a + 2 = p + 1$, where p is an odd prime, and set $b_{ij} = 0$ if $i = j$, $b_{ij} = 1$ if $j - i$ is a square (modulo p), and $b_{ij} = -1$ if $j - i$ is not a square (modulo p). The resulting matrix (Equation 28) is a conference matrix. For example, if $p = 5$, the squares modulo 5 are $1^2 = 1$ and $2^2 = 4$, and we obtain

$$C_6 = \begin{pmatrix} 0 & 1 & 1 & 1 & 1 & 1 \\ 1 & 0 & 1 & - & - & 1 \\ 1 & 1 & 0 & 1 & - & - \\ 1 & - & 1 & 0 & 1 & - \\ 1 & - & - & 1 & 0 & 1 \\ 1 & 1 & - & - & 1 & 0 \end{pmatrix}$$

where $-$ stands for -1. Note that this construction gives a cyclic matrix B_{n-1}.

The same construction works if p is replaced by any odd prime power p^m, and the rows and columns of B_{n-1} are labeled with the elements of the Galois field $GF(p^m)$. Now we set $b_{ij} = 1$ if $j - i$ is a square in $GF(p^m)$, etc. This resulting matrix B_{n-1} is no longer cyclic.

The construction gives symmetric, cyclic, conference matrices of orders 6, 14, 18, 30, 38, 42, 54, 62,$\cdots$, and symmetric conference matrices of orders 10, 26, 50,$\cdots$. Choosing the mask $W = C_n$ we obtain a weighing design that has mean squared error

$$\epsilon = \epsilon_i = \frac{\sigma^2}{n-1}$$

This is only slightly inferior to a Hadamard matrix--compare Equation 19.

This concludes our discussion of matrices. For additional information see the references, especially refs. 4, 10, 11, 43, 45, 46, 51, and 59.

FURTHER TOPICS

We have said nothing about the actual construction of a multiplexing spectrometer, but for this the reader is referred to Chapter 5 of ref. 2. In practice it is more convenient to move the mask (for example that shown in Figure 6) continuously across the dispersed light rather than in steps. This produces a linear distortion of the final spectrum and can therefore be compensated for. The analysis of this and a number of other systematic errors is given in ref. 24 and Chapter 6 of ref. 2.

A different type of error is analyzed in ref. 26: this is a remarkable constellation of blips that are produced when the slits in the mask are all slightly wider (or all slightly narrower) than they should be. The explanation and correction of these blips involves an unusual application of the theory of Galois fields to spectroscopy.

The technique described in this chapter, of improving accuracy by measuring several unknowns simultaneously, can be applied to other problems besides weighing and spectroscopy. For example, instead of the unknowns being the components of a beam of light at different wavelengths, they could be the spatial components representing the different points of a black-and-white picture. A device of this kind is called an <u>imager</u>.[2,19,25] More generally the unknowns could be a combination of spatial and wavelength components, as in the case when we are constructing a colored picture. Such an instrument is called a <u>spectrometric imager</u>. In each case the theory, and the choice of the optimal masks, is exactly the same as for the multiplexing spectrometers described here. Other applications will be found in Chapter 7 of ref. 2. The technique is potentially applicable whenever it is possible to measure several unknowns at once, weights and lengths being characteristic examples. Recent applications to X-ray astronomy are described in refs. 60 and 61.

Finally, many variations of the basic design are possible. For example it is possible to have masks at both the entrance and the exit of the diffraction grating, thus allowing even more light to reach the detector. Once again the reader is referred to ref. 2.

REFERENCES

1. Yates, F. 1935, J. Roy. Statist. Soc. Supp. $\underline{2}$, 181-247.
2. Harwit, M., and Sloane, N. J. A. 1979, <u>Hadamard Transform Optics</u>, Academic Press, N.Y.
3. Banerjee, K. S. 1975, <u>Weighing Designs for Chemistry, Medicine, Economics, Operations Research, Statistics</u>, Marcel Dekker, N.Y.
4. Geramita, A. V., and Seberry, J. 1979, <u>Orthogonal Designs</u>, Marcel Dekker, N.Y.
5. Geramita, A. V., and Wallis, J. S. 1974, Utilitas Math. $\underline{6}$, 209-236.
6. Hotelling, H. 1944, Ann. Math. Statist. $\underline{15}$, 297-306.
7. Kiefer, J. 1959, J. Roy. Statis. Soc., Ser. B. $\underline{21}$, 272-319.
8. Kishen, K. 1945, Ann. Math. Stat. $\underline{16}$, 294-300.
9. Mood, A. M. 1946, Ann. Math. Stat. $\underline{17}$, 432-446.
10. Raghavarao, D. 1971, <u>Constructions and Combinatorial Problems in Design of Experiments</u>, John Wiley, N.Y.
11. Sloane, N. J. A., and Harwit, M. 1976, Appl. Opt. $\underline{15}$, 107-114.
12. Golay, M. J. E. 1949, J. Opt. Soc. Amer. $\underline{39}$, 437-444.
13. Golay, M. J. E. 1951, J. Opt. Soc. Amer. $\underline{41}$, 468-472.
14. Fellgett, P. 1951, Ph.D. Thesis, Cambridge Univ., Cambridge.

15. Fellgett, P. 1967, J. de Physique, Colloque C2, 28, 165-171.
16. Ibbett, R. N., Aspinall, D., and Grainger, J. F. 1968, Appl. Opt. 7, 1089-1093.
17. Decker, J. A., Jr., and Harwit, M. 1968, Appl. Opt. 7, 2205-2209.
18. Nelson, E. D., and Fredman, M. L. 1970, J. Opt. Soc. Amer. 60, 1664-1669.
19. Harwit, M. 1971, Appl. Opt. 10, 1415-1421; 1973, 12, 285-288.
20. Harwit, M., Phillips, P. G., Fine, T., and Sloane, N. J. A. 1970, Appl. Opt. 9, 1149-1154.
21. Harwit, M., Phillips, P. G., King, L. W., and Briotta, D. A., Jr. 1974, Appl. Opt. 13, 2669-2674.
22. Sloane, N. J. A. 1979, Mathematics Magazine 52, 71-79.
23. Sloane, N. J. A., Fine, T., Phillips, P. G., and Harwit, M. 1969, Appl. Opt. 8, 2103-2106.
24. Sloane, N. J. A., Harwit, M., and Tai, M.-H. 1978, Appl. Opt. 17, 2991-3002.
25. Swift, R. D., Wattson, R. B., Decker, J. A., Jr., Paganetti, R., and Harwit, M. 1976, Appl. Opt. 15, 1595-1609.
26. Tai, M.-H., Harwit, M., and Sloane, N. J. A. 1975, Appl. Opt. 14, 2678-2686.
27. Decker, J. A., Jr. 1971, Appl. Opt. 10, 510-514.
28. Ben-Israel, A., and Charnes, A. 1963, J. Soc. Indust. Appl. Math. 11, 667-699.
29. Ben-Israel, A., and Greville, T. N. E. 1974, Generalized Inverses: Theory and Applications, John Wiley, N.Y.
30. Nashed, M. Z., ed. 1976, Generalized Inverses and Applications, Academic Press, N.Y.
31. Penrose, R. 1955, Proc. Cambridge Philos. Soc. 51, 406-413.
32. Penrose, R. 1956, Proc. Cambridge Philos. Soc. 52, 17-19.
33. Price, C. M. 1964, SIAM Review 6, 115-120.
34. Rao, C. R., and Mitra, S. K. 1971, Generalized Inverse of Matrices and Its Applications, John Wiley, N.Y.
35. Cramér, H. 1946, Mathematical Methods of Statistics, Princeton Univ. Press, Princeton.
36. Deutsch, R. 1965, Estimation Theory, Prentice-Hall, Englewood Cliffs, N.J.
37. Efron, B. 1975, Advances in Math. 16, 259-277.
38. Efron, B., and Morris, C. 1977, Scientific American 236, Number 5, 119-127.
39. James, W., and Stein, C. 1961, Proc. Fourth Berkeley Symp. Math. Stat. Prob., Univ. Calif. Press, Berkeley, Calif. 1, 361-379.
40. Stein, C. 1955, Proc. Third Berkeley Symp. Math. Stat. Prob., Univ. Calif. Press, Berkeley, Calif. 1, 197-206.
41. Fischer, J., private communication.
42. Hadamard, J. 1893, Bull. Sci. Math. (2) 17, 240-248.
43. Hall, M., Jr. 1967, Combinatorial Theory, Blaisdell, Waltham, Mass.
44. Turyn, R. J. 1974, J. Combinatorial Theory 16A, 313-333.

45. Wallis, W. D., Street, A. P., and Wallis, J. S. 1972, Lecture Notes in Math. 292, Springer-Verlag, N.Y.
46. Golomb, S. W., ed. 1964, Digital Communications with Space Applications, Prentice-Hall, Englewood Cliffs, N.J.
47. van Lint, J. H., MacWilliams, F. J., and Sloane, N. J. A. 1979, SIAM J. Appl. Math. 36, 62-72.
48. MacWilliams, F. J., and Sloane, N. J. A. 1976, Proc. IFFF 64, 1715-1729.
49. MacWilliams, F. J., and Sloane, N. J. A. 1977, The Theory of Error-Correcting Codes, North-Holland, Amsterdam.
50. Baumert, L. D. 1967, JPL Space Programs Summary 37-43-IV, 311-314.
51. Baumert, L. D. 1971, Lecture Notes in Math. 182, Springer-Verlag, N.Y.
52. Thoene, R., and Golomb, S. W. 1966, JPL Space Programs Summary 37-40-IV, 207-208.
53. Berlekamp, E. R. 1970, Bell Syst. Tech. J. 49, 969-986.
54. Belevitch, V. 1950, Electronics Comm. 27, 231-244.
55. Belevitch, V. 1956, Proc. Symp. Modern Network Synthesis, Polytechnic Inst. Brooklyn, Brooklyn, N.Y., pp. 175-195.
56. Belevitch, V. 1968, Ann. Soc. Scientifiques Bruxelles 82(I), 13-32.
57. Delsarte, P., Goethals, J.-M., and Seidel, J. J. 1971, Canad. J. Math. 23, 816-832.
58. van Lint, J. H. 1974, Lecture Notes in Math. 382, Springer-Verlag, N.Y., Chapter 14.
59. Brenner, J., and Cummings, L. 1972, Amer. Math. Monthly 79, 626-630.
60. Miyamoto, S., Tsunemi, H., and Tsuno, K. 1981, "Some Characteristics of the Hadamard Transform X-ray Telescope," preprint.
61. Skinner, G. K. 1980, J. British Interplanetary Soc. 33, 333-337.

PROCESSING SOFTWARE FOR FOURIER TRANSFORM SPECTROSCOPIES

C. L. Dumoulin and G. C. Levy

Department of Chemistry
The Florida State University
Tallahassee, Florida 32306

INTRODUCTION

Data processing techniques centered around the Fourier Transform (FT) have greatly expanded the power and flexibility of many conventional data manipulation methods. In addition, the availability of inexpensive mini- and micro-computers has made digital data processing techniques cost-effective for most scientific and engineering laboratories. Instrumental methods such as FT-NMR have evolved as a direct consequence of advances in computer hardware and the creation of the Fast Fourier Transform (FFT) algorithm. While the Fourier transform itself is not a user-interactive technique, the use of interactive software in all other phases of Fourier transform data processing is advantageous. This chapter will use examples from a modular software package nearing completion at Florida State University[1] to illustrate how interactive software can be written and used. The package, FSUNMR, was written for, but is not limited to, NMR spectral data. Considerable attention will be given to one of the most difficult aspects of programming many small computers: the implementation of very large data arrays. Finally, a Fourier transform algorithm for large (virtual) arrays will be presented.

SOFTWARE SUPPORT FOR THE FOURIER TRANSFORM

The power of the Fourier transform as a processing tool is greatly enhanced by the judicious use of support routines. Most processing systems provide the user with a collection of mathematical functions that can be applied to the data before and after a Fourier transform. These functions can greatly affect the data obtained from various experiments. Mathematical functions exist for signal-to-noise enhancement, resolution enhancement, etc. Other mathematical operations can be performed on the data in conjunction with the

Fourier transform to provide integration, differentiation, convolution, and correlation. These operations are straightforward and normally require little user intervention.

Some operations are too complex for non-interactive execution. For example, many data processing operations require multiple input parameters prior to computation. Frequently, as in FSUNMR, the user chooses the level of control he/she will have over a processing operation (total, partial, or none). The choice of control level is often made on the basis of data quality and/or user familiarity with the data. Output in particular should be highly interactive. Users need total control over a rapid "soft" copy of the data (graphics) for interactive processing. Once the data has been manipulated, a permanent record of the data usually needs to be made on "hard" copy devices (printers and plotters). A single plot format for data would be too restrictive for most laboratories and users are generally given a choice of format. Tabular listings of numeric data are often generated during the course of processing. Most modern computer operating systems support spooling of print (and perhaps plot) output permitting user interaction to proceed concurrently with the output operation. Several interactive processing routines from FSUNMR are described below. Modules such as these can be valuable processing tools, increasing user throughput and versatility in most laboratories.

Peak analysis

Once spectral data is acquired and processed in its final form, two steps remain for the spectroscopist: quantification and interpretation. To quantify the spectrum, the spectroscopist typically tabulates pertinent information which may include peak positions, intensities, widths, and/or integrals. Manual quantification of spectra is usually straightforward but limited in accuracy and often tedious. The untrained eye has little difficulty distinguishing peaks from interfering factors such as noise, baseline distortion, and spectral offsets. Automatic quantification by computer software, on the other hand, is difficult to implement because human judgment must be imitated when compensating for interfering factors. Furthermore, peak analysis software should be constructed to be totally general, quantifying a broad range of spectral types.

Since the user will never be totally satisfied with automatic analysis all the time, it is very important to allow him to assume control over all parts of the quantification whenever necessary. For example, the peak analysis routine of FSUNMR performs an automatic quantification by default. The user is given a wide range of options when the computations are complete. These include insertion and deletion of peaks, manual declaration of a threshold, peak position calibration, and a complete set of display options. Good peak analysis routines can save the spectroscopist a lot of time without

Processing Software for Fourier Transform Spectroscopies

compromising the quality of the results. One scenario for automatic peak analysis is outlined below:

    i. Determine the noise level and offset.

    ii. Locate peaks.

    iii. Measure peak intensities, widths, and integrals.

Noise and offset determination

The smallest value for a significant signal against a noise background is generally given as three times the standard deviation of the noise.[2] Calculation of the standard deviation of a set of numbers is a common technique which can be used for spectral noise calculation only if the set of chosen points contains only noise. Conceivably, the user could indicate a signal-free region to the calculation. Automatic noise and offset determination can be achieved, however, if the standard deviation calculation is made to iterate. Suppose the standard deviation and average of all the data (both signal and noise) is calculated. A first approximation of the noise level would then be three times the standard deviation. Likewise, the offset is approximated by the average. The standard deviation can then be recalculated for all points suspected of being noise. A data point is defined as noise when:

$3*($standard deviation$) > $ absolute value of $($DATA $-$ OFFSET$)$

The old and new standard deviations can be compared and if their difference is not within a predetermined limit, the process can be repeated with the new values for the offset and noise level. An iteration count should be kept to stop the calculation in case subsequent standard deviations do not converge. If this happens, control should be returned to the user so that the offset and noise level can be calculated manually.

Peak identification

Armed with the noise level and offset, location of the peaks is now possible. Peaks can be defined in a variety of ways, but regardless of the criterion used, it is prudent to distinguish noise from possible peaks. A threshold can be defined as the offset plus the noise level. Only data points whose magnitudes are greater than the noise level need be considered in the search for possible peaks. The actual peak-picking algorithm should be able to pick unresolved peaks and ignore noise spikes on peak shoulders. Cooper[3] has outlined a good peak identification technique. His method searches for a local maximum in the data above the threshold. A peak is defined as a significant rise in the data followed by a significant fall. Only a point whose difference from the previous point is greater than the noise level is considered significant. This technique works well for any type of peak as long as the threshold is chosen properly.

Intensities, widths, and integrals

The procedure outlined above merely finds the data point of greatest intensity. Rarely will this point be located at the center of the peak. Accurate measurements of peak position and peak intensity can be obtained by interpolating the points in the immediate vicinity of the maximum point with an appropriate function. Linear interpolations yield good approximations for peak position but usually overestimate the peak intensity. Parabolic interpolations, on the other hand, provide good approximations for both peak intensity and location. An accurate measurement of peak height is essential if the peak width at half height is to be calculated empirically; small deviations in measured peak intensity can cause disproportionate errors in measured widths. A parabolic interpolation of peak height coupled with linear interpolation of peak sides generally gives satisfactory results.

Integrals can be automatically calculated using a number of standard techniques. The integrating window can be determined by searching for the points of a peak that cross the threshold determined for the peak selection. A first approximation can be obtained by summing all the points in the integration range. More precise results can be obtained by linearly interpolating the data points (trapezoidal rule). Integration by Simpson's rule (described in most calculus texts) is the method of choice, however, because it is a polynomial interpolation. Regardless of the method employed, two aspects of the data can introduce large systematic errors in integrals: baseline offset and baseline distortion. Baseline problems and integration will be discussed in subsequent sections.

If the theoretical line shape is known for a peak, the function can be fitted to the data, and the intensity, width, and integral determined. In practice, however, experimental line shapes can be too complex to characterize conveniently. NMR lines, for example, have been shown by the Bloch equations to be Lorentzian in shape. A number of experimental factors, however, can contribute non-Lorentzian components to the observed line shape. Furthermore, unresolved peaks will have to be fitted to a sum of individual Lorentzians rather than to a single curve.

Baseline flattening

Baseline distortions can be caused by a variety of instrumental and processing conditions. Ideal spectral baselines have an average mathematical value of zero over all segments of data. Distorted baselines, on the other hand, have varying local averages. Truncation of discrete data can cause serious distortions in the Fourier transform co-domain. Experimental conditions such as overloaded amplifiers and overflowing digitizers can also distort baselines. Often, a distortion is not really a distortion at all, but an additional signal due to the sample matrix or an interfering substance. If the distortion cannot be prevented, it is useful to be able to compensate for it during processing. Well-characterized distortions due to spectral interference or matrix background can often be measured in the absence of the signal of interest. This spectral "blank" can then be subtracted from the distorted spectrum. Baseline distortions due to processing factors and instrumental problems can be highly variable, however, changing from spectrum to spectrum. Under these conditions, a spectral blank is not available, but one can be constructed from the distorted data.

Pearson[4] has written a routine to remove broad spectral features such as baseline distortions in the presence of narrow features such as signals. His algorithm fits a function to the signal-free components of a spectrum. The function is then subtracted from the data set to give a partially corrected spectrum. The process is repeated until the baseline is flat. Almost any function can be used for the fitting procedure. Since most distortions manifest themselves as continuous curves, fitting a multi-order polynomial is appropriate. The choice of data points for the fitting procedure is similar to the technique used in the automatic noise determination algorithm described earlier: only data points whose values are less than three times the standard deviation of the entire data set are considered in the fit.

Pearson's method can be made more versatile if the user is allowed to specify the order of the correction. A zero-order polynomial correction, for example, is more useful than a sixth-order correction if the user wants to change the spectral offset without risking modification to the line shapes. Furthermore, the extra computation required for higher-order corrections is unnecessary and time-consuming when the data has simple distortions. Very complex distortions can be corrected if an additional improvement is made to the algorithm so that the fitting procedure acts on small portions or blocks of the data set. Thus, a fourth-order correction on four individual blocks of data can be as effective as a sixteenth-order correction on the entire spectrum. It is important that the user place the endpoints of each block in a signal-free region. Otherwise, part of a signal may contribute to the fitted polynomial at the block edge, causing discontinuities between adjacent blocks. Default settings for the polynomial order and number of blocks are useful, but the user should be allowed to optimize the execution conditions of the correction since bad distortions can make the routine slower than a Fourier transform.

Integration

Spectroscopic measurements can usually be made under analytical conditions such that the amount of signal due to a component in the sample is proportional to the concentration. If all the line widths and line shapes are identical, then the signal intensity is proportional to concentration. Frequently, line widths and line shapes are subject to a variety of chemical and instrumental influences, however, and peak integrals must be measured for quantitative results. The precision or reproducibility of integrations obtained on discrete data points increases with the number of integrated points. Accuracy also increases with the number of integrated points. The choice of integration limits determines the number of data points to be integrated and thus has a large effect on the integral.

NMR lines theoretically have a Lorentzian distribution:

$$f(\nu) = A \frac{a}{a^2 + (2\nu)^2}$$

in which a is the full width at half maximum height and A is the peak intensity times a. Integrating between the limits $-\nu_1$ and $+\nu_1$ gives:

$$\int_{-\nu_1}^{+\nu_1} f(\nu) \, d\nu = 2Aa \int_0^{\nu_1} \frac{1}{a^2 + (2\nu)^2} \, d\nu$$

$$= A \tan^{-1}\left(\frac{2\nu}{a}\right)$$

When ν_1 is infinite, the total integrated intensity becomes:

$$\int_{-\infty}^{+\infty} f(\nu) \, d\nu = \frac{A\pi}{2}$$

It is unreasonable to integrate digitally over all frequencies, so an approximation must be made by choosing the limits. Fortunately, if this approximation is consistently implemented, the ratio of individual integrals within a spectrum will remain constant. Table 1 shows how the limits can be chosen as a function of peak width to obtain various accuracies for a Lorentzian line.

Since most of the integrated intensity lies within a few multiples of the line width, small errors in the determination of the line width or in the line shape will have very little effect on the integral if the limits are consistently chosen as several line widths.

Table 1. Accuracy of integration for Lorentzian lines as a function of integration range.

Accuracy (fraction of total integral)	Integration Range (in units of line width)
99.99%	6366
99.90	636
99.00	63.6
90.00	6.31
80.00	3.08
50.00	1.00

Automatic integrations can be implemented in automatic peak analysis software if peak widths are determined. The user should be given the option to change the default limit criterion, however, in order to integrate partially resolved peaks. The user can increase the number of integrated points without changing the limit criterion by zero-filling the data set while in the Fourier co-domain. This action increases the digital resolution, or number of points per peak, thus increasing the precision of a peaks integration. Convolution techniques, such as line broadening, will not affect integrals when the integration limits are chosen as a function of line width. Sloping, offset, or distorted baselines can exert disastrous effects on integrals, however, and must be compensated for prior to integration. The signal-to-noise ratio of a peak will affect the precision of an integration, but the effect is usually negligible if enough data points are integrated.

PROGRAM DATA MANAGEMENT

Virtual memory

One of the most important aspects of any software package is the data structure. Most interactive data processing software consists of a mass storage data base (files on a disk subsystem) under the control of a program or set of programs acting on only one data set at a time. A data set is typically transferred to computer memory to be modified and/or analyzed. As long as a data set fits into the allotted memory space, the design of the software is straightforward. Many general-purpose mini- and micro-computers, however, have a limited memory address space (typically 64 Kbytes), making data structure design for large data sets quite complex. If the size of the data set is limited to a power of 2, then the largest data set conveniently handled by a typical mini-computer will occupy 32 Kbytes of memory (the program and other data tables will consist of the rest of memory). Typical floating point format stores one number in 4 bytes of memory, thus limited the maximum data size to 8192 floating point data elements. If the data set needs to consist of both real and imaginary parts (a consequence of FT spectroscopy), then the maximum data size becomes 4096 points. 4096 points is a comfortable maximum for many spectroscopies and is usually not a limitation. Unfortunately, certain applications require much larger data sets and the design of software for these data sets is far from trivial.

Obviously, large data structures are most easily accommodated by computers than can address all memory locations of a data set. Most mini- and micro-computers, however, are incapable of addressing such a large range and must rely on a hardware mechanism to switch in various parts of memory. Thus, a small computer can access large amounts of memory (one megabyte, for example), even though it can only address 64 Kbytes at any given instant. Memory not currently addressable by the CPU is frequently called extended memory. Multi-user operating systems for most mini- and micro-computers use this mechanism to give each user of the system the maximum amount of memory addressable by the CPU. This memory space is usually independent of all other memory spaces and is switched in and out of the CPU's address range very quickly. Some of the more sophisticated operating systems treat extended memory as portions of a disk file that are heuristically read to and written from memory. This concept is known as virtual memory and it permits the use of arrays that are larger than the entire computer memory.

The details of the memory redefinition process depend on the hardware design of the computer. Simultaneous redefinition of all addressable memory (bank switching) is found on some micro-computers. This is not well-suited to the implementation of large data structures because the program must be re-mapped along with the data. Most mini-computers, on the other hand, allow portions (pages) of memory to be redefined. This arrangement allows the data to be called into the CPU address space without disturbing the program. Typically, a group of pages in the memory address space is treated as a window that can be made to contain desired memory pages. These pages can be part of a collection of pages forming a large virtual array. The size of the virtual array is thus limited to the total amount of memory (primary storage) and disk space (secondary storage) in the computer and not to the CPU's addressing range. Parts of the program can also be moved in and out of the address space in a similar manner, providing very fast program overlaying. In general, virtual memory data bases can be more efficiently accessed than secondary data bases because of the slower speed associated with I/O (input/output). Nevertheless, the conscious use of virtual memory is not commonplace among most programmers.

Processing Software for Fourier Transform Spectroscopies

Transparent re-mapping

From a programmer's point of view, the use of virtual data arrays can take one of two forms. The data array re-mapping can be under the direct control of the programmer or it can be transparent to him. The latter case can be accomplished easily with a set of assembly level routines designed to make high level language subroutine calls look like memory references. Thus the programmer is actually calling a subroutine to perform data array operations. These virtual array manager subroutines can be quite sophisticated and usually can be written to be totally transparent to the junior programmer. Unfortunately, the overhead of this method is quite heavy, and memory access times are slowed by orders of magnitude over ordinary memory addressing. The heuristic re-mapping of multiple windows will reduce this overhead significantly for cases where individual array positions are not addressed sequentially but in some coherent pattern. For example, the following code would run much more quickly with two virtual windows than it would with a single window simply because fewer re-map operations would be executed.

```
      DO 100 I=1,MAXINDEX/2

      K=MAXINDEX-(I-1)

      TEMP=DATA(I)

      DATA(I)=DATA(K)

      DATA(K)=TEMP

100   CONTINUE
```

The efficiency of ordinary Fast Fourier Transform (FFT) algorithms for transparently virtual data arrays also benefits somewhat from multiple windows. Unfortunately, the algorithm does not lend itself very well to this method of memory addressing. Use of a transparently virtual FFT is practical only if the total window size represents a significant fraction of the data array.

Explicit re-mapping

If the programmer is given control over the virtual array windows, then the data array can be addressed directly and the subroutine overhead present in the transparent system is avoided. Of course, algorithms should be designed to minimize the amount of re-mapping to maintain efficiency. As in the case of the transparent virtual array manager, multiple array windows yield the greatest efficiency. The algorithm shown below is functionally equivalent to the previous one, but is more than an order of magnitude more efficient despite its apparent greater complexity.

```
              PAGECOUNT=MAXINDEX/512

              WINDOWA=1

              WINDOWB=PAGECOUNT


              DO 100 I=1,PAGECOUNT/2

              CALL ALIGNWINDOW(WINDOWA,DATA_A)

              CALL ALIGNWINDOW(WINDOWB,DATA_B)

                 DO 50 J=1,512

                 TEMP=DATA_A(J)

                 DATA_A(J)=DATA_B(512-J+1)

                 DATA_B(512-J+1)=TEMP

        50       CONTINUE

              WINDOWA=WINDOWA+1

              WINDOWB=WINDOWB-1

   100        CONTINUE
```

In this case, the programmer is responsible for maintenance of two virtual array windows, each 512 data points long. The subroutine, "ALIGNWINDOW" aligns a given page of the data array with the appropriate window (DATA_A or DATA_B). Thus, the 513th position in the data array can be addressed directly by aligning page 2 with a window, (e.g., DATA_A) and directly addressing the first position of that window. FFT algorithms can be constructed that can make efficient use of this type of mechanism, as can other pre- and post-FT operations. The timing statistics for FFT routines using various data array mechanisms are given in Table 2.

THE FFT ALGORITHM AND VIRTUAL DATA ARRAYS

Almost every Fast Fourier Transform algorithm written to date has been written for data sets residing entirely within the CPU address space. Performing a floating point Fourier transform on a very large data set with a conventional FFT algorithm therefore requires a computer capable of directly addressing large amounts of memory. Several variations of the FFT algorithm have been devised in which the data set is not entirely memory resident. Singleton[5]

Table 2. Timing statistics for various floating point FFT algorithms on a virtual memory mini-computer.

	Size					
	512-512	1K-1K	2K-2K	4K-4K	8K-8K	16K-16K
Direct	1 sec	2 sec	4 sec	8 sec	xxx[a]	xxx[a]
Transparent	1 sec	2 sec	140 sec	300 sec	460 sec	xxx[b]
Explicit	1 sec	2 sec	5 sec	11 sec	25 sec	59 sec

[a] 4K-4K is the maximum data size for directly addressed data.
[b] The routine deadlocked.

has written an FFT algorithm that acts on sequentially stored data (magnetic tape or disk files), requiring little memory. Unfortunately, his technique requires multiple intermediate files and is most efficient when 8 files are used. Brenner[6] has presented techniques by Ryder and Granger that are better suited to block oriented disk files than Singleton's technique. All of the above algorithms, however, are slowed by peripheral I/O, which comprises a significant part of the overall execution time. Virtual array FFT's, on the other hand, perform little I/O because the data is usually in memory. Consequently, virtual array FFT algorithms have the potential to be only slightly slower than equivalent directly addressed FFT algorithms.

If the data set to be Fourier transformed is expressed as $d(j)$, $j = 0,1,2,\cdots,N-1$, in which $d(j)$ consists of real and imaginary parts, then the discrete Fourier transform is given by:

$$D(k) = \sum_{j=0}^{N-1} d(j) \, W$$

in which $k = 0,1,2,\cdots,N-1$

and $W = \exp(2\pi i j k/N)$

The inverse transform is similar, and can be expressed as:

$$d(j) = \frac{1}{N} \sum_{k=0}^{N-1} D(k) W$$

in which $\quad j = 0, 1, 2, \cdots, N-1$

and $\quad W = \exp(-2\pi i jk/N)$

Very efficient Fourier transform algorithms can be written if N is limited to a power of 2 (i.e., $N = 2^M$, $M = 1,2,3,\cdots$). The Cooley-Tukey algorithm[7] performs the above computation by first resequencing the data so that each data point d(i) is exchanged with data point d(k), where k is simply the binary reversal of i. M passes are then made over the data such that each pass performs N/2 two-by-two transforms on different pairs of data points, and by the end of the final pass, every data point has been mathematically mixed with every other data point. Figure 1 shows how the individual two-by-two transforms are carried out during three passes of an 8-point transform. Note that the first pass performs 8 two-by-two transforms on adjacent pairs of data. By the end of the second pass, however, every four points have been transformed. Thus, for an N-point transform, the I'th pass will result in 2^{M-I} transformed blocks of 2^I points. Very large data sets that are only partially within the CPU's address space at any given time can therefore be transformed by an algorithm making use of smaller internal transforms.

Potentially, the most inefficient part of any algorithm acting on virtual data is the memory redefinition operation. Virtual arrays residing entirely in extended memory can be manipulated much more quickly than virtual arrays residing wholly or partially on disk. Therefore, it is desirable to keep as much of the virtual array as possible in the primary storage. The re-map operation, while orders of magnitude more efficient than disk I/O, is still relatively slow, however, and care should be taken to minimize the number of these operations. With this constriction in mind, the Cooley-Tukey algorithm can be modified to act on a virtual data set.

A virtual array transform using a window that can accommodate 2^I complex data points can perform an N-point transform ($N = 2^M$) in the following four steps:

    i. Resequence the entire data set by exchanging data points having bit-inverted addresses.
   ii. Do 2^I-point transforms (I passes each) on the 2^{M-I} blocks of data by the most expedient means possible.
  iii. Do the remaining passes of the transform as efficiently as possible.
  iv. Resequence the data to present it in the most familiar manner.

Processing Software for Fourier Transform Spectroscopies   81

The bit reversal

The use of multiple re-mapping windows for the real and imaginary arrays is necessary because individual data points to be exchanged often reside in different blocks of the array. An algorithm using two windows is flow-charted in Figure 2. With this algorithm, a data set consisting of X blocks will be re-mapped R times, where R is:

$$R = \left[\sum_{y=1}^{X} (X-Y+2) \right] - 1$$

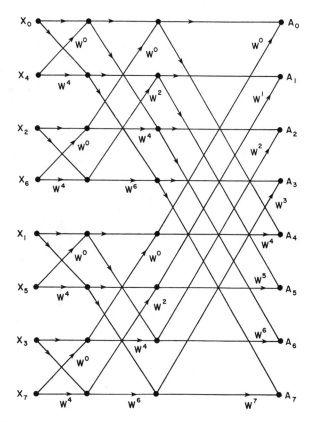

Figure 1. The effect of fast Fourier transformation on an 8-point data set.[8] (a) Overview.

Algorithms utilizing more than two windows per array are conceivable. Greater efficiency and simplicity are best attained, however, by larger windows rather than more windows. If the blocks of the virtual array are not entirely in extended memory, but are instead inserted heuristically distributed between memory and disk, then the approach outlined in Figure 2 has some definite advantages in limiting the amount of disk I/O. A heuristic operating system will tend to keep the most recently used blocks of the virtual array intact in memory. This algorithm acts on a continually decreasing set of blocks as it progresses, and thus at some point during the execution of this algorithm, all remaining blocks to be acted on will be in extended memory. At that point, no further disk I/O is necessary, and the remainder of the routine proceeds at maximum speed.

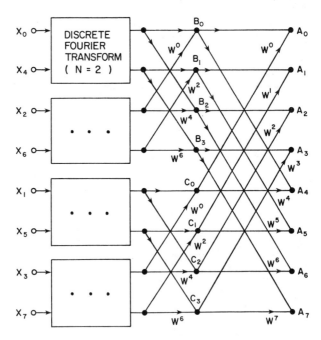

Figure 1. The effect of fast Fourier transformation on an 8-point data set. (b) After the first pass.

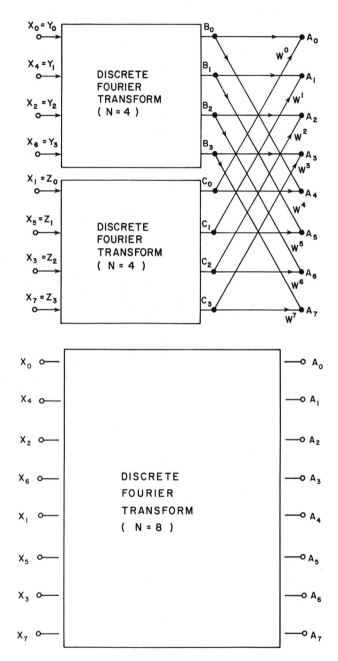

Figure 1. The effect of fast Fourier transformation on an 8-point data set. (c) After the second pass (top diagram). (d) After the third pass (bottom diagram).

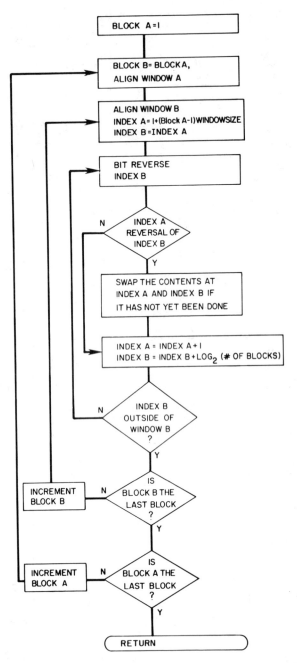

Figure 2. Flow chart of the bit reversal algorithm.

The internal transforms

Unlike the bit reversal step, the internal transforms never need to access data outside the block being transformed. Thus only one window per array is needed. If two windows used in the bit reversal step are adjacent in the CPU's address space, it should be quite easy to treat the windows as one large window whenever necessary. The single window can then be re-mapped in one operation rather than two. Since the internal transforms completely transform the data in the window before re-mapping, the first I passes over the entire data set can be accomplished by re-mapping each block into the window only once. For example, a window 1024 data points wide is large enough to allow the first 10 of 14 passes of a 16K-16K point transform to occur with only 16 re-map operations per array.

The internal transforms can be optimized to be most efficient. For the internal transforms, the transform data are always accessible (within the window). The speed-limiting step, therefore, is the calculation of the sine and cosine terms. Since these terms repeat within each pass of the transform, the most efficient algorithm minimizes the number of calculations by performing the individual 2-by-2 transforms for all points requiring the same sine and cosine terms before recalculating those terms. This approach also minimizes the need for a sine look-up table. Figure 3 shows how an FFT can be written in this manner. If an array processor is available, the internal transforms can be delegated to it for a significant increase in speed. It should be noted, however, that the bit reversal of the array processor FFT would need to be defeated (or nullified), since the bit reversal will have already been performed on the entire data set.

The final passes

After the last block has been transformed, the data has been subjected to I passes of a Cooley-Tukey algorithm. All that remains to complete the transform are the final M-I passes. During these passes, the data points for the 2-by-2 transforms are always in different blocks of the array. Thus, dual windows are necessary. Furthermore, subsequent passes act on different blocks of data, and an approach in which multiple passes are performed for discrete blocks of data (as in the internal transforms) is impossible. Consequently, the final passes are characterized by extensive re-mapping; each block in the array is re-mapped once during each pass. To make matters worse, two windows are in use and the block size is reduced to half the size of the window used for the internal transform. As in the bit reversal routine, the speed-limiting step, therefore, is the re-map operation.

A relatively simple method for the final passes is illustrated in Figure 4. Note that the two innermost loops correspond to the inner loop of a non-virtual Cooley-Tukey arrangement. The extra loop re-maps the window and resets the window index whenever the innermost loop has finished acting on a block of data. No attempt can be made to minimize the number of sine and cosine calculations within a block because every 2-by-2 transform requires different trigonometric values. A sine look-up table may increase efficiency for the final passes, but this has not yet been verified.

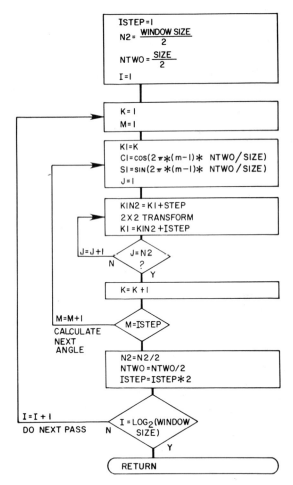

Figure 3. Flow chart of the internal transforms.

Processing Software for Fourier Transform Spectroscopies 87

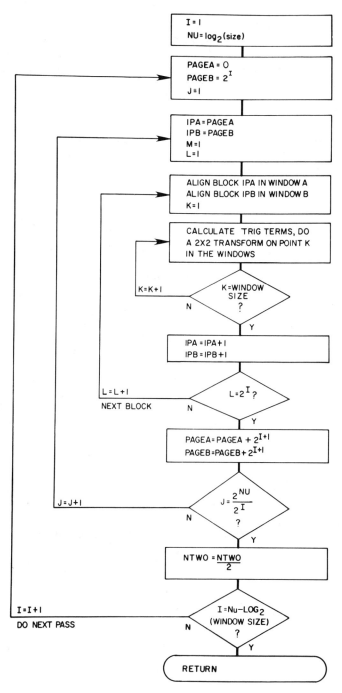

Figure 4. Flow chart of the final passes.

Reordering the data

In some applications, such as quadrature detection nuclear magnetic resonance spectroscopy, it is customary to present the transformed data so that zero is in the center of the frequency domain. Proper spectral presentation can be achieved by reversing the order of the first half of the data set, followed by a reversal of the second half. For other applications not requiring re-sequencing, this step should be ignored. The re-sequencing operation is straightforward, requiring two re-map windows, and does not deserve further discussion other than that found in the sample subroutine in Appendix A.

Performance

In principle, this explicitly re-mapped algorithm can calculate the Fourier transform for any size data set greater than the largest window size. Greatest efficiency can be attained by maximizing the window size, thereby minimizing the number of re-map operations. The timing statistics presented in Table 2 were obtained from a Data General Eclipse S/130 mini-computer running the Advanced Operating System (AOS). The data were single precision, floating point (firmware). Two 4K-byte (1024 data point) windows were used, one each for the real and imaginary arrays. Each window was further divided into two indpendent windows for the bit reversal, the final passes, and the reordering. All the major subroutines were written in FORTRAN 5, with the exception of the bit reversing calculation, which was written in assembly language. Assembly-level drivers were also written to interface the FORTRAN routines to the memory remapping system calls. The operating system (AOS) heuristically manages the distribution of the virtual arrays between memory and disk; the times in Table 2 are for data sets residing entirely within memory.

Table 3 shows how the various parts of the transform contribute to the total number of re-map operations. The total number of re-map operations roughly triples for each doubling of the data size. If a re-map operation takes 2 msec to re-define the CPU's address space, then approximately 3 sec or 5% of the total time of a 16K-16K transform is used for this purpose. If parts of the virtual array reside on disk, however, disk I/O will substantially increase the time required for an individual re-map operation. Clearly, the bit reversal routine becomes the least efficient of all the routines if memory re-mapping is slow. The execution times for the parts of the transform are shown in Table 4. Computationally, the least efficient routine is the final passes, because the algorithm used in the final passes is slower (by a factor of more than two) than the algorithm used for the internal transforms.

Table 3. Total number of re-map operations (both arrays) for various parts of the FFT algorithm.

Size	Bit Reversal	Internal Transform	Final Passes	Reorder	Total
2K-2K	26	4	8	4	46
4K-4K	86	8	32	8	142
8K-8K	302	16	96	16	446
16K-16K	1118	32	256	32	1470

Table 4. Execution times for the various parts of the FFT algorithm (sec).

Size	Bit Reversal	Internal Transform	Final Passes	Re-order	Total
2K-2K	<1	4	1	<1	5
4K-4K	1	6	4	<1	11
8K-8K	2	13	10	<1	25
16K-16K	5	28	24	2	59

The algorithm presented here is by no means the most efficient one possible. Any modification designed to increase speed, however, is likely to make the routine larger and more complex. Although this algorithm was written for virtual data, it is readily adaptable for disk-based data arrays: only the calls to the window alignment routine need be replaced. Should the data be totally and explicitly disk-based, a multi-tasking FFT should be possible in which the transform calculations on one block of data occur simultaneously with the disk I/O of another block.

ACKNOWLEDGMENTS: The National Science Foundation provided primary support for this project (Grant CHE 77-26473). We would like to acknowledge Professor Frank Anet of U.C.L.A. for his collaboration, contributions, and advice during the development of our Fourier transform software.

REFERENCES

1. Dumoulin, C. L. and Levy, G. C. 1980, Comps. and Chem., in press.
2. DeGalan, L., Analytical Spectrometry, Higler Ltd., London, p. 19.
3. Cooper, J. W. 1977, "The Minicomputer in the Laboratory," Wiley Interscience, New York, p. 248.
4. Pearson, G. A. 1977, J. Magn. Reson. 27(2), 265.
5. Singleton, R. C. 1967, IEEE Trans. Audio and Electroacoustics Au-15(2), 91.
6. Brenner, N. M. 1969, IEEE Trans. Audio and Electroacoustics Au-17(2), 128.
7. Cooley, J. W. and Tukey, J. W. 1965, Math. Comput. 19, 297.
8. GAE Subcommittee on Measurement Concepts, IEEE Trans. Audio and Electroacoustics, Au-15(2), 45.

APPENDIX A. The source code of the virtual array transform and its subroutines.

```
C
C      FAST FOURIER TRANSFORM (COOLEY-TUKEY FORM)--MODELED AFTER ONE
C      WRITTEN BY E. O. BRIGHAM
C
C      REVISION HISTORY:
C
C         WRITTEN BY J. Y. READ AND C. L. DUMOULIN...31 AUGUST, 1978
C         MODIFIED BY F. A. L. ANET..................13 June, 1979
C            MODIFICATION: THE PASSES OF THE TRANSFROM WERE RESTRUC-
C                          TURED SO THAT THE SINE AND COSINE TERM
C                          CALCULATIONS ARE PERFORMED ONCE FOR A
C                          GIVEN ANGLE DURING AN ENTIRE PASS.  THIS
C                          RESULTS IN AN INCREASE OF SPEED BY ABOUT
C                          A FACTOR OF 2.
C
C         MODIFIED BY C. L. DUMOULIN.................10 June, 1980
C            MODIFICATION: THE TRANSFORM WAS RESTRUCTURED TO HANDLE
C                          VIRTUAL ARRAYS. THE BITREVERSAL FUNCTION
C                          AND THE DATA REORDERING FUNCTION WERE
C                          SUBROUTINED.
C
C      ********************PERFORMANCE*************************
C
C               SIZE                    TIME
C               _____
C               2K-2K                   5 SEC
C               4K-4K                   11 SEC
C               8K-8K                   25 SEC
C               16K-16K                 59 SEC
C
```

Processing Software for Fourier Transform Spectroscopies

```
C                NOTE: THESE TIMES WERE OBTAINED ON AN ECLIPSE S/130 RUN-
C                      NING AOS WITHOUT OTHER SYSTEM ACTIVITY.  SUFFICIENT
C                      SHARED PAGES WERE AVAILABLE SO THAT BOTH THE REALS
C                      AND IMAGINARIES WERE ALWAYS IN MEMORY AND NEVER ON
C                      DISK.
C
C     ***************COMMON VARIABLES USED********************
C
C     SIZE=NO. OF REALS=NO. OF IMAGINARIES IN TRANSFORM (INTEGER)
C     XV1 = WINDOW TO REAL ARRAY (DEFINED IN "VMEM")
C     XV2 = WINDOW TO IMAGINARY ARRAY (DEFINED IN "VMEM")
C     FDATA(107)=0    IF QUAD
C               =1    IF SING. UNSORTED
C               =2    IF SING. SORTED
C     FDATA(108)=0    IF DATA IS FID
C               =1    IF DATA IS SPECTRA
C     KILL=.TRUE. TO SUPPRESS OUTPUT MESSAGES
C
C     ****************INTERNAL VARIABLES******************
C     NU=LOG(BASE2) OF SIZE
C     N2 USED TO DETERMINE THE SPACING BETWEEN DUAL NODES IN THE
C          COMPUTATIONAL ARRAYS.
C     NTWO STARTS AS SIZE/2 AND IS HALVED IN EVERY PASS.
C     NU1 USED TO DETERMINE THE BINARY SHIFT NECESSARY FOR THE COM-
C          PUTATION OF THE P FACTORS IN THE SIGNAL FLOW DIAGRAM.
C     K=INDEX OF ELEMENT IN THE COMPUTATIONAL ARRAY
C     K1N2=INDEX OF DUAL NODE FOR K+1 ELEMENT IN COMPUTATIONAL ARRAY
C     L=THE NUMBER OF THE COMPUTATIONAL ARRAY IN QUESTION
C     PAGE=# OF THE FIRST SHARED PAGE TO PUT IN THE WINDOW (INTEGER)
C
C     ****************************************************************
C
      SUBROUTINE FFT1
C
      INCLUDE "NMRCOMMON"
      INCLUDE "VMEM"
      INTEGER PAGE,PAGEA,PAGEB
      REAL REALS(1024),IMAGS(1024),AREALS(512),AIMAGS(512),
      BREALS(512),BIMAGS(512)
      EQUIVALENCE (XV1(0),REALS(1))
      EQUIVALENCE (XV2(0),IMAGS(1))
      EQUIVALENCE (XV1(0),AREALS(1))
      EQUIVALENCE (XV2(0),AIMAGS(1))
      EQUIVALENCE (XV1(512),BREALS(1))
      EQUIVALENCE (XV2(512),BIMAGS(1))
C
      INVERSE=1
      IF(FDATA(108).EQ.1.0) INVERSE=-1
      IF(FDATA(108).EQ.1.0) CALL REORDER(INVERSE)
      IF(.NOT.KILL) TYPE "CALCULATING FT"
```

```
C         CALCULATE NU
C
          NU=1
          ITEMP=SIZE
          DO 5 I=1,17
            ITEMP=ITEMP/2
            IF(ITEMP.EQ.1)GO TO 7
            NU=NU+1
     5    CONTINUE
          TYPE "FT ERROR"
C
C
C         CALL BIT REVERSE SUBROUTINE TO REARRANGE THE DATA
C
     7    CALL BITREVERSAL(NU)
C
          ISTP=1
          ARG=6.283185/FLOAT(SIZE)
C
C
C         DO 2**(NU-10) 1K-1K TRANSFORMS ON ADJACENT BLOCKS OF DATA.
C         THESE TRANSFORMS ARE DONE IN PLACE AND USE THE TECHNIQUE IM-
C         PLEMENTED BY F.A.L. ANET IN PREVIOUS VERSIONS OF THIS PRO-
C         GRAM.
C
C         VLGN1 AND VLGN2 ALIGN 1K REAL AND 1K IMAGINARY DATA POINTS,
C         RESPECTIVELY.  PAGE=PAGE+2 BECAUSE THERE ARE TWO MEMORY PAGES
C         FOR EACH 1K WINDOW.
C
          PAGE=0
          DO 100 L=1,2**(NU-10)
            CALL VLGN1(PAGE)
            CALL VLGN2(PAGE)
            ISTP=1
            N2=512
            NTWO=SIZE/2
            DO 10 I=1,10
              K=1
              DO 20 M=1,ISTP
                K1=K
                THETA=ARG*(M-1)*NTWO
                C1=COS(THETA)
                S1=INVERSE*SIN(THETA)
                DO 30 J=1,N2
                  K1N2=K1+ISTP
                  TREAL=REALS(K1N2)*C1 + IMAGS(K1N2)*S1
                  TIMAG=IMAGS(K1N2)*C1 - REALS(K1N2)*S1
                  REALS(K1N2)=REALS(K1)-TREAL
                  IMAGS(K1N2)=IMAGS(K1)-TIMAG
```

```
              REALS(K1)=REALS(K1)+TREAL
              IMAGS(K1)=IMAGS(K1)+TIMAG
              K1=K1N2+ISTP
 30        CONTINUE
           K=K+1
 20      CONTINUE
        N2=N2/2
        NTWO=NTWO/2
        ISTP=ISTP*2
 10    CONTINUE
      PAGE=PAGE+2
100  CONTINUE
C
C    DO THE REMAINING NU-10 PASSES OF THE FOURIER TRANSFORM
C
C    VPA1X ALIGNS A REAL DATA WINDOW
C    VPA2X ALIGNS AN IMAGINARY DATA WINDOW
C    VPAXA ALIGNS A REAL DATA PAGE AT THE START OF A WINDOW
C    VPAXB ALIGNS AN IMAGINARY PAGE AT THE SECOND PAGE OF A WINDOW
C
     DO 500 I=1,NU-10
       ARG1=ARG*NTWO
       PAGEA=0
       PAGEB=(2**I)
       DO 400 J=1,(2**(NU-10))/(2**I)
         IPA=PAGEA
         IPB=PAGEB
         M=0
         DO 350 L=1,2**I
           CALL VPA1A(IPA)
           CALL VPA2A(IPA)
           CALL VPA1B(IPB)
           CALL VPA2B(IPB)
           DO 300 K=1,512
             THETA=ARG1*(M)
             M=M+1
             C1=COS(THETA)
             S1=INVERSE*SIN(THETA)
             TREAL=BREALS(K)*C1+BIMAGS(K)*S1
             TIMAG=BIMAGS(K)*C1-BREALS(K)*S1
             BREALS(K)=AREALS(K)-TREAL
             BIMAGS(K)=AIMAGS(K)-TIMAG
             AREALS(K)=AREALS(K)+TREAL
             AIMAGS(K)=AIMAGS(K)+TIMAG
300        CONTINUE
           IPA=IPA+1
           IPB=IPB+1
350      CONTINUE
         PAGEA=PAGEA+2*(2**I)
         PAGEB=PAGEB+2*(2**I)
```

```
      400     CONTINUE
              NTWO=NTWO/2
      500     CONTINUE
C
C       IF THE FILE CONSISTED OF SINGLATURE DATA PERFORM AN ORTHO
C       NORMALIZATION
C
        IF(FDATA(107).EQ.2.0 , CALL ORTHO
C
C       THE FOLLOWING SUBROUTINE MOVES THE DATA TO A STANDARD NMR CON-
C       FIGURATION (ZERO FREQUENCY IN THE MIDDLE) IF THE DATA IS IN
C       THE TIME DOMAIN.
C
        IF(FDATA(108).EQ.0.0) CALL REORDER(INVERSE)
C
C
        RETURN
        END
C
        SUBROUTINE BITREVERSAL(NU)
C
C       THIS SUBROUTINE PERFORMS THE BIT REVERSAL FUNCTION OF THE
C       FOURIER TRANSFORM.  THE DATA IS STORED IN TWO VIRTUAL ARRAYS
C       (REALS & IMAGINARIES) EACH HAVING A MAPPABLE WINDOW.  EACH
C       WINDOW HAS TWO PARTS CONSISTING OF AT LEAST ONE PAGE EACH.  A
C       PAGE IS DEFINED AS 1024 MEMORY LOCATIONS AND THEREFORE CON-
C       TAINS 512 REAL NUMBERS.  THUS, A 2K WINDOW CONSISTS OF TWO
C       PAGES THAT CAN BE INDEPENDENTLY MAPPED, EACH CONTAINING 512
C       REAL NUMBERS. A 4K-4K DATA FILE WILL CONSIST OF 16 PAGES.
C
C
        INCLUDE "VMEM"
        INCLUDE "NMRCOMMON" C
        INTEGER XEDNI,PAGEA,PAGEB,WINDOWCOUNT,OLDINDEX,BLOCKNU,BEGAP
        REAL AREALS(512),AIMAGS(512),BREALS(512),BIMAGS(512)
C
        EQUIVALENCE (XV1(0),AREALS(1))
        EQUIVALENCE (XV2(0),AIMAGS(1))
        EQUIVALENCE (XV1(512),BREALS(1))
        EQUIVALENCE (XV2(512),BIMAGS(1))
C
C
        BLOCKNU=NU-9
        WINDOWCOUNT=2**(BLOCKNU)
        LASTPAGE=SIZE/512-1
        I=1
        PAGEA=0
        PAGEB=0
        OLDINDEX=0
        INDEX=0
```

```
C
C
      DO 10 K=PAGEA,LASTPAGE
      CALL VPA1A(PAGEA)
      CALL VPA2A(PAGEA)
C
      DO 20 L=PAGEB,LASTPAGE
      CALL VPA1B(PAGEB)
      CALL VPA2B(PAGEB)
C
      CALL BREV(PAGEB,BEGAP,BLOCKNU)
      INDEX=OLDINDEX+BEGAP
      DO 30 I=BEGAP+1,512,WINDOWCOUNT
      CALL BREV(INDEX,XEDNI,NU)
      IF(INDEX .GE. XEDNI) GO TO 40
C
      IF(XEDNI/512 .NE. PAGEB) GO TO 40
C
      IF(I .GT. 512) GO TO 30
      J=IAND(XEDNI,777K)+1
C
      TREAL=AREALS(I)
      TIMAG=AIMAGS(I)
      AREALS(I)=BREALS(J)
      AIMAGS(I)=BIMAGS(J)
      BREALS(J)=TREAL
      BIMAGS(J)=TIMAG
C
C
   40 INDEX=INDEX+WINDOWCOUNT
   30 CONTINUE
C
      PAGEB=PABEB+1
   20 CONTINUE
C
      OLDINDEX=OLDINDEX+512
      PAGEA=PAGEA+1
      PAGEB=PAGEA
C
   10 CONTINUE
C
      RETURN
C
      SUBROUTINE REORDER(INVERSE)
C
C     THIS SUBROUTINE PUTS THE RESULTS OF THE FOURIER TRANSFORM
C     INTO THE FORMAT NORMALLY USED BY NMR SPECTROSCOPISTS.  IT
C     ROTATES THE FIRST HALF OF THE DATA AND THEN THE SECOND HALF.
C
```

```
C     INVERSE= 1 IF FORWARD TRANSFORM (FACTOR=1)
C     INVERSE=-1 IF REVERSE TRANSFORM (FACTOR=SIZE)
C
      INCLUDE "VMEM"
      INCLUDE "NMRCOMMON"
C
      INTEGER PAGEA,PAGEB
      REAL AREALS(512),AIMAGS(512),BREALS(512),BIMAGS(512)
C
      EQUIVALENCE (XV1(0),AREALS(1))
      EQUIVALENCE (XV2(0),AIMAGS(1))
      EQUIVALENCE (XV1(512),BREALS(1))
      EQUIVALENCE (XV2(512),BIMAGS(1))

      FACTOR=1
      IF(INVERSE .EQ. -1) FACTOR=SIZE
C
      PAGEA=0
      PAGEB=SIZE/(512*2)-1
C
      DO 100 I=1,2
C
         DO 75 J=1,SIZE/(512*4)
         CALL VPA1A(PAGEA)
         CALL VPA1B(PAGEB)
         CALL VPA2A(PAGEA)
         CALL VPA2B(PAGEB)
C
            DO 50 K=1,512
            L=512+1-K
            TREAL=AREALS(K)/FACTOR
            TIMAG=AIMAGS(K)/FACTOR
            AREALS(K)=BREALS(L)/FACTOR
            AIMAGS(K)=BIMAGS(L)/FACTOR
            BREALS(L)=TREAL
            BIMAGS(L)=TIMAG
   50       CONTINUE
C
         PAGEA=PAGEA+1
         PAGEB=PAGEB-1
   75    CONTINUE
C
      PAGEA=SIZE/(512*2)
      PAGEB=SIZE/512-1
  100 CONTINUE
C
      RETURN

           *******************************************
```

```
;BITREV              CHUCK DUMOULIN           11 JULY,1978
;
;MODIFIED FOR AOS 21 AUG, 1979
;BIT REVERSAL OF 16 BIT INTEGER
;
;CALLING SEQUENCE:
;               CALL BITREV(INTEGER,NUM)
;
        TITLE BITREV
        DEFARGS                 ;START DEFINING ARGUMENTS
;
        DEF INTEGER             ;FIRST ARG
        DEF NUMBER              ;SECOND ARG
;
        DEFTMPS                 ;DEFINE TEMPORARIES (THERE AREN'T ANY)
;
        FENTRY BITREV
        LDA 2,@NUM,3            ;LOAD # OF BITS TO REVERSE
        NEG 2,2                 ;MAKE NEG
        SUB 0,0                 ;ACC0=0
        LDA 1,@INTEGER,3        ;LOAD INTEGER
        MOVR 1,1                ;LSB INTO CARRY
        MOVL 0,0                ;CARRY INTO RESULT
        INC 2,2,SZR             ;ALL BITS REVERSED?
        JMP .-3                 ;NO
        STA 0,@INTEGER,3        ;LOAD RESULT
        FRET                    ;RETURN
        END
```

DISPERSION VERSUS ABSORPTION (DISPA): HILBERT TRANSFORMS IN SPECTRAL LINE SHAPE ANALYSIS

Alan G. Marshall

Departments of Chemistry and Biochemistry
The Ohio State University
Columbus, OH 43210

INTRODUCTION

Chapter 1 describes the origin and shape of a hypothetical spectral absorption signal arising from a <u>single</u> driven oscillator with a single natural frequency and a single <u>relaxation</u> time. Experimental spectra, on the other hand, often exhibit signals composed of a sum of <u>two or more</u> peaks of different natural frequencies and/or line widths. Most generally, one can distinguish between a superposition of lines of different position (Fig. 1a) and a superposition of lines of different width (Fig. 1b).

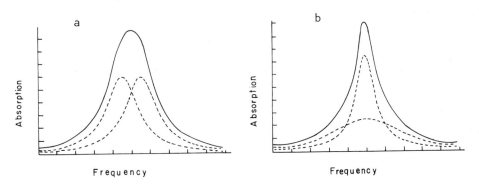

Figure 1. Schematic diagram of the composite absorption spectrum for the two most general sources of inhomogeneous spectral line broadening. (a) Sum of two Lorentzians of equal area and width, but different position; (b) Sum of two Lorentzians of the same area and position, but different width.

One of the most basic (and most difficult) problems in spectroscopy is how to analyze a broad absorption signal into its individual components. Once the line-broadening mechanism has been established, it is then straightforward to "fit" the observed absorption line shape according to that mechanism, by varying the parameter(s) of the component lines: line widths and/or splittings, exchange rate, and the like. Unfortunately, since it is often possible to "fit" a single broad experimental line according to any of several proposed mechanisms (e.g., Fig. 1a or 1b), it becomes necessary to find reliable criteria for identifying the correct line-broadening mechanism. Prior approaches have typically required multiple experiments designed to preferentially shift (as by change in solvent), broaden (as by change in temperature), or disperse (as by change in applied magnetic field strength in the case of NMR) the various signal components of the composite "inhomogeneously broad" line. However, multiple experiments necessitate multiple controls (e.g., changing the magnetic field strength in NMR changes relaxation times as well as chemical shifts), and the problem can easily get out of hand.

As noted in Chapter 1, spectroscopic "absorption" and "dispersion" signals each provide equivalent information, provided that the system is "linear" (i.e., that the response amplitude is proportional to the excitation amplitude, so that no "saturation" occurs). Historical the absorption spectrum has been preferred over dispersion, because the absorption line shape is narrower and more symmetrical. However, the various Fourier transform spectrometers[1] produce digitized absorption and dispersion spectra with equal vertical scaling, each containing half the available signal information.[2] Moreover, even when only the absorption-mode signal is available, the dispersion-mode spectrum may be generated by a Hilbert transform.[3]

The advantages of combining both the absorption and dispersion line shapes into a single display (DISPA) will now be demonstrated. The new plot produces a circular reference curve for a simple Lorentzian line shape.[4] A related semicircular display[5] has long been used to detect and characterize multiple relaxation processes for the simpler dielectric relaxation case, for which the "peaks" are always centered at the same (zero) frequency. In this chapter, theoretical simulations and experimental evidence will show how the DISPA display can distinguish reliably between many different line-broadening mechanisms, based on a graphical display of data from a single spectrum, without any prior assumptions about the spectral line shape.

THEORY AND SIMULATIONS

Single Lorentzian line: the DISPA reference circle

The fundamental normalized Lorentzian absorption and dispersion spectral line shapes, $A(\omega)$ and $D(\omega)$, are given in Equations 1:[6]

$$A(\omega) = \frac{\tau}{1 + (\omega_0 - \omega)^2 \tau^2} \quad [1a]$$

$$D(\omega) = \frac{(\omega_0 - \omega)\tau^2}{1 + (\omega_0 - \omega)^2 \tau^2} \quad [1b]$$

in which ω_0 is the "natural" motional frequency of the sample, ω is the irradiating frequency of the excitation source, and τ is the "relaxation time" for that transition. For the simple mechanical analog [i.e., an object of mass, m, suspended from a spring of force constant, k, subject to frictional resistance, f, driven by an oscillating force, $F(t) = F_0 \cos(\omega t)$], the natural frequency is $\omega_0 = \sqrt{k/m}$, and the relaxation time is $\tau = 2m/f$.

The present treatment begins from the simpler limit that $m \to 0$, (massless weight on a spring), for which the normalized absorption and dispersion line shapes become:[6]

$$D(\omega) = \frac{1}{1 + \omega^2 \tau^2} \quad [2a]$$

$$A(\omega) = \frac{\omega \tau}{1 + \omega^2 \tau^2} \quad [2b]$$

$\omega > 0$

in which $\tau = f/k$ in the mechanical analog. Equations 2a and 2b give the frequency-dependence of the dielectric or ultrasonic parameters (e.g., ϵ' and ϵ'', the in-phase and 90°-out-of-phase components of the a.c. dielectric constant) in dielectric or ultrasonic relaxation experiments.

For the line shapes of Equations 2, Cole and Cole first noted that since[5]

$$[D(\omega) - \tfrac{1}{2}]^2 + [A(\omega)]^2 = \tfrac{1}{4} = [\tfrac{1}{2}]^2 \quad [3]$$

a plot of $A(\omega)$ versus $D(\omega)$ for positive must give a semicircle, centered at $+1/2$ on the positive abscissa. Empirically, it was found that a "Cole-Cole" plot of ϵ'' vs. ϵ' could often be represented by a semicircle whose center was displaced <u>below</u> the abscissa. <u>The extent of the downward displacement could then be used to characterize the breadth</u> of the distribution in dielectric relaxation times for that sample. Only very recently have such downward displacements been related to physically reasonable microscopic motional models.[7,8]

Some 37 years later,[4] it was noted that a similar property results from the (different) Lorentzian line shape of Equations 1:

$$[A(\omega) - \tfrac{\tau}{2}]^2 + [D(\omega)]^2 = \tfrac{\tau^2}{4} = [\tfrac{\tau}{2}]^2 \qquad [4]$$

Equation 4 predicts that a plot of normalized dispersion, $D(\omega)$, vs. normalized absorption, $A(\omega)$, for a single Lorentzian line will yield a circle of radius, $\tau/2$, centered at $A(\omega) = (\tau/2)$ on the A-axis (see Figure 2). Since a single Lorentzian line is two-fold symmetric about its center, the "reference" DISPA plot is two-fold symmetric about the A-axis, so that only the top half of the circle need be displayed in Figure 2.

Two or more Lorentzians: deviations from the DISPA reference circle

Figure 2 shows that a single Lorentzian line yields a DISPA "reference" circle. For a DISPA plot for any other experimental line shape, any deviation from a reference circle having the same absorption peak height will then reflect non-Lorentzian composite line shape. We will next examine DISPA plots for several types of line-broadening, to determine the direction and magnitude of the displacement from the corresponding "reference" circle.

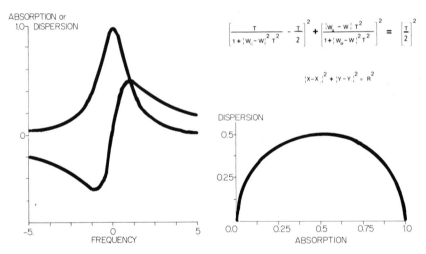

Figure 2. (Left): Normalized absorption and dispersion (left) for a single Lorentzian line (frequency in units of $1/\tau$). (Right): Plot of dispersion versus absorption (DISPA) for the right-hand half of each (symmetrical) spectrum at left, to give the upper half of the DISPA "reference" circle for a single Lorentzian line.

Figure 3 shows a family of DISPA plots for unresolved spectral doublets, in which the two component Lorentzians (each of the same area and width) are separated by up to about one-half the width of either component line. Also shown are the composite absorption and dispersion spectra from which the DISPA plots were constructed. Although each of the composite absorption spectra gives just a single unresolved peak, each of the DISPA plots shows a well-defined displacement <u>outside</u> and to the <u>right</u> of the reference circle. Moreover, the <u>magnitude</u> of the displacement is directly related to the magnitude <u>of the doublet separation</u>.

The diagnostic value of the DISPA plot becomes further evident from the plots of Figure 4. Each DISPA plot is constructed from a line shape consisting of a superposition of Lorentzians of equal widths, whose resonant frequencies satisfy a Gaussian distribution in frequency. For this line-broadening mechanism, the DISPA plots are displaced <u>outward</u> and to the <u>left</u> of the reference circle. Again, the <u>magnitude</u> of the DISPA displacement is directly related to the magnitude of the <u>spread</u> in resonant frequencies of the component lines.

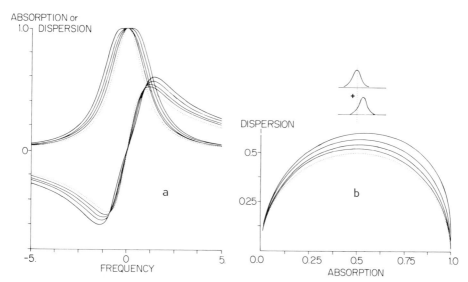

<u>Figure 3</u>. Composite absorption (top) and dispersion (bottom) spectra (a), and corresponding dispersion:absorption (DISPA) plots (b), for a spectrum consisting of the sum of two Lorentzian lines of equal width and area, separated by $0.6/\tau$, $0.8/\tau$, $1.0/\tau$, and $1.2/\tau$ in resonant frequency (see text). [Taken from reference 4.]

Figure 4 also shows that it is not easy in practice to distinguish in the absorption-mode between a Gaussian-broadened envelope (innermost solid curve at left of Figure 4) and a single Lorentzian absorption peak of the same height and width (filled circles in Figure 4), because the two line shapes diverge only at frequencies far from resonance. The DISPA displacements, on the other hand, are visually obvious. For the two mechanisms of Figures 3 and 4, then, the DISPA display can already distinguish between line-broadening due to two resonant frequencies, or to a distribution in resonant frequency.

Figure 5 shows DISPA plots for composite line shapes consisting of a superposition of Lorentzian lines centered at a common resonant frequency, whose line widths vary as a log-Gauss distribution (i.e., a Gaussian distribution in $\log(\tau/\tau_0)$). In this case the DISPA plots are displaced centrally inward from the reference circle, and the magnitude of the displacement is directly related to the width of the log-Gauss distribution of relaxation times. This situation is analogous to a distribution in dielectric relaxation times in the Cole-Cole plot, and a similar displacement is observed in that case.[5,7,8]

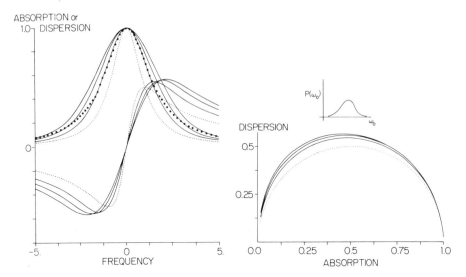

Figure 4. Composite absorption and dispersion spectra (left) and corresponding DISPA plots (right) for a spectrum consisting of a superposition of Lorentzian lines of equal widths, whose resonant frequencies satisfy a Gaussian distribution, with σ = 0.5, 1.0, or 1.5 s^{-1}, respectively. [Taken from ref. 4.]

Dispersion versus Absorption (DISPA) Line Shape Analysis

For the examples so far (Figures 3-5), it appears that distributions of two or more Lorentzians of different position produce DISPA displacement outside the reference circle, while distributions of two or more Lorentzians of different width lead to DISPA displacement inside the reference circle. In fact, this conclusion has recently been proved true under quite general conditions.[9] The DISPA plot can thus readily distinguish between the two most general spectroscopic line-broadening mechanisms (Figure 1), providing a solution to the the problem posed at the outset of this Chapter.

The diagnostic potential of the DISPA display is not limited to the two mechanisms of Figure 1. For example, the direction of the DISPA displacement is outside and to the right for two lines of different position, but outside and to the left for a Gaussian distribution in line position. Thus, the DISPA plot can distinguish between a multiplet consisting of just two lines, or more than two adjacent lines.

Another line-broadening mechanism is illustrated in Figure 6: chemical "exchange" between two sites of different resonant frequency. Although the absorption spectrum again shows just a single unresolved peak, the DISPA plot exhibits substantial displacement outside the reference circle, and the magnitude of the displace-

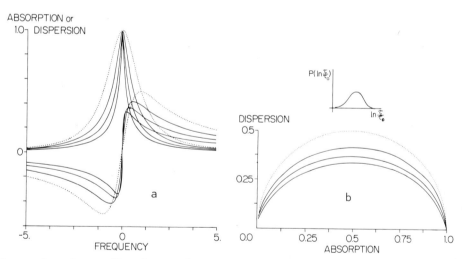

Figure 5. Composite absorption and dispersion spectra (left) and corresponding DISPA plots (right), for a superposition of Lorentzians centered at a common resonant frequency, whose relaxation times satisfy a log-Gaussian distribution, with σ = 0.5, 1.0, and 1.5. [Taken from ref. 4.]

ment is again directly related to the line-broadening parameter (in this case, the lifetime for chemical exchange).

By now it should be clear that the DISPA plot is extraordinarily sensitive to small deviations from simple Lorentzian line shape. Some 16 different line-broadening mechanisms have now been subjected to DISPA analysis,[3,4,10-12] and the results are summarized in Figure 7. Of course, it is almost never necessary to consider <u>all</u> these mechanisms in analyzing a single experimental spectrum; <u>it</u> is nevertheless heartening to note that the <u>direction of displacement</u> of a given DISPA plot from its reference <u>circle usually</u> suffices to discriminate between any two mechanisms. Moreover, in each case the <u>magnitude</u> of the displacement is directly related to the line-broad<u>ening parameter</u> (doublet splitting, chemical exchange lifetime, modulation amplitude, etc.), so that the DISPA plot can be used <u>quantitatively</u> to determine that parameter. Alternatively, once the <u>line-broadening</u> mechanism has been determined visually from the DISPA plot, the <u>line-broadening parameter</u> can be extracted by fitting to usual absorption-mode line <u>shape</u> to that mechanism.

Since the DISPA plot depends on proper instrumental "phasing" adjustments to give "pure" absorption and "pure" dispersion signals,

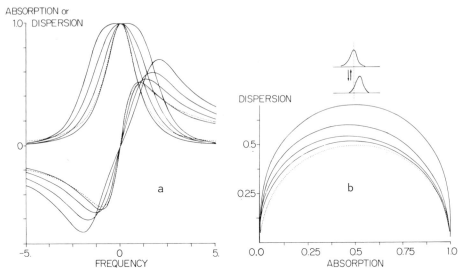

Figure 6. Absorption and dispersion spectra (left) and corresponding DISPA plots (right) for a spectrum resulting from exchange between two Lorentzian lines of equal width and area, but different resonant frequency. Exchange rate constant $k = \sqrt{2}$, $1.2\sqrt{2}$, $1.4\sqrt{2}$, and $1.6\sqrt{2}$ s^{-1} (proceeding from outermost to innermost solid curves), for line separation of 4 s^{-1}. [Taken from ref. 4.]

it is necessary to consider the effect of incorrect phasing upon the DISPA display. Plot "f" in Figure 7 shows that incorrect phasing acts simply to <u>rotate</u> the DISPA circle; since it is the only mechanism producing <u>such an</u> effect, it should be readily recognizable in practice (see Experimental section).

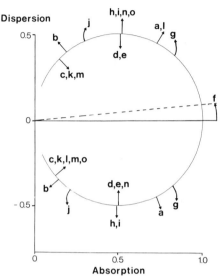

Figure 7. Reference circle (diameter equal to maximum observed absorption peak height) and direction of displacement of a DISPA plot for various line-broadening mechanisms: (a) unresolved pair of Lorentzians of equal width and different position;[4] (b) Gaussian distribution (in position) of Lorentzians of equal width;[4] (c) pair of Lorentzians of equal position but different width;[4] (d) log-Gauss distribution in relaxation time for Lorentzians of equal resonant frequency;[4] (e) log-Gauss distribution in correlation time for Lorentzians of equal resonant frequency;[4] (f) phase misadjustment;[4] (g) spectrum obtained by Fourier transformation of a truncated time-domain transient;[10] (h) power broadening;[11] (i) over-modulation;[3] (j) chemical exchange between two peaks of different position;[4] (k) chemical exchange between two peaks of different width;[4] (l) distortion from one adjacent peak of equal intensity and width;[11] (m) distortion of the central peak of a triplet, produced by the two outer peaks;[11] (n) (n) effect of too-long recorder time constant on EPR derivative spectrum;[3] (o) effect of slight baseline drift on EPR derivative spectrum.[3] Finally, for a simultaneous distribution in both peak position and peak width, the peak position distribution dominates DISPA behavior.[11] [Taken from ref. 11.]

Dispersion from absorption: Hilbert transforms for EPR spectra

The DISPA plot is easily generated for nuclear magnetic resonance (NMR) spectra, because discrete Fourier transformation of a time-domain response sampled at N equally spaced time increments yields N/2-point absorption and dispersion data sets, each containing half the spectral information,[2] and each scaled with the same vertical gain. The most obvious problem in extending the DISPA method to an electron spin resonance (ESR) spectrum is that the spectrum is usually recorded using magnetic field modulation, with the modulation frequency much less than the spectral line width. The resultant ESR frequency-domain spectrum is thus an absorption-mode derivative, with no direct detection of the dispersion mode. It is therefore necessary to generate the dispersion-mode mathematically, in much the same spirit that an X-ray diffraction pattern is converted mathematically into the desired electron-density map.

The steady-state ESR absorption spectrum, $A(\omega)$, is equivalent to the Fourier transform of the time-domain response, $f(t)$, to an impulse excitation, if the magnetic resonance system is linear (i.e., absence of saturation). Since $f(t)$ is causal (i.e., $\overline{f(t)} = 0$ for $t < 0$), there is a simple mathematical relation between the absorption spectrum, $A(\omega)$, and its corresponding dispersion spectrum, $D(\omega)$:

$$D(\omega) = -\frac{1}{\pi} \int_{-\infty}^{+\infty} \frac{A(\omega')}{\omega - \omega'} d\omega' \qquad [5]$$

In spectroscopy, Equation 5 is known as a Kramers-Kronig relation,[12]; in the context of servomechanisms as a Bode relation;[13] and in mathematics as a Hilbert transform.[14] Although numerical Hilbert transformation of a digitized absorption spectrum can be carried out directly, it is more efficient to perform the Hilbert transform using the much more rapid fast Fourier transform algorithm.[15] The operation can then be performed by a small on-line minicomputer, allowing the user to take advantage of the many apodization and other data manipulation and display available in most minicomputer software libraries. A flow chart for one successful scheme is shown in Figure 8.

In the Figure 8 scheme, a simulated ESR absorption derivative data set (Figure 8a) is first integrated using a numerical integration procedure[16] particularly well-suited to functions sampled at non-equal abscissa increments.[3] Since the discrete Fourier transform requires data points sampled at equal abscissa increments, the absorption data set is next interpolated using a piecewise cubic spline function, and the baseline flattened to first order to give an absorption spectrum sampled at equally spaced increments (Figure 8b). Since the fast Fourier transform algorithm requires 2^n data points, n = integer, the absorption data set is then padded with zeroes at each side of the peak, to increase the data size to 2^n points. The interpolated, baseline-flattened, zero-filled absorption data set is then subjected to an inverse discrete fast Fourier transform to give the (complex) function shown in Figure 8c. When the real and imaginary data sets of Figure 8c are multiplied by the weight function shown in Figure 8d (i.e., the sign of the right-hand half of each data set is reversed), and the resulting (complex) data set is subjected to a discrete fast Fourier transform, the desired dispersion spectrum of Figure 8e is produced. Finally, an ESR DISPA plot may then be generated, using the digitized absorption (Figure 8b) and digitized dispersion (Figure 8e) thus generated.

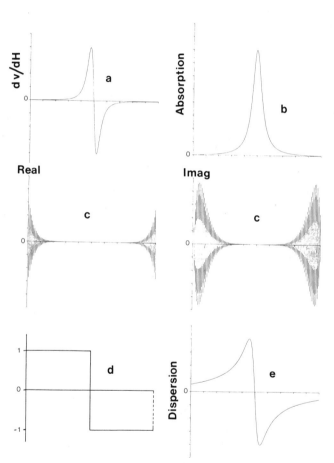

Figure 8. Flow diagram for a data reduction procedure that generates digitized absorption and dispersion data sets, (b) and (e) from a digitized ESR absorption derivative spectrum (a). See text for details. [Taken from ref. 3.]

Linearizing the DISPA circle

The DISPA analysis described above is based on comparison of an experimental curve to a reference circle. Although that display is qualitatively useful in identifying the correct line-broadening mechanism, (the analogous Cole-Cole plot[5] in dielectric relaxation continues as a popular display mode after nearly 40 years), a display in which the same experimental DISPA data is compared to a straight line could better help to determine quantitatively the line-broadening parameter(s) of that mechanism.

Bruce and Marshall[17] have considered five possible algorithms, each of which converts the DISPA reference circle to a straight line. However, these linearizations differ in their suitability for quantitating experimental DISPA data with respect to that reference line.

For example, Figure 9 shows one possible linearization suggested by the form of the absorption and dispersion line shapes of Equations 1, namely a plot of $D(\omega)$ versus $(\omega_0-\omega) \cdot A(\omega)$. The related plot of ϵ' versus ϵ''/ω for the in-phase and 90°-out-of-phase components of the steady-state a.c. dielectric constant offers distinct advantages in the analysis of dielectric relaxation data.[18] Although a single Lorentzian gives a simple straight line for this display, other line shapes lead to a double-valued plotted function with a tight loop rather than a single closed curve (e.g., the unresolved doublet line shape shown in Figure 9). The noise from actual experimental DISPA data would thus obscure the shape of the plotted curve, and limit its diagnostic value in discriminating between different line-broadening mechanisms. Moreover, this particular plot is not especially sensitive to deviations from Lorentzian line shape. For example, the line shape resulting from power broadening[11] produces a plot of $D(\omega)$ versus $(\omega_0-\omega) \cdot A(\omega)$ that is identical to that obtained for the reference Lorentzian line shape.

A second class of linearized DISPA displays is suggested by the constant radius of the reference DISPA circle. Thus, a plot of (the square of) the radius of an experimental DISPA display,

$$[R(\omega)]^2 = [A(\omega) - (A_{max}/2)]^2 + [D(\omega)]^2 \qquad [6]$$

as a function of the (frequency) distance away from the absorption maximum, A_{max}, should accurately reflect the sign and magnitude of deviations from the reference circle. Also, by plotting R^2 rather than R, we further magnify small differences between different line shapes. Among several such linearizations, the one that appears to provide optimal scaling and simplest interpretation is shown in Figure 10. Finally, by scaling the abscissa according to the observed absorption-mode line width at half-maximum height, the plot can be constructed with no prior assumptions about line shape.

Dispersion versus Absorption (DISPA) Line Shape Analysis   111

Figure 10 shows linearized DISPA plots for two simulated non-Lorentzian line shapes. For example, Figure 10a shows the DISPA deviations for a line shape consisting of an unresolved doublet of two Lorentzians of different resonant frequency. The maximum (upward) displacement from the reference line clearly occurs at frequencies <u>less than</u> one half-width away from the observed absorption maximum. In contrast, Figure 10b shows that for a log-Gauss distribution in relaxation time, the maximum displacement (downward this time) occurs at a frequency approximately <u>equal</u> to the half-width at half-maximum height of the observed absorption peak.

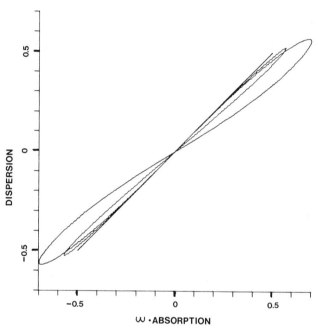

<u>Figure 9.</u> Plot of dispersion vs. (frequency times absorption), in which frequency is measured in units of $(1/\tau)$, where τ is the relaxation time of the Lorentzian line shape of Equations 1. The straight diagonal line is for a single Lorentzian line. The narrow and wide loops correspond to a spectrum consisting of the sum of two equally intense Lorentzians of the same width, separated by $0.6/\tau$ and $1.0/\tau$, respectively. The frequency of the center of the absorption peak is taken as zero (i.e., $\omega_0 = \omega$ in Equations 1). [Taken from ref. 12.]

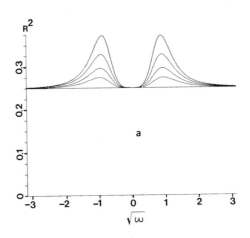

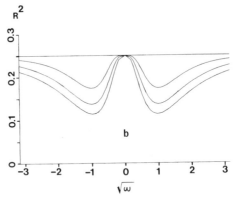

Figure 10. Plots of DISPA radius squared versus square root of frequency. Frequency is measured in units of half-width at half-maximum observed absorption peak height, so that no assumptions about line shape need be made. The horizontal line is for a reference Lorentzian line shape. (a) Composite spectrum consisting of two Lorentzians of equal width and intensity, separated by $0.6/\tau$, $0.8/\tau$, $1.0/\tau$, and $1.2/\tau$, respectively. (b) Composite spectrum resulting from a log-Gauss distribution in relaxation time, with distribution parameter $\sigma = 0.5$, 1.0, and 1.5, respectively. The (circular) DISPA plots for (a) and (b) are shown in Figures 3 and 5, respectively. [Taken from ref. 12.]

Dispersion versus Absorption (DISPA) Line Shape Analysis

EXPERIMENTAL RESULTS

Nuclear magnetic resonance

Since the diagnostic value of the DISPA analysis depends on detection of deviations from a reference circle or straight line, it is first necessary to show that an experimental Lorentzian line yields a well-defined reference curve. Figure 11 shows a DISPA plot for the ^{1}H NMR spectrum of the residual HDO in a sample of 99.7% D_2O. Absorption and dispersion data sets were generated by fast Fourier transformation of the HDO free induction decay following a 90° excitation pulse, and the absorption-mode display was phase-adjusted until visually symmetrical. The time-domain signal was weighted by a decreasing exponential of time constant 0.04 sec (sensitivity enhancement apodization) to produce more points per frequency-domain line width. The solid line in Figure 11 is a semicircle whose diameter is set equal to the maximum absorption peak height, and only half the spectral data are plotted, since the experimental line shape was symmetrical. Finally, although deliberate mis-phasing can lead to detectable distortion of the DISPA plot, the usual visual criterion of phasing to a symmetrical absorption-mode is in practice sufficient to make such distortion undetectable experimentally.[19]

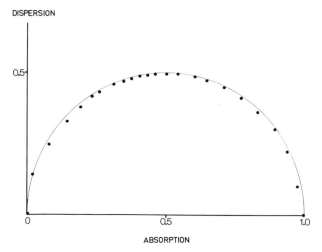

Figure 11. Dispersion versus absorption (DISPA) plot based on the ^{1}H NMR spectrum of the residual HDO in commercial 99.7% D_2O. Experimental data are shown as filled circles, and the solid curve is a semicircle whose diameter is set equal to the observed absorption peak height. [Taken from ref. 19.]

The excellent agreement between the experimental data and the reference semicircle expected for a single Lorentzian line in Figure 11 shows that any deviations from the reference circle for subsequent data sets will reflect non-Lorentzian line shape from the sample rather than artifacts from the instrument or the data reduction.

The reliability of any line shape analysis depends on the signal-to-noise ratio (S/N) of the data. In order to establish the minimum S/N need to generate useful DISPA data, pseudo-random noise, Gauss-distributed about zero, has been added independently to simple Lorentzian absorption and dispersion line shapes. The root-mean-square deviation for the Gaussian distribution was adjusted to give a ratio of (maximum absorption-mode signal):(rms noise) of 20:1, 10:1, or 5:1 (Figure 12). The DISPA plots constructed from these spectra (also shown in Figure 12) suggest that a rule-of-thumb S/N ratio of at least 10:1 (and preferably 20:1 or better) is necessary in order that the experimental DISPA curve can be distinguished from its reference circle (or straight line).

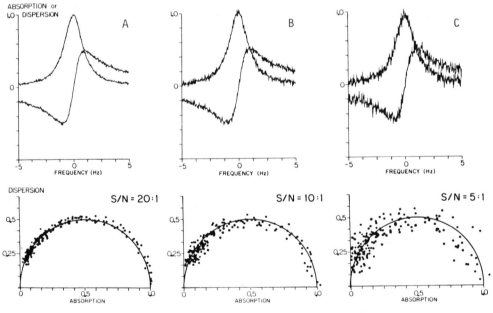

Figure 12. Absorption and dispersion spectra (upper row) and corresponding DISPA plots (lower row) for Lorentzian line shape with S/N (measured for absorption-mode) of (A) 20:1, (B) 10:1, and (C) 5:1. [Taken from ref. 10.]

Figure 13a shows a DISPA plot for a ^{19}F NMR spectrum of 5-fluorouracil at 94.1 MHz. The free induction decay was apodized with a decreasing exponential of time constant 0.04 sec, so that the resulting absorption spectrum consists of two Lorentzian lines of width 8.0 Hz, separated by 5.6 Hz (i.e., an unresolved doublet). As predicted by Figure 3, the experimental DISPA data are displaced <u>outward</u> and to the <u>right</u> of the reference circle. Figure 13b is the DISPA plot for the ^{1}H NMR signal of HDO in a non-spinning sample, subjected to a large z-gradient in applied static magnetic field. Figure 13b thus corresponds to an absorption spectrum consisting of a superposition of Lorentzians of many different resonant frequencies. The DISPA data points are displaced <u>outward</u> and to the <u>left</u>, as predicted qualitatatively by correspondence to the theoretical Gaussian distri-

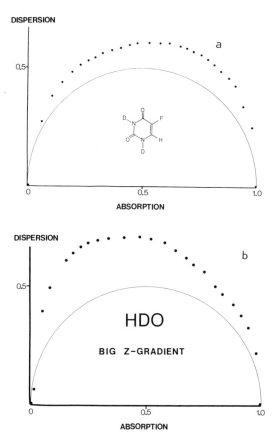

Figure 13. Experimental DISPA plots of NMR spectral data. (a) ^{19}F unresolved doublet from 5-fluorouracil. (b) ^{1}H distribution in chemical shift from HDO subjected to a large z-gradient. Compare (a) and (b) to Figures 3 and 4, respectively. [Taken from ref. 19.]

bution in ω_0 of Figure 4. These two examples show that the predicted diagnostic value of the DISPA plot in discriminating between various possible line-broadening mechanisms is realizable in practice.

A final interesting "calibration" example is shown in Figure 14. Sufficiently rapid chemical exchange between two sites of different chemical shift can produce coalescence of the two absorption signals to give a non-Lorentzian line shape. Figure 14 shows DISPA plots at two temperatures for N,N-dimethyltrichloroacetamide (DMTCA) in $CDCl_3$. Internal rotation between the magnetically nonequivalent conformers is fast enough to produce coalescence at a temperature slightly below 20°C.[20] The DISPA plot from data acquired at 20°C (i.e., just above coalescence temperature) shows clearly the predicted (Figure 6) displacement outside the reference circle, whereas the data set at 33°C (well above coalescence) approaches Lorentzian behavior (experimental data points located on the reference circle) expected in the fast-exchange limit.

The important conclusion from Figure 14 is that the DISPA plot can readily show whether a system undergoing chemical exchange has reached the "slow" or "fast" exchange limit (i.e., Lorentzian line shape, with DISPA data points on the reference circle) or not, based on data at a single temperature. It is always inconvenient, and not always practical (as with heat-labile compounds) to vary the temperature to discover whether the "slow" or "fast" limit has been reached; here the DISPA plot gives an immediate answer from a single spectrum.

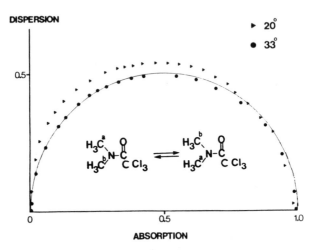

Figure 14. DISPA plots based on ^1H NMR spectra of an approximately 10% solution of N,N-dimethyltrichloroacetamide in $CDCl_3$, at 20°C (solid triangles) and at 33°C (solid circles). [Taken from ref. 19.]

Dispersion versus Absorption (DISPA) Line Shape Analysis

For each of the experimental NMR spectra plotted in Figures 11, 13 and 14, the line-broadening mechanism was known in advance. These examples thus serve to "calibrate" the DISPA approach, by showing that the predicted deviations from the DISPA reference curve are in fact observed experimentally.

Figure 15 shows a DISPA plot for a system for which the line-broadening mechanism was not previously known: ^{1}H NMR spectrum of benzene adsorbed onto dry silica gel. In this case, the free induction decay was acquired for 0.4 sec, and apodized with a decreasing exponential of time constant 0.1 sec to enhance S/N ratio without distorting the line shape. Since the DISPA data points in Figure 15 lie on a smooth curve displaced outside and to the left of the reference circle, DISPA analysis indicates that line-broadening here is dominated by a distribution in chemical shift among adsorbed benzene molecules, rather than by a heterogeneity in line width. Further analysis of this line shape should therefore either take into account this chemical shift distribution, or employ a detection scheme using some sort of spin-echo pulse train to eliminate differences in chemical shift before analyzing the line shape.

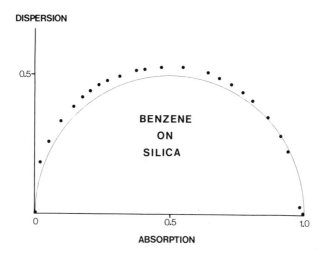

Figure 15. DISPA plot for ^{1}H NMR spectrum of benzene adsorbed onto dry silica gel (non-spinning sample). The displacement of the DISPA data from the reference circle serves to identify the dominant line-broadening mechanism (see text). [Taken from ref. 19.]

In Figure 15, small molecules were observed binding to a polymer; Figure 16 shows DISPA analysis for the polymer itself. In this case, the single broad ^{31}P NMR signal from the phosphate resonances of RNA in purified 70S ribosomes is recorded. A priori, it is not known whether the line shape is dominated by a distribution in chemical shift or a distribution in line width.

The DISPA plot of Figure 16 immediately resolves the issue: the displacement of data points inside the reference circle indicates that the dominant line-broadening mechanism is a distribution in line width. In corroboration, later ^1H NMR results have confirmed the presence of regions of varying flexibility in the ribosomal RNA.[21]

In yet another NMR example, Sykes et al. used a DISPA analysis of the ^{19}F NMR spectrum for M13 coat protein; the DISPA data points were located on the reference circle, within experimental error, showing that the two fluorotyrosines exhibited the same chemical shift.[22] Subsequent experiments based on solvent shifts confirmed that both fluorinated amino acid residues were "buried" and not accessible to solvent.

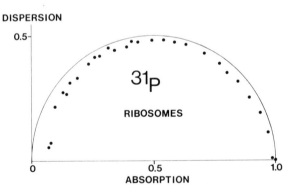

Figure 16. DISPA plot for the ^{31}P NMR signal from 70S ribosomes. The free induction decay of 512 points was padded with another 512 zeroes to increase digital resolution. Data were provided by Drs. T. R. Tritton and I. M. Armitage of Yale University. [Taken from ref. 9.]

Dispersion versus Absorption (DISPA) Line Shape Analysis

Electron Spin Resonance

Absorption-mode spectral peaks located less than about 10 line widths from the peak from which a DISPA plot is constructed can produce significant distortion of that DISPA plot (see Figure 7 and ref. 11). Therefore, DISPA analysis in ESR spectroscopy should be restricted either to systems with just a single signal (as for melanin or coal), or to systems with well-separated peaks (as for nitroxide "spin-labels" in relatively rapid motion). For a symmetrical ESR triplet, such as the nitroxide, the central line will give the least distortion from the other peaks[11]).

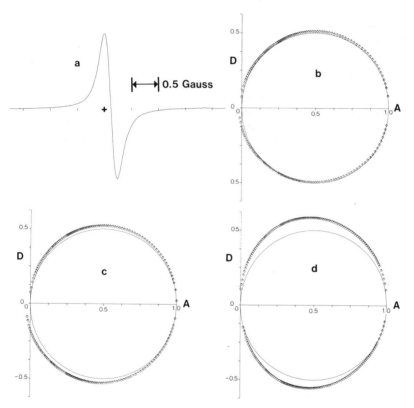

Figure 17. Effect of modulation broadening on experimental ESR DISPA plots. (a) v-mode derivative ESR spectrum ($m_I=0$ line) from 0.0005M peroxylamine disulfonate, using 13 mG modulation amplitude, 100 MHz modulation frequency, 5 mW power, swept at 1.0 G/min with a time constant of 0.125 sec. The "+" designates 3225 G. (b) DISPA plot from the data of (a). (c) DISPA plot as for (a), but with modulation amplitude = 0.2 G. (d) DISPA plot as for (a), but with modulation amplitude = 0.4 G. [Taken from ref. 3.]

Figure 17 shows an ESR peak (Figure 17a) and corresponding DISPA plot (Figure 17b) for peroxylamine disulfonate, generated by the algorithm of Figure 8. The close fit between the experimental points and the reference circle shows that the natural line shape is a nearly perfect Lorentzian. In agreement with theory (Figure 7 and ref. 3), the DISPA plots of Figures 17c,d show that the effect of excess modulation amplitude is to displace the data points <u>outside</u> the reference circle. The magnitude of the displacement is directly related to the modulation amplitude. Moreover, since the displacement can be scaled as the <u>ratio</u> of the modulation amplitude, H_m, to the natural absorption-mode line width, $\Delta H_{1/2}$, a <u>single</u> DISPA plot suffices to determine the absolute modulation amplitude when the unbroadened absorption-mode line width is known.[3] The DISPA plot thus provides a simple determination of H_m. In contrast, conventional determination of modulation amplitude requires either a direct measurement of H_m or a series of measurements[23] of derivative peak-to-peak separation as a function of H_m.

The DISPA plot readily distinguishes between <u>inhomogeneous</u> line-broadening (Figure 17), and <u>homogeneous</u> line-broadening (Figure 18). Figure 18 is a DISPA plot constructed from the ESR derivative spectrum of peroxylaminedisulfonate at a higher concentration than in Figure 17 (ca. 0.01M compared to 0.0005M). Although the experimental line width is about 15% larger than at the lower concentration (due to exchange broadening), the circular shape of the DISPA curve confirms that the line shape is still Lorentzian, and that the line-broadening is therefore a <u>homogeneous</u> process.

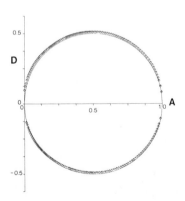

Figure 18. DISPA plot corresponding to the ESR v-mode derivative spectrum of peroxylamine disulfonate. Conditions are as for Figure 17a, except that the concentration has been increased to ca. 0.01M, in order to produce additional (homogeneous) exchange broadening. [Taken from ref. 3.]

Dispersion versus Absorption (DISPA) Line Shape Analysis

A final ESR experimental DISPA example is shown in Figure 19. The spin-labeled cardiolipin molecule is a phospholipid, containing a polar "head" group and four fatty acid hydrophobic "tails". It has been incorporated into lipid bilayers as a probe of molecular flexibility and mobility in model membranes.[24]

An important component of the observed ESR line width in such systems is the unresolved hyperfine splitting from the 12 methyl protons adjacent to the nitroxide group. The effect of proton hyperfine coupling is clearly evident in the DISPA plot constructed from the central line of the ESR triplet spectrum, showing why the coupling cannot be ignored in simulating the observed spectrum with a theoretical (homogeneous) line-broadening mechanism.

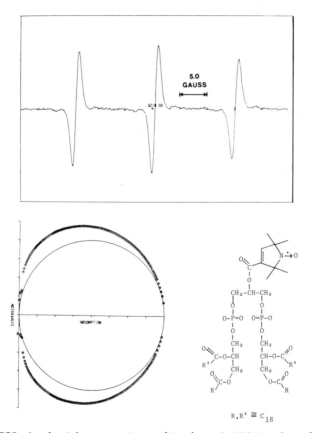

Figure 19. ESR derivative spectrum (top) and DISPA plot (bottom) constructed from the central $m_I = 0$ line, for a spin-labeled cardiolipin. The fatty acid groups (R, R') denote C_{18} chains. [F. G. Herring, private communication.]

SUMMARY AND FUTURE APPLICATIONS

The examples in this Chapter clearly demonstrate that a plot of spectroscopic dispersion versus absorption (DISPA) is extraordinarily sensitive to any deviation from Lorentzian line shape. More precisely, the direction of displacement of an experimental DISPA plot from its reference circle is often diagnostic of the line-broadening mechanism, and the magnitude of that displacement is directly related to the magnitude of the line-broadening parameter for that mechanism. Finally, all this information is extracted from a single experimental spectrum. How then should such plots best be applied?

It is suggested that the circular or linearized DISPA display first be used to identify the dominant line-broadening mechanism for the broad spectral peak of interest. Once the mechanism has been established, the line-broadening parameter for that mechanism can be extracted either by fitting the original absorption (NMR) or absorption derivative (ESR) spectrum, or by fitting the linearized DISPA shape. Alternatively, the DISPA result may suggest that the experiment be modified before further interpretation: e.g., changing the temperature if the DISPA plot indicates that the "fast" or "slow" exchange limit has not yet been reached (Figure 13), or proceeding to multiple-pulse refocussing experiments in case the line shape is dominated by a distribution in chemical shift (Figure 14). In any case, the DISPA plot should be the first step in analyzing the shape of an isolated spectral line, because the plot maximizes the mechanistic information available from a single spectrum.

It will be interesting to test the generality of the results obtained so far. For example, the NMR spectra of adsorbed species were dominated by a distribution in chemical shift rather than a distribution in line width. On the other hand, the NMR spectra of polymers themselves were dominated by a distribution in line width. ^{13}C NMR spectra of polymers often offer well-separated broad lines, for which DISPA analysis should be especially informative, and for which conventional line shape analysis experiments may be extremely time-consuming.

Another class of applications involves spectra consisting of a single broad line, such as NMR spectra of quadrupolar nuclei or the ESR spectra of melanin or coal. For example, DISPA experiments are currently in progress[25] to detect proposed "dynamic frequency shifts" in the NMR spectra of spin-3/2 nuclei such as ^{23}Na. The proton NMR spectrum of water is different for normal than for cancerous tissues: exploring the mechanism for this difference offers another possible DISPA application.

For each example shown here, a single line-broadening mechanism was dominant. Future applications must extend to cases involving two or more comparably important possible contributions.[11,19]

ACKNOWLEDGMENTS: This work was supported in part by the American Chemical Society (Petroleum Research Fund 11458-AC6), the Ohio State University, and the Alfred P. Sloan Foundation (1976-80).

REFERENCES

1. Marshall, A. G. 1981, in "Physical Methods of Modern Analytical Chemistry," Vol. 3., ed. Kuwana, T., Academic Press, N. Y., in press.
2. Bartholdi, E., and Ernst, R. R. 1973, J. Magn. Reson. 11, 9.
3. Herring, F. G., Marshall, A. G., Phillips, P. S., and Roe, D. C. 1980, J. Magn. Reson. 37, 293-303.
4. Marshall, A. G., and Roe, D. C. 1978, Anal. Chem. 50, 756-763.
5. Cole, K. S., and Cole, R. H. 1941, J. Chem. Phys. 9, 341.
6. Marshall, A. G. 1978, Biophysical Chemistry: Principles, Techniques, and Applications, Wiley, N. Y., Chapters 13-14.
7. Salefran, J. L., and Bottreau, A. M. 1977, J. Chem. Phys 67, 1909-1917.
8. Colonomos, P., and Gordon, R. G. 1979, J. Chem. Phys. 71, 1159-1166.
9. Marshall, A. G. 1979, J. Phys. Chem. 83, 521-524.
10. Marshall, A. G., and Roe, D. C. 1979, J. Magn. Reson. 33, 551-557.
11. Marshall, A. G., and Bruce, R. E. 1980, J. Magn. Reson. 39, 47-54.
12. Slichter, C. P. 1978, "Principles of Magnetic Resonance," 2nd ed., Springer-Verlag, Berlin, pp. 47-50.
13. Champeney, D. C. 1973, "Fourier Transforms and their Physical Applications," Academic Press, London, pp. 244-246.
14. Bracewell, R. 1965, "The Fourier Transform and Its Applications," McGraw-Hill, New York, pp. 267-272.
15. Cooley, J. W., and Tukey, J. W. 1965, Math. Comp. 19, 297.
16. Squire, W. 1976, Int. J. Numeric. Methods Eng. 10, 478.
17. Bruce, R. E., and Marshall, A. G. 1980, J. Phys. Chem. 84, 1372-1375.
18. Cole, R. H. 1955, J. Chem. Phys. 23, 493-499.
19. Roe, D. C., Marshall, A. G., and Smallcombe, S. H. 1978, Anal. Chem. 50, 764-767.
20. Allerhand, A., and Gutowsky, H. S. 1964, J. Chem. Phys. 41, 2115.
21. Tritton, T. R. 1980, FEBS Lett. 120, 141-144.
22. Hagen, D. S., Weiner, J. H., and Sykes, B. D. 1979, Biochemistry 18, 2007.
23. Poole, C. P., Jr. 1967, "Electron Spin Resonance," InterScience, New York, Chapter 10.
24. Herring, F. G. 1981, private communication.
25. Werbelow, L. G., and Marshall, A. G. 1981, submitted for publication.

FOURIER TRANSFORM ION CYCLOTRON RESONANCE SPECTROSCOPY

Melvin B. Comisarow

Department of Chemistry
University of British Columbia
Vancouver, B.C. V6T 1Y6
CANADA

PRINCIPLES OF ICR EXCITATION AND DETECTION

This chapter discusses the principles of Fourier transform spectroscopy as applied to a type of mass spectrometry called ion cyclotron resonance (ICR) spectroscopy.

As in any mass spectrometer, the ICR spectrometer has provision for ionizing a gaseous sample and then determining the masses of the ions that are formed. The ICR experiment is typically conducted at very low pressures, usually in the range 10^{-8} to 10^{-4} torr. The operating features of ICR spectrometers have been extensively reviewed[1-7] and will be discussed only briefly here.

A moving ion in a homogeneous magnetic field will be constrained to a circular path which is perpendicular to the magnetic field and will orbit an an angular frequency, ω, called the cyclotron frequency, given by Equation 1, in which q is the ion charge, m is the ion mass and B is the magnetic field strength.

$$\omega = \frac{q\,B}{m} \qquad [1]$$

According to Equation 1, an ensemble of ions of differing masses will have a spectrum of cyclotron frequencies which is characteristic of that ensemble. For a magnetic field strength of 1 tesla and a mass range of 15 amu to 1500 amu, the cyclotron frequency spectrum (Equation 1) extends from 10 kHz to 1 MHz and thus falls in the radiofrequency region of the electromagnetic spectrum.

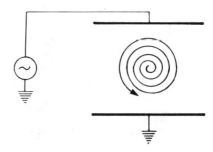

Figure 1. The cyclotron resonance principle as applied to mass spectrometers. An alternating electric field whose frequency equals the cyclotron frequency (Equation 1) for a particular ion mass, excites the cyclotron motion of that ion. An oscillator is connected to the plates of a capacitor, whose dimensions define the sample volume, and gives rise to an alternating electric field within the capacitor. If the frequency of the oscillator equals the cyclotron frequency (Equation 1) of an ion located within the capacitor, the radius of the ion's cyclotron orbit will be increased (i.e., the ion cyclotron motion is excited). This phenomenon is called cyclotron resonance. The kinetic energy of the ion increases as the ion follows the spiral path shown, and the presence of cyclotron resonance is detected by measuring the signal that is induced in the plates of the capacitor by the excited ion motion.

It follows from Equation 1 that measuring the cyclotron frequencies of a sample of ions will determine the mass of the sample ions--i.e., determine the mass spectrum of that sample. The measurement of the cyclotron frequencies of a sample of ions is accomplished in the following manner. A radio frequency electric field is applied to the sample of ions by connecting the output of a radiofrequency oscillator to the plates of a parallel plate capacitor whose dimensions define the volume of the sample (see Figure 1). If the frequency of the oscillator equals the cyclotron frequency for a particular ion, that ion will absorb energy from the electric field and will follow the spiral path shown in Figure 1. This absorption of energy from the alternating electric field is called "ion cyclotron resonance." The increase of the ion's orbital radius is called "exciting the cyclotron motion" of the ion. If the oscillator frequency does <u>not</u> equal the cyclotron frequency for a particular ion, the cyclotron motion of that ion will not be excited.

Once excited cyclotron motion has been achieved, the excited cyclotron motion at a particular ion mass will induce an alternating

Fourier Transform Ion Cyclotron Resonance Spectroscopy

charge on the plates of the ICR cell.[8] If the plates of the ICR cell are connected together with a resistor, then an alternating charge on the plates of the ICR cell implies an alternating current between the plates of the cell and an alternating voltage across this resistor. The magnitude of the alternating voltage is given by[8]

$$V_S(t) = \frac{N q^2 r B R}{m d} \sin(\omega t) \qquad [2]$$

in which N is the number of ions, q is the ion charge, m is the ion mass, r is the radius of the ion orbit, B is the magnetic field strength, R is the magnitude of the resistance, d is the spacing between the ICR cell plates and ω is the cyclotron frequency (Equation 1). Monitoring the "ICR signal voltage" (Equation 2) as the frequency of the applied voltage is swept over the range of cyclotron frequencies permits ICR mass analysis.

The most commonly used electronic circuit for conventional ICR mass analysis is the marginal oscillator. This circuit has the capability of providing the alternating electric field required for ICR excitation and simultaneous detection[9] of the ICR signal (Equation 2). The ICR mass spectrum is obtained by plotting the ICR voltage signal strength (Equation 2) as a function of magnetic field strength, as the magnetic field is swept to equate the ion cyclotron frequencies with the fixed frequency of the marginal oscillator. Typically, about 30 min are required to sweep over the mass range m = 15 to m = 200, the upper mass limit of most conventional ICR spectrometers.

CONCEPTUAL DIFFERENCES BETWEEN FT AND CONVENTIONAL ICR

In the Fourier transform ion cyclotron resonance (FT-ICR) spectrometers,[10-24] the cyclotron motion of ions of many different masses is excited essentially simultaneously (see Figure 2). The presence of excited cyclotron motion is then detected as the alternating voltage (Equation 2) after the exciting oscillator is turned off. Thus, unlike the conventional ICR spectrometers in which ion excitation and ion detection are simultaneous, ion excitation and ion detection in the FT-ICR spectrometer are temporally distinct.

As in the application of Fourier methods to other spectroscopies, the application of Fourier methods to ICR spectroscopy permits the whole ICR spectrum to be obtained very quickly. For example, an FT-ICR spectrum is given below which extends from m = 50 to m = 1200 amu. That spectrum resulted from transient FT-ICR signals that were observed for only 1.6 sec. In contrast, a conventional ICR spectrometer would require about 30 min to scan up to m = 200 and would be unable to scan above that mass.

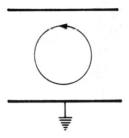

Figure 2. Ion motion during the detection period of a Fourier transform ion cyclotron resonance (FT-ICR) experiment. The cyclotron motions of the ions in the sample are excited along spiral paths as shown in Figure 1 by an oscillator which is then turned off. The excited ions then proceed on the circular paths shown and induce a time domain signal in the plates of the capacitor which is amplified and detected. Fourier transformation of this time domain signal yields the frequency domain spectrum

From the description of the FT-ICR experiment of the previous paragraph, it is clear that the FT-ICR experiment is operationally similar to the Fourier transform nuclear magnetic resonance (FT-NMR) experiment, in that many spectral positions are first excited during an excitation period, after which the excitation of the spectrum persists for a period of time that is long with respect to the excitation time. The excitation gives rise to a time domain signal which contains components from all spectral positions that have been excited. This time domain signal which, when sampled and subjected to Fourier transformation, results in the frequency domain spectrum that is characteristic of the sample. The FT-ICR experiment is operationally unlike the Fourier transform infrared (FT-IR) experiment, in which the infrared spectrum is obtained by Fourier transformation of a spatially dispersed interferogram.

Ion cyclotron resonance experiments (both conventional and Fourier) are similar to NMR experiments (both conventional and Fourier) in that a magnetic field is required for the experiment and in that the width of the spectrum is proportional to the magnetic field strength. ICR experiments differ from NMR experiments in two fundamental aspects. First, ion cyclotron resonance results from the interaction of alternating electric fields with electrically charged ions,[8] whereas nuclear magnetic resonance involved the interaction of alternating magnetic fields with magnetic dipoles. Second, the relaxation time of the time domain signal in the FTNMR experiment is fixed, whereas the relaxation time of a time domain ICR signal can be made arbitrarily long.[14] For both NMR and ICR experiments, spectral line width is usually determined by the relaxation time of the excited sample. This is true for both conventional detection methods and Fourier detection methods. However, the relaxation time of an excited NMR sample is determined primarily by intramolecular couplings and thus is largely a function of the sample molecule and beyond control by the experimentalist. In ICR experiments, however, the relaxation time is determined by the rate of intermolecular ion-molecule collisions.[15,17,18] Therefore, if the sample is diluted (i.e., the pressure is lowered), the ion-molecule collision rate will be lowered and the relaxation time of the excited ICR sample will become arbitrarily long.[15,18]

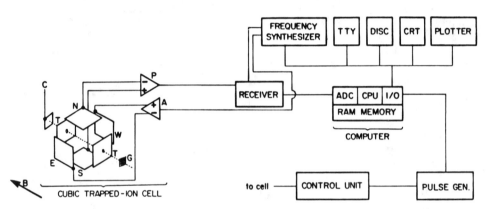

Figure 3. Block diagram of Fourier transform ion cyclotron resonance (FT-ICR) spectrometer. The operation of the spectrometer is described in the text.

EXPERIMENTAL FT-ICR OPERATION

Preliminary FT-ICR experiments[10-15] were performed on a prototype instrument which was developed by modification of an existing conventional trapped-ion cell ICR spectrometer. A new FT-ICR spectrometer has been constructed recently in our laboratories and all experiments described herein have been performed on this instrument. A block diagram of the present instrument is shown in Figure 3. The spectrometer consists of a cubic trapped-ion cell, a vacuum system (not shown), and the associated analog and digital electronics. The operation of the instrument is as follows: the sample is introduced into the cell and is ionized by a short pulse from the electron beam. After a delay time that permits parent ions to react with the background neutral gas and form product ions, the cyclotron motion of all ions in the cell is coherently excited. This excitation is accomplished by turning on a frequency synthesizer and applying a swept alternating electric field to the cell via the transmitter amplifier, A, and the transmitter plates, E, W. The synthesizer is then turned off. The exited cyclotron motion persists and induces an analog signal[8] in the receiver plates, N, S, which is amplified by the preamplifier, P, converted to a digital signal by the analog to digital converter, ADC, and stored in the memory of a digital computer. The pulse sequence is then repeated for a chosen number of times. The resultant summed, digital signal is then subjected to Fourier transformation to produce an ICR frequency domain spectrum. This can be converted to a mass spectrum and then plotted. Such a spectrum is shown in Figure 4. The spectrum in Figure 4 was obtained from 100 transient signals, each of which lasted for 16 msec. The total acquisition time for collecting the raw data is thus only 1.6 sec. In contrast, a conventional scanning ICR spectrometer would require about 20 min to produce a mass spectrum up to m/e = 200 and would be unable to resolve mass peaks above this mass with at least unit mass resolution.

The trapped-ion cell shown in Figure 3 differs from that of the original design[25] in two respects. First, the cell is cubic in geometry rather than orthorhombic. The second important difference between the cubic trapped-ion cell and the conventional cell is that in the cubic cell the excitation is applied to the cell in a differential manner and the signal[8] is detected in a differential manner. The advantages of the cubic cell are: first, the maximum mass resolution which is obtainable from the cell is about 2-4 times greater than that obtained[10,11] with the conventional cell; and second, that the reliability of the cell is greater.

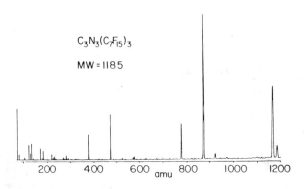

Figure 4. FT-ICR mass spectrum of tris perfluoroheptyl azine. The total data acquisition time for collecting the time domain data, which when transformed gave the spectrum, was 1.6 sec.

Fourier Transform Ion Cyclotron Resonance Spectroscopy

Signal accumulation of the broad-band signals that exist in FT-ICR spectroscopy presents some challenging problems in computer technology. The acquisition of broad-band signals must satisfy the Nyquist sampling theorem which states that the signal must be sampled at a rate which exceeds twice the highest frequency component in the broad-band signal. For an FT-ICR spectrometer operating at 20 kilogauss (2 tesla), the cyclotron frequencies extend from about 1 MHz to d.c. for a mass range of m/e = 30 to ∞. A transient signal containing the signals for this mass range must then be sampled at a rate that is in excess of 2 million samples/sec. Each sampling (i.e., each digital word) must be stored and added to the corresponding word from the previous run. It is to be noted that no random access computer memory (RAM) is fast enough to accept data at this rate. For the present generation of minicomputers, the maximum rate at which signal accumulation can be carried out is about 300 KHz, which corresponds to a spectral bandwidth of only 150 KHz. By use of the spectral segment extraction technique,[10-15] this band can be positioned anywhere in the 1 MHz ICR spectrum. The receiver configuration for spectral segment extraction with the present spectrometer is shown in Figure 5. Functionally, it is identical to that as previously described.[10-15] The receiver configuration for direct spectral acquisition is also shown in Figure 5. The output of the ADC is

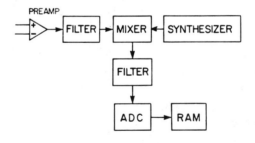

Figure 5. Configuration of the FT-ICR spectrometer for spectral segment extraction and for direct data acquisition.

routed to a shift register memory (SR) which serves as an intermediate storage device. This memory is accessible via a single input port and a single output port. Each new input word causes a successive transfer of the contents of a given location to the adjacent location. Currently, shift register memories are available with clock rates of up to 100 MHz. Once the complete digital transient signal for an FT-ICR pulse sequence has been acquired in the SR memory, the SR memory is read out at a rate of 300 KHz to the general purpose random access memory of the computer. By using the SR as an intermediate storage device, direct spectral acquisition of an FT-ICR transient containing frequency components from the entire spectrum can be achieved.

FT-ICR PERFORMANCE

High-resolution mass spectra

It was clear from the beginning of FT-ICR development[14] that unlike conventional ICR spectrometers, FT-ICR spectrometers would be capable of high or even ultrahigh resolution mass spectrometry. The reasons for this have been discussed previously.[10,11,14] Some examples of ultrahigh resolution FT-ICR mass spectra obtained with the present instrument are shown in Figure 6. According to the theory,[10,15,18] for a fixed acquisition time and a fixed relaxation time, FT-ICR resolution is predicted to be inversely proportional to ion mass. The data of Figure 6 were obtained under such conditions. Figure 6f is a plot of the experimental mass resolution vs. ion mass for the data of Figures 6a-6e, which confirms this theoretical expectation. It should be noted that the carbon monoxide peak in Figure 6 corresponds to a CO pressure of only 50 femtoatm. The FT-ICR technique is thus capable of simultaneously producing spectra with ultrahigh mass resolution and high sensitivity in a fairly short data acquisition time (10 sec in the case of Figure 6a).

Mass calibration

The problem of mass calibration is an important consideration for any mass spectrometer. It was recognized early[15] that FT-ICR mass calibration could be particularly convenient since the actual physical measurement was the cyclotron frequency. More so than for most physical measurements, frequency can be determined to great accuracy with extreme ease and thus mass calibration of FT-ICR spectra is potentially very convenient. Furthermore, FT-ICR spectrometers can be operated so as to produce very high mass resolution and the mass calibration accuracy is potentially very high. For example, by using the N_2 peak in Figure 6a to determine the magnetic field strength from the cyclotron equation (Equation 1), the CO and C_2H_4 peaks in Figure 6a can be calibrated to an accuracy of better than 10 micromass units.

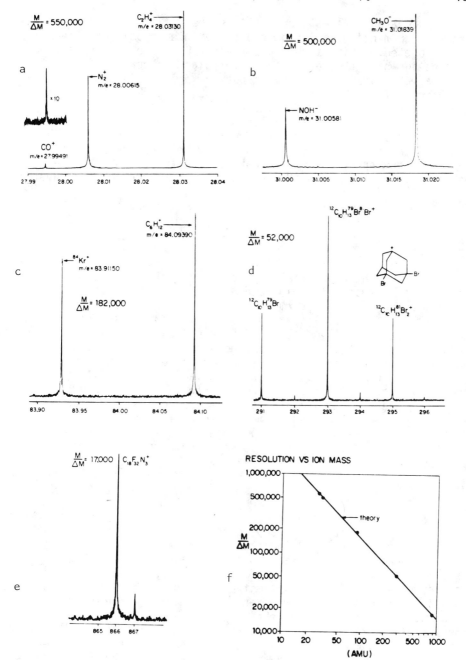

Figure 6. Ultrahigh resolution FT-ICR mass spectra. Each transient was acquired for 1.02 sec and had 250 msec relaxation time. The 867 peak in (e) is the carbon-13 satellite of the 866 peak. (f) is a plot of the mass resolution versus ion mass for the spectra of (a)-(e).

Of greater interest, however, is the ability to calibrate the mass spectrum using calibrant peaks that are remote from the mass to be calibrated. This is very important for ion-molecule reaction studies in which the reactant ion formulae are certain (being parent and fragment ions from neutrals) but the product ion formulae have to be inferred from the mass spectrum. Table 1 gives the results of an FT-ICR mass calibration experiment which was conducted on the spectrum in Figure 4. The calculated masses in Table 1 were obtained using the equation of Beauchamp and Armstrong[26] as rewritten by McIver and Leford.[27] These equations require two known masses to calibrate the spectrum, but do not require an accurate knowledge of magnetic field strength. McIver and Ledford[27] derived the equation

$$\frac{m}{e} = \frac{-B + (B^2 + 4A\omega^2)^{1/2}}{2\omega^2} \quad [3]$$

in which ω is the experimentally measured cyclotron frequency, m/e is the ion mass to charge ratio, and A and B are two constants. Once A and B are known, mass determination requires only a knowledge of the cyclotron frequency. Table 1 gives the major ions seen in Figure 4 and the experimentally determined frequencies. Also included are the calculated and actual masses for each ion. Various combinations of masses have been used for the determination of the constants A and B. It is clear from the errors columns in Table 1 that FT-ICR mass spectra can be calibrated with considerable accuracy using calibrant ions that are far removed in mass from the ion to be calibrated. For example, by using the ions at m = 69 and m = 131 to determine A and B, the ion mass at m = 119 can be determined within an error of 3 millimass units and the ion at 866 can be measured with accuracy of 0.1 amu.

Wide mass range

Ion cyclotron resonance spectroscopy is perhaps the most versatile single technique for studying ion-molecule reactions. The mass range of conventional ICR spectrometers extends to an upper mass limit of about m/e = 200, as above this mass the resolution is usually less than one mass unit. Most ICR ion-molecule reaction studies have therefore been limited to ions whose mass is below 200. This is perhaps the most important limitation of conventional ICR spectroscopy. Furthermore, the detection sensitivity of any ICR spectrometer decreases as mass increases[8] and in practice long filter time constants and accordingly slow scan speeds are needed when observing higher masses. An important feature of the FT-ICR technique is that all of the capability of conventional ICR spectrometers is retained in Fourier operation. The high speed, high sensitivity, wide mass range, and high mass resolution resulting from Fourier operation overcomes the above limitations of conventional ICR spec-

Table 1. Calibration Data for the FT-ICR Spectrometer.

ION	EXACT m/q	FREQUENCY (Hz)	CALCULATED[a] m/q	ERROR
CF_3^+	68.9952	454148	68.9952	0.000
$C_2F_5^+$	118.9920	263183	118.989	-0.003
$C_3F_5^+$	130.9920	239037	130.991	0.000
$C_3F_7^+$	168.9888	185219	168.979	-0.010
$C_8F_{14}N^+$	375.9807	83053.8	375.950	-0.030
$C_{10}F_{17}N_2^+$	470.9790	66230.7	470.930	-0.049
$C_{16}F_{29}N_2^+$	770.9598	40321.4	770.856	-0.104
$C_{18}F_{32}N_3^+$	865.9582	35865.7	865.668	-0.290
$C_{24}F_{44}N_3^+$	1165.9390	26542.7	1165.64	-0.299
$C_{24}F_{45}N_3^+$	1184.9374	26100.2	1185.13	0.193

ION	CALCULATED[b] m/q	ERROR	CALCULATED[c] m/q	ERROR
CF_3^+	68.9952	0.000	68.9952	1.9788×10^{-5}
$C_2F_5^+$	118.992	0.000	118.991	2.3932×10^{-4}
$C_3F_5^+$	130.995	0.003	130.994	-2.0599×10^{-3}
$C_3F_7^+$	168.986	-0.003	168.984	4.0564×10^{-3}
$C_8F_{14}N^+$	375.998	0.017	375.984	-3.8061×10^{-3}
$C_{10}F_{17}N_2^+$	471.009	0.030	470.986	-6.9742×10^{-3}
$C_{16}F_{29}N_2^+$	771.083	0.123	771.020	-6.0129×10^{-2}
$C_{18}F_{32}N_3^+$	865.958	0.000	865.877	0.08085
$C_{24}F_{44}N_3^+$	1166.18	0.241	1166.03	-0.09371
$C_{24}F_{45}N_3^+$	1184.9374	0.753	1185.53	0.39940

[a] m/q = 68.9952 and 130.9920 used to fit Equation 3.
[b] m/q = 68.9952 and 865.9582 used to fit Equation 3.
[c] All points fitted to Equation 3 to find best values of A and B.

troscopy and opens up an enormous range of higher molecular weight compounds whose ion-molecule chemistry can now be investigated. An example of the sort of FT-ICR data that can be obtained from an ion-molecule reaction study is shown in Figure 7, which gives FT-ICR spectra of cyclopentadienyl chromium dicarbonylthionitrosyl[28] as a function of reaction time. The spectra in Figure 7 show the decrease in reactant ion intensity and the increase in product ion intensity as the reaction time is made progressively longer.

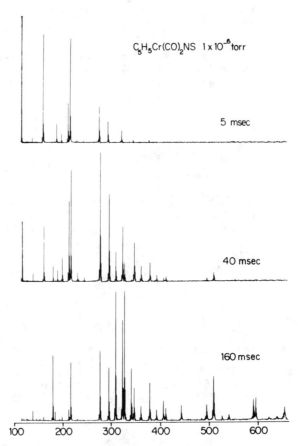

Figure 7. FT-ICR mass spectra of cyclopentadienyl chromium dicarbonylthionitrosyl as a function of reaction time. The chemistry derived from this data is discussed in the text.

Fourier Transform Ion Cyclotron Resonance Spectroscopy

FT-ICR DOUBLE RESONANCE

One feature of FT-ICR spectroscopy which is particularly valuable for ion-molecule reaction studies is the double resonance experiment.[16] In the ICR double resonance experiment, a reactant ion is ejected from the ICR cell as soon as it is formed. Product ions that result from that reactant ion are then absent from the spectra, and this absence can be used as a definitive diagnostic tool to identify ion-molecule reaction pathways. An attractive feature of the double resonance experiment when performed on FT-ICR spectrometers[16] is that all product ions which are derived from a particular reactant ion are identified in a <u>single</u> double resonance experiment. Conventional ICR spectrometers require many double resonance experiments, one for each product ion, to discover the same information about the ion-molecule reaction system. Examples of double resonance experiments performed on an FT-ICR spectrometer are shown in Figure 8, taken from the ion-molecule study described below. Note that in Figure 8, eight different ions are simultaneously identified as ion-molecule reaction products that have the cyclopentadienylchromium ion as their chemical precursor.

Example: FT-ICR studies of gas-phase ion-molecule chemistry of transition metal carbonyl complexes

As mentioned above, ICR spectrometers have been used as research tools for the investigation of gas phase ion-molecule reactions. An example of the sort of ion-molecule study which can be carried out by FT-ICR techniques but which would be beyond the capability of any other technique is the study of the ion-molecule condensation chemistry of the three organometallic compounds, η^5-cyclopentadienyl manganesetricarbonyl (CpMn(CO)$_3$), <u>1</u>, η^5-cyclopentadienyl chromiumdicarbonylnitrosyl (CpCr(CO)$_2$NO), <u>2</u>, and η^5-cyclopentadienyl chromiumdicarbonylthionitrosyl (CpCr(CO)$_2$NS), <u>3</u>. Compounds <u>1</u> and <u>2</u> form an isoelectronic pair; <u>2</u> and <u>3</u> are a congener pair.

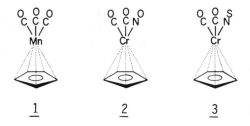

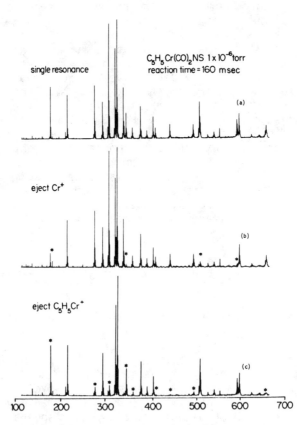

Figure 8. FT-ICR single resonance and double resonance spectra of CpCr(CO)$_2$NS, 3, from m/e = 100 to m/e = 700. (a) is the single resonance spectrum of 3 which was ionized with 25 eV electrons and allowed to self-react for 160 msec prior to FT-ICR mass analysis. (b) was obtained under the same experimental conditions as for (a) but with double resonance ejection of Cr$^+$ at zero reaction time. The differences between (a) and (b), each identified by an asterisk in (b), indicate the contribution of Cr$^+$ to the overall ion chemistry of 3. (c) was obtained under the same experimental conditions as for (a), but with ejection of CpCr$^+$ at zero reaction time. The differences between (a) and (c), each identified by an asterisk in (c), indicate the contribution of CpCr$^+$ to the overall ion chemistry of 3. Note that in (c), nine product ions are simultaneously identified as ion-molecule reaction products which have CpCr$^+$ as their original ancestor ion.

Compounds 1 and 2 are well known organometallic compounds. Compound 3 has recently been synthesized by Legzdins and Kolthammer.[28] It is the first organometallic thionitrosyl compound.

From data of the type given in Figures 7 and 8 for 3, the following chemical reactions can be identified as occurring in the $CpMn(CO)_3$ system:

$$Mn^+ + CpMn(CO)_3 \longrightarrow CpMn_2(CO)_2^+ + CO$$

$$Cp(Mn)_2(CO)_2^+ + CpMn(CO)_3 \longrightarrow Cp_2Mn_3(CO)_5^+$$

$$CpMn^+ + CpMn(CO)_3 \longrightarrow Cp_2Mn_2^+ + 3CO$$

$$CpMn^+ + CpMn(CO)_3 \longrightarrow Cp_2Mn_2CO^+ + 2CO$$

$$CpMn^+ + CpMn(CO)_3 \longrightarrow Cp_2Mn_2(CO)_2^+ + CO$$

$$CpMn^+ + CpMn(CO)_3 \longrightarrow Cp_2Mn_2(CO)_3^+$$

$$CpMn(CO)^+ + CpMn(CO)_3 \longrightarrow Cp_2Mn_2(CO)_3^+ + CO$$

$$CpMn(CO)_2^+ + CpMn(CO)_3 \longrightarrow Cp_2Mn_2(CO)_3^+ + 2CO$$

$$Cp_2Mn_2^+ + CpMn(CO)_3 \longrightarrow Cp_2Mn_2(CO)_2^+ + CO$$

$$Cp_2Mn_2CO^+ + CpMn(CO)_3 \longrightarrow Cp_3Mn_3(CO)_2^+ + 2CO$$

$$Cp_2Mn_2CO^+ + CpMn(CO)_3 \longrightarrow Cp_3Mn_3(CO)_3^+ + CO$$

Similar experiments give the condensation chemistry of 2 and 3. This chemistry is summarized in Figures 9-13.

Although space limitations preclude a complete discussion of the condensation reactions of 1, 2, and 3, certain overall observations are immediately evident from even a cursory examination of the above data. First, the formation of ions of very complex formulae in all three systems indicates the stability of the complex ions and further indicates that complex neutral compounds of similar formula could be stable isolable species. This work thus suggests the existence of many complex organometallic compounds which have not yet been prepared.

The condensation chemistry of 1, 2, and 3 is quite similar in that in all systems fragment ions will react with the parent neutral to form more complex ions. In the $CpMn(CO)_3$, 1, system, these condensation reactions stop after the formation of tertiary ions. In contrast, in the chromium nitrosyl, 2, and thionitrosyl, 3, systems, the tertiary ions are reactive.

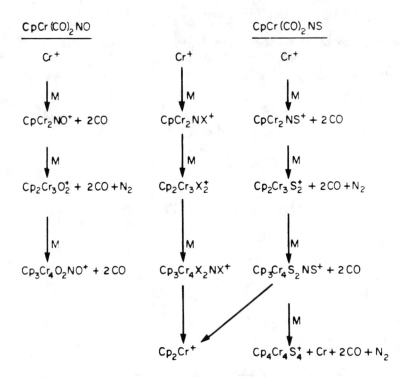

Figure 9. Positive ion condensation chemistry of $CpCr(CO)_2NX$, $X = O$, S. For the left column of data, $M = CpCr(CO)_2NO$ and for the right column of data, $M = CpCr(CO)_2NS$. This figure displays the ion chemistry resulting from Cr^+ as the original reactant ion.

In the case of the thionitrosyl system, the ion $Cp_4Cr_5S_3(NS)^+$ (m = 662) is formed via a sixth kinetic order reaction: i.e., as the product of five sequential reactions (cf. Figure 10). Each of the reactions in the $CpMn(CO)_3$ system is a reaction of the reactant ion with the parent neutral to form a more complex ion with the expulsion of CO molecules. While many similar reactions occur with the nitrosyl and thionitrosyl systems 2 and 3, systems 2 and 3 also react via other reaction types. Thus even though 1 and 2 are isoelectronic, the replacement of a CO ligand in 1 with an NO ligand in 2 opens up several new reaction channels. Of particular interest are the reactions in which the elements of N_2 are eliminated as neutral products. This reaction is noteworthy because one nitrogen atom must come from each of the reactant ion and the reactant neutral. Another usual reaction is the collision-induced dissociation reaction which eliminates C_5H_5N (the elements of pyridine).

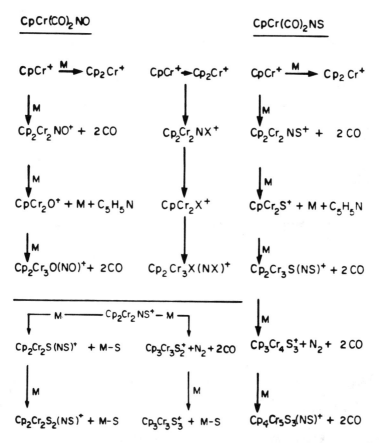

Figure 10. Positive ion condensation chemistry of $CpCr(CO)_2NX$, X = O, S. For the left column of data, M = $CpCr(CO)_2NO$; for the right column and for reactions (19)-(22), M = $CpCr(CO)_2NS$. This figure displays the ion chemistry resulting from $CpCr^+$ as the original reactant ion.

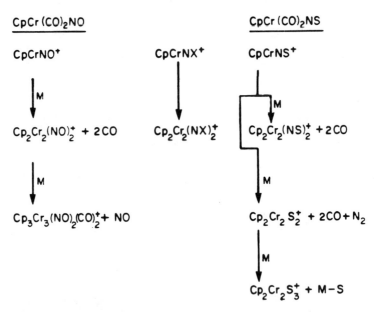

Figure 11. Positive ion condensation chemistry of $CpCr(CO)_2NS$, X = O, S. For the left column of data, M = $CpCr(CO)_2NO$, and for the right column, M = $CpCr(CO)_2NS$. This figure displays the ion chemistry resulting from $CpCrNX^+$ as the original reactant ion.

Although the present data does not contain any solid evidence for any particular ion structures, the structures of certain known compounds and quantum mechanical[29-32] bonding theories suggest two possible types of structures for the ions. Many organometallic compounds are known which contain metal-metal bonds.[31-33] The stability of these cluster compounds containing metal-metal bonds has been thoroughly examined using molecular orbital theory.[29-31] Accordingly, one can postulate for example that the dimetallic and trimetallic ions in the $CpMn(CO)_3$ system contain manganese-manganese bonds with the cyclopentadienyl and CO ligands attached to the metals. Similar structures may be postulated for many of the ions in the $CpCr(CO)_2NS$ systems. Alternatively, one can postulate that the complex ions have a sandwich structure with alternating metal atoms and cyclopentadienyl rings. The CO and NO ligands, if present, would be attached to the metal(s). Such structures would be analogous to compounds such as ferrocene and the well-known nickel "triple-decker sandwich" ion.[34] One attractive aspect of the

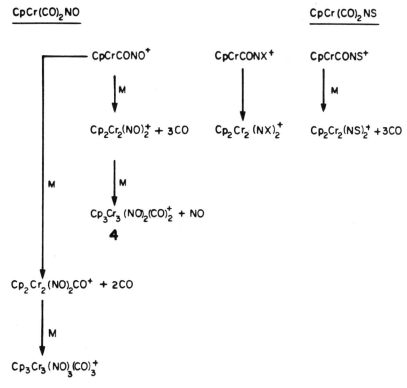

Figure 12. Positive ion condensation chemistry of $CpCr(CO)_2NX$, X = O, S. For the left column of data, M = $CpCr(CO)_2NO$ and for the right column, M = $CpCr(CO)_2NS$. This figure shows the ion chemistry resulting from $CpCrCONX^+$, X = O, S as the original reactant ion.

sandwich structures is that many of the reactions could be viewed as a nido to close conversion in the Wade formalism.[31] It should be noted that Wade[31] has predicted the stability of complexes formed by the bonding of a metal ion to the cyclopentadienyl ring in compounds such as $CpMn(CO)_3$, **1**.

The sulfur abstraction reactions in the chemistry of **3** are noteworthy for two reasons. First, the corresponding oxygen abstraction reactions in **2** do not occur. Of perhaps greater interest however is the fact that all of the ions which abstract sulfur contain more than one Cr atom. The ions which contain only one Cr atom do not abstract sulfur, but react in other ways. One explanation for this behavior would be that the reactant ions contain metal-metal bonds and that the abstracted sulfur is a "bridging" sulfur in the product ion. Compounds with metal-metal bonds and sulfur bridges are known.[35]

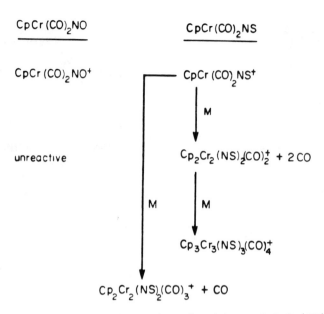

Figure 13. Positive ion condensation chemistry of $CpCr(CO)_2NX$, $X = O, S$. This figure displays the ion chemistry resulting from the parent ion $CpCr(CO)_2NX^+$, $X = O, S$ as the original reactant ion.

CONCLUSION

While the above work illustrates that the capability of the ICR technique for studying ion-molecule reactions is greatly enhanced by the use of Fourier techniques, it is possible that other applications of FT-ICR will also be important. One area where FT-ICR spectroscopy may be useful is analytical mass spectrometry.[23,24] The slow speed and small mass range of conventional ICR spectrometers have been severe limitations to the use of ICR spectrometers for analytical mass spectrometry. These limitations are minimized by the application of Fourier methods and it is possible that FT-ICR spectroscopy may be a useful method for certain analytical mass spectrometric problems.[23,24]

ACKNOWLEDGMENT: This research was supported by the Natural Sciences and Engineering Research Council of Canada and by the Research Corporation.

REFERENCES

1. Henis, J. 1972, in "Ion-Molecule Reactions," ed. J. L. Franklin, Vols. 1 and 2, Plenum Press, N.Y., Chapter 9.
2. Beauchamp, J. L. 1971, Ann. Rev. Phys. Chem. 22, 527.
3. Drewery, C. J., Goode, G. C., and Jennings, K. R. 1972, in "MTP International Review of Science, Mass Spectroscopy, Physical Chemistry," Series One, Vol. 5, ed. A. D. Buckingham and A. Maccoll, Butterworth, London, p. 183.
4. Brauman, J. I., and Blair, L. K. 1973, in "Determination of Organic Structures by Physical Methods," Vol. 5, ed. F. C. Nachod and J. J. Zuckerman, Academic Press, N. Y., p. 152.
5. Gray, G. A. 1971, Adv. Chem. Phys. 19, 141.
6. Lehman, T. A., and Bursey, M. M. 1976, "Ion Cyclotron Resonance Spectrometry," Wiley-Interscience, N.Y.
7. Baldeschwieler, J. D. 1968, Science 159, 263.
8. Comisarow, M. B. 1978, J. Chem. Phys. 69, 4097.
9. Comisarow, M. B. 1978, Int. J. Mass Spec. Ion Phys. 26, 369.
10. Comisarow, M. B. 1978, in "Transform Techniques in Chemistry," ed. P. R. Griffiths, Plenum Press, N.Y., Chapter 10.
11. Comisarow, M. B. 1978, in "Advances in Mass Spectrometry," Vol. 7B, ed. N. R. Daly, Heydon, London, p. 1042.
12. Comisarow, M. B., and Marshall, A. G. 1974, Chem. Phys. Lett. 25, 282.
13. Comisarow, M. B., and Marshall, A. G. 1974, Chem. Phys. Lett. 26, 489.
14. Comisarow, M. B., and Marshall, A. G. 1975, J. Chem. Phys. 62, 293.
15. Comisarow, M. B., and Marshall, A. G. 1976, J. Chem. Phys. 64, 110.
16. Comisarow, M. B., Grassi, V., and Parisod, G. 1978, Chem. Phys. Lett. 57, 413.
17. Parisod, G., and Comisarow, M. B. 1979, Chem. Phys. Lett. 62, 303.
18. Marshall, A. G., Comisarow, M. B., and Parisod, G. 1979, J. Chem. Phys. 71, 4434.
19. Comisarow, M., and Melka, J. 1979, Anal. Chem. 51, 2198.
20. Marshall, A. G., and Roe, D. C. 1980, J. Chem. Phys. 73, 1501.
21. Marshall, A. G. 1979, Anal. Chem. 51, 1710.
22. Marshall, A. G. 1979, Chem. Phys. Lett. 63, 575.
23. Ledford, E. B., Ghaderi, S., White, R. L., Wilkins, C. L., and Gross, M. L. 1980, Anal. Chem. 52, 469.
24. Ledford, E. B., White, R. L., Ghaderi, S., Gross, M. L., and Wilkins, C. L. 1980, Anal. Chem. 52, 1090.
25. McIver, R. T., Jr. 1970, Rev. Sci. Instrum. 41, 555.
26. Beauchamp, J. L., and Armstrong, J. T. 1969, Rev. Sci. Instrum. 40, 123.
27. Ledford, E. B., and McIver, R. T., Jr. 1976, Int. J. Mass Spec. Ion Phys. 22, 399.

28. Kolthammer, B., and Legzdins, P. 1978, J. Amer. Chem. Soc. 100, 2247.
29. Schilling, B. E. R., and Hoffman, R. 1979, J. Amer. Chem. Soc. 101, 3456.
30. Lauher, J. W. 1978, J. Amer. Chem. Soc. 100, 5305.
31. Wade, K. 1975, Chem. in Brit., 177.
32. Lauher, J. W., Elian, M., Summerville, R. H., and Hoffman, R. 1976, J. Amer. Chem. Soc. 98, 3219.
33. Band, E., and Muetterties, E. L. 1978, Chem. Rev. 78, 639.
34. Salzer, A., and Werner, H. 1972, Angew. Chem. 84, 949.
35. Mayerle, J. J., Denmark, S. E., Depamphilis, B. V., Ibers, J. A., and Holm, R. H. 1975, J. Amer. Chem. Soc. 97, 1032.

FOURIER TRANSFORM NUCLEAR QUADRUPOLE RESONANCE SPECTROSCOPY

Stanley M. Klainer[*], Tomas B. Hirschfeld[†]
and Robert A. Marino[§]

[*] Lawrence Berkeley Laboratory, 90/1140
Berkeley, CA 94720

[†] Lawrence Livermore National Laboratory, L325
Livermore, CA 94550

[§] Hunter College of CUNY
695 Park Avenue
New York, NY 10021

INTRODUCTION

FT-NQR (Fourier Transform Nuclear Quadrupole Resonance) spectroscopy is one of the newer analytical tools. Although nuclear quadrupole coupling constants were first observed in atoms by Schmidt and Schuler[1] in 1935 and in molecules by Kellog[2] and coworkers in 1936, it was the discovery of pure quadrupole coupling transitions in solids by Dehmelt and Kruger[3,4] in 1950 and 1951 that initiated the understanding of the NQR phenomenon as presently utilized.

During the past 30 years, the growth of NQR has been anything but phenomenal. A great quantity of theoretical work has been reported, but the experimental investigations have been restricted primarily to nitrogen-containing and halogenated compounds. How, then, can an application paper, such as this be justified? The answer lies in responding to two specific questions:
  (a) What has hindered the growth of NQR to date and can this be overcome with present day technology?
  (b) Why is it important that NQR be developed as a practical analytical technique at this time?

In responding to these queries, it is possible to show that FT-NQR is both needed and implementable.

This chapter is directed at the applied scientist, and, therefore, it does not contain the in-depth theoretical and mathematical information of interest to the research scientist. Texts by Schempp and Bray,[5] Lucken,[6] Slichter,[7] and Biryukov, Voronov and Safin[8] can supplement the information on NQR presented in this chapter. There is, however, no primer on FT-NQR, and the basic concepts of this specific area will be addressed in this discussion.

NQR SPECTROSCOPY

Nuclear Quadrupole Resonance (NQR) is a branch of radiofrequency spectroscopy. The NQR spectrometer detects the interaction of a nuclear quadrupole moment with the electric field gradient (EFG) produced by the charge distribution in a solid state compound. The quadrupole moment arises because the nuclear charge distribution is non-spherical. Resonance occurs when transitions from one spin state to another are excited by radiofrequency electromagnetic oscillations. In this way, NQR is quite similar in principle to nuclear magnetic resonance (NMR). In NQR, however, the energy levels depend on the coupling of the nuclear moment to the internal electric field gradient, whereas in NMR they are primarily dependent on the coupling of the nuclear magnetic moment with an external magnetic field. Since the EFG is a very sensitive function of the molecular and crystal structure, the resonance frequencies and band shapes are specific to each compound, and NQR data can be used for determining unambiguous sample identification, local electronic structure, atomic arrangement, order/disorder phenomena, and crystal phase transformations. In addition, molecular dynamics in the solid state can be studied. Furthermore, since NQR data are sensitive to changes in temperature and pressure, there is the possibility of obtaining strain information.

Basic concept of NQR

In order to understand the origin of nuclear quadrupole resonance, it is usual to visualize the nucleus as a classical distribution of positive charges, $\rho_N(\vec{r})$, over a volume of characteristic dimensions on the order of nuclear radii, i.e., 10^{-13} cm (Figure 1). On the other hand, a charge cloud extending over a volume on the order of several Angstroms (10^{-8} cm) generates an electrostatic potential $\phi(\vec{r})$ which can be considered as varying slowly over the region occupied by the nucleus. The electrostatic energy W of the system can be expressed[7] as

$$W = \int \rho_N(\vec{r}) \, \phi(\vec{r}) \, d\vec{r} \qquad [1]$$

which can be expanded in terms of the moments of the nuclear charge distribution,

Fourier Transform Nuclear Quadrupole Resonance Spectroscopy

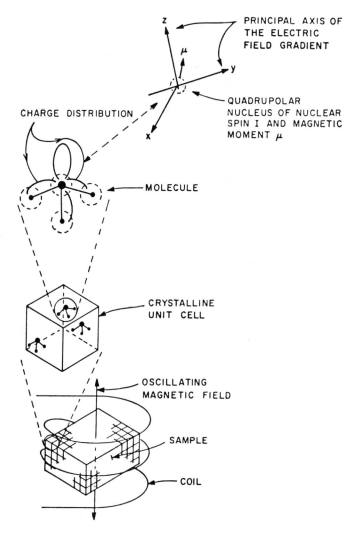

Figure 1. Graphic display showing origin and detection principles of NQR.

$$W = -\phi(0)q_N + \vec{p}\cdot\vec{E}(0) - \frac{1}{6}\sum_{ij} q(0)_{ij} Q_{ij} + \cdots \quad [2]$$

in which (0) means evaluated at the origin, the center of mass of the nucleus, and

$q_N = \int \rho_N(\vec{r})d\vec{r}$ nuclear charge

$\vec{p} = \int \rho_N(\vec{r})\vec{r}d\vec{r}$ nuclear dipole moment vector

$Q_{ij} = \int \rho_N(\vec{r})x_i x_j d\vec{r}$ nuclear quadrupole moment tensor

$\vec{E} = -\vec{\nabla}\phi$ external electric field vector

$q_{ij} = (\partial E_i/\partial x_j)_0$ external electric field gradient tensor

From quantum mechanical considerations, it can be proven that the only non-vanishing terms in Equation 2 are the first, third, and other odd-ordered ones. The first term is simply a constant, and, therefore, the orientationally-dependent term in the energy is the quadrupolar energy (hexadecapole interactions are extremely small if they exist at all):

$$W_Q = \frac{1}{6} \sum_{ij} q_{ij} Q_{ij} \qquad [3]$$

Q_{ij} is diagonal and is related to the components of the nuclear spin I_i according to

$$Q_{ii} = \frac{eQ}{I(2I-1)}[3I_i^2 - I(I+1)] \qquad [4]$$

in which $I = (\sum I_i^2)^{1/2}$ is the nuclear spin and eQ is defined as the nuclear quadrupole moment. Since $\nabla^2\phi = 0$ by the Laplace equation, we have

$$\sum_i q_{ii} = 0$$

The quantum mechanical expression for the quadrupole energy (the Hamiltonian) takes the particularly simple form

$$H_Q = \frac{eQ}{2I(2I-1)} \sum_i q_{ii} I_i^2 \qquad [5]$$

which can also be expressed in the form

$$H_Q = \frac{e^2qQ}{4I(2I-1)}[(3I_z^2 - I^2) + \eta(I_x^2 - I_y^2)] \qquad [6]$$

in which I and I_j are spin operators.

Fourier Transform Nuclear Quadrupole Resonance Spectroscopy 151

The quantity e^2qQ is called the quadrupole coupling constant of the system and η is defined as the asymmetry parameter of the electric field gradient (EFG)

$$\eta \equiv \frac{q_{xx} - q_{yy}}{q_{zz}}$$

and $eq \equiv eq_{zz}$, so $e^2qQ = (eQ)(eq_{zz})$ [7]

when the axes are labelled $|q_{zz}| \geq |q_{yy}| \geq |q_{xx}|$ by convention.

For nitrogen-14, with nuclear spin I = 1, the solution of the Schroedinger equation, $H_Q\Psi_N = E_N\Psi_N$, in which Ψ_N is the nuclear wave function, can be shown to lead to a three-level system of energies given by

$$E_z = e^2qQ/2$$

$$E_x = -E_z(1-\eta)/2$$

$$E_y = -E_z(1+\eta)/2 \quad [8]$$

Transitions between these levels can be induced with oscillating magnetic fields of the proper (resonant) frequencies (Figure 2). The frequencies of these transitions are

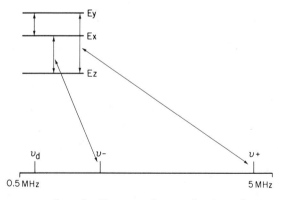

Figure 2. NQR energy level diagram for spin I = 1 and approximate range of nitrogen-14 transition frequencies.

$$\nu_+ = \frac{e^2qQ}{4h}(3+\eta)$$

$$\nu_- = \frac{e^2qQ}{4h}(3-\eta)$$

$$\nu_d = \frac{e^2qQ\eta}{2h} = \nu_+ - \nu_- \qquad [9]$$

Determining any two of the three NQR frequencies completely describes the magnitude of the EFG in the vicinity of the nitrogen nucleus:

$$\frac{e^2qQ}{h} = \frac{2}{3}(\nu_+ + \nu_-)$$

$$\eta = \frac{2(\nu_+ - \nu_-)}{e^2qQ/h} \qquad [10]$$

As another example, for nuclei of spin I = 5/2, such as aluminum-27, the solution of the Schroedinger equation can be shown to lead to a three-level system of energies[6] given by solutions of the secular equation[9]

$$E^3 - 7(3+\eta^2)E - 20(1-\eta^2) = 0 \qquad [11]$$

in which E is in units of $e^2qQ/20$. For the case, $\eta = 0$, the energies of the three eigenstates are

$$E(\pm\tfrac{1}{2}) = -4\,\frac{e^2qQ}{20}$$

$$E(\pm\tfrac{3}{2}) = -1\,\frac{e^2qQ}{20}$$

$$E(\pm\tfrac{5}{2}) = +5\,\frac{e^2qQ}{20} \qquad [12]$$

The frequencies of the resultant allowed transitions are given by

$$\nu_1 = \nu(\tfrac{3}{2} \to \tfrac{1}{2}) = \frac{3}{20}\,\frac{e^2qQ}{h}$$

$$\nu_2 = \nu(\tfrac{5}{2} \to \tfrac{3}{2}) = \frac{6}{20}\,\frac{e^2qQ}{h} \qquad [13]$$

For the general case, $\eta \neq 0$, no solution in closed form can be written down. However, tables of the results of numerical solution of the eigenvalue problem for $0 \leq \eta \leq 1$ in steps of 0.001 are available,[10] so that observation of two NQR frequencies uniquely determines both the quadrupole coupling constant, e^2qQ, and the asymmetry parameter, η, of the site under study. For large values of η, it becomes theoretically possible to observe the "forbidden" transition corresponding to the sum of ν_1 and ν_2 in Equation 10. This may be especially useful to do in situations of low quadrupole coupling.

As an additional example, for nuclei of spin $I = 7/2$, such as uranium-235, the solution of the Schroedinger equation can be shown to lead to a five-level system of energies given by solutions of the secular equation[5]

$$E^4 - 42(1+\eta^2/3)E^2 - 64(1-\eta^2)E + 105(1+\eta^2/3)^2 = 0$$

in which E is in units of $e^2qQ/28$.

For the general case, $\eta \neq 0$, no solution in closed form can be written. However, tables of the results of numerical solution of the eigenvalue problem for $0 \leq \eta \leq 1$ in steps of 0.001 are also available,[10] so that observation of two identified NQR frequencies again uniquely determines both the quadrupole coupling constant, e^2qQ, and the asymmetry parameter, η, of the site under study.

Experimental techniques

Experimental techniques for the determination of NQR spectra can be divided into three categories:
  a. Continuous-wave (CW) methods
  b. Transient methods
  c. Double resonance techniques.

CW methods have enjoyed great popularity for they are simple and inexpensive. The limitations of these techniques are so severe, however, that in the low MHz region they are rarely used anymore. Transient methods include the superregenerative technique (unproductive at the low frequencies of interest in many applied problems), and pulsed methods. Double resonance techniques are essentially pulsed methods that monitor the resonance of one type of nuclear species while another is being perturbed.[7] When applicable, these methods are extremely sensitive and are very convenient for locating unknown resonances, although line shapes are not always reliable. Interested readers are referred to the excellent review by Edmonds.[11]

Pulsed techniques that operate in the so-called spin-echo mode are particularly good for applied NQR, and they naturally lend themselves to remote detection.[12]

To understand the reasons for the advantage of the pulsed method,[13-17] it is necessary to begin by defining some basic magnitudes that describe an NQR line. In the frequency domain, an NQR line is fully described by its frequency and by a normalized shape function $S(\omega)$. For example,

$$S(\omega) = \frac{2}{\pi \Delta \omega} \frac{1}{1 + (\frac{\omega - \omega_0}{\Delta \omega})^2} \qquad [14]$$

However, spin-echo techniques operate in the time domain; here there are three parameters that describe an NQR absorption line:
    a. Spin-lattice relaxation time, T_1, the characteristic time with which a bulk magnetization is established and energy can flow between the spin system and the lattice.
    b. Spin-spin relaxation time, T_2, the characteristic time which describes the coupling between nuclei and which establishes the time scale for the observation of spin-echoes.
    c. A spin-echo shape function, $G(t)$. The "width" of this function is defined as $2T_2^*$. $G(t)$ is the Fourier transform of the line shape function, $S(\omega)$; thus $T_2^* \propto (1/\Delta\omega)$.

Continuous-wave (CW) and superregenerative methods are poor techniques for lines with large T_1 and/or large $\Delta\omega$ values (short T_2) associated with the NQR line (to avoid saturation in CW, the input power level has to be reduced to very small values). Broad lines, expected to be important in applied NQR, are weak since their areas are constant. Spin-echo wide-line methods are not only unaffected by long T_1 values, but since $G(t)$ and $S(\omega)$ are related by a Fourier transformation, the equality

$$G_{MAX} = \int_0^\infty S(\omega) \, d\omega = 1 \qquad [15]$$

holds (Figure 3). Thus, broad lines in the frequency domain do not affect the maximum intensity of the echo signal.

The NQR spectrum of a substance is determined by placing about 25 grams of sample inside the inductor of a tank circuit, which is then subjected to a series of radiofrequency pulses of frequency f. Whenever the frequency of these pulses satisfies the resonance condition, $f = \nu_Q$, where ν_Q is one of the quadrupole frequencies, absorption of energy takes place and is retransmitted as a series of signals [free induction decay (FID) or spin-echo]. Therefore, by monitoring and detecting the transmitted signals as a function of the frequency of the pulses, the energy levels of the quadrupolar nucleus are completely determined. A number of different pulse sequences are possible; this results in more or less efficient signal production, depending on the values of the relaxation times.

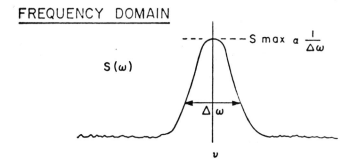
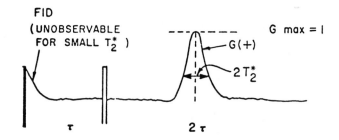

Figure 3. Parameters for spin-echo Fourier transforms.

NQR INSTRUMENTATION AND DATA HANDLING

For applied NQR spectroscopy, pulsed techniques are used in conjuction with a variety of pulse sequences and FT data processing. This approach provides for maximum sensitivity and versatility.

Pulsed NQR spectrometer

The most advanced pulsed NQR spectrometer in the nitrogen frequency region presently in operation was described by Harding et al.[18] in 1979. A block diagram of this instrument is shown in Figure 4. This is an FT-NQR spectrometer that operates from 0.5 to 5 MHz. The features of this instrument are:
(a) The use of heterodyne techniques throughout to eliminate carrier feed-through.
(b) A matching network design that allows one-knob tuning of both the transmitter and receiver.

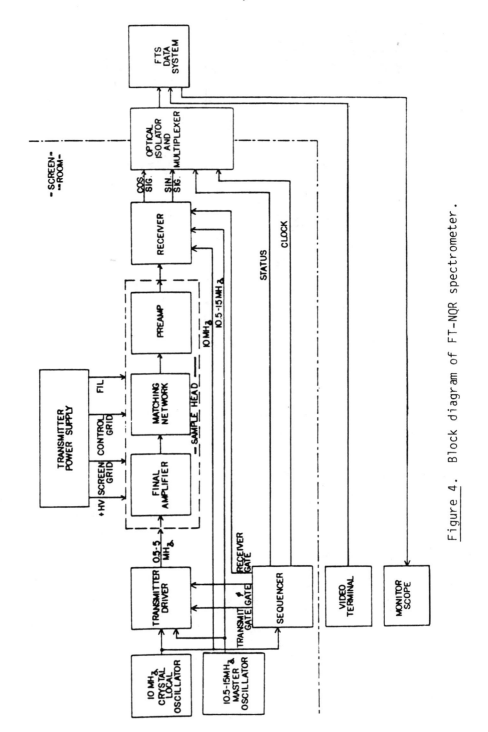

Figure 4. Block diagram of FT-NQR spectrometer.

Fourier Transform Nuclear Quadrupole Resonance Spectroscopy

(c) The choice of several excitation sequences (Carr-Purcell, Meiboom-Gill, modified Carr-Purcell, spin-locked spin-echo, and the standard sequence of $\pi/2$ pulses) for generation and collection of signals at high data rates.

(d) Fast Fourier transform routines for studying line shapes, for facilitating spectral searches, and for improved sensitivity.

(e) Ability to handle small samples (0.1 cc).[19]

The specifications for this instrument are given in Table 1, column 1. This instrument, although state-of-the-art and much more versatile and sensitive than previous spectrometers, is not optimally suited to many applied NQR uses. A new system suited for practical NQR use is presently under design and scheduled for construction in late 1981. Its anticipated specifications are shown in Table 1, column 2. When this instrument is operational, many of the obstacles which hinder NQR growth should be overcome.

Table 1. Pulsed FT-NQR spectrometer.

Specification	Existing System[18]	System Under Design
Frequency range	0.5 to 5 MHz	0.5 to 64 MHz
Sample volume	40 to 0.1 cc[19]	100 to 5×10^{-4} cc
System sensitivity (SLSE)[21]; S/N = 3, 1 sec integration	200 mg	1 mg
System recovery time (Sample Q = 120)	150 μsec @ 3 MHz	<15 μsec @ 1 MHz
$\pi/2$ pulse width	50 μsec	10 μsec
Sample coils	1 per octave	Same
Matching networks	1 per octave	Same
Sample operating temperature	77-350 K	4-350 K
Remote detection	Yes, non-directional	Yes, directional
Automatic spectral search	No	Yes, programmed sequence

Pulse sequence

In order to optimize the NQR data format, it is necessary to choose the pulse sequence most suited for the measurement. Several of these exist and each has its benefits.

(1) Free induction decay

The NQR response of a single crystal to a resonant pulse of rf irradiation is completely analogous to the spin 1/2 NMR case. One difference is that the intensity of the free induction decay (FID) response depends on the orientation of the radiofrequency field H_1 with respect to the EFG principal axis system. For a spin $I = 1$ system, the three resonance lines ν_+, ν_- and ν_d can be observed only when the field H_1 is oriented, respectively, along the x, y and z principal axes of the EFG tensor. As an example, for a ν_- line, the expected value of the magnetization along the y axis is proportional to

$$\frac{h\nu_-}{3kT} \sin(\sqrt{2}\omega_1 t_w) \cos(2\pi\nu_- t) \qquad [16]$$

in which $\omega_1 = \gamma H_1$ measures the intensity of the irradiation, and t_w is the duration of the irradiating pulse. The maximum response is obtained when

$$\sqrt{2}\omega_1 t_w = \sqrt{2}\gamma H_1 t_w = \pi/2 \qquad [17]$$

in analogy with the NMR case.

In NQR, however, the sample usually consists of a polycrystalline powder. Then a convolution for all orientations must be made. The result[20] is that the $\sin(\sqrt{2}\omega_1 t_w)$ function in the expression for the expectation value of the magnetization becomes a Bessel function, $J_1(\sqrt{2}\omega_1 t_w)$. This function has its first maximum, analogous to a "90° pulse", at $\sqrt{2}\omega_1 t_w = 0.66\pi$ and not at 0.5π like the sine function. Similarly, the first null, corresponding to a "180° pulse", occurs for a value of the argument equal to 1.43π rather than simply π.[20]

(2) Spin-echo and Carr-Purcell (CP) sequence

Following an FID experiment in a case where T_2 (spin-spin relaxation time) $> T_2^*$ (spin-echo shape function), it is possible to recall part of the magnetization not lost through T_2 processes by applying a "180° pulse" at a time τ after the first pulse. As is well known, an echo will form at a time 2τ and this echo can be recalled repeatedly at integral multiples of this time, $2n\tau$, by the application of additional "180° pulses" at times $(2n-1)\tau$. The amplitude of the resultant echo train decays with time constant T_2 (Figure 5).

Fourier Transform Nuclear Quadrupole Resonance Spectroscopy

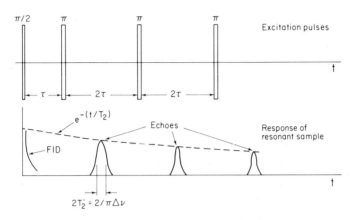

Figure 5. Parameters for Carr-Purcell sequence.

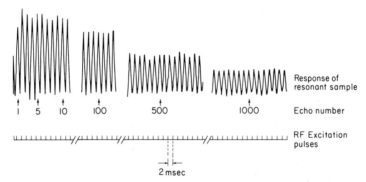

Figure 6. Spin-Locked-Spin-Echo (SLSE) train for ν_- line of $NaNO_2$ at 77 K. Transmitter frequency set at center of resonance line.

Figure 7. Spin-Locked-Spin-Echo train for ν_- line of $NaNO_2$ at 77 K. The transmitter frequency was chosen a few kHz away from resonance.

Signal-to-noise enhancement can be obtained by coherently adding successive echoes in the sequence. The optimum time for co-addition is easily shown to be 1.26 T_2. Due to the generally small value of T_2 in solids, however, this method does not result in an appreciable enhancement of the signal-to-noise ratio.

(3) Spin-lock spin-echo (SLSE)

A major advance in S/N enhancement was made by Marino and Klainer[21] when an adaptation of the Ostroff-Waugh sequence[22] was made to NQR. The sequence needed is essentially a Meiboom-Gill-modified-CP sequence in which all the pulses have the same "flip angle" of "90°," which means 0.66π in the NQR of an I = 1 nucleus. In Figure 6 ($NaNO_2$ at 77 K, on resonance) and Figure 7 ($NaNO_2$ at 77 K, slightly off resonance), it can be seen that the spin-echo train in this sequence persists for times of order T_1 (spin-lattice relaxation time, actually $T_1\rho$) and not the much shorter T_2. Coherent addition of the echoes in this case results in considerable enhancement of the S/N, since $T_1\rho \gg T_2$ is the typical situation in these solids. Marino and Klainer showed that the optimum enhancement is $0.64(T_{2\epsilon}/2\tau)^{1/2}$, where $T_{2\epsilon}$ is the effective decay constant of the echo train and 2τ is the spacing between echoes, or equivalently the spacing between pulses of the excitation sequence. It was further shown that the NQR effect was completely analogous to the spin 1/2 NMR case discussed by Waugh[23] in that the decay constant $T_{2\epsilon}$ tends to $T_1\rho$ as τ is reduced to values less than T_2. Furthermore, for intermediate values of $T_{2\epsilon}$, this parameter is proportional to τ^{-5}, again in analogy to NMR. Figure 8 shows this functional dependence for $NaNO_2$ at 77 K. Recently, Cantor and Waugh[24] have developed a theory to explain the main features of this NQR effect, using a model of a polycrystalline solid with each nitrogen site having one nearest neighbor.

(4) Strong Off-Resonance Comb (SORC)

Recently one of the authors (RAM) has used a new pulsed NQR experiment which can have considerable advantages in enhancing the S/N ratio of weak lines. This represents new and as yet unpublished information.

The steady-state response of an ensemble of nuclear spins, I = 1/2, in high magnetic field H_0, to a strong radiofrequency field H_1, applied off-resonance by Δf, has long been known.[7] When all the conditions for the establishment of a spin temperature in the rotating frame are met,[7] the x-component of the magnetization, which is experimentally observable, is given by the expression

$$M_x = M_0 \left[\frac{H_1(2\pi\Delta f/\gamma)}{H_1^2 + H_{1oc}^2 + (2\pi\Delta f/\gamma)^2} \right] \quad [18]$$

Fourier Transform Nuclear Quadrupole Resonance Spectroscopy

in which M_0 is the equilibrium longitudinal magnetization, γ is the magnetogyric ratio of the nucleus, and H_{loc} is a measure of the local field at the nuclear site due to its neighbors. Results analogous to Equation 18 have also been derived and observed for a quadrupolar system[25] with nuclear spin I = 3/2 when subjected to the same strong, long, off-resonant H_1 irradiation.

The preliminary results obtained, when the irradiation field H_1 is applied in a long train of equally-spaced identical pulses, are presented here. Although the SORC experimental data reported here are for a quadrupolar I = 1 system, analogous effects in a magnetic system or a quadrupolar system with spin different from unity can be expected.

Figure 9a defines the parameters of the SORC sequence. Here a train of radiofrequency pulses of duration t_w and spacing τ is applied, Δf away from exact resonance, to a pure nuclear quadrupole system in zero external magnetic field.

The variation of the signal amplitude vs. Δf, the distance from exact resonance, shows two features as depicted in Figures 9b and 9c. First, the signal amplitude is modulated by a sinusoid of period $1/\tau$, the pulse repetition rate. This phenomenon is best understood by considering that the Fourier transform of the transmitter pulse has periodicity $1/\tau$. This leads to successive maxima and minima in the

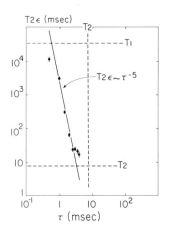

Figure 8. Double logarithmic plot of $T_{2\epsilon}$, the long decay constant (ϵ = effective) of the spin-echo train, vs. τ, the spacing of the first two pulses in the SLSE excitation sequence. The transition is the ν_- line of $NaNO_2$ at 77 K at 3757 kHz. The solid line is a fit to the data of $\tau^5 T_{2\epsilon}$ = constant. Also shown are the relaxation times T_1 and T_2 for the transition.

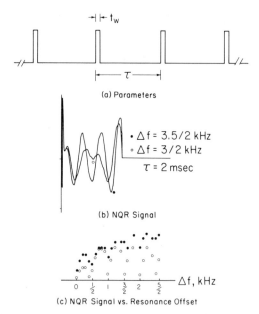

Figure 9a. Timing diagram for SORC sequence (Strong Off-Resonance Comb of rf pulses).
9b. Superposition of nuclear induction signals obtained in SORC observation window when Δf = 3/2 kHz and Δf = 3.5/2 kHz. Note destructive interference in the former case. All data were taken on the ν_- line of $NaNO_2$ at 77 K.
9c. Plot of signal height obtained in the middle and right-hand side of observation window using the SORC sequence.

NQR signal (Figures 9b,c) as the transmitter frequency is changed, i.e., Δf is varied. Alternatively, and more naively, this modulation can be interpreted as the destructive interference of type I signals (immediately following the rf pulse) and type II signals (immediately preceding the rf pulse) in their overlap region as the frequency, Δf, is slowly varied.

Another feature shown in Figure 9c is the shape of the envelope, possibly conforming to a function of the type, $(\Delta f)/(A^2 + (\Delta f)^2)$, such as Equation 18. Insufficient data have been taken so far to ascertain the degree of agreement with theory on this last point.

The nuclear induction signals present in the observation window between successive pulses of the SORC sequence are shown in Figures 10a-j as a function of the pulse separation τ. All data were taken on the ν_- line of $NaNO_2$ at 77 K. The magnitude of type I signal

Fourier Transform Nuclear Quadrupole Resonance Spectroscopy 163

is then plotted vs. τ in Figure 11. Note that for $\tau > 8$ msec, the magnitude of type I signals increases with τ as might be expected for an FID signal subject to spin-lattice relaxation. On the other hand, for $\tau < 5$ msec, signals at both ends of the observation window are of comparable size, and they grow exponentially with decreasing τ. This is the region of interest.

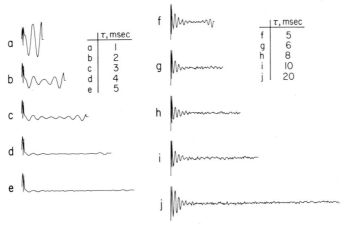

Figure 10. Trace of steady-state nuclear induction signal in an SORC sequence for various values (a-j) of the pulse spacing τ. Signals are to scale. All data were taken on the ν_- line of NaNO$_2$ at 77 K. Plots e and f are identical signals displayed on two different scales.

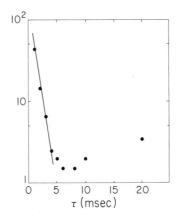

Figure 11. A semilogarithmic plot of the type I signal height (see text) vs. the pulse spacing parameter τ. The data are taken from Figure 10.

Figure 12 shows the variation of the type II signal vs. $\langle H_1 \rangle$ for $\tau = 3$ msec. The experimental points are obtained for four different values of the instantaneous field H_1 obtained by changing the pulse width t_w at constant "flip angle" $\overline{2\gamma t_w H_1}$. The dotted line is the curve $F = \langle H_1 \rangle / [\langle H_1 \rangle^2 + B^2]$ with $B = 0.05$ G. The fact that $\langle H_1 \rangle$, the average value, rather than H_1, the peak value, is the important parameter and that there is good agreement of the data with the form of Equation 18 is strong evidence that the ensemble of spins is responding to the time-averaged field of the SORC sequence in a manner analogous to the conventional long, strong, off-resonant pulse.

The size of the parameter B is found to be about 0.05 G, which is approximately two orders of magnitude too small for the value expected from the contribution of Δf to Equation 18. This discrepancy is reduced by a factor of 5 when the experiment is repeated for $\tau = 1$ msec, as shown in Figure 13. Comparison of Figs. 12 and 13 suggests that the pulse nature of the experiment is still very important for $\tau = 1$ msec and that quantitative agreement cannot be expected until τ is reduced further.

The potential of this technique appears to be great since signals can be obtained at essentially 100% duty cycle. However, further experimentation is needed to understand completely the operational parameters of SORC.

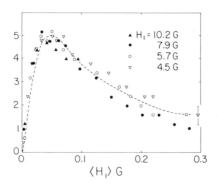

Figure 12. A plot of the magnitude of type II signals (see text) vs. the time-averaged value of the rf field, $\langle H_1 \rangle$, using a SORC sequence. All data were taken on the ν_- line of $NaNO_2$ at 77 K with $\Delta f = 3.5/2$ kHz; $\tau = 3$ msec.

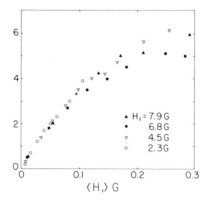

Figure 13. A plot of the magnitude of type II signals (see text) vs. the time-averaged value of the rf field, $\langle H_1 \rangle$, using a SORC sequence. Data were taken on the ν_- line of $NaNO_2$ at 77 K with $\Delta f = 1.5/2$ kHz; $\tau = 1$ msec.

FT considerations

There are three main reasons for doing FT spectroscopy:
(a) Enhancement of signal-to-noise over CW methods is given by the square root of the ratio of the total width of the spectrum to the typical line width, $\sqrt{T_2^*/\Delta}$.
(b) Pulsed methods are singularly well-suited to data processing.
(c) The line shape is readily obtained directly from the output.
Of these, the first reason, probably the most important in other disciplines, is not nearly so important in NQR, because the ratio defined above is not large and can often be close to unity. This is so because in solids, line widths are relatively large (a few kHz), while the bandwidths that can be suitably irradiated are in the 10-100 kHz range. The other two advantages, however, have provided the impetus toward the growth of FT-NQR.

Comparison to high-resolution NMR (HR-NMR)

In HR-NMR, the width of the spectrum Δ is much less than the carrier frequency f_0; $\Delta/f_0 \leq 10^{-3}$ can be expected even in worst cases. In NQR, however, the ratio Δ/f_0 is often of order unity.

This implies that the NQR spectrum must be obtained one frequency interval at a time and cannot usually be displayed in one operation as in the case of HR-NMR.

Processing techniques

Once FID or echo signals have been obtained, the proper FT treatment for each will yield the desired line shape spectrum.[26,27] In this section, the procedures the authors found most satisfactory are discussed.

The problem of phase correction for NQR spectra presents a particuar problem not found in fixed-frequency spectrometers. The fact that the NQR spectrometer operates at variable frequencies means that for NQR signals, instrumental phase shifts which are not corrected for before final data collection will be present. These phase shifts can be considerable, and are of course also present in echoes, where the receiver dead time problem present for FID signals does not occur. In Table 2, the mathematical results for Lorentzian line shapes are collected as guides to the solution of this problem.

The results in the Table show that the presence of phase shifts produces an admixture of absorption and dispersion modes in the case of FID signals, and a possible loss of intensity in echo signals. This problem can be avoided while at the same time preserving the true Lorentzian line shape, if the modulus squared transform is computed for FID signals, and the modulus transform is computed for echoes. Thus the proper line shapes are obtained in each case regardless of the degree and source of phase shift, without need for a separate "phase correction" subroutine.

In Figures 14-21, the results of computer-simulated spectra are shown. An echo and an FID signal have been simulated for both Lorentzian and Gaussian line shapes. Cosine, sine, the square of the modulus, and modulus transforms are computed and displayed. Figures 14-17 are for no phase shift, while Figures 18-21 have a phase shift of 30° in the time domain signals. Note that in all cases, the conclusions discussed for proper data processing are borne out, namely that the modulus squared transform should be used for FID signals, and the modulus transform for echo signals.

Figures 22-24 show experimental spectra that demonstrate the foregoing arguments. Figures 22a and 23a are the nitrogen-14 NQR FID signals at 77 K from hexamethylenetetramine (HMT) and urea, respectively. The cosine and sine transforms of HMT, Figures 22b,c, clearly show the admixture of absorption and dispersion expected when phase shifts exist in the time-domain data. This effect is much less evident in the cosine and sine transforms of urea, Figures 23b,c, which occurred with only a small phase shift. Finally, the proper line shapes are shown in Figures 22d and 23d, the modulus squared transforms of the time-domain signals. Note the fine structure on the HMT line, first reported by Colligiani and Ambrosetti.[28] The modulus transforms, Figures 22e and 23e, are shown for comparison, and they are visibly broader than the true line shapes.

Table 2. Lorentzian FID and echo complex signals and their Fourier transforms.

FID

Time domain: $f(t) = e^{-\alpha t} e^{i\omega_0 t} e^{-i\phi}$, $t \geq 0$

Frequency domain:

$$C = \text{cosine transform} = \frac{1}{\alpha^2 + (\omega-\omega_0)^2} [\alpha\cos(\phi) - (\omega-\omega_0)\sin(\phi)]$$

$$S = \text{sine transform} = \frac{1}{\alpha^2 + (\omega-\omega_0)^2} [\alpha\sin(\phi) + (\omega-\omega_0)\cos(\phi)]$$

$$C^2 + S^2 = \text{modulus squared transform} = \frac{1}{\alpha^2 + (\omega-\omega_0)^2}$$

$$[C^2 + S^2]^{1/2} = \text{modulus transform} = [\frac{1}{\alpha^2 + (\omega-\omega_0)^2}]^{1/2}$$

ECHO

Time domain: $f(t) = \begin{cases} e^{\alpha t} e^{-i\omega_0 t} e^{-i\phi}, & t < 0 \\ e^{-\alpha t} e^{i\omega_0 t} e^{-i\phi}, & t \geq 0 \end{cases}$

Frequency domain:

$$C = \text{cosine transform} = \frac{1}{\alpha^2 + (\omega-\omega_0)^2} 2\alpha \cos(\phi)$$

$$S = \text{sine transform} = \frac{1}{\alpha^2 + (\omega-\omega_0)^2} 2\alpha \sin(\phi)$$

$$C^2 + S^2 = \text{modulus squared transform} = [\frac{2\alpha}{\alpha^2 + (\omega-\omega_0)^2}]^2$$

$$[C^2 + S^2]^{1/2} = \text{modulus transform} = [\frac{2\alpha}{\alpha^2 + (\omega-\omega_0)^2}]$$

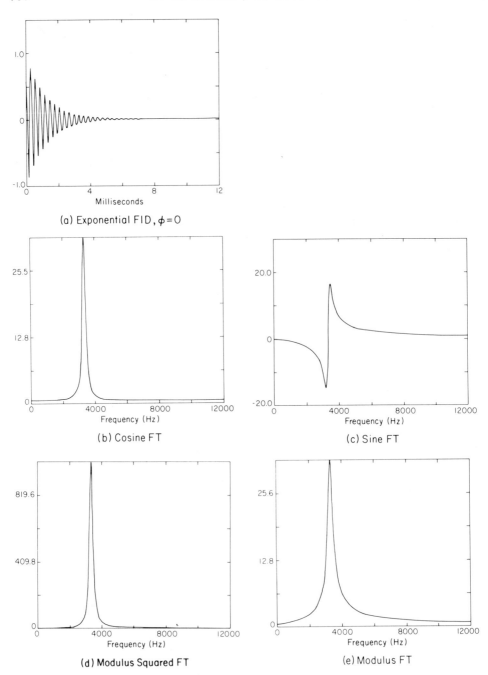

Figure 14. Computer-simulated exponential FID signal and its Fourier transforms. (a) time domain signal, $e^{-\alpha t}\cos(\omega t)$; (b) cosine FT of (a); (c) sine FT; (d) square of the modulus FT; (e) modulus FT.

Fourier Transform Nuclear Quadrupole Resonance Spectroscopy

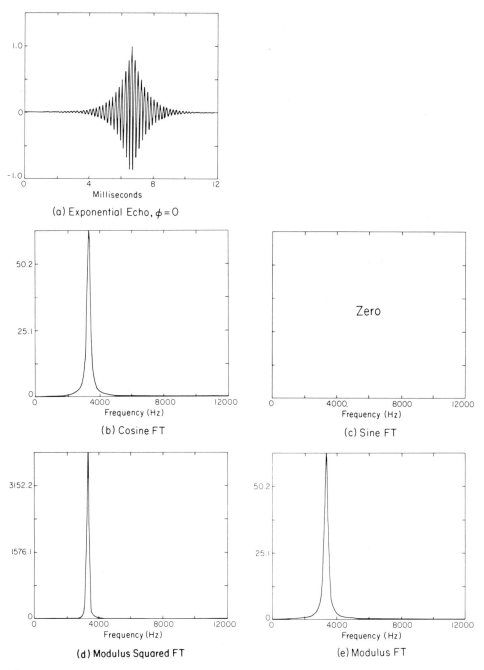

Figure 15. Computer-simulated exponential echo signal and its Fourier transforms. (a) time domain signal, $e^{-\alpha|t|}\cos(\omega t)$; (b) cosine FT of (a); (c) sine FT = 0; (d) square of the modulus FT; (e) modulus FT.

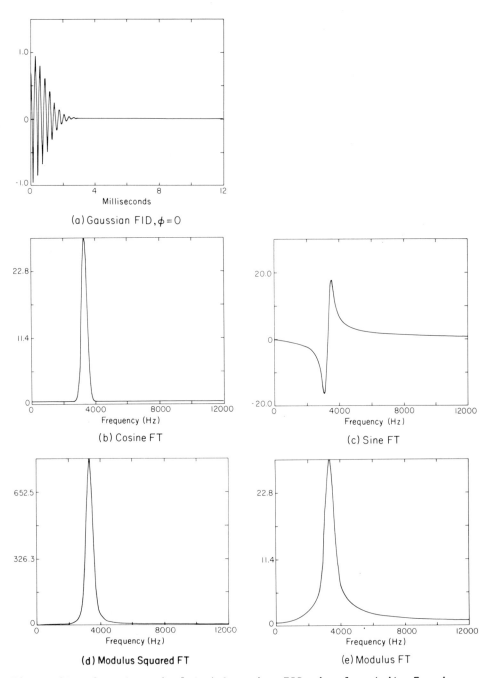

Figure 16. Computer-simulated Gaussian FID signal and its Fourier transforms. (a) time domain signal, $\exp[-(\alpha t)^2] \cos(\omega t)$, $t \geq 0$; (b) cosine FT; (c) sine FT; (d) square of the modulus FT; (e) modulus FT.

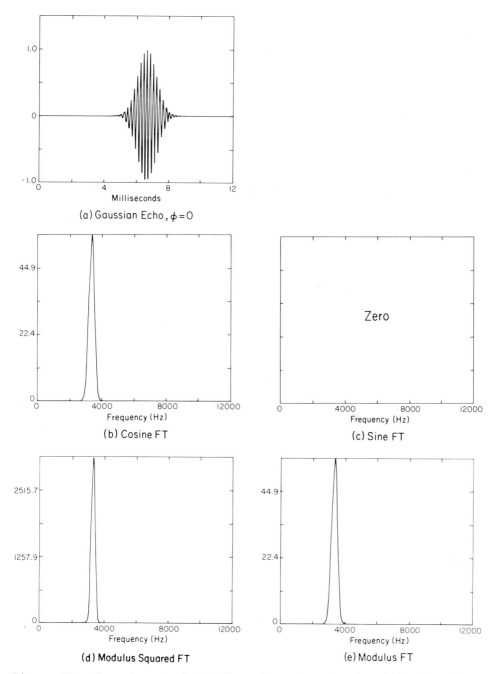

Figure 17. Computer-simulated Gaussian echo signal and its Fourier transforms. (a) time domain signal, $\exp[-(\alpha t)^2]\cos(\omega t)$; (b) cosine FT; (c) sine FT = 0; (d) square of the modulus FT; (e) modulus FT.

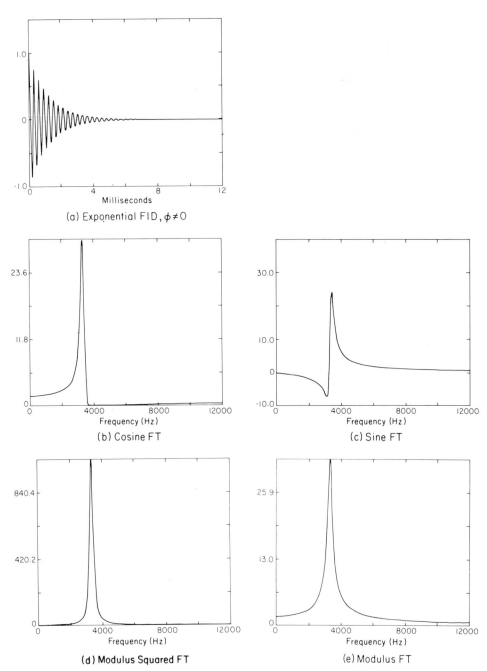

Figure 18. Computer-simulated phase-shifted exponential FID and its Fourier transforms. (a) time domain signal, $e^{-\alpha t}\cos(\omega t - \phi)$; (b) cosine FT; (c) sine FT; (d) square of the modulus FT; (e) modulus FT.

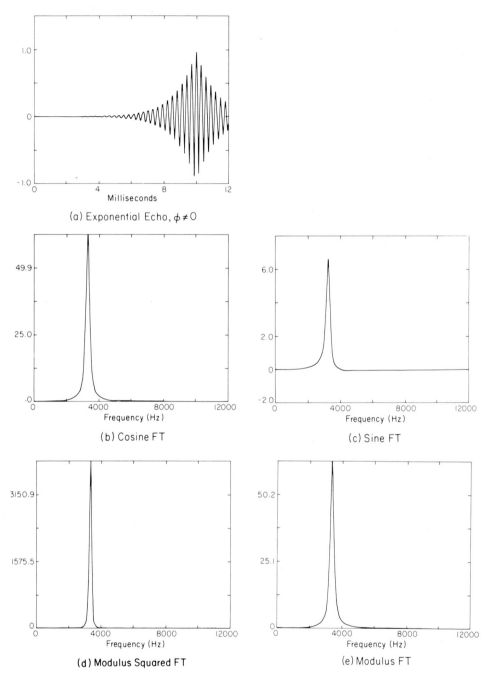

Figure 19. Computer-simulated phase-shifted exponential echo and its Fourier transforms. (a) time domain signal, $e^{-\alpha|t|}\cos(\omega t-\phi)$; (b) cosine FT; (c) sine FT; (d) square of the modulus FT; (e) modulus FT.

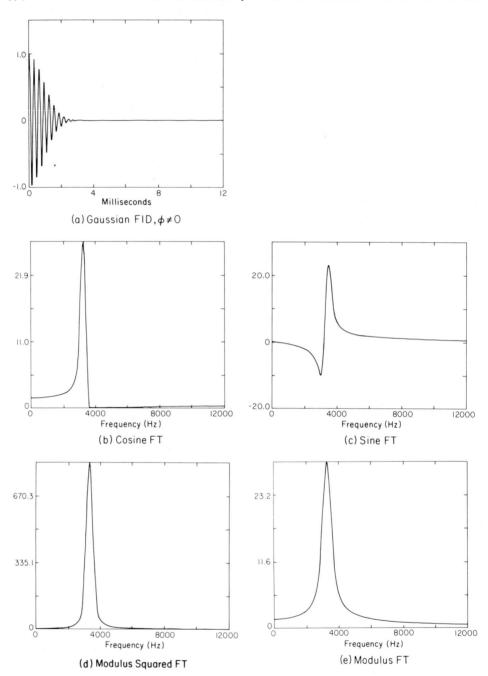

Figure 20. Computer-simulated phase-shifted Gaussian FID and its Fourier transforms. (a) time domain signal, $\exp[-(\alpha t)^2]\cos(\omega t - \phi)$; (b) cosine FT; (c) sine FT; (d) square of the modulus FT; (e) modulus FT.

Fourier Transform Nuclear Quadrupole Resonance Spectroscopy

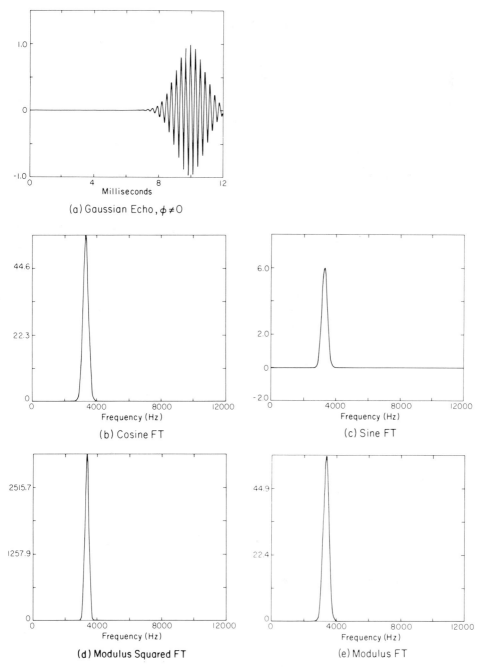

Figure 21. Computer-simulated phase-shifted Gaussian echo and its Fourier transforms. (a) time domain signal, $\exp[-(\alpha t)^2]\cos(\omega t - \phi)$; (b) cosine FT; (c) sine FT; (d) square of the modulus FT; (e) modulus FT.

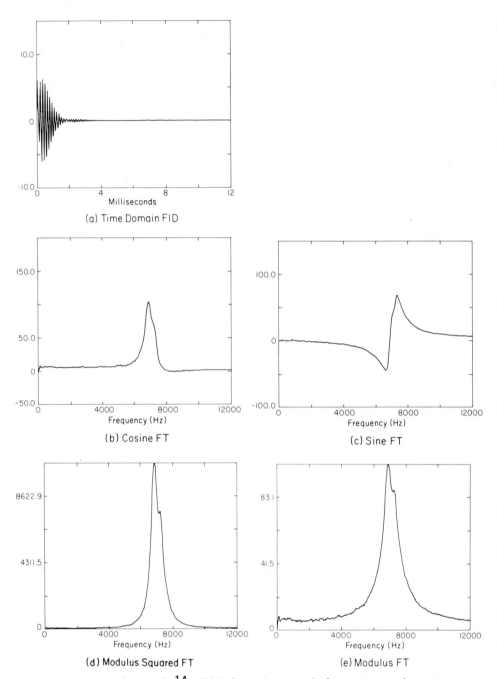

Figure 22. Experimental ^{14}N FID from hexamethylene-tetramine at room temperature and its Fourier transforms. (a) time domain signal; (b) cosine FT; (c) sine FT; (d) square of the modulus FT; (e) modulus FT.

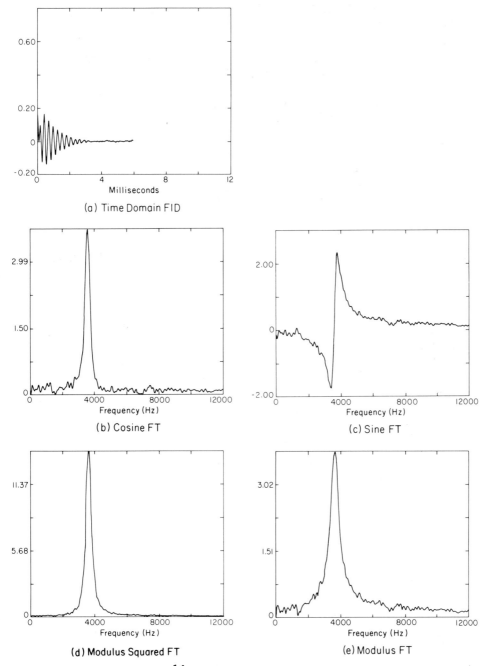

Figure 23. Experimental ^{14}N FID from the ν_+ line in urea at 77 K and its Fourier transforms. (a) time domain signal; (b) cosine FT of (a); (c) sine FT; (d) square of the modulus FT; (e) modulus FT.

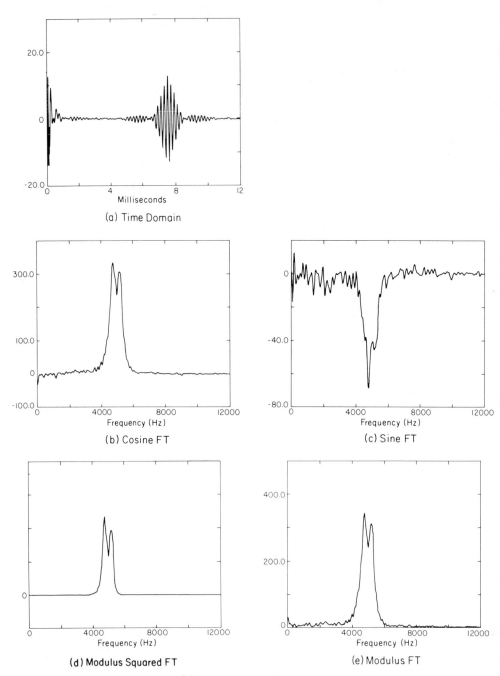

Figure 24. Experimental ^{14}N spin echo from 727 kHz ν_- doublet in monoclinic TNT at 77 K and its Fourier transforms. (a) time domain; (b) cosine FT; (c) sine FT; (d) square of modulus FT; (e) modulus FT.

An application of an echo signal is shown in Figure 24a, a doublet of ν_- lines from the monoclinic phase of TNT at 77 K. Figures 24b,c are, respectively, the cosine and sine transforms of this signal. Note that, as expected, both of these transforms yield valid line shapes with fractional amplitudes, and that the <u>modulus</u> transform, Figure 24e, yields the correct line shape.

APPLICATIONS

The current intense interest in the chemistry, physics, and crystallography of solids is responsible for the renewed activity in NQR. In 1964, Grechishkin and Soifer[29] suggested specific applied categories into which NQR could be divided. These have been updated and include:

(a) Investigation of the nature of chemical bonds in solids (the NQR frequency is directly dependent upon the type of hybridization and the degree of ionization of the chemical bond).
(b) Establishment of nonequivalence of the location of resonant atoms in crystal lattices and molecules.
(c) Qualitative analyses (each chemical compound has a definite NQR spectrum).
(d) Inspection of purity of chemical synthesis products (the NQR line intensity is directly dependent upon the amount of soluble impurities in a sample).
(e) Structural analysis of crystals (NQR is a valuable complement to x-ray methods.)
(f) Detection of phase transitions in crystals and the investigation of their kinetics.
(g) Measurement of average rotary vibration frequencies and average moments of inertia of molecules in crystals, from the temperature dependence of the NQR lines.
(h) Measurement of strain.
(i) Determination of temperature.

Table 3 lists some of the capabilities of NQR. These examples have been chosen because they represent real, current-day problems which have been presented for possible solution using NQR.

CONCLUSIONS

It is now possible to answer the two questions posed in the introduction. NQR development has been hindered by:

(a) Lack of proper instrumentation.
(b) Inefficient sample excitation techniques.
(c) Primitive data handling and processing methods.

Table 3. Suggested applications of FT-NQR.*

Characterization of new compounds (drugs, chemicals, explosives, liquid crystals, etc.).

Degree of crystal order (in clays, solid solutions, etc.).

Locus of aluminum atoms in plagioclase minerals.

Shelf life of materials (drugs, chemicals, explosives, etc.).

Determination of <u>in-situ</u> strain (salt domes, hard rock, etc.).

Detection of contraband (drugs, explosives).

Identification of contaminants (i.e., nature of N, S, and O in coal).

Measurement of temperature.

Radiation damage in solids.

Crystalline polymorphism studies (i.e., relationship of TNT crystalline form to impact stability, etc.).

Measurement of organic complexes (i.e., V, Cu, Ni in crude oil).

Identification of host minerals in bore-holes (i.e., measurement of uranium grade and type).

Characterization of phase transitions (i.e., in rock-forming minerals such as feldspars, pyroxenes, phyllosilicates, etc.).

Interpretation of crystal order with respect to thermodynamic properties (in minerals).

Determination of non-stoichiometric behavior (i.e., minerals, alloys, single crystals, etc.).

Identification of small solid samples (i.e., exsolution in aluminosilicates, dangerous materials, rare compounds, etc.).

---

* In some instances, NQR techniques other than FT may prove to be more suitable, i.e., double resonance NQR, acoustic NQR, etc.

It can be shown that solutions have been found for each of these drawbacks:

(a) Sensitive, high-resolution, automated, pulsed NQR spectrometers which utilize state-of-the-art electronic, radar and computer techniques fulfill the instrumentation needs.
(b) The availability and usability of a variety of pulse sequences such as spin-lock spin-echo (SLSE) or strong off-resonance comb (SORC) provide superior sample excitation.
(c) Fourier transform data processing, using modulus squared or modulus transforms, represents the state-of-the-art procedure for the special NQR conditions.

The need for FT-NQR is clearly demonstrated by the examples given in Table 3. The growing interest in understanding the behavior of solids has provided an impetus for analytical techniques suitable for use with solid state systems. Particular emphasis has been directed at natural systems, such as minerals, and military and security applications. Furthermore, the possibility of measuring in-situ strain is of major interest to the rock mechanics community. A technology that can measure both single crystals and polycrystalline materials, which can respond to contaminants or homogeneous small samples, which can operate both in the laboratory or under semi-remote conditions while providing high resolution spectral data for solid samples has all of the versatility to develop into an important analytical method.

ACKNOWLEDGMENTS: This work was supported in part by the U. S. Department of Energy under Contract No. W-7405-ENG-48, and the U. S. Army Research Office under Contract No. DAAG-29-79-0025. The authors wish to thank Mr. R. Connors of Block Engineering, Cambridge, MA for his assistance in running the NQR spectra. They would also like to recognize Mr. N. Henderson, Office of Nuclear Waste Isolation, Battelle-Columbus, Columbus, OH, and Dr. C. Boghosian, U. S. Army Research Office, Research Triangle, NC for their continued support. Technical editing was done by Dr. Ellory Schempp of Lawrence Berkeley Laboratory, Berkeley, CA.

REFERENCES

1. Schuler, H., and Schmidt, T. 1935, Z. Phys. $\underline{94}$, 457.
2. Kellog, J. M. B., Rabi, I. I., Ramsey, N. F., and Zacharias, J. R. 1936, Phys. Rev. $\underline{55}$, 728.
3. Dehmelt, H. G., and Kruger, H. 1950, Naturwiss. $\underline{37}$, 111.
4. Dehmelt, H. G., and Kruger, H. 1951, Naturwiss. $\underline{38}$, 921.
5. Schempp, E., and Bray, P. J., 1970, "Nuclear Quadrupole Resonance Spectroscopy," Physical Chemistry, an Advanced Treatise, Vol. 4, Academic Press, New York.
6. Lucken, E. A. C. 1969, Nuclear Quadrupole Coupling Constants, Academic Press, New York.

7. Slichter, C. P. 1978, Principles of Magnetic Resonance, 2nd ed., Springer-Verlag, Berlin, Heidelberg, New York.
8. Biryukov, I. P., Voronkov, M. G., and Safin, I. A. 1969, Tables of Nuclear Quadrupole Resonance Frequencies, Israel Program for Scientific Translations, Jerusalem.
9. Das, T. P., and Hahn, E. L. 1958, Nuclear Quadrupole Resonance Spectroscopy, Solid State Physics, Supplement 1, Academic Press, New York. Note that the error in the secular equation for spin 5/2 given on page 13 has been corrected.
10. Livington, R., and Zeldes, H. 1955, "Tables of Eigenvalues for Pure Quadrupole Spectra, Spin 5/2 and 7/2," Oak Ridge National Laboratory Report ORNL-1913.
11. Edmonds, D. T. 1977, Physics Reports 29, 233.
12. Hirschfeld, T., and Klainer, S. M. 1980, J. Mol. Struc. 58, 63.
13. Zussman, A., and Alexander, S. 1968, J. Chem. Phys. 49, 3792.
14. Petersen, G., and Oja, T. 1974, Advances in Nuclear Quadrupole Resonance, ed. J. A. S. Smith, Vol. 1, Heyden, London, p. 179.
15. Abe, Y., Ohneda, Y., Hirota, M., and Kojima, S. 1974, J. Phys. Soc. Japan 37, 1061.
16. Colligiani, A., and Ambrosetti, R., 1976, Gazz. Chim. It. 106, 439.
17. Gibson, A. A. V., Goc, R., and Scott, T. A. 1976, J. Magn. Reson. 24, 103.
18. Harding, J. C., Wade, D. A., Marino, R. A., Sauer, E. G., and Klainer, S. M. 1979, J. Magn. Reson. 36, 21.
19. Marino, R. A., Harding, J. C., and Klainer, S. M. 1980, J. Mol. Struc. 58, 79.
20. Petersen, G. L. 1975, Ph.D. thesis, Brown University.
21. Marino, R. A., and Klainer, S. M. 1977, J. Chem. Phys. 67, 3388.
22. Ostroff, E. D., and Waugh, J. S. 1966, Phys. Rev. Lett. 16, 1097.
23. Waugh, J. S. 1970, J. Mol. Spec. 35, 298.
24. Cantor, R. S., and Waugh, J. S. 1980, J. Chem. Phys. 73, 1054.
25. Pratt, J. C., Raghunathan, P., and McDowell, C. A. 1974, J. Chem. Phys. 61, 1016; 1975, J. Magn. Reson. 20 313.
26. Lenk, R., and Lucken, E. A. C. 1974, Pure Appl. Chem. 40, 199.
27. Colligiani, A., and Ambrosetti, R. 1976, Gazz. Chim. It. 106, 439.
28. Colligiani, A., and Ambrosetti, R. 1974, J. Chem. Phys. 60, 1871.
29. Grechishkin, V. S., and Soifer, G. B. 1964, Pribory i Tekhnika Eksperimenta 1, 5 (in Russian).

FOURIER TRANSFORM DIELECTRIC SPECTROSCOPY

Robert H. Cole and Paul Winsor, IV

Department of Chemistry
Brown University
Providence, RI 02912

INTRODUCTION

This paper is primarily concerned with the techniques usually described as time domain spectroscopy (TDS) or time domain reflectometry (TDR). These have been most commonly applied to studies of time or frequency dependent behavior of dielectrics with negligible ohmic or d.c. conductance, but can be used for substances with appreciable conductance and indeed for studies of any electrical properties which can be characterized by an effective admittance or impedance.

The methods really amount to nothing more than obtaining the system properties from real time measurements of transient currents following application of a voltage pulse. Techniques of this kind have been in use for many years when the times of interest are milliseconds, seconds or longer, but the special interest in TDS is due to the use of pulse generation and observation techniques with picosecond time resolution which makes possible the equivalent of steady state a.c. measurements at frequencies from a few MHz to several GHz, a region that has traditionally been difficult and tedious by frequency domain methods.

The use of Fourier transform methods for analysis of TDS observations is dictated by several problems that do not arise for slower phenomena. The first is that although fast rising voltage or electric field pulses can be applied, they cannot readily be made of simple enough form to avoid dealing with a dielectric or other response which is a convolution of a dielectric response function with the voltage pulse form in real time.

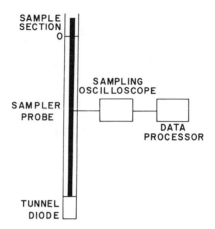

Figure 1. Schematic drawing of TDS arrangement for total reflection measurements.

The second is that propagation effects must be dealt with if sample dimensions are an appreciable fraction of a wavelength, and this situation is not readily avoided at frequencies for which the method is otherwise useful. Both problems are better handled by use of onesided Fourier (Laplace) transforms, rather than direct time domain solutions, as a result of the convolution theorem for the former, and solution of the field equations in the frequency domain for the latter.

There are of course problems, common to any analyses requiring numerical Fourier transformations, of aliasing, truncation, and time referencing. Happily, these can be handled quite readily for dielectric problems because the pulse forms to be transformed are relatively simple and because effects of apparatus response characteristics and unwanted reflections can be eliminated or minimized by judicious apparatus design and use. In the sections that follow, we present basic analysis of TDS measurements, useful sample cell designs and working equations, methods for evaluation of transforms involved, and representative results.

BASIC ANALYSIS

The principal elements in commercial TDS equipment are a tunnel diode generator of repetitive voltage pulses with rise time of ca. 30 ps at a rate of ca. 100 KHz, probe circuits for observation of these pulses and their reflections in a coaxial line, and sampling oscilloscope circuits to "stretch" real observation times--tens of picoseconds to nanoseconds--to several milliseconds by repetitive scanning and summing.

The schematic in Figure 1 shows the essentials for measurements of reflections from a sample section terminating the observation line. Referred to the beginning of this section as origin, the sampler probe at $x = -\ell$ receives the incident pulse $V_0(t + \ell/c)$ and sum of reflected pulses $R(t-\ell/c)$ of the voltage $V(t) = V_0(t) + R(t)$ at the sample input. From transmission line theory, the sample input current $I(t)$ is given by $I(t) = G_c[V_0(t) - R(t)]$, where $G_c = (L_c/C_c)^{1/2}$ is the characteristic conductance of the coaxial line with inductance L_c and capacitance C_c per unit length. For Fourier component $v(i\omega)$ at frequency given by

$$v(i\omega) = L_{i\omega}V(t) = \int_0^\infty \exp(-i\omega t)V(t)\, dt$$

and corresponding $i(i\omega)$, one then has the complex input admittance $y_{in}(i\omega)$ given by

$$y_{in}(i\omega) = \frac{i(i\omega)}{v(i\omega)} = G_c \frac{v_0(i\omega) - r(i\omega)}{v_0(i\omega) + r(i\omega)} \qquad [1]$$

From continuity of $i(i\omega)$ and $v(i\omega)$, this is also the ratio of i and v in the sample section at $x = 0$ as determined by the dielectric material of interest, its geometrical configuration, and boundary conditions for the sample section. The dielectric response characteristics can be described macroscopically at frequencies of interest by the relation of Maxwell displacement and field vectors $\underline{D}(t,\underline{r})$ and $\underline{E}(t,\underline{r})$

$$\underline{D}(t,\underline{r}) = \int_{-\infty}^{t} [-\frac{d\phi(t-t')}{d(t-t')}]\underline{E}(t',\underline{r})\, dt' \qquad [2]$$

in which $\phi(t)$ is the macroscopic dielectric relaxation function of interest, related in linear response theory, for example, to molecular dipole correlation functions.[1-3] It is here that the use of transforms is advantageous, as for a pulse voltage with $\underline{E}(t,\underline{r}) = 0$ for $t < 0$ one has

$$\underline{d}(i\omega,\underline{r}) = \epsilon(i\omega)\, \underline{e}(i\omega,\underline{r}) \qquad [3]$$

with the complex permittivity ϵ of steady state a.c. response given by

$$\epsilon(i\omega) = L_{i\omega}[-\dot{\phi}(t)] \qquad [4]$$

The remaining problem of relating displacements and fields in the sample section to the admittance y_{in}, and so of obtaining ϵ as a function of the observables v_0 and r or their equivalents, must now be obtained from the solution of the electromagnetic wave equations (reducing to the Laplace or Poisson equation at low frequencies) for appropriate sample boundary conditions. Before giving results for useful realistic cases, it is helpful to consider the idealized simple case of a sample in a capacitor with geometric capacitance C_g and dimensions small compared to a wavelength. Neglecting any effects of the transition between the sample and the end of the coaxial line, the counterpart of Equation 2 reduces to

$$I(t) = C_g \frac{d}{dt} \int_0^t [-\dot{\phi}(t-t')] V(t') \, dt'$$

Taking the transform and using equation 1 gives

$$\epsilon = \left(\frac{G_c}{C_g}\right) \frac{v_0 - r}{i\omega(v_0 + r)} = \left(\frac{G_c}{C_g}\right) \frac{p}{i\omega q} \qquad [5]$$

where for simplicity the argument $(i\omega)$ of transforms (lower case) of real-time functions (capital letters) is omitted here and henceforth unless there is risk of confusion.

As an illustration of the observed behavior and types of functions to be transformed, Figure 2 shows the reflection $R(t)$ from Equation 5 for a step-like voltage pulse and dielectric with "Debye" relaxation expressed by

$$\epsilon = \epsilon_\infty + (\epsilon_s - \epsilon_\infty)/(1 + i\omega\tau_D) \qquad [6]$$

in which τ_D is the relaxation time for frequency dependent permittivity of (real) magnitude $\epsilon_s - \epsilon_\infty$ and ϵ_∞ (also real) represents instantaneous polarization compared to the time resolution. The curve $R(t)$ has an initial negative peak of shape and breadth determined by ϵ_∞ and rise time of the incident pulse. This is followed by an approximately exponential rise to the final level $V_0(\infty)$ of $V_0(t)$ with time constant τ_{app} nearly equal to the sum of τ_D and the time constant $(\epsilon_s - \epsilon_\infty) C_g/G_c$ of the coaxial line.[4,5]

The forms of the functions $P(t) = V_0(t)$ and $Q(t) = V_0(t) + R(t)$, whose transforms appear in Equation 5, are also shown in Figure 2. Both are relatively simple here and in more realistic cases, and present no unusual difficulties in evaluation of their transforms numerically (but see Section VI). A particularly simple result is for the zero frequency limit, as then the transform $p = v_0 - r$ is

Fourier Transform Dielectric Spectroscopy

simply the area under the P(t) curve, strictly to infinite time, and the transform $i\omega(v_0-r)$ of the derivative dQ/dt is the limiting value $Q(\infty) = V_0(\infty) + R(\infty)$, again strictly at infinite time.

The ideally simple result, Equation 5, is not practically realizable without modification, but it and the forms of the functions involved give a useful illustration of how the observable quantities can be related to the desired dielectric behavior as represented by ϵ or its equivalent. Solutions for y_{in} in terms of ϵ can be obtained for other sufficiently simple geometries by transmission line or electromagnetic field methods, as discussed in the books by Johnson[6] and Markuvitz[7] for example, and in other cases by semiempirical methods.

SAMPLE CELL ARRANGEMENTS

Two basically different kinds of cell arrangement can be distinguished: those in which the dielectric filled section terminates the coaxial pulse generating and sampling line, and those in which this section is itself terminated by a further admittance. The former are usually simpler and more useful, but the latter can have advantages for some purposes. If the sample section is symmetrical with respect to interchange of its input and output connections, and is terminated by an arbitrary admittance y_d, its input admittance is given from network theory by

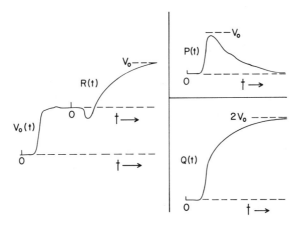

Figure 2. Left: Incident pulse $V_0(t)$ and multiple reflection pulse R(t) from open coaxial line sample as seen by sampling probe. Upper right: Sample input current function $P(t) = V_0(t) - R(t)$. Lower right: Sample input voltage $Q(t) = V_0(t) + R(t)$.

$$y_{in} = \frac{y_0 + y_d}{1 + Z_s y_d} \quad [7]$$

in which y_0 is the open circuit admittance, for $y_d = 0$. The short circuit input impedance, Z_s (i.e., the value of $1/y_{in}$ for $y_d = \infty$), is determined at low frequencies by the effective inductance of the sample section. We now describe useful arrangements of both kinds.

Lumped capacitance cell

An approximation to the "ideally simple" cell discussed above, shown schematically in Figure 3a, has been used by Iskander and Stuchly[8] and others. One complication is from the fringing field at the sample boundary, particularly if a retaining shell of some kind is used to confine a liquid sample. At low frequencies, this is essentially an electrostatic problem of solving Laplace's equation for the appropriate boundary conditions. At sufficiently high frequencies, more general solutions of Maxwell's equations are required to take account of non-stationary electric fields and wave propagation in the region between the end of the coaxial line and the sample. Solutions for the first problem are given in Markuvitz[7] and elsewhere. The second seems not to have been considered, at least in TDS applications, but might be solved in useful approximation by assuming an effective series inductance between the line and sample, and obtaining its value by measurements of a liquid, or liquids, of known permittivity.

End capacitance cell

In this even simpler cell[6] shown in Figure 3b, the volume beyond the end of the coaxial line center conductor (supported by a disk with dimensions to match the coaxial line admittance) is filled with dielectric. The sample is thus in the fringing field at the end of the line, or more generally can be regarded as filling a section of circular wave guide used at frequencies below cutoff for wave propagation.

Experimental tests, supported by theoretical calculations for similar but not identical geometry,[5] have established a useful approximate expression for the input admittance, and calculations of Somla[9] as discussed by Bussey[10] give a similar result. This is

$$y_{in} = \frac{i\omega C_c d\epsilon}{1 - a(\omega d/c)^2 \epsilon}$$

and hence from Equation 1

Fourier Transform Dielectric Spectroscopy

$$\epsilon = (\frac{c}{d}\frac{p}{i\omega q}) [1 + a(\omega d/c)^2(\frac{c}{d}\frac{p}{i\omega q})]^{-1} \qquad [8]$$

In these equations, d is the length of coaxial line with capacitance C_c for unit length and the same geometric capacitance (d is approximately 0.6 mm for 7 mm 50 ohm line), and c = 0.300 mm/ps is the speed of propagation in the line, related to C_c and G_c by $c = G_c/C_c$. The high frequency correction term, with coefficient a of order 1 determined by calibration with known samples, is adequate for frequencies such that $\omega < 0.3\, c/d(\epsilon a)^{1/2}$.

Cells of this type are very simple to construct and use, and because of their small capacitance are particularly suitable for measuring liquids of high permittivity ($\epsilon_s > 30$) at frequencies up to several GHz.

Coaxial line cell

Another simple and more versatile cell consists of filling a length d of coaxial line with the sample of interest, as shown in Figure 3c. If fringing effects at the end of this length of line are neglected, the input admittance of the cell is given from transmission line theory by

$$y_{in} = i\, C_c d\, /f(\tilde{z}) \qquad [9]$$

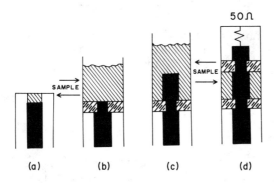

Figure 3. Simplified schematic drawings of dielectric sample cells for reflection measurements. (a) Lumped capacitance cell, (b) End capacitance cell, (c) Open coaxial line cell, (d) Coaxial line cell with 50 ohm termination. Bottom sections in all cases are 50 ohm 7 mm precision coaxial lines.

in which the function $f(Z)$, accounting for propagation effects in the line with $Z = (\omega d/c)\epsilon^{1/2}$, is

$$f(Z) = Z\cot Z = 1 - \frac{1}{3}(\omega d/c)^2 \epsilon - \frac{1}{45}(\omega d/c)^2 \epsilon^2 + \cdots$$
[10]

and hence

$$\epsilon = (c/d)\frac{v_0 - r}{i\omega(v_0-r)} f(Z) = (c/d)\frac{p}{i\omega q}(Z\cot Z) \quad [11]$$

The series expansion converges for $Z < \pi/2$, the upper limit being the condition of quarter wave resonance if ϵ is real. In practice, Equations 9 and 10 are useful for $Z^2 < 1$, in which case the Z^4 term in $Z\cot Z$ can be neglected, leading after use of Equation 1 to an explicit solution for ϵ in terms of the ratio $(p/i\omega q)$ of transforms of the observed pulses.

The effect of fringing fields at the end of the coaxial sample line can usually be made small but not negligible. Two ways of accounting for it are to use Equation 7 with an appropriate expression for y_d or to use an effective electrical length d' that is larger than the physical length d and determined by calibration with samples of known ϵ. The solutions for ϵ including these effects are discussed in reference 11.

Coaxial sample line with matched termination

This arrangement, shown in Figure 3d, has been used and discussed by several investigators (for reviews, see references 13, 14, 26) and consists in continuing the sample section with coaxial line terminated by its characteristic conductance G_c.

On using Equation 9 for y_0, $y_d = G_c$, and $Z = i\omega L_c d/f(Z)$ (where the inductance L_c per unit length of line is given by $L_c = 1/c^2 C_c$ from transmission line theory) in Equation 7 for y_{in} together with Equation 1, one obtains[11]

$$\epsilon - 1 = \frac{2c}{d}\frac{r}{i\omega(v_0+r)}(f(Z) + \frac{i\omega d}{c})$$

which again gives an explicit solution if the Z^2 series approximation for $f(Z)$ is used. The observed reflection $R(t)$ for a Debye dielectric in this arrangement is shown in figure 4. The difference from $R(t) = 0$ for no sample present is seen to be smaller than the corresponding differences for sample termination arrangements as a result of shunting the sample by $G_c = 20$ mmho (for 50 ohm coaxial line). The arrangement can be useful if thin samples with well defined boundary conditions are needed for measurements of high permittivity at high frequencies, but sample termination designs are easier to use and in the writers' opinion usually preferable.

An alternative method of using this cell is to place the sampling probe in the line following the cell to measure the pulse transmitted by the sample to this line. For a properly matched empty cell ($\epsilon = 1$), this pulse is a copy of V_0 delayed by the transit time d/c through the cell, and the difference between this pulse and the one with sample present is a measure of $\epsilon-1$. This arrangement has been analyzed in detail by Gestblom and Noreland,[12] but again in the writers' opinion does not have significant advantages over reflection methods.

Short circuited sample

Using a short circuit to terminate a length of coaxial line filled with dielectric is at first sight attractive, but has very limited usefulness. On putting $y_d = \infty$ in Equation 7, one sees that

$$y_{in} = 1/Z_s = f(Z)i\omega L_c d = \frac{1}{G_c}\frac{V_0+r}{V_0-r}$$

Thus the only dielectric information is contained in the argument $Z = (\omega d/c)\sqrt{\epsilon}$ of $f(Z)$ and first appears in terms of order ω^2. The arrangement has potential value for magnetic materials with permeability μ differing from unity, however, as one then has $Z_s = i\omega L_c d\mu/f(Z)$, with Z in $f(Z)$ given by $Z = (\omega d/c)\sqrt{\epsilon\mu}$. This possible use has been discussed in reference 13, but has not been developed.

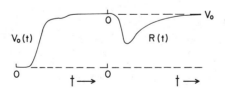

Figure 4. Incident and reflected voltage pulses from sample section in 50 ohm coaxial line with 50 ohm termination.

MODIFIED METHODS AND USES

Multiple reflection expressions

The admittance analysis used in the preceding derivations[11] is usually a better way of obtaining useful expressions for permittivity in terms of transforms of incident and reflected pulses from samples in coaxial lines than the commonly used alternative of starting with transmission line formulas for reflections in terms of the incident pulse and sample reflection coefficient $\rho = (1 - \sqrt{\epsilon})/(1 + \sqrt{\epsilon})$. The latter approach does have its uses, however, particularly if one is interested in the sequence of multiple reflections in the sample which are, so to speak, concealed in the propagation factor $f(Z)$ of admittance formulas.

Using the open-circuit terminated coaxial cell as an example, Equation 1 can be solved for $r(i\omega)$ in terms of $v_0(i\omega)$ and ρ, giving after some rearrangement the result,

$$r(i\omega) = v_0(i\omega) \frac{\rho + \exp(-i2Z)}{1 + \rho \exp(-i2Z)}$$

$$= v_0(i\omega)[\rho + (1-\rho^2)\exp(-i2Z) - \rho(1-\rho^2)\exp(-i4Z) + \cdots] \qquad [12]$$

For a non-dispersive dielectric with ϵ real and independent of ω, the series expansion is the frequency domain counterpart of a series of reflections separated by the round trip travel time $2d\epsilon^{1/2}/c$ and of progressively smaller amplitude for $\rho > -1$, as shown schematically in Figure 5. One sees also that the approximate solution for ϵ by taking $Z^2 \ll 1$ corresponds to a smoothed version of this steplike structure.

For real dielectrics with ϵ complex and frequency dependent, inverse transformation of Equation 12 to obtain $R(t)$ is not possible in any simple way, even for a Debye relaxation function, because of the awkward square root of ϵ in ρ and $\exp(-i2Z)$, but one expects a rounded and distorted version of the idealized progression of steps for ϵ and ρ real, and v_0 a step function.

Solutions of Equation 12 or equivalent closed forms for ϵ in terms of r/v_0 are possible by successive numerical approximations, such as the Newton-Raphson method for determining the complex root ϵ of $F(\sqrt{\epsilon}) - r(i\omega)/v_0(i\omega) = 0$. This procedure has been used successfully, as discussed by Suggett,[14] but has the disadvantages that an initial estimate of ϵ is necessary and that conditions for rapid or acceptable convergence to the true value are not easily

Fourier Transform Dielectric Spectroscopy

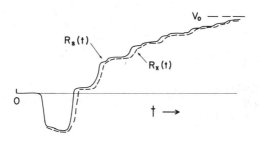

Figure 5. Multiple reflections from 20 mm cell. $R_s(t)$ is for solvent toluene; $R_x(t)$ is for 0.5 weight percent propylene carbonate solution in toluene. Interval between reflections is ca. 210 ps.

recognized. Both these difficulties are overcome quite simply by using the alternative expressions obtained more directly by admittance analysis.

Evaluation of v_0 from reflected pulse transforms

An advantage common to all but (D) of the methods described above is that the incident pulse waveform $V_0(t)$ needed to evaluate $v_0(i\omega)$ can be obtained quite simply from the reflection $R_1(t)$ for the empty cell, rather than by direct observation at a time $2\ell/c$ earlier. This simplifies the problem of consistent time referencing discussed in section V, and also eliminates the problem of correcting for differences in sampling probe response to forward and return waves.

Again taking the open-circuited coaxial cell as an example, the transform r_1 of the reflection for $\epsilon = 1$ is (from Equation 11) related to v_0 for the incident pulse by

$$v_0 = r_1 \frac{1 + \frac{i\omega d}{c}(\tan Z_1/Z_1)}{1 - \frac{i\omega d}{c}(\tan Z_1/Z_1)} \qquad [13]$$

in which $Z_1 = \omega d/c$. In practice, the value of $\tan Z_1/Z_1$ can be approximated by $1 + Z_1^2/3$ and usually by unity (for frequencies such that $(\omega d/c)^2/3 \ll 1$). The transform v_0 is then obtainable from r_1 by simple algebra.

An alternative, valid for $Z_1^2 \ll 1$, is to observe that Equation 13 is then equivalent to

$$v_0 = r_1 \exp(i\omega 2d/c)$$

and the real time $V_0(t)$ is, on taking the inverse transform, related to $R_1(t)$ by

$$V_0(t) = R_1(t + 2d/c)$$

Thus to this approximation, $V_0(t)$ at the cell input can be obtained simply by shifting the observed $R_1(t)$ forward by the time interval $2d/c$ of delay in the open circuit reflection.

Samples with d.c. conductivity

If the dielectric of interest has a low frequency conductance that is not negligible in comparison to the coaxial line conductance G_c, the observed reflections change and some modification of the working formulas is desirable to make explicit the differences. If the low frequency specific conductance is denoted by σ_s, the formulas remain valid if ϵ is replaced by ϵ_t given by

$$\epsilon_t = \epsilon + \sigma_s/i\omega k \qquad [14]$$

in which the conversion factor k depends upon the choice of units. For σ_s and ϵ in esu and ω in sec^{-1}, $k = 4\pi = 1/0.0884$, and for the convenient units of σ_s in mho/m, $1/\omega$ in ps, and ϵ the relative or electrostatic permittivity, $k = 0.113$. The effect of σ_s added to ϵ for a coaxial cell (C) is shown schematically in Figure 6. The final levels of $P(t)$ and $Q(t)$ are no longer zero and $2V_0$; instead $P(t)$ has a finite value and $Q(t)$ is less than $2V_0$ by amounts determined by a ratio of sample to coaxial line conductance in relations which are easily obtained from Ohm's law.[5,11]

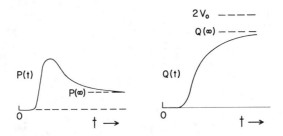

Figure 6. Input pulse forms for reflection from a conducting dielectric sample in open coaxial line. Left: sample input current function $P(t)$. Right: sample input voltage $Q(t)$. The ratio of sample conductance G to coaxial line conductance G_c is equal to the ratio $P(\infty)/Q(\infty)$.

Fourier Transform Dielectric Spectroscopy

In using ϵ_t defined in this way, the burden of representing any relaxation effects in conductivity as a function of ω is thrown on the derived values of ϵ. As is clear from Maxwell's equations, there is no way of separating σ and ϵ relaxation effects experimentally by electromagnetic measurements alone, and any such separation must be made by arbitrary assignment or comparison with a theoretical model.

Difference methods

The device of replacing $V_0(t)$ or its transform by a reflection pulse, or its transform, from a reference sample of known permittivity can be very useful when one is interested in small relaxation effects produced, for example, by adding impurities, ions, or polar solute molecules in low concentration to a solvent medium. Using measurements of solution and solvent reflections $R_s(t)$ and $R_x(t)$ for permittivities ϵ_x and ϵ_s, one can obviously combine the two admittance equations for any of the methods to eliminate v_0 and obtain ϵ_x or $\epsilon_x - \epsilon_s$ in terms of ϵ_s and the transforms r_x and r_s. The important quantity for good precision in $\epsilon_x - \epsilon_s$ is then the transform $(r_s - r_x)$ of $R_s(t) - R_x(t)$, and this difference can be obtained with improved precision from digital processing of separate records of $R_s(t)$ and $R_x(t)$. Methods of this kind have been described in some detail by Cole, Mashimo, and Winsor,[15] and should be useful for a variety of systems.

DESIGN CONSIDERATIONS

Obtaining satisfactory results with any of the methods described above requires attention to a number of factors. In this section, the more important ones are discussed briefly, with references to sources of more detailed information.

The high frequency limit of TDS measurements is ultimately set by the form of the incident pulse, response characteristics of the sampling probes and circuitry, and transmission characteristics of the coaxial measuring line. The fastest tunnel diode and sampling circuits in current commercial TDS equipment generate step-like 250 mV pulses with rise times of order 50 ps as recorded from the output. For these, the frequency components of the transform $i\omega v_0$ are nearly independent of frequency to about 5 GHz and fall to nearly zero at 15 GHz, with half amplitude at about 10 GHz,[15] thus setting the last figure as an approximate upper limit for adequate information in the reflected pulses. The transmission characteristics of precision 7 mm coaxial lines are essentially constant, with negligible losses as used, at frequencies below 18 GHz. Ordinarily, these upper limits are well above limits imposed by limitations of cell design and magnitude of permittivity to be measured; the latter have been indicated in preceding discussion.

An important advantage of the repetitive real time sampling techniques used in TDS equipment is that their outputs can be signal averaged by digital signal analyzers or minicomputers. To be effective, however, the circuit outputs must be sufficiently free of drift in time and amplitude scales, and commercial equipment has limitations in both respects. Drift and jitter of the time of output pulses relative to the initiating trigger pulses evidently blurs out their average, with loss of high frequency information as a result of the low-pass filter effect.[16] For measurements of samples with moderately large permittivities and differences as a function of frequency, averaging of 100-200 pulses in a few seconds at most is often sufficient to reduce noise and jitter to acceptable levels and drift is not a serious factor, particularly if triggering of the signal averaging equipment synchronously with the pulse of interest can be employed.[15]

For measurements of effects so small as to require several minutes of signal averaging, drift can nullify the advantages and stabilization is necessary. Andrews[17] has described improved tunnel diode power supply and triggering circuits which help considerably. Elliott[18] has developed feedback methods using lock-in amplifiers, by which the scanning and triggering of the pulses are locked into synchronism for indefinite periods of time. As an indication of the sensitivity then realizable, Elliott was able to detect and measure a capacitance of 1.8 femtofarad (1 ff = 10^{-3} pf) by signal averaging for 15 minutes, with a sensitivity of 0.018 ff which is about 200 times that of the commercial TDS equipment without signal averaging.

For less demanding applications, adequate records of reflection-time curves can be obtained more simply with reasonable attention to such matters as careful adjustment, adequate warm-up time, and the like. There is, however, a further important consideration: in any method of obtaining permittivities, one needs to evaluate the ratio of two transforms of different pulses or combinations of them. To obtain the ratio with sufficient accuracy, it is essential that the origin of time for all the records involved be the same or known relative to each other with sufficient precision.

The nature and magnitude of resulting errors can easily be deduced from the fact that if the zero of time for one record is in error by a time δt from the zero for a second one, then its transform relative to the second will be in error by a factor $\exp(i\omega\delta t) \cong 1 + i\omega\delta t$, corresponding to a phase shift $\omega\delta t$. The resultant error in derived quantities can then be worked out from the appropriate formulas by using this factor, and ordinarily results in some such requirement as that $\omega\delta t$ should be less than, say, 0.01 at the highest frequency of interest. For measurements to 5 GHz as an example, this requires that $\omega\delta t$ be less than, or known to less than, 0.3 ps in real time.

The necesary accuracy in time referencing can be realized in several ways. A simple one if appropriate dual channel TDS sampling circuits are available is to add to the reflection pulse signal a timing marker that is derived from, and synchronized with, the incident voltage pulse. Methods of doing this are described in reference 15, which also has references to other methods.

There are various other artifacts and quirks of behavior peculiar to particular pieces of equipment too numerous and specific to be considered here. In the writers' experience, the consequences are usually, but not always, only too apparent and the origins diagnosed quite readily (in part because of time resolution), with remedies possible by fairly simple additions to, or changes in, the equipment.

EVALUATIONS OF TRANSFORMS

The price paid for the convenience and simplicity of the transform expressions, if it really is one, is the necessity of evaluating the transforms of reflection or transmission pulses. With few exceptions, this must be done numerically, as only rarely is it possible to devise analytic real time functions which both fit the data and have known, manageable transforms. It may of course be instructive to work out results of simple approximations to the pulse shapes or do the inverse of finding what these will be for simple forms of ϵ.

The problems of numerical transformation are not greatly different from those in other applications and much of the extensive development of methods for other problems is immediately useful. Their application does need to be considered in the context of how the dielectric reflection records are or can be obtained, and their characteristic features. Before considering these in detail, some common features should be recognized.

As already pointed out, all the working formulas given above involve only ratios of transforms, either of observed pulses or of sums or differences of them. Because of this, one need not consider response characteristics of the sampling circuits used to obtain the pulses if the circuits are linear. This is because the output $G_S(t)$ is the convolution of the pulse $G(t)$ with the sampler response function $S(t)$ and the transform of $G_S(t)$ is $g_S(\omega) = s(\omega)g(\omega)$; hence the factor $s(\omega)$ cancels in any ratio of similarly obtained transforms. [One must of course have adequate response $s(\omega)$ for desired frequency ω.]

The principal errors in numerically approximating the transform integral to infinite time by a discrete Fourier transform (DFT), or series of a finite number of values at discrete points, are from aliasing and truncation. The magnitudes of the errors for the functions $G(t)$ usually encountered can be readily assessed and kept satisfactorily small by satisfying simple criteria. Examples of G(t) are the functions P(t) and Q(t) shown schematically in Figure 2: P(t) has an initial peak followed by a decay to zero or a finite limiting value, while Q(t) rises monotonically to a long time limit, and both have rounded step-like progressions as a result of multiple reflections in the sample.

The dielectric function ϵ to be determined also has rather simple frequency dependence. As contrasted with a sharp spectral line or lines, for example, the loss curve $\epsilon''(\omega)$ typically exhibits a broad maximum with only gradual inflections in the wings and is more or less symmetrical on a logarithmic frequency scale. The corresponding $\epsilon'(\omega)$ dispersion curve is a smoothly decreasing function of ω, which is also best represented on a logarithmic frequency scale. The Debye relaxation function, Equation 6, is the classic prototype; experimentally observed deviations are in the direction of producing still broader and more gradually changing curves for ϵ' and ϵ''.

The form of numerical evaluation almost invariably used is the simple summation,

$$q(i\omega) = \int_0^\infty \exp(-i\omega t) G(t)\, dt \cong \Delta \sum_{n=0}^{N} \exp(-i\omega n\Delta) G(n\Delta) \qquad [15]$$

with internal Δ and range $n\Delta$. An alternative, first proposed by Tuck,[11,19] is to approximate the integral by a series of triangular pulses: the resultant transform is the same series as in Equation 15 multiplied by the square of the "sinc" function $[\sin(\omega\Delta/2)/(\omega\Delta/2)]^2$. Whether or not this gives a better approximation to the true transform for various choices of interval Δ is debatable, but the question is academic when only the ratio of transforms is needed, as the extra factor then cancels. A Simpson's rule (parabolic) approximation could also be used, but there seems to be little point in using the more complicated formulas[20] in preference to Equation 15 with smaller interval Δ.

The fast Fourier transform (FFT) algorithm[21] now readily available is economical when many real time points and derived values at many frequencies are needed. It is, however, not ideally suited to evaluations of dielectric relaxation functions, because as already mentioned one usually wants these for frequencies that are uniformly spaced on a logarithmic time scale, rather than for the constant intervals and harmonically related frequencies required by the FFT. As a result, use of the FFT gives too coarse intervals for small ω and an unnecessarily large number of points for large ω. This does not preclude running a readily available program two or three times wiht different base frequencies. However, the necessary

Fourier Transform Dielectric Spectroscopy

number N of data points for reasonable description of reflection pulses such as P(t) and Q(t) is small enough that the simple brute force or "slow" Fourier transform is practical with quite modest computer facilities, and then the frequencies ω for the sums can be specified quite arbitrarily as desired.

Aliasing errors

The use of sums of points at finite intervals to approximate the pulse function and its integral transform can be described as the process of simulating the true function, strictly of infinite duration, by a periodic function which is the part of the function from $t = 0$ to $t = N\Delta$ repeated indefinitely, with the numerical summations giving the Fourier series coefficients of the periodic function at frequencies $n/N\Delta$. The frequency spectrum so derived has errors in describing the true spectrum at high frequencies if it has components at frequencies above a maximum f_m given by $f_m = 1/2\Delta$. Shannon's sampling theorem[22] gives an indication of the largest acceptable value Δ for the summing interval: if the true real time pulse to be transformed has no Fourier components above a frequency f_m, then it is completely determined by its values at the times separated by the interval $\Delta = 1/2f_m$ and there will be no aliasing errors in the derived spectrum at frequencies less than f_m.

We have already remarked that the pulse waveforms in TDS equipment have small amplitudes at and above f_m = 15 GHz, and the appropriate Δ would be no greater than 30 ps by this criterion if aliasing errors are to be avoided. For arbitrarily chosen frequencies rather than harmonics of the base frequency $f_0 = 1/N\Delta$ the situation is less simple, but in practice values of Δ smaller than given by this criterion are usually appropriate for a reasonable simulation of the pulse itself. As a rule of thumb, it is prudent to take Δ less than $1/4f_m$, where f_m is the highest frequency of interest.

Truncation errors

The errors resulting from necessarily terminating the numerical summations at a maximum time $T_N = N\Delta$ depend on the long time behavior of the function $G(t)$ to be transformed. If $G(t) \to 0$ as $t \to \infty$, the time transform has a finite, non-zero value, but if $G(t)$ approaches a non-zero limit $G(\infty)$ as $t \to \infty$ the true transform diverges as $1/i\omega$ as $\omega \to 0$, with obvious problems in obtaining it numerically for low frequencies. We consider first the effect of cutting off the summation for the simpler case.

Functions such as P(t) in Figure 2 with zero long time limit and finite area often have a long time behavior that is at least roughly an exponential decay: $P(t) = P(t_N)\exp[-(t-t_N)\tau_{app}]$ for $t > t_N$. (A case in point is Debye relaxation with $\tau_{app} = \tau_D + (\epsilon_s - \epsilon_\infty)d/c$ as discussed in section II.) If τ_{app} is estimated from the decay at times before t_N in the interval of summation, the error $\Delta p(i\omega)$ from truncating the numerical transform at $t = t_N$ is

$$\Delta p(i\omega) = \int_{t_N}^{\infty} \exp(i\omega t) \, P(t_N) \, \exp[-(t - t_N)/\tau_{app}] \, dt$$

$$= P(t_N) \, \tau_{app} \, \exp(-i\omega t_N)/(1 + i\omega \tau_{app}) \qquad [16]$$

In the zero frequency limit, the error is obviously just the area of the $P(t_N)$ curve for $t > t_N$, namely $P(t_N)\tau_{app}$, and at high frequencies it is modified by the phase factor $\exp(-i\omega t_N)$ which produces a sinusoidal oscillation around the true $p(i\omega)$.

This pattern is shown in the example given in Figure 7, of results for ϵ of methanol at 25 °C with Debye relaxation time τ_D = 52 ps using an end capacitance cell with effective length d = 0.694 mm. In the complex plane plot, values obtained by truncation of $P(t)$ at t_N = 516 ps are plotted as open circles and corrected values obtained by Equation 16 as filled circles. The uncorrected static permittivity ϵ_s is too small by 6 per cent, and with increasing frequency the difference oscillates around the correct values lying on a Debye semicircle (solid curve).

Considerations of this kind suggest a simple correction for errors of truncating the numerical integration at time t_N: t_N should be long enough that $P(t_N)$ is no more than a few per cent of its maximum value, so that the remaining error can then be evaluated with sufficient accuracy by adding the transform of a long time tail which is an extrapolation of the behavior as t_N is approached from shorter times.

Leakage errors

This term is used in the literature[23] to refer to truncation errors for any function that is not zero at the final point used in numerical integration; here we take it to mean the errors for a function which has a non-zero limit as $t \to \infty$ and a divergent transform as $\omega \to 0$, the function $Q(t)$ for example.

At least three procedures have been described in the literature for correcting the discrete numerical transform to give the proper $1/\omega$ limiting behavior:

(1) Nicolson[24] proposed the procedure of representing $G(t)$ as the sum of a "ramp-step" function $G_{RS}(t)$ with linear rise in arbitrary time t_R to the long time limit $G(\infty)$ of $G(t)$, with this constant value thereafter, plus the difference $\Delta G(t) = G(t) - G_{RS}(t)$. The function $G_{RS}(t)$ defined by

$$G_{RS}(t) = G(\infty) \frac{t}{t_R}, \quad 0 < t < t_R$$
$$= G(\infty), \quad t > t_R$$

has the transform $g_{RS}(i\omega) = G(\infty)[1 - \exp(-i\omega t_R)]/(i\omega)^2$ with the proper $(i\omega)^{-1}$ asymptotic behavior, while the transform $\Delta g(i\omega)$ of $\Delta G(t)$ is of a finite function going to zero at $t \to \infty$, which can be evaluated without difficulty using appropriate correction for truncation. The desired transform, $g(i\omega) = g_{RS}(i\omega) + \Delta g(i\omega)$, can be evaluated for any frequency ω, but as Nicolson pointed out is very simple for frequencies $\omega_n = n(2\pi/t_R)$ with n an integer, as then $g_{RS}(i\omega_n) = 0$.

(2) Gans[23] proposed a different algorithm, consisting of taking the transform of a new function $G_{WA}(t)$, obtained from the original $G(t)$ in the interval $0 < t < t_N$ by the prescription

$$G_{WA}(t) = G(t), \quad 0 < t < t_N$$
$$= G(t) - G(t - t_N), \quad t_N < t < 2t_N$$

This function now has a zero or small value at $t = 2t_N$ and its numerical Fourier transform over the interval $0 < t < 2t_N$ gives $g(\omega_r)$ for frequencies $\omega_r = r(2\pi/2t_N)$ with $r = 1,3,5,\ldots,N-1$.

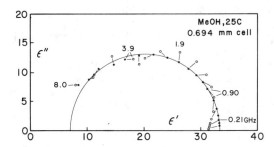

Figure 7. Complex dielectric permittivity plots for methanol at 25°C measured in end capacitance cell with d = 0.69 mm. Open circles are from transforms truncated at t_N = 516 ps. Filled circles and solid semicircle result after truncation corrections described in the text.

(3) A seemingly quite different procedure is to use the Samulon[25] formula

$$g(i\omega) = \frac{1}{i\omega} \frac{\omega\Delta/2}{\sin(\omega\Delta/2)} \exp(-i\omega\Delta/2)$$

$$\times \sum_{n=0}^{N} \exp(-i\omega\Delta) [G(n\Delta+\Delta) - G(n\Delta)] \qquad [17]$$

For a function $G(t)$ approaching a constant $G(\infty)$, the differences $G(n\Delta+\Delta) - G(n\Delta)$ go to zero and the sums remain finite, with the $(i\omega)^{-1}$ divergence of $g(i\omega)$ explicitly represented. The reason for this form is easily understood: the differences are approximations to the derivative dG/dt and the transform $L[dG/dt] = i\omega LG(t) = i\omega g(i\omega)$ if $G(0) = 0$, as is true for functions of interest, with the amplitude and phase factors accounting for errors from taking finite differences to approximate the derivative.

The use of this formula is restricted only by the zeroes of $\sin(\omega\Delta/2)$, so the result is satisfactorily usable for frequencies less than $1/\Delta$ at least. It is also to be noticed that if both transforms in a ratio are evaluated by the formula with the same Δ, the amplitude and phase factors cancel.

A choice between these alternatives for transforming functions $G(t)$ with the $(i\omega)^{-1}$ divergence might seem to require considerable analysis, but Dutuit[25] has demonstrated analytically and by numerical examples that they all give the same result. The Samulon formula has been criticized as being sensitive to noise resulting in fluctuating difference, but summation is a smoothing process and signal averaging also makes this concern unimportant.

The writers' experience with the Samulon formula is that it is easy to use. It obviously gives such functions as $i\omega q(i\omega)$ occurring in working formulas of section III directly. It is also effective for obtaining complex conductances $\sigma = \sigma_s + k(i\omega\epsilon)$ from such ratios as $(v_0-r)/(v_0+r)$ of transforms which both diverge as $(i\omega)^{-1}$, as both can be evaluated by the Samulon formula with all amplitude and phase factors canceling.

An alternative procedure for evaluating dielectric properties of conducting systems can be illustrated by considering the use of Equation 11 for the coaxial line cell. For frequencies such that $f(Z) = 1$ and using $\epsilon_t = \epsilon + \sigma_s/i\omega k$ as in IV.C., Equation 11 gives

$$\epsilon = \frac{c}{d} \frac{v_0 - r}{i\omega(v_0 + r)} - \frac{\sigma_s}{i\omega k} \qquad [18]$$

and

$$\frac{\sigma_s}{k} = \frac{c}{d} \lim_{\omega \to 0} \frac{v_0 - r}{v_0 + r}$$

$$= \frac{c}{d} \lim_{t \to \infty} \frac{V_0(t) - R(t)}{V_0(t) + R(t)} \quad [19]$$

where the last equality follows from a limit theorem for Laplace transforms. Combining these gives

$$\epsilon = \frac{c}{d} \frac{[(v_0 - r) - (v_0 + r)\sigma_s/k]}{i\omega(v_0 + r)} \quad [20]$$

Both $(v_0 - r)$ and $(v_0 + r)$ diverge as $(i\omega)^{-1}$ but the difference in the numerator of Equation 18 does not, and can be evaluated as the transform of $[V_0(t) - R(t)] - [V_0(t) + R(t)]\sigma_s/k$, with σ_s/k given by equation 19.

Either of the procedures just described works well for obtaining dielectric properties of systems with appreciable conductivity, with the choice determined primarily by convenience of available methods for data processing.

(4) Signal smoothing. The various algorithms for digital signal smoothing available in signal analyzers and small computers can be useful in data processing to reduce the number of points needed for acceptable numerical transforms. The simplest of these suffices to illustrate the kind of procedure and effect of using it.

The smoothing algorithm defined for point $G(n\Delta)$

$$S_1 G(n\Delta) = \frac{1}{4} G(n\Delta - \Delta) + \frac{1}{2} G(n\Delta) + \frac{1}{4} G(n + \Delta)$$

gives on iteration

$$S_2 G(n\Delta) = \frac{1}{16} G(n\Delta - 2\Delta) + \frac{1}{4} G(n\Delta - \Delta) + \frac{3}{8} G(n\Delta)$$

$$+ \frac{1}{4} G(n\Delta + \Delta) + \frac{1}{16} G(n\Delta + 2\Delta)$$

The numerical transform of S_1G for G defined at points $n\Delta$ is

$$L_{i\omega}S_1G = \Delta \sum_{n=0}^{\infty} \exp(-i\omega n\Delta) \, S_1 \, G(n\Delta) = \cos^2(\omega\Delta/2) \, L_{i\omega} G$$

$$\cong [1 - (\omega\Delta)^2/8] L_{i\omega} G$$

if the discrete transform is for the interval Δ small enough to approximate the true one with negligible error. It is then easy to see that the discrete transform evaluated for the smoothed points $S_1G(2n\Delta)$ at intervals 2Δ only has essentially the same correction factor, $\cos^2(\omega\Delta/2)$. It contains, however, contributions from all points $G(n\Delta)$, with the correction factor resulting from phase errors in the calculated contributions of the points $G(2n\Delta + \Delta)$ which would not be included in the discrete transform of $G(2n\Delta)$. As a result, transformation of the smoothed function with the coarser interval can reduce effects of irregularities without serious degradation of the information if the factor $\cos^2(\omega\Delta/2)$ is not appreciably different from unity. Moreover, this factor again cancels out of ratios of transforms of functions smoothed in the same way.

The helpful effect of judicious smoothing is particularly useful when transforms of the difference of two nearly identical reflection pulses are required, as in difference methods described in section IV.D. An example shown in Figure 8 is of the difference of the multiple reflection $R_x(t)$ from a 0.5 weight per cent solution of propylene carbonate in toluene and the reflection $R_s(t)$ for solvent toluene, both using a coaxial cell with effective length $d = 20$ mm. The differences in travel time and amplitude of successive reflections produce the comb-like structure of $R_s(t) - R_x(t)$ after both are smoothed twice with original interval $\Delta = 8$ ps of signal analyzer channels. This structure and its transform are degraded considerably if fewer than one in five of the data points are used, but the transform of the difference can be adequately approximated to 5 GHz by using an interval 4Δ and twice smoothed differences.

CONCLUSION

This review of the uses of Fourier transform methods in obtaining dielectric permittivities and related quantities in the MHz to GHz frequency range from time domain reflection measurements has been written primarily to show how transforms and numerical transform methods can be used to obtain useful results in otherwise difficult frequency regions in relatively simple ways. A comprehensive review of the extensive literature prior to 1977 on many approaches to such problems has not been attempted, as most of these have been thoroughly treated in earlier reviews.[13,14,25]

Fourier Transform Dielectric Spectroscopy

The great advantage of time domain methods over steady state frequency methods is of course that a single time domain pulse record with good resolution obtained in a few seconds contains in principle the equivalent of information from measurements at an indefinitely large number of frequencies in a band several decades wide. The precision with which this information can be extracted from such records is steadily being improved by effective use of digital data acquisition and processing techniques requiring only modest computer capabilities. With the development of difference methods in particular, the precision and accuracy now possible appears to be at least comparable to that from most ultrahigh frequency and microwave methods which are not highly specialized and limited in applicability. This state of progress has been reached only recently, and much of the earlier TDS work gave disappointing results because of inadequate development of suitable methods and of failure to recognize pitfalls in applying them.

The discussion given here has been primarily in the context of using TDS equipment with pulse durations limited to a microsecond at most and pulse rise times of ca. 50 ps with resolution of less than 1 ps in sampling interval possible. Much of the analysis could equally well be used for instrumentation with other time ranges when and if this is developed. Regrettably, there seems to be little prospect of such equipment becoming available for much shorter times, to help bridge the gap between microwave frequencies and far infrared Fourier transform spectroscopy, for example. On the low frequency side, the prospects are brighter for time domain methods of obtaining permittivities at frequencies from, say, 100 KHz to 20 MHz quickly and simply to supplement the powerful transformer bridge and transient methods available for lower frequencies.

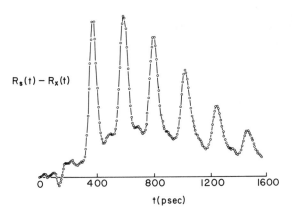

Figure 8. Difference $R_s(t) - R_x(t)$ of solvent toluene and propylene carbonate (0.5 wt. % solution reflections of Fig. 5. Maximum peak height approximately 7% of V_0; interval between peaks ca. 210 ps.

REFERENCES

1. Kubo, R. 1957, J. Phys. Soc. Japan 6, 570.
2. Glarum, S. H. 1960, J. Chem. Phys. 33, 639.
3. Cole, R. H. 1965, J. Chem. Phys. 42, 637.
4. van Gemert, M. J. C. 1974, J. Chem. Phys. 60, 3963.
5. Cole, R. H. 1975, J. Phys. Chem. 79, 1459.
6. Johnson, W. C. 1950, "Transmission Lines and Networks," McGraw-Hill, New York.
7. Marcuvitz, N. 1951, "Waveguide Handbook," McGrawHill, New York.
8. Iskander, M. F.,and Stuchly, S. S. 1972, IEEE Trans. Instrum. Meas. IM 21, 425.
9. Somlo, P. I. 1967, Proc. Inst. Radio Elec. Eng. Australia 28, 7.
10. Bussey, H. E. 1980, IEEE Trans. Instrum. IM 29, 120.
11. Cole, R. H. 1975, J. Phys. Chem. 79, 1469.
12. Gestblom, B.,and Noreland, E. 1977, J. Phys. Chem. 81, 782.
13. Cole, R. H. 1977, Ann. Rev. Phys. Chem. 28, 283.
14. Suggett, A. 1972, in "Dielectric and Related Molecular Processes," Vol. I, Chemical Society, London, p. 100.
15. Cole, R. H., Mashimo, S., and Winsor, P. 1980, J. Phys. Chem. 84, 786.
16. Elliott, B. J. 1970, IEEE Trans. Instrum. Meas. IM 19, 391.
17. Andrews, J. R. 1970, IEEE Trans. Instrum. Meas. IM 19, 171.
18. Elliott, B. J. 1976, IEEE Trans. Instrum. Meas. IM 25, 376.
19. Tuck, E. O. 1967, Math. Comput. 21, 239.
20. Filon, L. N. G. 1928, Proc. Roy. Soc. Edin. XLIX, 38.
21. Cooley, J. W., and Tukey, J. W. 1965, Math. Comput. 19, 297.
22. Shannon, C. 1949, Proc. I.R.E. 37, 10.
23. Gans, W. L., and Andrews, J. R. 1975, NBS Technical Note 672.
24. Nicolson, A. M. 1973, Electron. Lett. 9, 317.
25. Dutuit, Y. 1979, Revue de Phys. App. 14, 939.
26. Clarkson, T. S., Glasser, L., Tuxworth, R. W. and Williams, G. 1977, Adv. Mol. Relaxation Processes 10, 173.

PULSED FOURIER TRANSFORM MICROWAVE SPECTROSCOPY

Willis H. Flygare

Noyes Chemical Laboratory
University of Illinois
Urbana, Illinois   61801

INTRODUCTION

In this chapter, we describe the technique of Fourier transform microwave spectroscopy. We distinguish here two rather different types of sample absorption cells which require somewhat different theoretical descriptions. First, we describe the theory for the relatively broad-band waveguide absorption cell in which the radiation is described as a <u>traveling wave</u>. Second, we describe the narrow-band Fabry-Perot cavity absorption cell in which the radiation is described as a <u>standing wave</u>.

In the first case, a microwave oscillator at frequency ν is phase stabilized to a harmonic of a lower-frequency oscillator that is monitored by a frequency counter. This primary oscillator at frequency ν is phase stabilized to a local oscillator at a frequency ν - 30 MHz which is subsequently used in the superheterodyne detection of the molecular signal. The primary oscillator is formed into a pulse with switches; the microwave pulse then transmits as a <u>traveling wave</u> through a section of waveguide that contains the molecular sample. The microwave pulse (typically a $\pi/2$ pulse) polarizes a band of frequencies around ν that overlaps one or more rotational transitions. After the polarizing pulse reaches the end of the sample cell, it rapidly dies away leaving a rotationally polarized gas which then commences to emit a coherent <u>traveling wave</u> at the molecular resonance frequencies. The coherent <u>emission</u> is then detected by mixing with the local oscillator in a balanced mixer. The theory and practice of these traveling wave experiments are now well understood and have been applied in a number of static gases which illustrate the advantages in sensitivity and resolution by working in the time domain relative to the frequency domain.

In the second case, the radiation is contained as a standing wave in a Fabry-Perot cavity. There are, however, significant differences in this latter standing wave experiment relative to the above traveling wave experiment in a waveguide. The primary and local oscillators are set up as described above. The primary microwave source at ν is then formed into a pulse with a switch, after which the pulse is impedance matched to a Fabry-Perot cavity. Maximum polarization of the gas is achieved following a nearly $\pi/2$ pulse, for $T_2 \gg \tau_p$, where T_2 is the polarization relaxation time, related to the low power steady state transition half-width at half height by $\Delta\nu = 1/(2\pi T_2)$, and τ_p is the microwave pulse length. Following polarization of a band of molecular frequencies centered at ν, the microwave pulse is switched off and it dissipates with a relaxation time τ_c, the cavity relaxation time. We now require the molecules to maintain their coherent polarization for a period of time long relative to τ_c; $T_2 \gg \tau_c$. The $T_2 \gg \tau_c$ requirement allows the original microwave pulse to die away before the coherent polarization is lost. After the microwave pulse has died away, the coherent emission from all polarized transitions is observed in the superheterodyne receiver as a beat between the local oscillator field and the field from the coherent emission.

In this chapter, we summarize the theory of rotational transition polarization for both the traveling wave in a waveguide and the standing wave in a Fabry-Perot cavity through a solution of the Bloch equations. We also develop the equations necessary to understand the nature of the coherent emission following the decay of a short pulse of radiation into the appropriate sample cell. Coupling of the emitted traveling or standing wave into a detector is also considered. Several examples are given. The traveling wave Fourier transform experiment is well suited to observe a stable static gas. The standing wave Fabry-Perot cavity Fourier transform experiment is well suited to observe transient or otherwise weakly bound or short-lived molecular species. In this chapter, we summarize from our previous publications in this field.[1-9]

BASIC THEORY

Consider an ensemble of non-degenerate two-level quantum systems interacting with an electric field, $\underline{E}(r,t)$ through the electric dipole interaction. In the absence of collisions, the quantum mechanical Hamiltonian for the j'th molecule is[1,4,7]

$$\underline{H}(r_j,t) = H_0 - (\hbar^2/2m)\underline{\nabla}^2 - \underline{\mu}\cdot\underline{E}(r_j,t) \qquad [1]$$

in which r_j is the center-of-mass of the j'th molecule, H_0 is the time-independent free molecular Hamiltonian whose eigenfunctions include energy levels E_a and E_b corresponding to the upper and lower states, respectively, $\underline{\nabla}$ is the gradient with respect to r_j, m is

Pulsed Fourier Transform Microwave Spectroscopy

the molecular mass, and μ is the dipole moment operator. Since the external field interaction does not affect the molecular center-of-mass motion, the density matrix $\sigma^j(r_j,p_j,t)$ in classical phase space, where r_j and p_j are the classical variables for the position and momentum of the j'th molecule, satisfies

$$i\hbar(\frac{\partial}{\partial t} + m^{-1}p_j \cdot \nabla)\sigma^j_{\alpha\alpha'}(r_j,p_j,t)$$
$$= [H_0 - \mu \cdot E(r_j,t), \sigma^j(r_j,p_j,t)]_{\alpha\alpha'} \quad [2]$$

in which the brackets on the right-hand side indicate the commutator. Provided that each molecule interacts independently with the field, we may work with a macroscopic density matrix having elements defined by

$$\sigma_{\alpha\alpha'}(r,p,t) = \sum_j \int d^3r_j\, d^3p_j\, \sigma^j_{\alpha\alpha'}(r_j,p_j,t)\, \delta(r-r_j)\, \delta(p-p_j) \quad [3]$$

Using $v = p/m$, Equation 2 becomes

$$i\hbar(\frac{\partial}{\partial t} + v \cdot \nabla)\, \sigma_{\alpha\alpha'}(r,v,t)$$
$$= [H_0 - \mu \cdot E(r,t), \sigma(r,v,t)]_{\alpha\alpha'} \quad [4]$$

Using the development in McGurk, Schmalz, and Flygare,[1] we take

$$E(r,t) = 2\hat{z}\,\mathcal{E}\cos(\omega t - ky), \quad [5]$$

a plane polarized traveling wave, where $\hat{z}$ is a unit vector perpendicular to the direction of propagation, and $\mathcal{E}$ may in general have the form $\mathcal{E}(r,t)$. Using this, Equation 4 becomes

$$i\hbar(\frac{\partial}{\partial t} + v \cdot \nabla)\sigma_{aa} = 2\mathcal{E}(\mu_{ba}\sigma_{ab} - \mu_{ab}\sigma_{ba})\cos(\omega t - ky)$$

$$i\hbar(\frac{\partial}{\partial t} + v \cdot \nabla)\sigma_{ab} = \sigma_{ab}(E_a - E_b) + 2\mu_{ab}\mathcal{E}(\sigma_{aa}-\sigma_{bb})\cos(\omega t - ky)$$

$$i\hbar(\frac{\partial}{\partial t} + v \cdot \nabla)\sigma_{bb} = 2\mathcal{E}(\mu_{ab}\sigma_{ba} - \mu_{ba}\sigma_{ab})\cos(\omega t - ky) \quad [6]$$

Following McGurk et al.[1], we transform to the density matrix in the interaction representation defined by

$$\rho = \exp(\frac{iS}{\hbar}(t - \frac{y}{c})) \; \sigma \exp(-\frac{iS}{\hbar}(t - \frac{y}{c})),$$

$$S = \begin{bmatrix} E_a & 0 \\ 0 & (E_a + \hbar\omega) \end{bmatrix} \quad [7]$$

Substituting ρ for σ in Equation 6, and making the rotating wave approximation, one obtains

$$i\hbar(\frac{\partial}{\partial t} + \underline{v}\cdot\underline{\nabla})\rho_{aa} = \mathcal{E}(\mu_{ba}\rho_{ab} - \mu_{ab}\rho_{ba}) \quad [8]$$

$$i\hbar(\frac{\partial}{\partial t} + \underline{v}\cdot\underline{\nabla})\rho_{ab} = -\hbar\Delta\omega\rho_{ab} + \mathcal{E}\mu_{ab}(\rho_{aa} - \rho_{bb}) \quad [9]$$

$$i\hbar(\frac{\partial}{\partial t} + \underline{v}\cdot\underline{\nabla})\rho_{bb} = \mathcal{E}(\mu_{ab}\rho_{ba} - \mu_{ba}\rho_{ab}) \quad [10]$$

where
$$\Delta\omega = \omega_0 - \omega(1 - v/c), \quad [11]$$

$$\omega_0 = \frac{E_b - E_a}{\hbar},$$

and we have assumed a dispersion relationship, $k = \omega/c$. To make the connection with Equations 17-19 in McGurk et al.,[1] we specify $\mathcal{E}(\underline{r},t) = \mathcal{E}(t)$, with no spatial dependence. With no $\underline{r}$ dependence in Equations 8-10 above, the $\underline{v}\cdot\underline{\nabla}$ terms may be handled trivially by specifying that $\rho_{\alpha\alpha'}(\underline{r},\underline{v},t) = \rho_{\alpha\alpha'}(\underline{v},t)$. One is left with

$$i\hbar \frac{\partial \rho_{aa}}{\partial t} = \mathcal{E}(\mu_{ba}\rho_{ab} - \mu_{ab}\rho_{ba})$$

$$i\hbar \frac{\partial \rho_{ab}}{\partial t} = \mathcal{E}\mu_{ab}(\rho_{aa} - \rho_{bb}) - \hbar\Delta\omega\rho_{ab} \quad [12]$$

$$i\hbar \frac{\partial \rho_{bb}}{\partial t} = \mathcal{E}(\mu_{ab}\rho_{ba} - \mu_{ba}\rho_{ab})$$

Pulsed Fourier Transform Microwave Spectroscopy

These may be compared to Equations 17-19 and 65 in McGurk et al.[1]

Consider now the standing wave case. The electric fields in a Fabry-Perot cavity[9] in the TEM_{mnq} modes described here will be polarized Gaussian standing waves of the form

$$\underline{E}(\underline{r},t) = 2\hat{\underline{z}}\, \mathcal{E}(\underline{r},t)\cos(\omega t) \qquad [13]$$

We transform to the density matrix in the interaction representation defined by

$$\rho = \exp\left(\frac{iS}{\hbar}t\right)\sigma \exp\left(-\frac{iS}{\hbar}t\right) \qquad [14]$$

where S is defined in Equation 7. Again making the rotating wave approximation, one obtains

$$i\hbar\left(\frac{\partial}{\partial t} + \underline{v}\cdot\underline{\nabla}\right)\rho_{aa} = -\mathcal{E}(\underline{r},t)(\mu_{ab}\rho_{ba} - \mu_{ba}\rho_{ab})$$

$$i\hbar\left(\frac{\partial}{\partial t} + \underline{v}\cdot\underline{\nabla}\right)\rho_{ab} = \mathcal{E}(\underline{r},t)\mu_{ab}(\rho_{aa} - \rho_{bb}) - \hbar\Delta\omega\rho_{ab} \qquad [15]$$

$$i\hbar\left(\frac{\partial}{\partial t} + \underline{v}\cdot\underline{\nabla}\right)\rho_{bb} = -\mathcal{E}(\underline{r},t)(\mu_{ba}\rho_{ab} - \mu_{ab}\rho_{ba})$$

where

$$\Delta\omega = \omega_0 - \omega \qquad [16]$$

with no v/c term, since the transformation $\sigma \to \rho$ involves no spatial dependence. The effect of molecular motion on the standing wave solutions enters via the $\underline{v}\cdot\underline{\nabla}$ terms.

Following closely the development in McGurk et al.[1] for the traveling wave, we rewrite Equation 15 in terms of the polarization and population differences of the gas. The polarization is given by

$$P = Tr(\mu\sigma) \qquad [17]$$

where Tr represents the trace of the matrix representation. Using Equation 14 gives

$$P = Tr\left(\mu \exp\left(-\frac{iS}{\hbar}t\right)\rho \exp\left(\frac{iS}{\hbar}t\right)\right) \qquad [18]$$

Expanding Equation 18, one obtains

$$P = \mu_{ba}\rho_{ab}\, e^{i\omega t} + \mu_{ab}\rho_{ba}\, e^{-i\omega t} \qquad [19]$$

$$= (P_r + iP_i)\, e^{i\omega t} + (P_r - iP_i)\, e^{-i\omega t} \qquad [20]$$

with

$$P_r + iP_i = \mu_{ba}\rho_{ab} \qquad [21]$$

by definition, and where P_r and P_i are real quantities. Noting that $N_a = \rho_{aa}$, $N_b = \rho_{bb}$, the number densities of molecules in the a and b states, respectively, from Equations 12 or 15, 19, and 20, one obtains the Bloch-type equations for both cases:

<u>Traveling wave:</u>

$$(\partial/\partial t)P_r + \Delta\omega P_i + \frac{P_r}{T_2} = 0$$

$$(\partial/\partial t)P_i - \Delta\omega P_r + \kappa^2 \mathcal{E}\left(\frac{\hbar\Delta N}{4}\right) + \frac{P_i}{T_2} = 0 \qquad [22]$$

$$(\partial/\partial t)\frac{\hbar\Delta N}{4} - \mathcal{E}P_i + \frac{\hbar}{4}\frac{(\Delta N - \Delta N_0)}{T_1} = 0$$

<u>Standing wave:</u>

$$\left(\frac{\partial}{\partial t} + \underline{v}\cdot\underline{\nabla}\right) P_r + \Delta\omega P_i + \frac{P_r}{T_2} = 0$$

$$\left(\frac{\partial}{\partial t} + \underline{v}\cdot\underline{\nabla}\right) P_i - \Delta\omega P_r + \kappa^2 \mathcal{E}(\underline{r},t)\left(\frac{\hbar\Delta N}{4}\right) + \frac{P_i}{T_2} = 0 \qquad [23]$$

$$\left(\frac{\partial}{\partial t} + \underline{v}\cdot\underline{\nabla}\right)\left(\frac{\hbar\Delta N}{4}\right) - \mathcal{E}(\underline{r},t)P_i + \frac{\hbar}{4}\cdot\frac{(\Delta N - \Delta N_0)}{T_1} = 0$$

Here $\Delta N = N_a - N_b$, ΔN_0 is the equilibrium value of ΔN in the absence of an electric field, and

$$\kappa = 2\hbar^{-1}\, |<a|\mu_z|b>| \qquad [24]$$

The first-order relaxation terms involving T_1 and T_2 have been introduced phenomenologically in the usual way.[1]

These fundamental equations which govern all of the transient phenomena that we will discuss in this chapter are the electric dipole analogs of the Bloch equations in NMR.[10] The derivation of these equations has been given many times before.[4] Many times, a geometric interpretation of the Bloch equations is used.[11] When a two-level system interacts with an electric field of the form in Equation 5, for instance, the motion of the Bloch vector, $\vec{r}$, is governed by the vector equation

$$\frac{d\vec{r}}{dt} = \vec{\omega} \times \vec{r} \qquad [25]$$

in the absence of collisions. The Bloch vector $\vec{r}$ is given by the following combinations of density matrix elements:

$$\vec{r} = \begin{pmatrix} \sigma_{ba} + \sigma_{ab} \\ i(\sigma_{ba} - \sigma_{ab}) \\ \sigma_{aa} - \sigma_{bb} \end{pmatrix} \qquad [26]$$

while ω is given in terms of the perturbation that couples the lower state a and upper state b, $V_{ab} = -2\mu_{ab}\mathcal{E}\cos(\omega t - kz)$, plus the natural resonance frequency of the two-level system, $\omega_0 = (E_b - E_a)/\hbar$:

$$\vec{\omega} = \begin{pmatrix} (V_{ab} + V_{ba})/2\hbar \\ i(V_{ab} - V_{ba})/2\hbar \\ \omega_0 \end{pmatrix} \qquad [27]$$

Upon transforming to an interaction representation (i.e., a reference frame rotating at frequency ω) in which the density matrix is defined by Equation 7, and invoking the rotating wave approximation which consists of dropping all high-frequency motions with respect to ω, Equation 25 becomes

$$\frac{d\vec{r}'}{dt} = \vec{\omega}_{eff} \times \vec{r}' \qquad [28]$$

with

$$\vec{r}' = \begin{pmatrix} \rho_{ba} + \rho_{ab} \\ i(\rho_{ba} - \rho_{ab}) \\ \rho_{aa} - \rho_{bb} \end{pmatrix} \qquad [29]$$

and

$$\vec{\omega}_{eff} = \begin{pmatrix} -\omega_1 \\ 0 \\ \Delta\omega \end{pmatrix} \qquad [30]$$

The effective strength of the interaction with the radiation is $\omega_1 = \kappa\mathcal{E} = 2\mu_{ab}\mathcal{E}/\hbar$, where μ_{ab}, taken as real, is the dipole matrix element coupling states a and b, while $\Delta\omega = \omega_0 - \omega$ is the difference between the resonant frequency of the system and the applied frequency.

Interactions between molecules are taken into account phenomenologically by adding linear damping terms to the right-hand side of Equation 28, using some first-order relaxation theory for the density matrix such as Redfield theory.[12,13] This gives for the complete equation of motion of the system

$$\frac{d\vec{r}'}{dt} = (\vec{\omega}_{eff} \times \vec{r}') - \frac{\hat{i}' r_1' + \hat{j}' r_2'}{T_2} - \frac{\hat{k}'(r_3' - \bar{r}_3')}{T_1} \qquad [31]$$

$\hat{i}'$, $\hat{j}'$, and $\hat{k}'$ are unit vectors in the rotating frame, and $\bar{r}_3'$ is the thermal equilibrium value of r_3'. Equation 31 must be solved under appropriate conditions to describe microwave transient experiments.

Although the Bloch vector-rotating frame formalism is convenient for deriving the basic equation of motion and for understanding the similarities among transient phenomena in various fields, we find it more convenient to work with physically real quantities when describing specific experiments. The physical quantities of interest in microwave experiments are the polarization, or macroscopic induced dipole moment, per unit volume, and the population difference between the levels a and b per unit volume.[1]

Comparison of Equations 19 and 20 shows that

$$P_r = \frac{\mu_{ab}}{2}(\rho_{ab} + \rho_{ba})$$

$$P_i = \frac{\mu_{ab}}{2i}(\rho_{ab} - \rho_{ba}) \qquad [32]$$

and comparison with Equation 29 yields

$$P_r = \frac{\mu_{ab}}{2} r_1'$$

$$P_i = \frac{\mu_{ab}}{2} r_2' \qquad [33]$$

Thus, r_1' and r_2' are simply related to the real and imaginary (in-phase and out-of-phase) components of the macroscopic polarization.

The population difference is defined as before by

$$\Delta N = \rho_{aa} - \rho_{bb} \quad [34]$$

Thus, it is readily apparent that

$$\Delta N = r_3' \quad [35]$$

Similarly,

$$\Delta N_0 = \overline{r}_3' \quad [36]$$

These definitions for P_r, P_i, and ΔN may now be substituted into Equation 31 to give Equation 22. The roles of the phenomenological relaxation times T_1 and T_2 are now clear. The macroscopic polarization P, and hence the components P_r and P_i, relax to their equilibrium values of zero with a relaxation time T_2. The population difference N relaxes to its equilibrium value ΔN_0 with a relaxation time T_1.

In NMR, T_1 is called the longitudinal or spin-lattice relaxation time.[14] It represents the rate at which energy is exchanged between the spin system and the surroundings. Since in microwave experiments, energy is stored in the form of a population difference, the connection between these interpretations is clear. T_2 in NMR is called the transverse or spin-spin relaxation time.[14] It represents the rate at which individual spins lose coherence. In the experiments discussed here, T_2 can be interpreted as a loss of coherent polarization due to molecular interactions.

Although Equations 22 and 23 are written entirely in terms of physically real quantities, none of them are in fact directly observed experimentally. No matter what the details of the experimental apparatus used to observe the transient phenomena, all eventually observe the current generated by some sort of non-linear detector element. That current (in the square law regime) is proportional to the square of the total electric field incident on the detector, averaged over all times faster than the response time of the detector. It is therefore necessary to calculate the electric field produced at the end of the sample cell due to the polarization of the sample.

TRAVELING WAVE CASE: WAVEGUIDE SAMPLE CELL

In normal microwave spectroscopy, the frequency of the incident oscillator is swept through the resonance in a period of time long relative to the relaxation processes. The gas is then always in equilibrium with the radiation. Under these conditions, Equation 22 can be solved by setting $dP_r/dt = dP_i/dt = d\Delta N/dt = 0$, giving:

$$P_r = \frac{(1/2)\mu\omega_1(\Delta\omega)\Delta N_0}{(1/T_2)^2 + (T_1/T_2)\omega_1^2 + (\Delta\omega)^2}$$

$$P_i = \frac{-(1/2)\mu\omega_1(1/T_2)\Delta N_0}{(1/T_2)^2 + (T_1/T_2)\omega_1^2 + (\Delta\omega)^2} \qquad [37]$$

$$\Delta N = \frac{\Delta N_0[(1/T_2)^2 + (\Delta\omega)^2]}{(1/T_2)^2 + (T_1/T_2)\omega_1^2 + (\Delta\omega)^2}$$

$$\omega_1 = \kappa\mathcal{E} = \frac{\mu_{ab}\mathcal{E}}{\hbar}$$

P_i in Equation 37 gives the standard Lorentzian line shape for the absorption coefficient with the half-width at half-height given by:

$$\Delta\nu_{1/2} = \Omega_p/2\pi = \frac{1}{2\pi}[(1/T_2)^2 + (T_1/T_2)\omega_1^2]^{1/2} \qquad [38]$$

in which $\Omega_p/2\pi$ indicates the power saturated line width. If low power radiation is used, $(1/T_2)^2 \gg (T_1/T_2)\omega_1^2$, and the normal half-width at half-height of the Lorentzian is obtained, $\Delta\nu_{1/2} = 1/(2\pi T_2)$.

In a microwave pulse excitation of a sample in the on-resonant case where a two-level system is brought into resonance, $\Delta\omega = 0$, in a time short relative to the relaxation processes, we solve Equation 22, when $\Delta\omega = 0$, giving

$$P_r(t) = P_r(t_i) e^{-(t-t_i)/T_2}$$

$$P_i(t) = \frac{\mu\omega_1\Delta N_0}{2\Omega_p^2}\left\{e^{-(t-t_i)/T}\left[(1/T_2)\cos(\Omega_0(t-t_i))\right.\right.$$

$$\left.- [(T_1/T_2)\frac{\omega_1^2}{\Omega_0} + (1/T_2)(\frac{\phi}{\Omega_0})\sin(\Omega_0(t-t_i))]\right] - (1/T_2)\Big\}$$

$$+ e^{-(t-t_i)/T}\left\{P_i(t_i)[\cos(\Omega_0(t-t_i)) - \frac{\phi}{\Omega_0}\sin(\Omega_0(t-t_i))]\right.$$

$$\left.- \frac{\mu\omega_1}{2\Omega_0}[\Delta N(t_i) - \Delta N_0]\sin(\Omega_0(t-t_i))\right\}$$

$$\Delta N(t) = \frac{\Delta N_0}{\Omega_p^2} \{(T_1/T_2)\omega_1^2 \, e^{-(t-t_i)/T} [\cos(\Omega_0(t-t_i))$$

$$+ \frac{1}{\Omega_0 T} \sin(\Omega_0(t-t_i)) + (1/T_2)^2]\} + e^{-(t-t_i)/T}$$

$$\times \{P_i(t_i) \frac{2\omega_1}{\mu\Omega_0} \sin(\Omega_0(t-t_i)) + [\Delta N(t_i) - \Delta N_0][\cos(\Omega_0(t-t_i))$$

$$+ \frac{\phi}{\Omega_0} \sin(\Omega_0(t-t_i))] \qquad [39]$$

where $P_r(t_i)$, $P_i(t_i)$, and $\Delta N(t_i)$ contain the appropriate initial conditions at the initial time, t_i, where $t_i \leq t$, Ω_p is defined in Equation 38, and

$$\frac{1}{T} = \frac{1}{2}\left(\frac{1}{T_1} + \frac{1}{T_2}\right)$$

$$\phi = \frac{1}{2}\left(\frac{1}{T_2} - \frac{1}{T_1}\right) \qquad [40]$$

$$\Omega_0 = [\omega_1^2 - \frac{1}{4}\left(\frac{1}{T_2} - \frac{1}{T_1}\right)^2]^{1/2}$$

When $\omega_1^2 < (1/4)[(1/T_2)-(1/T_1)]^2$, Equations 39 must be modified by replacing the sine and cosine functions with hyperbolic functions and setting $\Omega_0 = [(1/4)[(1/T_2)-(1/T_1)]^2 - \omega_1^2]^{1/2}$.
Equations 39 are general and require only the initial conditions to specify completely the behavior of the system. Normally one finds that $(1/T_1) \cong (1/T_2)$, and therefore it is expected that

$$\Omega_0 \cong \omega_1 \qquad [41]$$

Usually, the initial conditions (before the on-resonant absorption) are obtained from the far off-resonance limit of the steady-state results in Equations 37, where $(\Delta\omega)^2 \gg (1/T_2)^2$ and $(\Delta\omega)^2 \gg (T_1/T_2)\omega_1^2$. Under these initial conditions, one has $P_r(t_i=0) = P_i(t_i=0) = 0$, and $\Delta N(t_i=0) = \Delta N_0$ from Equations 37. Substituting these results and the condition in Equation 41 into Equations 39 gives

<u>Limits</u>: $0 \leq t$; $\Delta\omega = 0$, $P_r(0) = P_i(0) = 0$, $\Delta N(0) = \Delta N_0$

$P_r(t) = 0$

$$P_i(t) = \frac{\mu\omega_1\Delta N_0}{2\Omega_p^2} \left\{ e^{-t/T}\left[(1/T_2)\cos(\omega_1 t) - [(T_1/T_2)\omega_1 \right.\right.$$
$$\left.\left. + (1/T_2)(\frac{\phi}{\omega_1})\sin(\omega_1 t)\right] - (1/T_2) \right\}$$

$$\Delta N(t) = \frac{\Delta N_0}{\Omega_p^2}\left\{(T_1/T_2)\omega_1^2\, e^{-t/T}[\cos(\omega_1 t)+(\frac{1}{\omega_1 T}\sin(\omega_1 t)] + (1/T_2)^2\right\}$$

[42]

Under the conditions of high power radiation where

$$\omega_1 T_1 \gg 1$$
$$\omega_1 T_2 \gg 1$$

[43]

Equations 42 reduce to

$$P_r(t) = 0$$
$$P_i(t) = -\frac{\mu}{2}\Delta N_0\, e^{-t/T}\sin(\omega_1 t)$$
$$\Delta N(t) = \Delta N_0\, e^{-t/T}\cos(\omega_1 t)$$

[44]

And if the interaction is terminated in a time short relative to the relaxation time, $t \ll T$, Equations 44 reduce to the normal equations that are used to describe π and $\pi/2$ pulse experiments.

$$P_r(t) = 0$$
$$P_i(t) = -\frac{\mu}{2}\Delta N_0\, \sin(\omega_1 t)$$
$$\Delta N(t) = \Delta N_0\, \cos(\omega_1 t)$$

[45]

The $\pi/2$ pulse, where

$$\omega_1 t = \pi/2$$

[46]

leads to maximum imaginary polarization, P_i, and zero population difference between the two levels

$$P_r(t_{\pi/2}) = 0$$
$$P_i(t_{\pi/2}) = -(\mu/2)\Delta N_0$$
$$\Delta N(t_{\pi/2}) = 0$$

[47]

Pulsed Fourier Transform Microwave Spectroscopy

At a later time, we have the pulse condition, where

$$\omega_1 t = \pi \qquad [48]$$

which leads to zero polarization and a population inversion.

Fourier transform methods have long been used in magnetic resonance to obtain higher signal-to-noise when dealing with weak spectra. Many of the same types of experimental situations are encountered in microwave spectroscopy. The technique uses a train of intense microwave pulses with carrier frequency $\nu_p = \omega_p/2\pi$, pulse length t_p, and period between pulses t_0. The radiation pulses travel along the z direction in a waveguide in the dominant TE_{10} mode. The short pulses are of sufficient duration and radiation power to produce a polarization in the sample gas which will begin to emit after the radiation power has passed.

To make the analysis tractable, certain conditions must be imposed on the exciting pulses. Consider the observation of all transitions in a band $\pm \Delta\omega$ around the carrier frequency ω_p. It is assumed that the pulse length is short relative to the relaxation times for the system, $t_p \ll T_2 \cong T_1$, and that the time between pulses is long relative to the polarization relaxation time, $t_0 \gg T_2$. It is also assumed that the pulse is intense enough to satisfy

$$\kappa_j \mathcal{E} = \frac{\mu_j \mathcal{E}}{\hbar} \gg |\Delta\omega_j| = |\omega_j - \omega_p| \qquad [49]$$

for all transitions j within the bandwidth $\Delta\omega$. ω_j denotes the natural resonance frequency of the j'th transition, and μ_j is the transition dipole for that transition. Equations 45 and 47 are now also valid if $|\Delta\omega_j| \neq 0$.

In these pulse experiments, polarization is created at different times in different parts of the waveguide as the pulse moves through the sample. The coherent field at the detector due to the emission from all these molecules has been evaluated by taking proper account of the phase of the emitted field from different parts of the cell.[4] The Fourier transform method shows its full power only for large bandwidths in long cells, and the effect of the waveguide on the form of the emitted field has also been considered explicitly, where the real ($\mathcal{E}_r$) and imaginary ($\mathcal{E}_i$) components for the amplitude of the emitted field are given by[4]

$$\mathcal{E}_r = 0$$
$$\mathcal{E}_i = \frac{4\pi\omega\ell}{v_g}\left(-\frac{\hbar\kappa_j}{4}\right)(\Delta N_0)_j J_1(\kappa_j \mathcal{E}_0 t_p) e^{-(t-\bar{t})/T_{2j}} \qquad [50]$$

in which $\bar{t} = t_p + 3z/v_g$, z is the distance traveled by the pulse, v_g is the pulse group velocity, and J_1 is the first-order Bessel function.

There is no loss of field intensity due to phase cancellation in this type of experiment even in long cells with large bandwidths, as long as the difference in transit time down the waveguide of the pulse and emission is shorter than the relaxation time of the system being studied.

The electric fields from all j transitions can be detected simultaneously by mixing them with a local oscillator at frequency ω_{LO} having a fixed phase relationship to ω_p, and integrating over the Doppler distribution (when necessary). The resultant signal is

$$\Delta S(t) = \frac{8\sqrt{\pi}\omega \ell q \beta}{v_g} \sum_j (-\frac{\hbar \kappa_j}{4})(\Delta N_0)_j J_1(\kappa_j \mathcal{E}_0 t_p) e^{-(t-\bar{t})/T_{2j}}$$

$$\times \int_{-\infty}^{\infty} e^{-q^2 \omega'^2} \cos[(\omega_j - \omega_{LO} + \omega')t - \phi_j(z)]\, d\omega' \qquad [51]$$

where the amplitude of the local oscillator has here been absorbed into the conversion efficiency of the mixer, β, and all phase factors have been collected together in $\phi_j(z)$. The remaining integral is over the Maxwell-Boltzmann velocity distribution, where $q = \sqrt{\ell n 2}/\Delta \omega_0$ and $\Delta \nu_0 = \Delta \omega_0/2\pi$ is the Doppler half-width at half-height.

The cosine Fourier transform of Equation 51 beginning at $t = t$ then gives the following signal as a function of frequency:

$$\Delta S(\omega) = \frac{4\sqrt{\pi}\,\omega\, q\beta}{v_g} \sum_j (-\frac{\hbar \kappa_j}{4})(\Delta N_0)_j J_1(\kappa_j \mathcal{E}_0 t_p)$$

$$\times \left\{ \cos(\phi_j(z)) \int_{-\infty}^{\infty} \frac{e^{-q^2\omega'^2(1/T_{2j})}}{(1/T_{2j})^2 + (\omega_j - \omega_{LO} - \omega + \omega')^2}\, d\omega' \right.$$

$$\left. + \sin(\phi_j(z)) \int_{-\infty}^{\infty} \frac{e^{-q^2\omega'^2}(\omega_j - \omega_{LO} - \omega + \omega')}{(1/T_{2j})^2 + (\omega_j - \omega_{LO} - \omega + \omega')}\, d\omega' \right\} \qquad [52]$$

This is a frequency-dependent (through $\omega_j(z)$ which depends on k_j and hence ω_j) linear combination of the absorption and dispersion spectra. The pure absorption spectrum can be regained from Equation 52 by numerical methods if the sine Fourier transform is also available.

It can be seen that the spectrum in the frequency domain can be obtained without power saturation in either the pressure or Doppler broadened limit.

Equation 52 shows that the intensities of each of the transitions depend on the pulse length, t_p. We have therefore to distinguish between two situations:

1. The first case is where we are interested in recovering one or more relatively weak lines with similar κ_j in a noisy spectrum. In this case, we do not need exact amplitude relations. t_p is then roughly adjusted to the first maximum of the Bessel function, and the spectrum can be obtained with optimum signal-to-noise ratio but slightly distorted amplitudes of the lines according to their κ_j.

2. In the second case, we are interested in the relation of the amplitudes of lines with different κ_j. The pulse length is then decreased sufficiently below the first maximum of the Bessel function for the strongest line, and the intensities of the lines will be proportional to κ_j^2, which is the same as in conventional steady-state spectroscopy for unsaturated transitions.

We will now describe in some detail the Fourier transform spectrometer constructed by Ekkers and Flygare.[5] The following requirements for their instrument working in the frequency range from 4 to 8 GHz with a maximum bandwidth of 50 MHz were carefully considered.

1. In order to saturate a transition with a dipole transition moment $\mu_{ab} \cong 1$ Debye somewhere in a band of 50 MHz, the pulse power must be sufficiently large to fulfill Equation 49 where $\Delta\omega/2\pi$ = 25 MHz. The waveguide formula

   $$P = E_0^2 \frac{a \cdot b}{480 \cdot \pi} \frac{\lambda}{\lambda_g}$$

   allows the calculation of the necessary power P in Watts. E_0 is the maximum field amplitude in Volts/cm, a and b are the dimensions of the waveguide in cm, and λ and λ_g are the free-space wavelength and the waveguide wavelength, respectively. For a C-band guide, we need pulse powers of greater than 10 Watts. To maximize the polarization, the pulse length has to be less than 10 nsec.

2. After the pulse is switched off, the power emitted from the excited molecules is (for a moderately strong transition) 80 to 100 dB below the pulse power. This means that the on/off ratio of the pulse has to be considerably higher than 100 dB.

3. After the signal is detected as a beat against the local oscillator frequency, the signal has to be brought into a suitable frequency range to make the analog-to-digital conversion possible. The maximum frequency of 50 MHz implies a sample rate of 10 nsec per point. The responses after each pulse have to be in phase to make averaging possible. The averaging process has to be fast compared to the time during which the signal is sampled in order to make the dead time between the pulses as short as possible.

The Ekkers-Flygare spectrometer is shown in Figure 1. The pulse sequence for the four PIN diode switches is shown in Figure 2. The delay of switch 2 is variable compared to switch 1. Power from the master oscillator (MO) reaches the 20 Watt CW TWT-amplifier only when the first two switches are open. At the input of the amplifier, the pulse has an on/off ratio of 80 dB, 10 nsec switching time, and a length that is continuously variable from 1 μsec to zero. At the output of the TWT-amplifier, a pulse of nearly 20 Watts peak power with a signal-to-noise ratio of about 50 dB is obtained. The TWT noise, which is also present when the pulse is off, is reduced by another 80 dB by switch 3. Any noise from the high power source is now negligible during the detection of the weak signals. Switch 4, with an isolation of 80 dB, protects the mixer crystal from the high power pulse. Switches 3 and 4 also have transition times of 10 nsec.

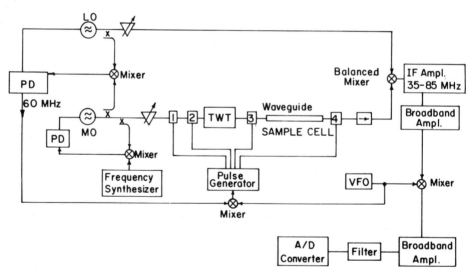

Figure 1. Pulsed microwave Fourier transform spectrometer from 4 to 8 GHz.[5] The numbers 1, 2, 3, and 4 designate the PIN-switches as described in the text. PD stands for phase detector and includes a phase locking system.

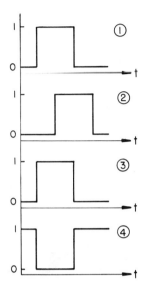

Figure 2. Pulse sequence for the PIN-switches 1 to 4 which produce a 10 nsec high-power microwave pulse and at the same time protect the detector during the pulse-on time.[5] Each of the four pulses is approximately 1 μsec long. The repetition rate for all four pulses is the period T. The triggering always occurs at the same phase of the incoming high frequency signal, but the number of cycles between the pulses is variable and determines T.

The sample cell consists of an empty 4 m C-band waveguide. The local oscillator (LO) is kept 60 MHz above the MO by a phase-lock loop. The detector is a balanced mixer with an IF-amplifier for 60 MHz and a bandwidth of 50 MHz. The signal band from 35 to 85 MHz is subsequently amplified and mixed down to 0-50 MHz with a variable frequency oscillator (VFO) set at 85 MHz. After mixing, the signal is amplified in a second stage, and all frequencies above 50 MHz are cut off by a low-pass filter. As noted above, the emission signal has a fixed phase relation to the MO. To allow averaging, the pulses have to be triggered from the phase-stable 60 MHz difference frequency between the MO and the LO. This 60 MHz must be mixed with the same signal from the VFO (85 MHz) which was used for the actual signal conversion. From this difference signal (25 MHz) the pulse gennerator is triggered. The repetition rate is adjustable to obtain a variable period between the pulses. Figure 3 shows the signal of 5 mTorr of formaldehyde (CH_2O) at 4823.1 MHz after a single pulse. In Figure 3a, the MO is exactly on-resonance, which leads to a beat at the mid-band frequency of 25 MHz. In Figures 3b and 3c, the MO frequency (the carrier frequency ω_p of the pulses) was moved 10 MHz and 20 MHz, respectively, off-resonance. The fact that the amplitude remains constant over the complete frequency band confirms that the assumption made in Eqution 49 is fulfilled. Figure 3 shows that a band of at least 40 MHz (20 MHz on each side of the carrier frequency) can be covered by the pulse train.

After the low-pass filter, the signals that lie in a frequency range of 0-50 MHz are fed into the analog-to-digital converter. Whereas AD-converters with 10 nsec resolution are presently available, commercial instruments have only very slow repetition rates because each recording has to be stored in a magnetic core memory before the gathering of new data. For any reasonable number of data points, this leads to repetition rates well below 500 Hz, or dead times of 20 msec between each recording. This leads to an unacceptable decrease in the signal-to-noise ratio.

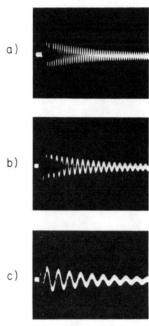

Figure 3. Oscilloscope traces of the emission signal at 5 mTorr of the CH_2O $1_{11} \rightarrow 1_{10}$ transition after a single pulse.[5] In Figure 3a, the carrier frequency from the MO is exactly on resonance (4823.1 MHz) with the absorption line. The beat frequency of 25 MHz is determined by the VFO (see Figure 1). In Figures 3b and 3c, the carrier was moved 10 and 20 MHz away from the resonance. It is evident that the amplitude of the emission remains constant. The beat frequency changes from 25 to 5 MHz from Figure 3a to 3c.

As mentioned previously, these signals are normally buried in the balanced mixer noise; therefore, one-bit accuracy analog-to-digital conversion can be used. One of several versions of one-bit analog-to-digital converters that we have developed will be described here. This is a typical method of near real-time averaging. The AD-converter consists of a one-bit comparator and a 512-bit bipolar random access memory for the storage of 512 points. The sampling rate is 51.2 MHz, which gives a resolution of slightly less than 20 nsec in the time domain. The resulting data collection time is 10 μsec. After the collection cycle, the instrument enters the averaging mode, in which the 512 one-bit words of the buffer memory are added to the previously collected data in 8 parallel chanels. The data are then stored in 16 RAM-memories of the same type. This allows the storage of the 512 points with a resolution of 16 bits. One averaging cycle takes 20 μsec, which leads together with the data collection time to a maximum repetition rate of 30 KHz. The actual repetition rate at which the instrument is run is determined experimentally by the variable repetition rate of the microwave pulses. Each data collection cycle of the analog-to-digital converter is therefore started by trigger pulses from the pulse generator. It is important to mention here that the 51.2 MHz clock for the converter has to be in phase with the incoming signals. Unless this is the case, the signal-to-noise ratio will rapidly decrease for signals with high frequencies. As a clock, we therefore use a simple RC-oscillator which is triggered at the beginning of each data collection cycle. The shortest possible time for the average memory to fill up to 16 bits is approximately 2 sec, after which the data are read into a PDP8/e laboratory computer where further averaging can be done.

After sufficient averaging, the signals in the time domain are transformed to the frequency domain in the computer. The resolution Δf of the obtained spectrum in the frequency domain is given by

$$\Delta f = (n \cdot \Delta t)^{-1} \qquad [53]$$

in which n is the number of points in the time domain and Δt their resolution. With a sample rate of 51.2 MHz and 512 points, we therefore obtain a resolution of 100 KHz per point in the frequency domain. The width, F, of the calculated spectrum in the frequency domain is given by

$$F = (2 \cdot \Delta t)^{-1} \qquad [54]$$

which means that with this sampling rate, the local oscillator frequency must be kept near the center of the 50 MHz band, since the width of the calculated spectrum is only 25 MHz. In a single experiment, one is left with the uncertainty of whether the detected line is above or below the reference frequency. This ambiguity can be resolved by repeating the experiment with a slightly different local oscillator frequency and noting the direction of the shift. In addition, incoming signals in the range of 25 to 50 MHz are folded back into the range from 0 to 25 MHz due to the slow sampling rate. This problem can be relieved by increasing the resolution of the AD-converter to 10 nsec.

According to Equation 52, the phase of the Fourier transformed signal in the frequency domain changes over the bandwidth of the instrument. This is, however, not the only phase problem that occurs. In addition, a frequency-dependent phase shift is introduced because the data collection can normally begin only a certain time (ca. 100 nsec) after the pulse is switched off. This and other phase shifts due to the use of band limiting filters are well-known in NMR and can easily be corrected with a digital computer. The true absorption spectrum, $S_a(\omega)$, is obtained as a linear combination of the cosine and sine Fourier transforms of the signal in the time domain: $S_{cos}(\omega)$ and $S_{sin}(\omega)$, respectively,

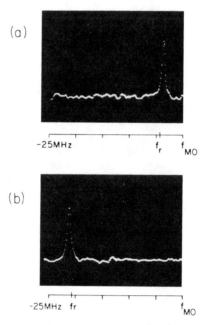

Figure 4. Fourier transform spectra of $C^{(13)}H_2O$ in natural abundance.[5] In both cases, 25 MHz are covered. f_r corresponds to the resonance frequency of the $1_{11} \rightarrow 1_{10}$ rotational transition at 4593.3 MHz. f_{MO} is the carrier frequency of the microwave pulses and is set to 4597 MHz for Figure 4a and 4614 MHz for 4b, respectively. The averaging time is 15 sec.

$$S_a(\omega) = S_{cos}(\omega)\cos(\phi) - S_{sin}(\omega)\sin(\phi) \qquad [55]$$

in which the phase factor ϕ, a linear function of the frequency, is normally fitted after both transforms have been calculated.

Figure 4 shows the frequency domain spectra obtained with our pulsed Fourier transform spectrometer for $C^{(13)}H_2O$ in natural abundance displayed on an oscilloscope. The formaldehyde pressure was approximately 1 mTorr. The spectra cover 25 MHz and each frequency point corresponds to 100 KHz. The displayed line corresponds to the $1_{11} - 1_{10}$ rotational transition of $C^{(13)}H_2O$ at 4593.3 MHz. The carrier frequency, f_{MO}, was kept at 4 MHz off-resonance in the upper spectrum and 21 MHz off-resonance in the lower spectrum. Both spectra were obtained after an averaging time of 15 sec and an optimum exponential filter was used in the digital conversion. The line has an absorption coefficient of 6×10^{-8} cm^{-1}. The obtained signal-to-noise ratio (peak signal amplitude to rms noise amplitude) is approximately 50:1.

For high-resolution studies, Δf has to be decreased according to Equation 53. This can be done by increasing Δt and/or adding zeroes at the end of the recorded spectra in the time domain.

Figure 5 shows the averaged signal in the time domain for the $1_{11} - 1_{10}$ rotational transition in CD_2O at 6083 MHz for very low pressures and T = -77°C. The signal was digitized with a sampling rate of 1 sec per point, and zeroes up to 1024 points were added.

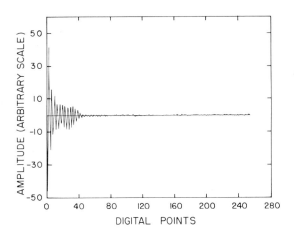

Figure 5. The transient emission signal of 0.2 mTorr of CD_2O averaged over 10^4 pulses converted at a rate of 1 μsec per point.[5] 256 points are shown here.

Figure 6 shows the Fourier transformed and phase-corrected spectrum in the frequency domain. The fast Fourier transform method yields in this case 512 points with a resolution of 1 KHz. The splitting of the lines occurs due to the nuclear quadrupole interaction.

At this point, it seems useful to consider the bandwidth, resolution, and sensitivity limits of Fourier transform spectroscopy in some depth. The commonly quoted expressions for bandwidth and resolution are, respectively,

$$B = 1/t_p$$
$$\Delta T = t_0$$
[56]

Here, it is assumed that the radiation pulse train is the limiting factor in the experiment. The first of Equations 56 merely represents the spectral width that can be polarized by the radiation when $\kappa_j \mathcal{E} t_p \cong 1$, which gives $B \cong \kappa_j \mathcal{E}$. The second of Equations 56 assumes real-time signal averaging, since the data collection time after one pulse, which determines resolution, will end with the onset of the next pulse.

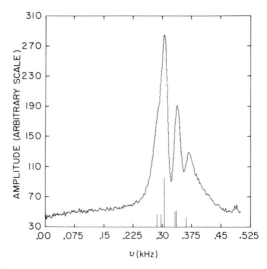

Figure 6. The Fourier transformed spectrum obtained from the emission in Figure 5 after a phase correction according to Equation 55.[5] The markers indicate the frequencies measured with a molecular beam spectrometer.

In practice, however, bandwidth and resolution are usually not given by Equations 56. One is almost always limited to a finite number of channels, n, in which data can be stored and averaged. The usable bandwidth F is limited by the sampling theorem, which requires that a periodic signal be sampled at least twice per cycle, leading to Equation 54, where Δt is the time resolution between points. The total time for which data may be accumulated is then $n \cdot \Delta t$, leading directly to Equation 53 for the attainable frequency resolution. There is thus a trade-off between resolution and bandwidth depending on the time interval, Δt.

There is also a trade-off between resolution and sensitivity. Since the total obtainable signal depends on how often the experiment is repeated, the longer the data collection time per experiment, the lower the sensitivity. This is analogous to the situation encountered in frequency domain spectroscopy with Stark modulation. The higher the Stark modulation frequency, the higher the sensitivity, but the greater the artificial broadening. Of course, increasing the modulation frequency beyond the width of the line does not lead to further improvement in sensitivity, just as repetition of the time domain experiment in a time shorter than the relaxation time does not increase the sensitivity because the molecules have not had a chance to relax from the previous polarization.

The optimum sensitivity of the Fourier transform spectrometer obtained by calculating the signal-to-noise ratio of a pulse Fourier transform spectrometer relative to a conventional absorption spectrometer has also been given.[4] Suppose that a spectral range F has to be investigated in a total time, T_t. Both experiments will use a superheterodyne detection system with a balanced mixer. The noise is assumed to be white with a power density P_0 per spectral unit. The S/N ratio is defined as the ratio of the peak signal amplitude to the rms noise amplitude.

It is intuitively easy to understand that the pulse Fourier transform spectrometer leads to an improvement in the signal-to-noise ratio as compared to the conventional steady-state experiment. In the steady-state experiment, the frequency is swept over a frequency range, and the spectrum is recorded as a function of the frequency. At a certain time, we look at one single frequency and recover only one point of the spectrum under investigation. Since, in many cases, a microwave spectrum is not very dense, most of the time is spent by recording the baseline between the spectral lines.

In contrast, the time domain collects data from all frequencies at all times. Of course, it also collects noise from all frequencies at all times as well, but since signals are coherent while the noise is random, the signal will increase faster than the noise. This intuitive argument is quantified as follows.

The expression for $(S/N)_s$ in the steady-state experiment is available in the literature (Ernst and Anderson, 1966). A single-scan, true slow passage experiment is assumed. It is also assumed that the balanced bridge is perfectly balanced and that $T_1 = T_2$. To reduce the noise bandwidth, a low-pass filter is used. The maximum S/N ratio is obtained with a linear matched filter and under conditions of partial saturation,

$$\kappa^2 \mathcal{E}^2 = 2(1/T_2)^2 \qquad [57]$$

Stark modulation, which must usually be used for baseline stabilization, leads to an additional reduction in the S/N ratio by a factor of $2/\pi$. The optimum S/N ratio for the single-scan experiment is then given by

$$(S/N)_s^2 = \frac{2\omega^2 \ell^2 \beta^2 \hbar^2 \kappa^2 \Delta N_0^2}{3\sqrt{3} \; RP_0 c^2} \left(\frac{T_t}{F}\right)\left(\frac{1}{T_2}\right) \qquad [58]$$

in which R is the resistance across which the signal is measured.

In the Fourier transform experiment, the total signal-to-noise is given as

$$(S/N)_P^2 = \left(\frac{T_t}{T} \cdot s_{max}\right)^2 / \left(RP\frac{T_t}{T}\right) \qquad [59]$$

$P(T_t/T)$ is the total noise accumulated during the time T_t and it is assumed that data is collected continuously in periods of time T. Thus T_t/T is the number of pulse experiments in T_t. s_{max} is the maximum Fourier component of the signal after proper filtering. To obtain the maximum possible S/N ratio, the use of a matched filter is again necessary. The filtering is done in the time domain by multiplying the pulse response by $\exp(-t/T_2)$. This corresponds to a matched filter in the frequency domain.

To obtain the maximum signal amplitude in the frequency domain, the signal in the time domain, Equation 51, must be multiplied by $\exp(-t/T_2)$ and Fourier transformed. Ignoring the Doppler distribution, maximizing the $J_1(\kappa \mathcal{E}_0 t_p)$ function in Equation 52, and selecting $\phi_j = 0$, all of which correspond to maximum signal, and selecting the resonance frequency component, gives

$$s_{max} = \frac{\sqrt{2}}{\sqrt{T}} \frac{T_2}{4} \left(\frac{4\pi\omega\ell\beta}{c}\right)\left(-\frac{\hbar\kappa\Delta N_0}{4}\right) \qquad [60]$$

The noise is also filtered, leading to

$$P = P_0 \frac{T_2}{2T} \qquad [61]$$

In both Equations 60 and 61, it has been assumed that $T \gg T_2$, so that $\exp(-T/T_2)$ can be ignored with respect to one. Substituting Equations 60 and 61 into Equation 59 gives the maximum attainable S/N in the pulse experiment:

$$(S/N)_p^2 = \frac{T_t T_2}{4RP_0 T}\left(\frac{4\pi\omega\beta}{c}\right)^2 \left(-\frac{\hbar\kappa\Delta N_0}{4}\right)^2 \qquad [62]$$

The signal-to-noise gain in the pulse experiment relative to the steady-state experiment can now be calculated.

$$\frac{(S/N)_p}{(S/N)_s} = \left[\frac{F}{2(1/(2\pi T_2))}\left(\frac{T_2}{T}\right)\left(\frac{3\pi(3)^{1/2}}{8}\right)\right]^{1/2} \qquad [63]$$

Equation 63 is valid only at large T. At smaller T, the equation will be modified by exponential factors which cause the last two terms to approach unity. A typical experimental value is $T = 3T_2$, which yields

$$\frac{(S/N)_p}{(S/N)_s} \cong \left(\frac{F}{2(1/(2\pi T_2))}\right)^{1/2} \qquad [64]$$

Equation 64 indicates that the gain in S/N ratio for the pulse method is essentially given by the square root of the ratio of the total sweep width and a characteristic line width in the spectrum. The actual gain in sensitivity in microwave spectroscopy depends on $|\langle a|\mu|b\rangle_j|$, the value of the dipole transition moment, because the degree of polarization obtained from Equation 51 depends on κ_j. Equation 64 assumes that the maximum of this function can be reached, but this is not always the case. With this limitation in mind, the following points may serve to summarize the advantages of the pulse method.[4]

(a) For an average dipole transition moment of 1 Debye, a spectral range of 50 MHz (+25 MHz on each side of the carrier) can be polarized. If we assume a typical line width of 500 KHz full width, then according to Equation 64, a factor of 10 can be gained in S/N. This may be decisive in the detection of transitions with very small population differences, i.e., rare isotopic species or vibrational satellites, etc.

(b) For transitions with very small dipole moments, the spectral range that can be polarized is also smaller. The pulse method, however, may still allow the detection of a weak line which requires a prohibitively high Stark voltage in the steady-state for modulation.

(c) For high-resolution spectra, the pulse method has the advantage that no power broadening occurs. The optimum signal-to-noise ratio for the steady-state experiment, on the other hand, was obtained under conditions which lead to a line broadening of $\sqrt{3}$.

A recent study of the Zeeman effect in vinyl formate[15] provides a comparison of time domain and frequency domain spectroscopy in terms of resolution and sensitivity. Figure 7 shows a time domain spectrum of transient emission signals for the $1_{11} \rightarrow 2_{12}$ transition in vinyl formate in a field of 17,590 Gauss. The interference pattern arises from transient emission signals from four resolvable $\Delta M = +1$ transitions. The spectrum was digitally filtered to improve the signal-to-noise ratio, and then Fourier transformed. The cosine and sine Fourier transforms were squared and added to yield the power spectrum in Figure 8. The same transition was examined using a 5 KHz Stark modulation spectrometer for similar conditions of averaging time and cell temperature. The resulting spectrum in shown in Figure 9. In comparing Figures 8 and 9, it is seen that the resolution and sensitivity is significantly better with the Fourier transform spectrometer.

The resolution of both the Stark modulation and Fourier transform spectrometers is limited by the comparable broadening from magnetic field inhomogeneity and Doppler broadening. However, to obtain an adequate signal-to-noise ratio (S/N) in the steady-state measurements, pressures around 10 mTorr were used. At these pressures, the line width is dominated by pressure broadening. Furthermore, enough microwave power had to be used to obtain good crystal rectification, which introduces additional broadening from partial power saturation. There is no power broadening with the Fourier transform spectrometer. In addition, the S/N ratio was good enough with the Fourier transform spectrometer that pressures around 3 mTorr were used. Therefore, typical line widths obtained with the Stark modulation spectrometer were two to three times larger. The consequence of the narrower line widths and better sensitivity of the Fourier transform spectrometer is evident in the data. The largest

Figure 7. Time domain spectrum of transient emission from the $1_{11} \rightarrow 2_{12}$ transition in vinyl formate for a field of 17590 G and $\Delta M = +1$ selection rules.[15] The data points are taken every 50 nanosec.

Pulsed Fourier Transform Microwave Spectroscopy   233

difference in frequency, $\Delta\nu$, between the calculated splittings and those observed with the Fourier transform spectrometer was 16 KHz. However, several $\Delta\nu$'s obtained with the Stark modulation spectrometer were larger.

There are two reasons for the increased sensitivity with the Fourier transform spectrometer. First, the steady-state spectrometer wiht 5 KHz modulation has significant 1/f noise. To improve the sensitivity, the modulation frequency could have been increased to 100 KHz to reduce the noise to white noise. However, the resolution

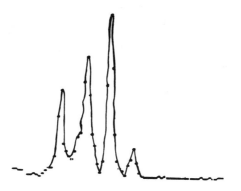

Figure 8. Power spectrum of Figure 7 after the time domain was digitally filtered.[15] The points are separated by 40 KHz. Four Zeeman transitions are easily resolved.

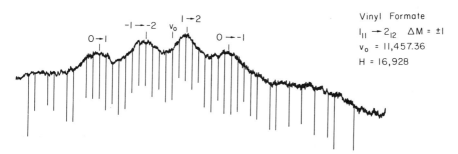

Figure 9. Spectrum of the $1_{11} \to 2_{12}$ transition in vinyl formate for a field of 16928 G and $\Delta M = +1$ selection rules using a 5 KHz Stark modulated spectrometer.[15] The markers under the peaks are every 50 KHz.

would have been significantly reduced by modulation broadening. The noise in the superheterodyned (60 MHz carrier) Fourier transform spectrometer is white noise. Therefore, the instrument has a lower noise figure and does not suffer this dilemma of sensitivity versus resolution. A second reason for the increased sensitivity is the bandwidth capabilities of the Fourier transform spectrometer. Instead of continuously scanning through the spectrum, the Fourier transform spectrometer samples the entire spectrum at once. For a typical scan of 1.2 MHz with vinyl formate, the bandwidth advantage is approximately a factor of 4. However, the available bandwidth of the Fourier transform spectrometer is approximately 10 times the frequency interval scanned in vinyl formate. Therefore, the bandwidth advantage will be even more significant for transitions with much larger Zeeman splittings. An exact value for the improved sensitivity with the Fourier transform spectrometer is difficult to ascertain, since the spectra were not run at identical operating conditions. However, we estimate the improvement in sensitivity of the Fourier transform spectrometer over the Stark modulation spectrometer in the vinyl formate study to be roughly a factor of 5 to 10 times better.

A final point worth noting is that most steady-state spectrometers use Stark modulation. This modulation gives rise to Stark lobes, which are very helpful for assignments and dipole moment measurements. However, Stark lobes complicate the Zeeman spectrum and can lead to distortion of the Zeeman transitions and the baseline. The Fourier transform spectrometer eliminates this complication. The microwave molecular Zeeman effect in trans-crotonaldehyde has also been reported.[16]

A Ku-band microwave spectrometer has also been constructed recently.[17] Careful attention to the details of optimal cell length, coherent noise reduction, and other important properties are apparent in this study. Remarkable resolution and sensitivities are achieved.

STANDING WAVE CASE: FABRY-PEROT CAVITY SAMPLE CELL

In this section, we replace the broadband waveguide absorption cell with a narrow-band Fabry-Perot cavity. The traveling wave is then replaced with a standing wave. We consider a static gas polarization and subsequent coherent emission in the Fabry-Perot cavity.[7,8] However, the use of a Fabry-Perot cavity and the pulsed Fourier transform microwave method is also well-suited for the measurement of the resonant transitions of transient or otherwise short-lived species.

This new development[6] combines the principles of a standing wave pulsed Fourier transform spectrograph in a Fabry-Perot cavity with a pulsed molecular source, where a high-pressure gas expands into a

Pulsed Fourier Transform Microwave Spectroscopy

vacuum through a supersonic nozzle. Furthermore, there are definite advantages in pulsing the nozzle.[6] The general sequence of events is first to pulse the molecular nozzle, allowing a large number of molecules or particles to expand adiabatically into a vacuum, thereby achieving a very low temperature gas. The expanding particles in the vacuum are directed between the mirrors of a microwave Fabry-Perot cavity. As the cold particles traverse the Fabry-Perot cavity, a microwave pulse is applied to the cavity through an input coupling through the cavity mirror. This microwave pulse polarizes a band of frequencies which may contain one of the resonant transitions in the molecules or other particles in the gas. After the polarizing pulse dies away, the polarized gas coherently emits at its resonant frequencies. These signals are then detected in a superheterodyne detector and subsequently Fourier transformed to give the spectrum.

We will describe briefly the principles of the Fabry-Perot cavity, the principles of the pulsed molecular nozzle and subsequent expansion into a vacuum, the simultaneous polarization of the gas by a microwave pulse, and finally the subsequent coherent emission at the appropriate resonance frequencies. Several examples of the operation of this apparatus will be given, followed by a thorough description of the apparatus.

Fabry-Perot cavity

In this section, we will describe the operation of a Fabry-Perot cavity in the microwave range.[9] The field distribution inside a Fabry-Perot cavity as shown in Figure 10 for a TEM_{nmq} mode is

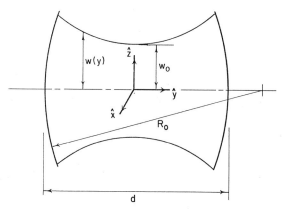

Figure 10. Geometry and coordinate systems used for the Fabry-Perot cavity.[7] The distance at which the electric field falls to 1/e of its on-axis value is $w(y)$, with $w(y=0) = 0$, the minimum value. The radius of curvature of both mirrors is $R_0 = 83.8$ cm.

$$\underline{E}(\underline{r},t) = 2\,\hat{\underline{z}}\,\mathcal{E}(\underline{r})f(t)\cos(\omega t) \qquad [65]$$

where

$$\mathcal{E}(\underline{r}) = \mathcal{E}_0 H_m\left(\sqrt{2}\frac{x}{w}\right) H_n\left(\sqrt{2}\frac{z}{w}\right) \frac{w_0}{w(y)} \exp(-\rho^2/w^2(y))$$

$$\times \cos[ky + (k\rho^2/2R) - \phi - \pi q/2] \qquad [66]$$

in which $\rho^2 = x^2 + y^2$

$$w_0 = \left[\frac{\lambda}{2\pi}(d(2R-d))^{1/2}\right]^{1/2}$$

$$w(y) = w_0\left[1 + \left(\frac{\lambda y}{\pi w_0^2}\right)^2\right]^{1/2}$$

$$\phi = \tan^{-1}(\lambda y/w_0^2)$$

and $\quad k = \omega_c/c$

Referring to Figure 10, $\hat{x}$, $\hat{y}$, and $\hat{z}$ are the unit vectors from the center of the cavity. $H_{m,n}$ are Hermite polynomials of order m,n. The radius of curvature of both mirrors is the same and equal to R. For the fundamental or TEM_{00q} mode, the Hermite polynomials are unity. λ is the free-space radiation wavelength, d is the mirror separation, and q + 1 the number of half-wavelengths between the mirrors. The beam waist, w_0, is the distance from the center of the cavity to the 1/e points of the field strength. The beam waist is a maximum when d = R, the confocal arrangement, and falls to zero when d = 0 or d = 2R. Using the radius of curvature of one of our mirrors, 84 cm, and a mirror separation of 70 cm, the beam waist at 10 GHz is 12.6 cm. The $k\rho^2/2R$ factor accounts for curvature of the wave front arising because of the curved mirrors. The phase front is planar at z = 0. The ϕ term is a phase shift difference between the Gaussian wave and a plane wave. The factor $\pi q/2$ is included to change $\cos(kz)$ into $\sin(kz)$ depending on whether the field is a maximum or zero at the center of the coordinate system.

The resonant frequencies, ν, of the resonator for the TEM_{mnq} mode are

$$\nu = \nu_0[(q + 1) + (1/\pi)(m + n + 1)\cos^{-1}(1 - d/r)] \qquad [67]$$

where $\nu_0 = c/2d$. Using a mirror separation of 70 cm, ν_0 is 214 MHz, so that the dominant modes, TEM_{00q}, are separated by that amount. The higher order modes m or n $\neq$ 0 can be easily seen depending on the coupling iris to the cavity. All modes are narrow and well separated in frequency for our operating conditions.

The Q of a resonant system is a very useful and important property. Q is defined by

$$Q = \omega \cdot W/P \quad [68]$$

in which ω is the angular frequency of the radiation, W is the total energy stored in the cavity, and P is the power dissipation. From the quasi-optical point of view of a resonator, the power in the wave reflecting back and forth between the two mirrors can be dissipated by diffracting out the sides of the mirrors, or it can be dissipated in ohmic losses in the metallic mirrors. The diffraction losses can be made arbitrarily small by increasing the mirror diameter. The power lost due to the finite conductivity of the metal mirrors can be calculated, although it is difficult to account for surface roughness or contamination.

Using Equation 68, four Q's will now be defined appropriate for a transmission Fabry-Perot cavity. The unloaded Q, Q_0, accounts for power dissipation in the cavity. Similarly, the input and output coupling Q's are labelled Q_{c1} and Q_{c2}. The sum of all dissipative elements defines the loaded Q, Q_L

$$1/Q_L = 1/Q_0 + 1/Q_{c1} + 1/Q_{c2} \quad [69]$$

The ratios of the unloaded Q to the two coupling Q's define two coupling coefficients β_1 and β_2

$$\beta_1 = \frac{Q_0}{Q_{c1}} \quad \text{and} \quad \beta_2 = \frac{Q_0}{Q_{c2}} \quad [70]$$

The power flow through the circuit can now be evaluated. For instance, the ratio of the power that is coupled into the cavity from the input generator, P_R, to the power available from the oscillator, P_0, is

$$\frac{P_R}{P_0} = \frac{4\beta_1}{(1+\beta_1+\beta_2)^2 + Q_0^2(\frac{\omega}{\omega_c} - \frac{\omega_c}{\omega})^2} \quad [71]$$

When the cavity is being driven on resonance ($\omega - \omega_c$) this expression reduces to

$$\left(\frac{P_R}{P_0}\right)_{\omega=\omega_c} = \frac{4\beta_1}{(1+\beta_1+\beta_2)^2} = \frac{4Q_L^2}{Q_{c1}Q_0} \quad [72]$$

The amount of power dissipated in the load, P_L, from the generator can be derived in the same way and is

$$\left(\frac{P_L}{P_0}\right)_{\omega=\omega_c} = \frac{4\beta_1\beta_2}{(1+\beta_1+\beta_2)^2} = \frac{4Q_L^2}{Q_{c1}Q_{c2}} \quad [73]$$

For the transmission cavity considered here, some power will always be coupled out through the output coupling iris. If we call P_F the power reflected from the input to the cavity, then

$$P_0 = P_F + P_R + P_L$$

or

$$1 = \frac{P_F}{P_0} + \frac{P_R}{P_0} + \frac{P_L}{P_0}$$

Using the values for P_R/P_0 and P_L/P_0 obtained before for the transmission cavity, we have

$$\frac{P_F}{P_0} = 1 - \frac{4\beta_1}{(1+\beta_1+\beta_2)^2} - \frac{4\beta_1\beta_2}{(1+\beta_1+\beta_2)^2}$$

$$= \frac{(1-\beta_1+\beta_2)^2}{(1+\beta_1+\beta_2)^2}$$

With $\beta_1 = \beta_2 = 1$, one ninth of the input power is reflected and four ninths each is dissipated in the load and in the cavity. Also for this case, $Q_0 = Q_{c1} = Q_{c2}$ and $Q_L = (1/3)Q_0$. In order to have no power reflected from the input coupling iris, $1 - \beta_1 + \beta_2 = 0$.

We will now examine in more detail the frequency dependence of the power transmitted through the cavity into the load. From before, we write the power dissipated in the load, P_L,

$$\frac{P_L}{P_0} = \frac{4\beta_1\beta_2}{(1+\beta_1+\beta_2)^2 + Q_0^2(\frac{\omega}{\omega_c} - \frac{\omega_c}{\omega})^2} \qquad [74]$$

Now, near resonance where $(\omega-\omega_c) \ll \omega$, we can write $(\omega+\omega_c) = 2\omega_c$, $\omega_c\omega = \omega_c^2$, and

$$\frac{\omega}{\omega_c} - \frac{\omega_c}{\omega} = \frac{\omega^2-\omega_c^2}{\omega_c\omega} = \frac{(\omega+\omega_c)(\omega-\omega_c)}{\omega_c\omega} \cong \frac{2(\omega-\omega_c)}{\omega_c}$$

Substituting gives

$$\frac{P_L}{P_0} = \frac{4\beta_1\beta_2(\omega_c/(2Q_0))^2}{(1+\beta_1+\beta_2)^2(\omega_c/(2Q_0))^2 + (\omega-\omega_c)^2} \qquad [75]$$

This is a familiar Lorentzian curve which has a full width, ω_c, at half height, $P_L/2P_0$, given by

$$\Delta\omega_c = 2(1+\beta_1+\beta_2)\left(\frac{\omega_c}{2Q_0}\right) = \frac{\omega_c}{Q_L} \qquad [76]$$

where Equations 69 and 70 are used in the last step.

The decay time constant, τ_c, of the energy W in a cavity can also be obtained from the definition of the Q. The energy stored in the cavity, W, will decay at a rate proportional to W,

$$-\frac{dW}{dt} = \frac{W}{\tau_c}$$

in which $1/\tau_c$ is the proportionality constant. The solution is

$$W = W_0\, e^{-t/\tau_c}$$

in which W_0 is the initial value. The rate of energy loss must equal the power dissipated, or

$$-\frac{dW}{dt} = P = \frac{\omega_c W}{Q_L}$$

so that

$$\frac{W}{\tau_c} = \frac{\omega_c W}{Q_L}$$

or

$$\tau_c = \frac{Q_L}{\omega_c} \qquad [77]$$

In summary, then, with $\omega = 2\pi$, $\Delta\omega = 2\pi\Delta\nu$,

$$Q_L = \frac{\nu_c}{\Delta\nu_c} = \frac{\omega_c}{\Delta\omega_c}$$

$$Q_L = \tau_c \omega_c$$

and $\nu_c = 1/(2\pi\tau_c)$

$$\tau_c = 1/(\Delta\omega_c) = 1/(2\pi\Delta\nu_c)$$

For a cavity with $Q_L = 10^4$ at $\nu_c = 10$ GHz, $\Delta\nu_c = 1$ MHz and $\tau_c = 0.16$ μsec.

We have used a set of mirrors designed for 8 GHz which operate from 4.5 to 18 GHz. Flat mirrors have also been tried, but due to the very critical parallelism adjustment, high Q's were never achieved.

The coupling iris is a round hole centered along the waveguide axis. Part of the back of the mirrors are machined out so that the waveguide can butt up against an area that is between ten to twenty thousandths of an inch from the front mirror surface. The coupling hole is centered in this area and centered in the front mirror surface. The inside dimensions of X-band (8-12 GHz) waveguide are 0.4 x 0.9 inch, and the coupling hole is about 0.38 inch in diameter. For lower frequencies, a coaxial connector device replaces the waveguide section, and antennas are inserted from the inside of the cavity into this coaxial connector. Simple straight wire configurations with the wire coming out of the connector and bending 90° along the mirror surface work well. The length of the wire and its distance from the mirror surface are adjusted to control the coupling, just as the circular diameter size controls the coupling in the waveguide feed mentioned above.

The mirrors are 14 inches in diameter and are suspended on four one-inch diameter stainless steel rods. The rods run the entire length of the vacuum chamber, which is an 18-inch outside diameter tube, 41 inches long. The rods connect two end plates that are bolted onto the main tube. A ten-inch port coming off at the middle of the main tube connects to a ten-inch diffusion pump. Directly above the ten-inch port is a six-inch flange where the pulse valve and associated electrical and gas feedthroughs are located. The mirrors ride on the four rods and are connected to the outside by waveguides that feed through end caps of the 18-inch tube. One mirror is held fixed while the position of the other mirror is adjusted by a rack-and-pinion gear reduction mechanism.

This mirror translation mechanism has to be made with some care in order to step the cavity resonant frequency 500 KHz as explained before. Considering the dominant mode, the resonant frequency was given as

$$\nu = (c/2d)[(q+1) + (1/\pi)\cos^{-1}(1 - d/r)]$$

The difference between two closely spaced frequencies $\nu = \nu_1 - \nu_2$ is given by

$$\Delta\nu = (c/2)(1/d_1 - 1/d_2)(q+1)$$

The mirror distance between modes is $\Delta d = d_2 - d_1$, and if $d_1 d_2 \cong d^2$,

$$\Delta d = \frac{2\Delta \nu d^2}{c(q+1)}$$

in which d is the distance between the mirrors and q+1 is the number of half-wavelengths between the mirrors. As an example, for d = 70 cm, λ = 3.5 cm, so that q+1 = 40, and $\Delta \nu$= 500 KHz, giving Δd = 41 μm. This small distance points to another concern: mechanical vibration can vary the cavity resonant frequency and can thereby amplitude-modulate the output signal. An amplitude modulated signal is demodulated in the mixer and contributes to low-frequency noise in the receiver. The diffusion pump with its boiling silicone oil contributes to noise in this way. However, proper filtering can block most of this noise.

Gas expansion into a supersonic nozzle; pulsed operation

We will now describe the principles of an adiabatic expansion of a gas into a vacuum. Normally, we will use a rare gas carrier, so most of the properties of the gases are dominated by the rare gas atoms. In our instrument, we operate the gas in the pulsed form. However, the pulse valve open-time is very short so that equilibrium expansion properties are very rapidly achieved. We are interested in the number of particles that enter the vacuum, the spatial distribution of these particles, and the temperature of the gas. We must understand the dynamics of the adiabatic expansion in order to simultaneously direct the polarizing microwave pulse into the gases at the optimum time.

A simplified density distribution is used for distances far from the nozzle,

$$\rho(r,\theta) = \rho_n (D^2/r^2) \cos\rho(\theta) \qquad [78]$$

in which D is the nozzle diameter, r and θ are the radial distance and polar angle from the nozzle orifice, as in Figure 10. The density of molecules at the nozzle is given by ρ_n.

The basic isoentropic relations for a perfect gas expanding through a supersonic nozzle are[9]

$$\frac{\rho}{\rho_0} = (1 + \frac{\gamma-1}{2} M^2)^{-1/(\gamma-1)}$$

$$\frac{a}{a_0} = (\frac{T}{T_0})^{1/2} = (1 + \frac{\gamma-1}{2})^{-1/2} \qquad [79]$$

$$v = Ma$$

In these equations, the $_0$ subscript denotes source conditions, M is the Mach number, γ is the ratio of specific heats ($\gamma = 5/3$ for a rare gas), a is the local speed of sound, and v is the flow velocity. The definition of a is

$$a_i = \sqrt{\gamma k T_i / m}$$

in which k is Boltzmann's constant, T_i is the temperature, and m is the particle's mass.

Returning to Equation 79, we can now compute ρ_n by noting that the Mach number is unity at the nozzle plane, giving

$$\rho_n = \rho_0 \left(\frac{\gamma+1}{2}\right)^{-1/(\gamma-1)} \qquad [80]$$

With 1 atmosphere and 300 K as in our normal source conditions, $\rho_0 = 2.69 \times 10^{19}$ molecules/cm^3, so that $\rho_n = 1.7 \times 10^{19}$ molecules/cm^3. Using a nozzle with a diameter of 1 mm, the number density on the nozzle axis and at the cavity center is $\rho(r = 17$ cm, $\theta = 0) = 6 \times 10^{14}$ molecules/cm^3. This number density would correspond to a static pressure of 17 mTorr.

The static gas in a reservoir behind the nozzle has, of course, zero flow velocity, with a Maxwellian distribution of velocities and an average speed, v_a, of

$$v_a = \sqrt{\frac{8}{\pi} \frac{kT}{m}}$$

During the expansion, the random translational kinetic energy and any internally stored energy (rotations and vibrations) is converted into mass flow through binary collisions. The effective rotational and vibrational temperature drops, the velocity distribution narrows and moves out along the flow velocity axis. Since local thermodynamic equilibrium was assumed to exist at all times in the derivation of Equations 79 that describe this process, we can calculate the flow velocity as a function of gas temperature, giving

$$v = \left[\frac{2\gamma}{\gamma-1} \frac{kT_0}{m} \left(1 - \frac{T}{T_0}\right)\right]^{1/2} \qquad [81]$$

If the expansion were to convert all internal energy to directed mass flow, the temperature would be zero, and Equation 81 simplifies to

$$v_t = \left(\frac{2\gamma}{\gamma-1} \frac{kT_0}{m}\right)^{1/2} \qquad [82]$$

the terminal velocity. Note that if the temperatures were zero, the sonic velocity would be zero, and since the flow velocity is still finite, the Mach number is infinity. Also, the terminal velocity is only a factor of 1.4 larger (for a rare gas) than the average velocity in the source. For Kr gas, the terminal velocity is 3.7×10^4 cm/sec.

The number of molecules that are released, N, in a single pulse of the nozzle is

$$N = C \rho_n v_n A_n t_v$$

in which v_n is the flow velocity at the nozzle, A_n is the nozzle area, t_v is the valve pulse time, and C is a discharge coefficient (C = 0.55 for $\gamma = 5/3$). At the nozzle, the Mach number is unity, giving from Equation 79

$$v_n = a_n = a_0 \left(\frac{\gamma+1}{2}\right)^{-1/2}$$

Using this and Equation 80 gives

$$N = \rho_0 a_0 A_n t_v (0.31)$$

For a pulse time of 3 ms, a circular nozzle of diameter 1 mm, 1 atmosphere and 300 K source conditions, there are 4×10^{18} particles released. Since the particles are traveling at approximately $v_t = 3.8 \times 10^4$ cm/s, there is about one tenth of this total inside the beam waist at any one time. Also, because we use a seeded beam, only 2%-5% of this number are dipolar molecules. If one is interested in the molecular dimers, the number of these dimers formed is then again some fraction of the number of molecules.

It is generally accepted that dimer formation in the expanding gas proceeds by a three-body collision, where the third body is needed to carry off excess kinetic energy, allowing the complex to stabilize. The forward rate constant is proportional to the termolecular collision rate and the back reaction is proportional to the bimolecular collision rate. The termolecular collision frequency is proportional to $p^2 D$, where the pressure and diameter are given by pD^q where q is 0.5. Other factors affecting dimer concentration are the source temperature, the geometry of the nozzle, and clearly the physical properties of the expanding particles.

As mentioned above, energy in the internal degrees of freedom is converted into increased mass flow by binary collisions which are proportional to pD. Because energy exchange between translation and rotation is very efficient, rotational temperatures should be equivalent to translational temperatures. An expression giving the terminal translational temperature of the gas can be derived from Equation 79. The terminal Mach number can be given by

$$M_t = \epsilon(\lambda'/D)^{(1-\gamma)/\gamma}$$

where ϵ is a collisional effectiveness constant. For argon, $M = 133(pD)^{0.4}$ giving

$$T = T_0[1 + 5896(pD)^{0.8}]^{-1}$$

where p is in atmospheres and D in cm. Using this formula for conditions of 1 atmosphere, D = 1 mm, and 300 K, the terminal translational temperature is 0.3 K. Measurements of rotational temperatures for various expansions are higher, but still in the 1-4 K range.

The advantage, in signal-to-noise, of using a pulsed nozzle source compared to a continuous source is now considered. There is a definite advantage to pulsing the sample gas into the cavity. If N particles can be pumped through the cavity during a long time t_1, and if $t_1 \gg T_2$, the number of radiation polarizing pulses (pulse time τ_p) that can be used effectively over time t_1 is $p \cong t_1/3T_2$, where we also estimate $\tau_p \ll T_2$. If the gas flows continuously through the cavity, the observed signal-to-noise ratio, S/N, for period of time, t_1, will be proportional to $N/\sqrt{p}$. However, if now all N particles are pulsed into the cavity allowing only a single polarizing pulse during the time period, t_1, the S/N will be proportional to N, assuming identical T_2 for both experiments.

Molecular polarization for near-resonant radiation stimulation and subsequent emission in the Fabry-Perot cavity

We will now solve Equation 23 for P_i, P_r, and ΔN, using functional forms of $\mathcal{E}(r,t)$ appropriate for the pulsed Fourier transform experiment carried out in a Fabry-Perot cavity.[7]

We are interested in applying these equations to both the static and pulsed gases. Consider first the pulsed nozzle experiment. In this case, it is possible to define for each coordinate r a unique velocity v(r), since it is known that all molecules travel at constant speed along straight-line paths originating at the nozzle opening. Multiplying the first of Equations 23 by $d\underline{v}$, and integrating, gives[7]

$$\frac{\partial P_r(\underline{r},t)}{\partial t} + \underline{\nabla} \cdot \int d\underline{v}\ \underline{v}\ P_r(\underline{r},\underline{v},t) + \Delta\omega P_i(\underline{r},t) + \frac{1}{T_2} P_r(\underline{r},t) = 0 \qquad [83]$$

Because each coordinate has a unique velocity associated with it, the second term is

$$\underline{\nabla} \cdot \int d\underline{v}\, \underline{v}\, P_r(\underline{r},\underline{v},t) = \underline{\nabla} \cdot [\underline{v}(\underline{r})\, P_r(\underline{r},t)]$$

Now define

$$P_r(\underline{r},t) = N(\underline{r},t)\, p_r(\underline{r},t)$$
$$P_i(\underline{r},t) = N(\underline{r},t)\, p_i(\underline{r},t) \qquad [84]$$
$$\Delta N(\underline{r},t) = N(\underline{r},t)\, \Delta n(\underline{r},t)$$

in which

$$\Delta N(\underline{r},t) = \sum_\alpha \int d\underline{v}\, \rho_{\alpha\alpha}(\underline{r},\underline{v},t)$$

is the number density of two-level systems at coordinate r at time t. Equation 83 becomes

$$N \frac{\partial p_r}{\partial t} + p_r \frac{\partial N}{\partial t} + p_r \underline{\nabla} \cdot (\underline{v}(\underline{r})N) + \underline{v}(\underline{r}N) \cdot \underline{\nabla} p_r + \Delta\omega N p_i$$
$$+ \frac{N p_r}{T_2} = 0 \qquad [85]$$

We will assume a conservation law for N of the standard form

$$\frac{\partial N(\underline{r},t)}{\partial t} + \underline{\nabla} \cdot [\underline{v}(\underline{r})N(\underline{r},t)] = 0$$

Dividing the remaining terms in Equation 85 by $N(\underline{r},t)$, and repeating this entire procedure for the second and third of Equations 23, one obtains

$$\left(\frac{\partial}{\partial t} + \underline{v}(\underline{r}) \cdot \underline{\nabla}\right) p_r + \Delta\omega p_i + \frac{p_r}{T_2} = 0$$

$$\left(\frac{\partial}{\partial t} + \underline{v}(\underline{r}) \cdot \underline{\nabla}\right) p_i - \Delta\omega p_r + \kappa^2 \mathcal{E}(\underline{r},t) \frac{\hbar \Delta n}{4} + \frac{p_i}{T_2} = 0$$

$$\left(\frac{\partial}{\partial t} + \underline{v}(\underline{r}) \cdot \underline{\nabla}\right) \frac{\hbar \Delta n}{4} - \mathcal{E}(\underline{r},t)\, p_i + \frac{\hbar(\Delta n - \Delta n_0)}{4 T_1} = 0 \qquad [86]$$

in which Δn_0 for a given molecular transition will depend only on the Boltzmann factors, and will be assumed to be a constant.

Now consider the static gas problem. Here $N(\underline{r},t) = N$, a constant. We may cast Equations 23 into a form similar to Equation 86, defining

$$P_r(\underline{r},\underline{v},t) = N \, W(\underline{v}) \, p_r(\underline{r},\underline{v},t) \qquad [87]$$

with analogous expressions for p_i and Δn. The normalized Boltzmann distribution of velocities is given by

$$W(\underline{v}) = \left(\frac{m}{2\pi kT}\right)^{3/2} e^{-m\underline{v}^2/2kT}$$

Using Equation 87, Equations 23 are, for the static gas problem

$$\left(\frac{\partial}{\partial t} + \underline{v}\cdot\underline{\nabla}\right) p_r + \Delta\omega p_i + \frac{p_r}{T_2} = 0$$

$$\left(\frac{\partial}{\partial t} + \underline{v}\cdot\underline{\nabla}\right) p_i - \Delta\omega p_r + \kappa^2 \mathcal{E}(\underline{r},t) \frac{\hbar\Delta n}{4} + \frac{p_i}{T_2} = 0 \qquad [88]$$

$$\left(\frac{\partial}{\partial t} + \underline{v}\cdot\underline{\nabla}\right) \frac{\hbar\Delta n}{4} - \mathcal{E}(\underline{r},t) p_i + \frac{\hbar(\Delta n - \Delta n_0)}{4T_1} = 0$$

This is identical to Equation 86, except $\underline{v}(\underline{r})$ is replaced by the phase-space vector $\underline{v}$, and $p_r(\underline{r},\underline{v}(\underline{r}),t)$ is replaced by $p_r(\underline{r},\underline{v},t)$, with similar exchanges for p_i and Δn. For the solutions to Equations 86 and 88 to be developed here, it is useful to note that for any velocity field satisfying the criterion

$$\underline{v}(\underline{r}) \cdot \underline{\nabla} v_i = 0, \qquad i = x,y,z \qquad [89]$$

solutions to Equations 86 and 88 may be exchanged by making a formal substitution, $p_r(\underline{r},\underline{v},t) \leftrightarrow p_r(\underline{r},\underline{v}(\underline{r}),t)$, etc. Since the pulsed nozzle molecular velocity field satisfies Equation 89, the terms $\underline{v}$ and $\underline{v}(\underline{r})$ in the solutions p_r, p_i, and n to be developed below may be freely interchanged according to whether one is dealing with the static gas or with the pulsed nozzle.

We first note the existence of formal solutions to Equations 86 and 88. One may combine the first two equations in Equation 88, for example, as

$$\left(\frac{\partial}{\partial t} + \underline{v}\cdot\underline{\nabla}\right)(p_r + ip_i) - i\Delta\omega(p_r + ip_i) + i\kappa^2 (\underline{r},t) \frac{\hbar\Delta n}{4}$$

$$+ \frac{(p_r + ip_i)}{T_2} = 0 \qquad [90]$$

The formal solution to this is[7]

$$(p_r + ip_i)(\underline{r},\underline{v},t) = \int_{-\infty}^{t} dt' \, \left(-i\frac{\kappa^2\hbar}{4}\right) e^{(i\Delta\omega - (1/T_2))(t-t')}$$

$$\times \mathcal{E}(\underline{r}-\underline{v}(t-t'),t') \, \Delta n(\underline{r}-\underline{v}(t-t'),\underline{v}(t'), t') \qquad [91]$$

with a similar result

$$(\Delta n(\underline{r},\underline{v},t) - \Delta n_0) = \int_{-\infty}^{t} dt' \left(\frac{4}{\hbar}\right) e^{-(t-t')/T_1} \quad [92]$$
$$\mathcal{E}(\underline{r}-\underline{v}(t-t'),t') \, p_i(\underline{r}-\underline{v}(t-t'),\underline{v},t')$$

for the population difference. Using Equations 91 and 92, the solutions for the polarization and population difference may be developed iteratively to any desired degree of accuracy, a process entailing considerable mathematical complexity. It will be possible to adopt here a simpler approach still capable of explaining the phenomena occurring in the Fabry-Perot cavity experiment.[7]

Consider an experiment in which the molecules in the cavity are brought into contact with a radiation field $2\mathcal{E}(\underline{r},t)\cos(\omega t)$ for a duration τ_p, where $\mathcal{E}(\underline{r},t)$ satisfies the cavity boundary conditions. We consider two sets of solutions to Equation 88. In the case where $\kappa\mathcal{E}\tau_p < \pi/2$ throughout the central part of the cavity, and where $kv\tau_p \ll 1$, where v is a characteristic molecular velocity component along the cavity axis, a typical molecule during the pulse time τ_p travels only a short distance compared to the characteristic distance k^{-1} of changes in the intensity of the standing wave electric field. One may then ignore the $\underline{v}\cdot\underline{\nabla}$ terms in Equation 88 to obtain

$$\frac{\partial p_r}{\partial t} + \Delta\omega p_i + \frac{p_r}{T_2} = 0$$

$$\frac{\partial p_i}{\partial t} - \Delta\omega p_r + \kappa^2 \mathcal{E}(\underline{r},t) \frac{\hbar \Delta n}{4} + \frac{p_i}{T_2} = 0 \quad [93]$$

$$\frac{\partial}{\partial t}\left(\frac{\hbar \Delta n}{4}\right) - \mathcal{E}(\underline{r},t)p_i + \frac{\hbar(\Delta n - \Delta n_0)}{4T_1} = 0$$

These equations are similar in form to Equation 46 in McGurk et al.[1], except that now p_r, p_i, and Δn are functions of $\underline{r}$ as well as t, and $\Delta\omega$ here contains no velocity dependence. When the two-level system is brought instantaneously into contact with a radiation field of constant amplitude, so that $\mathcal{E}(\underline{r},t)$ may be written $\mathcal{E}(\underline{r})$, one has[1] (assuming $T_1=T_2$)

$$p_i(\underline{r},t) = \frac{\kappa^2\hbar\mathcal{E}(\underline{r})\Delta n_0}{4T_2} \cdot \frac{e^{-t/T_2}(\cos\Omega t - \Omega T_2 \sin\Omega t) - 1}{(1/T_2)^2 + \Omega^2}$$

$$p_r(\underline{r},t) = \frac{-\kappa^2\hbar\mathcal{E}(\underline{r})\Delta n_0}{4T_2} \cdot \frac{\Delta\omega}{\Omega} \cdot \frac{e^{-t/T_2}(\Omega T_2\cos\Omega t + \sin\Omega t) - \Omega T_2}{(1/T_2)^2 + \Omega^2}$$

$$\Delta n(\underline{r},t) = \Delta n_0 \ [(\kappa\ \mathcal{E}(\underline{r}))^2\ e^{-t/T_2}\ (\cos\Omega t + (\Omega T_2)^{-1}\sin\Omega t)$$
$$+ (1/T_2)^2 + (\Delta\omega)^2]/[(1/T_2)^2 + \Omega^2] \qquad [94]$$

where

$$\Omega^2 = \kappa^2\ \mathcal{E}^2(\underline{r}) + (\Delta\omega)^2$$

These expressions can be simplified under certain experimental conditions.

The electric field $\mathcal{E}(\underline{r})$ inside the cavity appearing in these expressions can be related to the input power to the cavity. We assume critical coupling so that $Q_0 = Q_{c1} = Q_{c2}$. In the cavity, one has a maximum electric field strength given by[7]

$$E_0 = \left(\frac{128R\ Q_L^2}{\omega\ Q_{c1} d w_0^2}\right)^{1/2} \qquad [95]$$

in which

R = total available power at the input coupling

d = mirror spacing

w_0 = beam waist parameter.

Taking $R = 1$ mW, $Q_L = 1 \cdot 10^4$, $\omega = 2\pi \cdot 12 \cdot 10^9$ s^{-1}, $Q_{c1} = 3 \cdot 10^4$, $d = 47$ cm, and $w_0 = 6$ cm, values typical for our experiment, one obtains $E_0 = 6 \cdot 10^{-3}$ esu, or, picking up the factor of 2 from Equation 13, $\mathcal{E}_0 = 3 \cdot 10^{-3}$ esu, and $\kappa \mathcal{E}_0/2\pi = 1 \cdot 10^6$ Hz for a typical 1 Debye electric dipole moment. The bandwidth that one can work with in the Fabry-Perot experiment is limited by the cavity bandwidth, $\Delta\nu_c$. For a loaded Q of 10^4 at 12 GHz, one has

$$\Delta\nu_c = \frac{\nu}{Q_L} = \frac{12 \cdot 10^9}{10^4} = 1.2 \times 10^6 \text{ Hz} \qquad [96]$$

This fixes a maximum $\Delta\nu$ of 0.6×10^6 Hz. Our experimental work is actually generally limited to frequency offsets $\Delta\nu \leq 0.3 \times 10^6$ Hz, and this will always be the case in experiments discussed in this paper. Over most of the central region of the cavity, then, excluding the nodal surfaces of the standing wave field, one may invoke the condition ($\Delta\omega = 2\pi\Delta\nu$)

$$\frac{(\Delta\omega)^2}{(\kappa\mathcal{E})^2} \ll 1 \qquad [97]$$

Pulsed Fourier Transform Microwave Spectroscopy

It will be shown later that most of the molecular emission is produced in the high-field, antinodal, central region of the cavity, so this approximation is a good one.

In the case where $(\kappa \mathcal{E}_0)^2 \gg (\Delta\omega)^2$, $(\kappa \mathcal{E}_0)^2 \gg T_2^{-2}$, and where the polarizing pulse width τ_p satisfies

$$\tau_p \ll \frac{\kappa \mathcal{E}_0}{(\Delta\omega)^2}, \qquad \tau_p \ll T_2 \qquad [98]$$

conditions easily met in the pulsed nozzle experiment, equation 94 reduces to

$$p_i(\underline{r},t) = -\frac{\kappa \hbar \Delta n_0}{4} \sin(\kappa \mathcal{E}(\underline{r})t)$$

$$p_r(\underline{r},t) = -\frac{\kappa \hbar \Delta n_0}{4} \frac{\Delta\omega}{\kappa \mathcal{E}} [\cos(\kappa \mathcal{E}(\underline{r})t) - 1] \qquad [99]$$

$$\Delta n(\underline{r},t) = \Delta n_0 \cos(\kappa \mathcal{E}(\underline{r})t)$$

If $(\Delta\omega)/\kappa \mathcal{E}_0 \ll 1$, then p_r may be neglected compared to p_i. The resulting expressions for p_r, p_i, and Δn satisfy

$$\frac{\partial p_r}{\partial t} = 0$$

$$\frac{\partial p_i}{\partial t} + \kappa^2 \mathcal{E}(\underline{r}) \frac{\hbar \Delta n}{4} = 0 \qquad [100]$$

$$\frac{\partial}{\partial t}\left(\frac{\hbar \Delta n}{4}\right) - \mathcal{E}(\underline{r}) p_i = 0$$

with the initial conditions

$$p_r(\underline{r},t=0) = p_i(\underline{r},t=0) = 0$$

$$\Delta n(\underline{r},t=0) = \Delta n_0 \qquad [101]$$

Following the same considerations that led to Equation 100, a second set of solutions to the polarization Equations 86 and 88 are obtained by setting $\Delta\omega$ equal to zero, and dropping the terms involving T_1 and T_2. The resulting equations

$$(\frac{\partial}{\partial t} + \underline{v} \cdot \underline{\nabla})p_r = 0$$

$$(\frac{\partial}{\partial t} + \underline{v} \cdot \underline{\nabla})p_i + \kappa^2 \mathcal{E}(\underline{r},t) \frac{\hbar \Delta n}{4} = 0 \qquad [102]$$

$$(\frac{\partial}{\partial t} + \underline{v} \cdot \underline{\nabla}) \frac{\hbar \Delta n}{4} - \mathcal{E}(\underline{r},t)p_i = 0$$

will be valid for $(\kappa \mathcal{E}_0)^2 \gg (\Delta\omega)^2$, $\tau_p \ll (\kappa \mathcal{E}_0)/(\Delta\omega)^2$, and when $(\kappa \mathcal{E}_0)^2 \gg T_2^{-2}$ and $\tau_p \ll T_2$. Alternatively, Equations 102 may be regarded as a microscopic description of on-resonant polarization. The exact solutions to Equations 102, with the initial conditions in Equation 101 are[7]

$$p_r(\underline{r},\underline{v},t) = 0$$

$$p_i(\underline{r},\underline{v},t) = -\frac{\kappa \hbar \Delta n_0}{4} \sin(\int_{t_0}^{t} \mathcal{E}(\underline{r}-\underline{v}(\underline{r})(t-t'),t') \, dt')$$

$$\Delta n(\underline{r},\underline{v},t) = \Delta n_0 \cos(\int_{t_0}^{t} \mathcal{E}(\underline{r}-\underline{v}(\underline{r})(t-t'),t') \, dt') \qquad [103]$$

Here t_0 is defined as the time at which the electric field $\mathcal{E}(\underline{r},t)$ is switched on. We require that the velocity field $\underline{v}(\underline{r})$ satisfy $\underline{v} \cdot \underline{\nabla} v_i = 0$, where $i = x, y, z$. The compatibility of solutions 103 to the general result in Equations 91 and 92 may be verified by direct substitution. Comparison of Equations 103 to Equation 99 shows that the terms $\mathcal{E}(\underline{r})t$ in Equation 99 have been replaced by

$$\int_{t_0}^{t} \mathcal{E}(\underline{r}-\underline{v}(\underline{r})(t-t'),t') \, dt' \qquad [104]$$

When $\mathcal{E}(\underline{r},t) = \mathcal{E}(\underline{r})$, and a molecule travels only a distance short compared to the distance scale of changes in $\mathcal{E}(\underline{r})$, Equation 104 does reduce to the simpler expression. In general, subject to the previously listed conditions, the quantities $p_i(\underline{r},\underline{v},t)$ and $\Delta n(\underline{r},\underline{v},t)$ depend only on the total integrated electric field envelope experienced by the molecules as they move through the spatially complex, time-varying fields of the cavity.

Because it is always possible to work with optimum signal-to-noise in the short pulse limit, and because the additional algebraic complications of any long pulse results, which must be handled numerically, will not provide a significantly more helpful description of the operation of the spectrometer, we omit any attempt to provide a description of the long pulse limit. Accordingly, Equations 103 are simplified to

$$p_r(\underline{r},t) = 0$$

$$p_i(\underline{r},t) = -\frac{\kappa \hbar \Delta n_0}{4} \sin(\kappa \int_{t_0}^{t} \mathcal{E}(\underline{r},t')\, dt')$$ [105]

$$\Delta n(\underline{r},t) = \Delta n_0 \cos(\kappa \int_{t_0}^{t} \mathcal{E}(\underline{r},t')\, dt')$$

strictly valid in the short pulse limit defined by $kv\tau_p \ll 1$, for the case when $\kappa \mathcal{E}_0 \tau_p \leq 1$, or by $\tau_p \ll (kv\kappa\mathcal{E}_0)^{-1/2}$ when $\kappa \mathcal{E}_0 \tau_p > 1$.

After the system has been polarized, the input radiation is switched off and the molecules emit. Let the polarizing radiation at frequency ω be removed at time t_1. At that instant, one has a polarization (from Equation 20) given by

$$P(\underline{r},\underline{v},t_1) = [P_r(\underline{r},\underline{v},t_1) + iP_i(\underline{r},\underline{v},t_1)]\, e^{i\omega t_1} + \text{c.c.}$$ [106]

in which 'c.c.' stands for complex conjugate. The values $P_r(\underline{r},\underline{v},t_1)$ and $P_i(\underline{r},\underline{v},t_1)$ are determined by solving Equations 86 or 88 for $t \leq t_1$, and then using either Equation 84 or Equation 87. At times greater than t_1, the evolution of the quantities p_i, p_r, and Δn is described by Equations 86 or 88 with $\mathcal{E}(\underline{r},t)$ in those expressions set to zero. These are

$$(\frac{\partial}{\partial t} + \underline{v} \cdot \underline{\nabla})p_r + (\Delta\omega)p_i + \frac{p_r}{T_2} = 0$$

$$(\frac{\partial}{\partial t} + \underline{v} \cdot \underline{\nabla})p_i - (\Delta\omega)p_r + \frac{p_i}{T_2} = 0$$ [107]

$$(\frac{\partial}{\partial t} + \underline{v} \cdot \underline{\nabla})\frac{\hbar \Delta n}{4} + \frac{\hbar(n - n_0)}{4T_1} = 0$$

The solutions to these equations with arbitrary initial conditions at time t_1 are

$$p_r(\underline{r},\underline{v},t) = e^{-(t-t_1)/T_2} \{p_r(\underline{r}-\underline{v}(r)(t-t_1),t_1)\cos(\Delta\omega(t-t_1))$$
$$- p_i(\underline{r}-\underline{v}(r)(t-t_1),t_1)\sin(\Delta\omega(t-t_1))\}$$

$$p_i(\underline{r},\underline{v},t) = e^{-(t-t_1)/T_2} \{p_i(\underline{r}-\underline{v}(r)(t-t_1),t_1)\cos(\Delta\omega(t-t_1))$$
$$+ p_r(\underline{r}-\underline{v}(r)(t-t_1),t_1)\sin(\Delta\omega(t-t_1))\}$$ [108]

$$\Delta n(\underline{r},\underline{v},t) = \Delta n_0 + e^{-(t-t_1)/T_1}\{\Delta n(\underline{r}-\underline{v}(r)(t-t_1),t_1) - \Delta n_0\}$$

These expressions may be compared to Equation 132 in McGurk et al.[1]
Again we require that the velocity field satisfy Equation 89.

The wave equation for the electric field in the cavity produced by the polarized molecules is, for Maxwell's equations,

$$\left(\nabla^2 - \frac{1}{c^2}\frac{\partial^2}{\partial t^2} - \frac{4\pi\sigma}{c^2}\frac{\partial}{\partial t}\right)\underline{E} = \frac{4\pi}{c^2}\frac{\partial^2 \underline{P}}{\partial t^2} - 4\pi\underline{\nabla}(\underline{\nabla}\cdot\underline{P}) \quad [109]$$

in which σ is the conductivity of the medium in the cavity, and P is given by Equations 20 and 108, along with Equations 84 or 87, depending on which problem is being considered.

We are now ready to discuss pulsed time-domain spectroscopy carried out in a Fabry-Perot cavity. We will start with the static gas problem. This will allow us to draw out some features of the high-Q, standing wave phenomena common to both the static-gas and pulsed nozzle experiments, while still treating a familiar problem. A high-power microwave pulse of typical duration $\tau_p = 1\,\mu sec$ is applied to a gas sample in the cavity at thermodynamic equilibrium and at a pressure of several millitorr or less. Our discussion is of course limited to the region $\Delta\omega \ll \kappa\mathcal{E}_0$ as discussed earlier, although this restriction does not appear to be important experimentally. To help simplify some later results, we will for convenience also specify $\omega = \omega_c$ and $\Delta\omega \ll \Delta\omega_c$; that is, all polarization and emission processes are to be carried out well within the cavity bandwidth, with the cavity tuned to the carrier. Polarization of the gas is described by Equations 102 and 105. Using these and Equation 109, the field emitted from a static gas in the cavity is obtained.[7] Integrating over a Maxwell distribution of velocities for the static gas, one obtains

$$E(t) \propto e^{-t/T_2} \cos(\omega_0 t + \omega t_1)$$
$$\times \int dv_x dv_y dv_z\, e^{-mv^2/2kT} \int dx\,dy\,dz\, \sin(ky)\, e^{-\rho^2/w_0^2}$$
$$\times \sin[(\kappa\mathcal{E}_0 \tau_p\, e^{(-((x-v_x t)^2 + (z-v_z t)^2)/w_0^2)})\sin(k(y-v_y t))] \quad [110]$$

in which t is measured from the end of the polarization pulse. Molecular emission occurs at the transition frequency ω_0, modified by an envelope with a T_2 exponential decay and a time- and velocity-dependent six-dimensional integral containing all the Doppler dephasing information, as determined from the movement of molecules through the cell. This general arrangement of terms is independent of any approximations made in the expression for the cavity normal mode.

Making a change of variables $k(y-v_y t) \to u$, the integrals over y and v_y become

$$k^{-1} \int_{-\infty}^{\infty} dv_y \, e^{-mv_y^2/2kT} \int_{-\pi(q+1)/2}^{\pi(q+1)/2} \sin(u + kv_y t)$$

$$\times \sin[\kappa \mathcal{E}_0 \tau_p \, e^{-((x-v_x t)^2+(z-v_z t)^2)/w_0^2} \sin(u)] \, du \quad [111]$$

Here we assume that

$$\left(\frac{2kt}{m}\right)^{1/2} t \ll d \quad [112]$$

so that we may ignore edge effects due to the mirrors. Using $\sin(u + kv_y t) = \sin(u)\cos(kv_y t) + \cos(u)\sin(kv_y t)$ and the symmetry properties of two integrals, one obtains

$$4k^{-1} \int_0^{\infty} dv_y \, e^{-mv_y^2/2kT} \cos(kv_y t) \int_0^{\pi(q+1)/2} \sin(u)$$

$$\times \sin(u)\sin[\kappa \mathcal{E}_0 \tau_p \, e^{-((x-v_x t)^2+(z-v_z t)^2)/w_0^2} \sin(u)] du \quad [113]$$

Using the fact that $q+1$ is even, the integral over u is a representation of the ordinary Bessel function of order one, J_1. This last expression becomes

$$2\pi(q+1)k^{-1} \int_0^{\infty} dv_y \, e^{-mv_y^2/2kT} \cos(kv_y t)$$

$$\times J_1[\kappa \mathcal{E}_0 \tau_p \, e^{-((x-v_x t)^2+(z-v_z t)^2/w_0^2}] \quad [114]$$

$\pi(q+1)k^{-1}$ is just the cavity length, d. The integral over v_y is standard (McGurk et al.[1] Equation 145), giving

$$\left(\frac{\pi^{1/2}}{2ks}\right) e^{-t^2/4s^2} \quad [115]$$

in which $s = (\ln(2))^{1/2}(\Delta\omega_D)^{-1}$, and

$$\Delta\omega_D = \frac{\omega_0}{c} \left(\frac{2kT\ell\ln(2)}{m}\right)^{1/2} \quad [116]$$

the Doppler half-width. An entirely conventional Doppler-broadened envelope is obtained. It follows from Equation 115 that condition 112 may be written

$$2\left(\frac{2kT}{m}\right)^{1/2} \left(\frac{c}{\omega}\right)\left(\frac{m}{2kT}\right)^{1/2} \ll d ,\qquad [117]$$

or $\frac{\lambda}{\pi} \ll d$,

easily true for all cavity modes except possibly a fundamental TEM_{mno} mode. Implicit in Equation 115 is the necessity of having polarized the entire Doppler envelope of the transition of interest, and this can indeed be shown to follow from the conditions imposed in deriving that result. Choosing $v_{RMS} = (3kT/m)^{1/2}$ for a Maxwell-Boltzmann gas, one obtains from Equation 116

$$\Delta\omega_D \cong \frac{\omega_0}{c} v_{RMS} = k v_{RMS} \qquad [118]$$

Now if $\Delta\omega_D < \Delta\omega$, so that the carrier falls outside the Doppler envelope, then condition 96 guarantees that the entire line profile lies within $\kappa \mathcal{E}_0$. However, if $\Delta\omega_D > \Delta\omega$, so that the carrier lies within the envelope, and

$$k v_{RMS} \tau_p = \Delta\omega_D \tau_p \ll 1 \qquad [119]$$

or, $\Delta\omega_D \ll 1/\tau_p$,

which, when combined with $\Delta\omega < \Delta\omega_D$, ensures that the Fourier components of the carrier cover the entire line profile, even if $\kappa \mathcal{E}_0 < \Delta\omega_D$.

Using Equations 114 and 115 in Equation 110 gives

$$E(\underline{r},t) = 16\pi Q_L \frac{\kappa \hbar \Delta N_0}{4} \sin(ky)\, e^{-\dot{\rho}^2/w_0^2}\, e^{-t/T_2}\, e^{-t^2/4s^2}$$
$$\times \cos(\omega_0 t + \omega t_1)\, \left[\left(\frac{m}{2\pi kT}\right) \iint dv_x dv_z\, e^{-m(v_x^2 + v_z^2)/2kT} \left(\frac{2}{\pi w_0^2}\right)\right.$$
$$\left.\times \iint dx dz\, e^{-(x^2+z^2)/w_0^2}\, J_1(\kappa \mathcal{E}_0 \tau_p)\, e^{-((x-v_x t)^2 + (z-v_z t)^2)/w_0^2}\right],$$
$$[120]$$

still measuring from the end of the polarization pulse. If Q_L is written as

$$Q_L = \frac{\omega d}{\alpha c} \qquad [121]$$

where α is the fraction of energy lost by a wave in one transit of the cavity, then the effective length of the cavity may be taken to be d_α^{-1}, typically $10^2 d$ for a Q_L of 10^4 at 10 GHz. The four-dimensional integral in Equation 92 contains two kinds of information. First, there is the slowly varying time-dependent envelope resulting from the overlap of the exponential x^2, z^2, and $(x-v_x t)^2$ and $(z-v_z t)^2$ terms. Physically, this describes the gain or loss in signal as molecules move through the cavity beam waist. Second, one finds that the concept of a $\pi/2$ pulse, generally obtained by adjusting $\kappa \mathcal{E}_0 \tau_p$ to the first maximum of J_1, is not directly applicable to the cavity because of significant transverse variations in the electric field amplitude. Computer calculations carried out using more exact forms of the cavity normal modes show no changes in the line shapes from those predicted by this analytically derived result.

Expression 120 is sufficient for analysis of line shapes seen at the detector.

In Figure 11, we show the time domain signal from the OCS J = 0 → 1 transition at 12163 MHz. Data was taken at a gas pressure of less than 1 mTorr, with a polarization time τ_p = 2.5 μsec, and digitized at the rate of 0.5 μsec per point. Digital points have been connected by straight lines. A switch was closed at time t_s on the Figure, 0.35 μsec after the polarization pulse had ended. The carrier offset $\Delta \nu$ = 156.5 KHz. A low-Q mode, $Q_L \leq 5000$ was used in

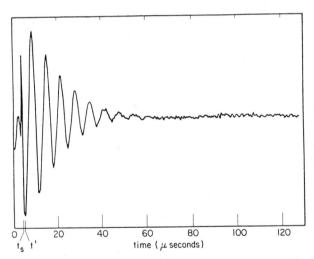

Figure 11. Transient emission signal obtained from the OCS J = 0 → 1 transition at 12163 MHz taken in the TEM$_{0037}$ cavity mode with a static gas sample.[7] A switch blocking the detector was opened at time t_s. t' is the first point at which the signal envelope can be measured. The signal was digitized at the rate of 0.5 μs/point and the points connected by straight lines.

this measurement to minimize ringing from the high-power pulse. Care was taken to obtain an emission envelope free from amplifier or filter distortion.

From Equation 120, the envelope should be a combination of Gaussian and exponential decays. If

$$\left(-\ln \frac{I(t-t')}{I(t')}\right)^{1/2} \qquad [122]$$

is plotted as a function of $(t-t')$, in which I is the envelope amplitude, and t' occurs a short time after t_s, one obtains a curve with a straight-line asymptote having slope $(2s)^{-1}$, and intercept sT_2^{-1} at $t = t'$, where s is defined by Equation 116. This is shown in Figure 12, choosing t' to be the first available maximum in the signal. We obtain $s = 13.12$ μsec, compared to an expected result of 13.7 μsec for OCS at 293 K, and a T_2 of 42 μsec, corresponding to a pressure of about 1/2 mTorr. Although Equation 120 has been derived subject to various restrictions on pulse length and carrier offset, we have never observed any significant deviations from this result over a wide range of variations of these parameters.

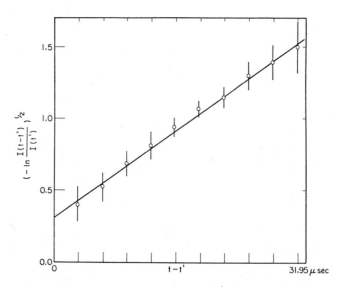

Figure 12. Plot of $(-\ln[I(t-t')/I(t')])^{1/2}$ versus $t - t'$ for the signal in Figure 11.[7] Here $I(t)$ is the signal envelope amplitude. The solid line is $(t-t')/26.24$ μsec $+ 0.320$, and may be used to determine the Doppler and T_2 de-phasing times.

Pulsed Fourier Transform Microwave Spectroscopy   257

In summary of the static gas experiment (Equation 120), the emission signal is centered at the molecular transition frequency ν_0, with an envelope combining an exponential T_2 de-phasing and a Gaussian Doppler de-phasing, identical to the well-known traveling wave results. The power reaching the detector in the critically coupled static gas-cavity experiment as a function of the molecular and experimental parameters has also been considered in detail elsewhere[8], where the signal in cavity experiment is found to exceed the corresponding waveguide experiment result in these two types of time-domain spectrometers.

Any discussion of the ratio of signal-to-noise in the two experiments must include several other considerations. The cavity is obviously a narrow-banded instrument, and although this makes searches for new resonances more difficult, it also reduces the noise power included in its normal working bandwidth. The bandwidth of the waveguide experiment, however, is generally made as large as possible by using a high-power TWT amplifier and wideband signal processing components, reaching values as large as 50 MHz for molecules with transition moments on the order of 1 Debye. In the Doppler broadened limit, from Figure 11, signal relaxation times will limit both experiments to pulse repetition rates of perhaps 20 KHz or less. Wall relaxation, definitely not a problem for the cavity experiment, may limit decay times in the waveguide cell to much less than that shown in Figure 11, particularly if a small cell cross section is used to increase the microwave power density and thus the bandwidth. In a carefully designed 20-foot waveguide system, the power-pulse ring-time can be made as short as 200 nsec, which, when combined with power pulse lengths of 10-30 nsec, make it possible to work well within the pressure-broadened regime without much loss in signal. The larger number of molecules approximately compensate for the shorter duration of the emission signal. In this case, repetition rates of 300 KHz are routinely possible. This approximately even trade-off between the number density and the signal duration changes if one is attempting to work with weakly bound dimer systems in a static gas. Here the signal duration is still inversely proportional to the first power of the gas pressure, but the number of species of interest, which we assume to depend on the product of the partial pressure of the two monomer components, can be made to be proportional to the square of the total pressure.

Of course, the tremendous increases in sensitivity and resolution of the Fabry-Perot instrument come when the large volume of the cavity is combined with the pulsed nozzle molecular source, and it is this combination that is taken up now.

In order to examine the radiation field from molecules in a Fabry-Perot cavity following a gas pulse into the cavity, we return to Equations 105. To obtain an appropriate form for $\mathcal{E}(r,t')$, we note that the frequency spacing between successive non-degenerate modes in the Fabry-Perot cavity is much greater than the width of any single mode. The electric field $\mathcal{E}(\underline{r},t')$ is separable as $\mathcal{E}(\underline{r})f(t')$, where, for the TEM$_{00q}$ mode (see Equation 66)

$$\mathcal{E}(\underline{r}) = \mathcal{E}_0 \frac{w_0}{w(y)} e^{-\rho^2/w^2(y)} \cos(ky - \pi q/2) \quad [123]$$

in which all terms are defined in Equation 66. The normal mode in Equation 123 has been simplified from the more exact expression (Equation 66) by neglecting small terms in the cosine argument. To be consistent, we will also use a simplified equation

$$d = \frac{c}{2\nu}(q+1)$$

to relate the frequency and mode number to the mirror separation.

After the polarizing radiation has been removed, p_r, p_i, and n continue to evolve as described by Equations 23 without the $\mathcal{E}(\underline{r},t)$ terms. The polarized gas now produces an electric field in the cavity. Using exact solutions to Equations 23 without the $\mathcal{E}(\underline{r},t)$ terms, and using Maxwell's equations to couple the polarization to the emitted electric field, one obtains for the emitted electric field in the cavity ($\omega_c = \omega_0$):[8]

$$E(\underline{r},t) = 8\pi Q_L (w_0/w) e^{-\rho^2/w^2} \cos(ky - \pi q/2) e^{-t/T_2}$$

$$\times \cos(\omega_0 t + (\omega - \omega_0)t_1) \frac{4}{\pi w_0^2 d} \int_V d^3r' \; p_i(\underline{r}' - \underline{v}(\underline{r}')t, t_1)$$

$$\times \frac{w_0}{w(y')} N(\underline{r}',t) e^{-\rho'^2/w^2(y')} \cos(ky' - \pi q/2) \quad [124]$$

Time is measured from the end of the polarization pulse at t_1. The quantity $p_i(\underline{r}' - \underline{v}(\underline{r}')t, t_1)$ is to obtained from Equations 102 and 123.

As discussed in detail before, Equation 124 has been derived under the assumption that the velocity field $\underline{v}(\underline{r})$ satisfies Equation 89, which may be verified directly for the pulsed nozzle velocity field. Physically, we require that all molecules travel at constant speeds along straight-line paths.

The normal mode form $(w_0/w)e^{-\rho^2/w^2} \cos(ky - \pi q/2)$ outside the integral in Equation 124 simply describes amplitude variations in the emitted electric field signal characteristic of the TEM_{00q} mode. The essential line shape information is contained in the time-dependent $e^{-t/T_2} \cos(\omega_0 t + (\omega-\omega_0)t_1)$ terms, and in the spatial integral, which contains all information related to the molecular motion, including Doppler de-phasing and de-phasing due to the movement of the molecules through the cavity.

To complete the derivation of the functional form of the time-domain line shapes seen in the pulsed nozzle experiment, we consider the characteristics of the pulsed nozzle gas expansion. We assume that all expanding particles travel at constant speed v_0 on radial paths originating at the nozzle. Using the geometry of Figure 10, the velocity field $\underline{v}(\underline{r})$ may be written as

$$\underline{v}(\underline{r}) = v_0 \frac{x\hat{\underline{x}} + y\hat{\underline{y}} - (h-z)\hat{\underline{z}}}{(x^2 + y^2 + (h-z)^2)^{1/2}} \qquad [125]$$

which satisfies Equation 89.

We will assume a molecular density distribution $N(\underline{r},t)$ of the form given in Equation 78

$$N(\underline{r},t) \propto \frac{\cos^p(\theta)}{r^2} = \frac{(h-z)^p}{(x^2+y^2+(h-z)^2)^{(p/2)+1}} \qquad [126]$$

in which r and θ are the radius and polar angle from the nozzle as in Figure 10. In the expanding gas, $T_2 \geq 10\ \mu sec$, justifying our neglect of relaxation processes during the polarization. Using the previous equations and dropping the normal mode form and numerical constants from in front of the integral in Equation 124, one obtains for the emitted electric field line shape

$$E(t) = e^{-t/T_2} \cos(\omega_0 t + (\omega-\omega_0)t_1) \int_V d^3r\ \sin(ky)$$
$$\times \sin[\kappa\ \mathcal{E}_0 \tau_p(\frac{w_0}{w(y-v_y t)})\ \exp(\frac{-(x-v_x t)^2-(z-v_z t)^2}{w^2(y-v_y t)})$$
$$\times \sin[k(y-v_y t)]]\ \frac{w_0}{w(y)}\ \exp(\frac{-x^2-z^2}{w^2(y)})\ \frac{\cos^p(\theta)}{r^2} \qquad [127]$$

in which v_x, v_y, and v_z are functions of x, y, and z. We have chosen $q+1$ to be even, and have dropped the primes inside the integral. Note that the integrand contains two Gaussian waist terms, one inside the sine expression and time-dependent, and one outside the sine and time-independent. Physically, these describe the loss of signal as the molecules move out of the beam waist.

It is convenient to separate Equation 127 into two parts. The $\cos(\omega_0 t+(\omega-\omega_0)t_1)\exp(-t/T_2)$ terms center the emission signal at ω_0 and provide a routine exponential damping. The interesting line shape information is contained in the remaining integral expression denoted by $I(t)$. $I(t)$ contains two types of signal damping. In the special case of a narrow beam traveling through the cavity perpendicular to the axis, there is a fall-off in signal due to the transverse motion of the molecules out of the cavity, mathematically expressed as the overlap of the $(x-v_x t)^2$, $(z-v_z t)^2$, and x^2, z^2 terms in the exponentials. A second type of damping, Doppler de-phasing, results from the movement of molecules from the region where they were polarized with one phase, to regions where they would have been polarized with a different phase, mathematically appearing in the $k(y-v_y t)$ argument. Numerical studies show that in the pulsed beam,[8] the damping due to transit of the molecules out of the cavity is negligible in comparison to the Doppler de-phasing. Because inclusion of this transit-time damping in our analysis would provide at best only a marginal improvement in our description of the gas expansion, we will not include this effect in our analysis. Accordingly, we will use here the line shape expression

$$I(t) = \int_{cavity} d^3r \, \sin(ky) \, \sin[\kappa \ell_0 \tau_p (\frac{w_0}{w(y)}) \exp(-\frac{x^2+z^2}{w^2(y)})$$

$$\times \sin[k(y-v_y t)]] (\frac{w_0}{w(y)}) \exp(-\frac{x^2+z^2}{w^2(y)}) \frac{\cos^p(\theta)}{r^2} \quad [128]$$

In Figure 13, we show the triple integral (Equation 128) calculated for the TEM_{0031} cavity mode as a function of t, for $p = 0.5$, corresponding to a $\cos^{0.5}(\theta)$ density distribution, a frequency $\nu_0 = 10$ GHz, and a molecular speed $v_0 = 4 \cdot 10^4$ cm/sec.[8] No relaxation processes except Doppler de-phasing are included in this calculation. The time, L_0, from the maximum of the curve at $t = 0$ to the first zero crossing z_1, and the distances L_i, $i = 1,2,\cdots$, between successive zero crossings are, in microseconds, 41.8, 47.0, 46.6, 46.4, and 46.5, respectively. These spacings conform to a general pattern, independent of any of the parameters in Equation 128, in which the spacings L_i following the first zero are nearly constant, but possibly quite different from the first spacing L_0. The ratio L_0/L_1 is an important parameter, independent of ν and v_0, depending only on the coefficient p. It varies monotonically from 0.85 for an isotropic distribution with $p = 0$, to 1.07 for $p = 2$, to arbitrarily large values as p increases. As p increases, the first zero crossing moves away from the origin, and the heights of the successive maxima in the curve become smaller in relation to the height of the curve at $t = 0$. A calculation is shown in Figure 14.[8] In the limit of large p, then, the envelope will show no oscillations, as is required for a tightly collimated beam traveling perpendicular to the cavity axis.

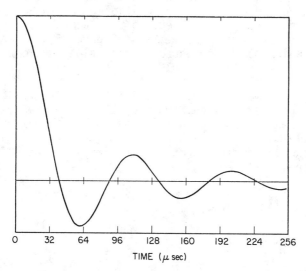

Figure 13. The envelope function in Equation 128, calculated for a density distribution $(\cos^{0.5}(\theta))/r^2$, $\nu_0 = 10$ GHz, molecular speed $v_0 = 4 \cdot 10^4$ cm/sec, and a mirror spacing of 48 cm. No T_2 is included in this calculation.[8]

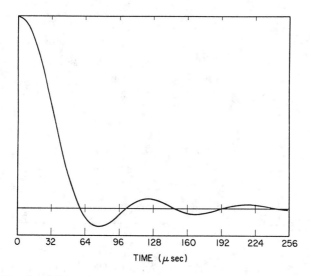

Figure 14. The envelope function in Equation 128, calculated for a density distribution $(\cos^{4.0}(\theta))/r^2$ and otherwise the same parameters as in Figure 13.[8]

In Figure 15, we show the power spectrum of the curve in Figure 13. The envelope in Figure 13 was multiplied by an arbitrary $\cos(\omega_0 t)$ before it was transformed, to displace the power spectrum along the frequency axis. The peaks in Figure 15 are separated by 2 x 10.23 KHz, which may be compared to the value $2\nu v_0/c$ of 2 x 13.33 KHz for ν = 10 GHz, and v = 4 x 10^4 cm/sec.

The splitting may be understood intuitively as follows. Consider the integral expression, Equation 128, when one has only a single narrow beam traveling on a radial path in the xy-plane, away from the nozzle at angle θ with respect to the z-axis. Ignoring edge effects and the beam waist terms in Equation 128, one obtains a line shape of (dropping the T_2 term)

$$E(t,\theta) \propto \cos(\omega_0 t + (\omega-\omega_0)t_1) \cos(\omega_0 \frac{v_0 \sin(\theta)}{c} t)$$
$$\times J_1(\kappa \mathcal{E}_0 \tau_p) \qquad [129]$$

The total emitted field is obtained by adding contributions from all θ, weighted by the angular density distribution. This gives

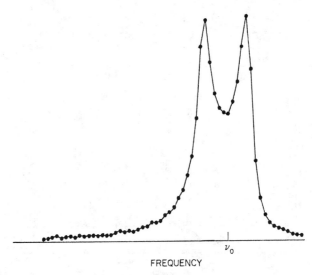

Figure 15. Power spectrum obtained by multiplying the envelope in Figure 13 by an arbitrary term, $\cos(\omega_0 t)$, and Fourier transforming. The peak separation is 20.46 KHz.[8]

$$E(t) \propto dJ_1(\kappa \mathcal{E}_0 \tau_p) \cos(\omega_0 t + (\omega-\omega_0)t_1) \int_0^{\pi/2} \cos(\omega_0 \frac{v_0 \sin(\theta)}{c} t)$$

$$\times \cos^p(\theta) \, d\theta = d(\frac{\pi}{2}) J_1(\kappa \mathcal{E}_0 \tau_p) \cos(\omega_0 t + (\omega-\omega_0)t_1)$$

$$\times \frac{1 \cdot 3 \cdot 5 \cdots (p-1)}{(\omega_0 v_0 t/c)^{p/2}} J_{p/2}(\omega_0 \frac{v_0}{c} t) \qquad [130]$$

Note the characteristic functional form, $[J_{p/2}(x)/x^{p/2}]$, which contains all of the information relating to the spatial distribution of molecules in the expansion to the observed line shape. We also note that the curve in Figure 14 bears a close resemblance to $\sin(x)/x$, the Fourier transform of a square absorption curve in frequency space. By considering the number of molecules having a particular velocity component $v_\parallel$ parallel to the cavity axis as a function of $v_\parallel$, for $p = 1$ for example, the connection between the time and frequency domains becomes straightforward.

The experimental system and results

A block diagram of the spectrometer is shown in Figure 16. The master oscillator (MO) is frequency-stabilized (FS) to two lower frequency standards. A one-GHz signal (1 GHz) is produced by multiplying a 20 MHz oven-controlled crystal oscillator which has a frequency stability of 1×10^9 parts per day. A variable frequency oscillator (VFO) is also used for continuous frequency adjustment

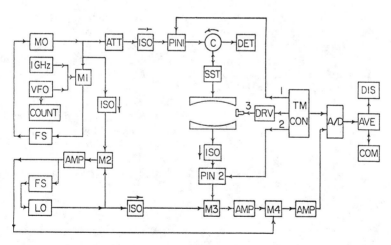

Figure 16. Block diagram of the apparatus.[9] See text for description.

within its range. We use a 10 MHz to 480 MHz (in 5 bands) tube oscillator that is stable to hundreds of Hz over minutes. This VFO is counted (COUNT) by a standard seven digit frequency counter. The mixer (M1) mixes the first harmonic of the VFO and the n'th harmonic of the 1 GHz with the master oscillator, giving an IF frequency of 30 MHz that is fed into the frequency stabilizer. This frequency stabilizer compares the IF frequency to its own 30 MHz crystal and gives an error voltage whose magnitude is proportional to the phase difference and whose polarity depends on whether the IF frequency is below or above 30 MHz. If the VFO's frequency is changed so that the mixed-down frequency is no longer 30 MHz, the frequency stabilizer will apply a voltage to the MO, changing its output frequency so as to cause the mixed-down signal to be 30 MHz once again. Now we can control not only the MO's frequency, but by counting the VFO's frequency and by knowing which harmonic of the 1 GHz we are using, we can calculate the MO's frequency to an accuracy of hundreds of Hz.

The local oscillator (LO) is locked by another frequency stabilizer to the master oscillator through another mixer (M2). Part of the 30 MHz signal that is fed into the second frequency stabilizer is split off to be used as a phase-coherent reference signal. Both the master and local oscillator are backward wave oscillators (BWO), and when locked, they have the frequency stability of the VFO. BWO's have good bandwidth, a relatively level power output, and can be swept easily. This last feature is convenient when working with cavities, as the cavity resonant shape and position are often adjusted. Cavity coupling information can also be obtained by sweeping across the cavity resonance.

Since both the master and local oscillator are operating continuously, the pulse to the cavity is formed by opening PIN diode 1, (PIN 1) for a time t_3 (see Figure 17 for sequence of pulses). This pulse of microwave energy (typically t_3 = 6 sec) passes through the circulator (C), slide screw tuner (SST), and impinges on the input coupling iris of the Fabry-Perot cavity. Any energy that is reflected from the cavity travels backward in the waveguide and is routed to a detector (DET) by the circulator. The output of this diode rectifier is fed into an oscilloscope which is triggered in synchronism with the PIN diode pulses. If the frequency of the MO is far off from the cavity resonant frequency, all the power incident upon the cavity is reflected and the output of the detector traces out the microwave pulse envelope. The PIN diodes we use have rise times (10-90%) of 10 nsec, and if one uses a fast detector such as a tunnel diode that can follow this rise time, the pulse envelope can be observed by this method. When the MO is at the cavity resonant frequency, the pulse envelope shape is different. Since the cavity energy builds up and decays with a time constant τ = 160 nsec, as given before, the pulse envelope will reflect this response.

Pulsed Fourier Transform Microwave Spectroscopy 265

Microwave energy during the pulse is also coupled out the opposite mirror. To protect our detector mixer (M3), we have another PIN diode (PIN 2) that blocks this energy. The TTL trigger pulses that operate these two PIN diodes are shown in Figure 17. The diodes conduct when the TTL level is +5 volts and have an 80 db isolation when the level is zero. As seen in Figure 17, PIN diode 2 reflects any power coupled out of the cavity when PIN diode 1 is conducting. The time t_3 is adjustable, and the time t_4 is equal to t_3 plus another adjustable amount to make sure that all of the original power is dissipated. The timing control (TM CON) forms these pulses and the pulse to the driver (DRV) that operates the molecular pulse valve.

The pulse length to the molecular valve, out of the driver, is about $t_1 = 3$ msec. This valve is a commercial in-line solenoid valve. t_2 in Figure 17 is the time between opening the mechanical valve to allow molecules to enter the field regions of the cavity and the initiation of the microwave pulse. This delay, t_2, has to be adjusted for each different gas mixture, to optimize the signal amplitude.

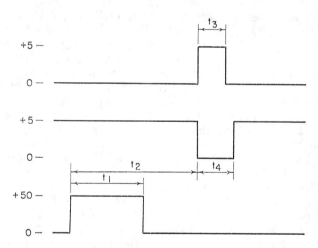

Figure 17. A time diagram of the various pulses to the spectrograph.[9] The lower pulse, which is on for a period of time t_1, is applied to the solenoid that controls the nozzle opening to introduce the sample into the Fabry-Perot cavity (see Figure 16). After an appropriate delay, the upper pulse, which is on for a period of time t_3, is applied to PIN 1 which opens the microwave pulse to the Fabry-Perot cavity. The middle pulse, which is on for a time t_4, is applied to PIN 2 in order to protect the balanced detector mixer (M3) from the high power microwave pulse.

The pulse to the molecular valve is repeated at a rate from one to ten Hz, depending on the carrier gas, back pressure, and nozzle size. Typically, 4×10^{18} particles are released per gas pulse. We use a ten-inch diffusion pump that can pump around 4×10^{19} particles per second, hence the 10 Hz repetition rate.

The pulse sequence to the PIN diodes occurs at twice the solenoid valve rate. One microwave pulse interacts with the molecules in the free expansion and the next microwave pulse interacts with the evacuated cavity.

If we call the frequency that the master oscillator is locked at, ν, then the local oscillator frequency is $\nu - 30$ MHz. The pulse of microwave energy creates a macroscopic polarization in the expanding molecules, which then emit radiation at the rotational transition frequency, ν_m. Both of the frequencies, ν and ν_m, have to be within the bandwidth of the cavity, so that $|\nu_m - \nu| = \Delta \ll 1$ MHz. Usually Δ is between 10 KHz and 600 KHz. After the power pulse dies away, PIN diode switch 2 opens and the molecular emission mixes with the local oscillator in another mixer (M3). The signal out of this mixer (M3) is at a frequency 30 MHz $+ \Delta$, depending on whether $\nu > \nu_m$ or $\nu < \nu_m$. This is amplified ($\overline{AMP}$) and mixed down again in another mixer (M4). The other input to this mixer (M4) is the 30 MHz signal we obtained before by mixing the master and local oscillators. This method assures identical phases in repeated emissions from the molecules in the cavity. The signal out of this mixer (M4) is at the frequency Δ, or the offset between the master oscillator and the molecular transition frequency. This signal is filtered and amplified and then fed into an analog-to-digital converter (A/D). The A/D is triggered to digitize a signal every time the microwaves are pulsed. The digitizer we currently use is a 6-bit converter with a dynamic range per pulse of only 1 in 64. This will shortly be replaced by a 10-bit A/D that gives a 1 in 1024 dynamic range.

After the signal has been digitized, the data is transmitted to an averager (AVE). This averager adds and subtracts alternate pulses so that all molecular emissions are added together and all alternate background scans are subtracted. The digitizer we currently use has 256 points, and by using 0.5 μsec per point, the entire data scan takes 128 μsec. Since the molecular pulse valve provides observable molecules in the cell for times longer than this, we also have the option of taking n adds and n subtracts in one molecular valve period. After the averager-memory fills, the data is transmitted to a VAX 11/780 computer (COM) which does a fast Fourier transform and returns this data to the averager. Either the time-domain or Fourier-transformed data can be displayed (DIS) from the averager. Because of the say we mix down the molecular emission signal, the start of the frequency domain data is at the master oscillator frequency. By counting the points over to a molecular peak, we find Δ and therefore we know the transition frequency ν_m.

Pulsed Fourier Transform Microwave Spectroscopy

Searching for unknown molecular lines consists of stepping the cavity mirror separation, hence its resonant frequency, and following along with the microwave oscillators. Since the cavity bandwidth is around 1 MHz, the step size is 500 KHz or less.

Figure 18 shows the time-domain record for the $J = 0 \rightarrow 1$ transition of $^{16}O^{12}C^{32}S$, which is known to consist of a single line at 12163 MHz. The spectrum was recorded by pulsing a 4% mixture of OCS in Ar through the nozzle. We use a thin plate flat orifice bolted to the bottom of the pulsed valve; we do not use a skimmer. The signal was digitzed at the rate of 0.5 μsec/point, and the points connected by straight lines. The corresponding power spectrum is shown in Figure 19, and has a frequency resolution of 3.9 KHz/point. The most prominent characteristic of this spectrum is the symmetric splitting apparent in both figures. As described above, this splitting is a general feature of all spectra taken with the pulsed nozzle in the Fabry-Perot cavity. We have found that the position of the center of this pattern is invariant under all changes in the spectrometer operating conditions, and corresponds to the molecular resonance frequency. The line shape itself provides several different kinds of information about the gas expansion. It is apparent, for instance, that the envelope of the experimental curve in Figure 18 falls somewhere between the theoretical envelopes given in Figures 13 and 14.

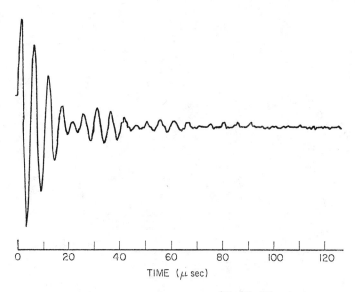

Figure 18. Time-domain signal from the $^{16}O^{12}C^{32}S$ $J = 0 \rightarrow 1$ transition, converted at 0.5 μsec/point.[8] The points have been connected by straight lines. The gas mixture is 4% OCS in Ar.

The gas dynamics of a pulsed supersonic nozzle molecular source have been investigated in detail by using a pulsed Fabry-Perot cavity microwave spectrometer.[8] The Doppler splitting phenomenon is discussed in detail and experimental line shapes are deconvoluted to give molecular velocities, de-phasing times, and density distributions. It was found that the density distribution of active molecules from the pulsed nozzle varies rapidly in time, starting with a depletion on the nozzle axis at short times after the nozzle is opened, and changing to an on-axis concentration at longer times.

Several examples of the use of this spectrograph in the observation of weakly bound van der Waals molecules are in the literature. Some of these results include assignment of the spectra and the determination of the spectroscopic constants and molecular structure of KrHCl.[6] The rotational spectra and molecular structures of ArHBr and KrHBr have also been given.[18] The ^{83}Kr nuclear quadrupole coupling has been measured in KrHF,[19] and the H-F spin-spin coupling, D nuclear quadrupole coupling, and the molecular structure of this molecule have been measured by observing the rotational spectra in KrHF and KrDF.[20] We have also examined the ^{131}Xe nuclear quadrupole coupling and molecular structure in XeHCl.[21] Also observed were the rotational spectra of the weakly bound dimer of carbon monoxide and the hydrogen halides HX (X = F, Cl, and Br).[22]

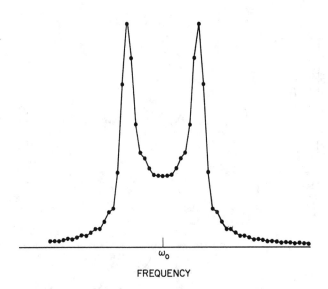

Figure 19. The power spectrum obtained by adding 256 zeroes to the data in Figure 18, and Fourier transforming.[8] The resolution is 3.9 KHz/point, with a splitting of 36.2 KHz.

Spectroscopy on heavy molecules that have low vapor pressure could be done by vaporizing them in a high-temperature oven that could easily be placed in the vacuum chamber. Another possibility that is being pursued is to use a high-temperature nozzle source to form rare-gas:metal-atom dimers.

By crossing the nozzle expansion with some excitation source such as a laser, electron beam or plasma, one might be able to see rotational transitions in excited states. Other types of nozzle sources might allow one to observe combustion or explosion products. Molecular radicals or ions would also be interesting species to study in the large cell by using the pulse techiques described here.

REFERENCES

1. McGurk, J. C., Schmalz, T. G., and Flygare, W. H. 1974, Adv. Chem. Phys. 25, 1.
2. McGurk, J. C., Mäder, H., Hofmann, R. T., Schmalz, T. G., and Flygare, W. H. 1974, J. Chem. Phys. 61, 3759.
3. Flygare, W. H., and Schmalz, T. G. 1976, Accts. Chem. Res. 9, 385.
4. Schmalz, T. G., and Flygare, W. H. 1978, in Laser and Coherence Spectroscopy, ed. J. I. Steinfeld, Plenum Press, N.Y., p. 125.
5. Ekkers, J., and Flygare, W. H. 1976, Rev. Sci. Instrum. 47, 448.
6. Balle, T. J., Campbell, E. J., Keenan, M. R., and Flygare, W. H. 1980, J. Chem. Phys. 71, 2723; 72, 922.
7. Campbell, E. J., Buxton, L. W., Balle, T. J., and Flygare, W. H. 1980, J. Chem. Phys. 74, 813.
8. Campbell, E. J., Buxton, L. W., Balle, T. J., Keenan, M. R., and Flygare, W. H. 1980, J. Chem. Phys. 74, 829.
9. Balle, T. J., and Flygare, W. H. 1981, Rev. Sci. Instrum. 52, 33.
10. Bloch, F. 1946, Phys. Rev. 70, 460.
11. Feynman, R. P., Hellwarth, R. W., and Vernon, F. L., Jr. 1957, J. Appl. Phys. 28, 49.
12. Redfield, A. G. 1965, Adv. Magn. Reson. 1, 1.
13. Liu, W. K., and Marcus, R. A. 1975, J. Chem. Phys, 63, 272.
14. Abragam, A. 1961, The Principles of Nuclear Magnetism, Clarendon Press, Oxford.
15. Voss, H. L., Krajnovich, D., Hoke, W. E., and Flygare, W. H. 1978, J. Chem. Phys. 68, 1439.
16. Hoke, W. E., Voss, H. L., Campbell, E. J., and Flygare, W. H. 1978, Chem. Phys. Lett. 58, 441.
17. Bestimann, G., Dreizler, H., Mäder, H., and Andresen, U. 1980, Z. Naturforsch. 35a, 392.
18. Keenan, M. R., Campbell, E. J., Balle, T. J., Buxton, L. W., Minton, T. K., Soper, P. D., and Flygare, W. H. 1980, J. Chem. Phys. 72, 3070.

19. Campbell, E. J., Keenan, M. R., Buxton, L. W., Balle, T. J., Soper, P. D., Legon, A. C., and Flygare, W. H. 1980, Chem. Phys. Lett. $\underline{70}$, 420.
20. Buxton, L. W., Campbell, E. J., Keenan, M. R., Balle, T. J., and Flygare, W. H. 1980, Chem. Phys. $\underline{54}$, 173.
21. Keenan, M. R., Buxton, L. W., Campbell, E. J., Balle, T. J., and Flygare, W. H. 1980, J. Chem. Phys. $\underline{73}$, 3523.
22. Legon, A. C., Soper, P. D., Keenan, M. R., Minton, T. K., Balle, T. J., and Flygare, W. H. 1980, J. Chem. Phys. $\underline{73}$, 583.

TWO-DIMENSIONAL FOURIER TRANSFORM NMR SPECTROSCOPY

Gareth A. Morris

Physical Chemistry Laboratory
South Parks Road
Oxford OX1 3QZ
ENGLAND

INTRODUCTION

Most of the chemical problems presented to an NMR spectroscopist can be solved by straightforward measurement and analysis of a conventional spectrum. A small but important proportion of the problems require a more flexible approach, either because the normal spectrum provides too little information, or because it contains so many signals that individual lines cannot be resolved. Many of these more awkward problems can be tackled with the aid of two-dimensional (or in popular parlance '2D') NMR.

In a normal Fourier transform NMR experiment, the free induction decay $S(t)$ generated by a single radiofrequency pulse is transformed to give a spectrum $S(f)$. In a double Fourier transform experiment, this principle is extended to the acquisition of a set of free induction decays $S(t_2)$ obtained using a sequence of pulses incorporating a variable delay t_1. A series of separate measurements is made with increasing values of t_1, until a complete matrix of NMR signals $S(t_1,t_2)$ is built up. Double Fourier transformation of this matrix, once with respect to t_1 and once with respect to t_2, yields a spectrum $S(f_1,f_2)$ which displays signal strength as a function of two independent frequencies. Just as the single Fourier transformation in a normal FT NMR experiment unravels the free induction decay to reveal the precession frequencies of the nuclear magnetizations, so in a 2D experiment the double transformation separates out signals according to their precession frequency f_2 during t_2 and their modulation frequency f_1 as a function of t_1.

There is a simple and direct analogy between such 2D NMR experiments and the well-known techniques of two-dimensional paper and gel chromatography. These improve the chromatographic separation of mixtures by using two different solvents to disperse different components in two orthogonal directions on a paper or gel surface, exploiting the different changes in R_f values on changing solvents. The final resting place of one constituent of a mixture is then characterized by two parameters R_{f1} and R_{f2}, analogous to the frequencies f_1 and f_2 in a 2D NMR spectrum.

Since the choice of pulse sequence for acquiring the matrix of free induction decays governs the distribution of signals in the resultant 2D spectrum, many different types of 2D NMR experiment are possible, with signal distribution in f_1 and f_2 reflecting a variety of different NMR parameters. Several previous reviews[1,2] have offered a simple introduction to 2D NMR methods; the aim of this chapter is to provide an overview of work in this field up to July, 1980, and to describe in a little more detail some of the technical features peculiar to these experiments. In the limited space available, experimental aspects will be emphasized at the expense of a full discussion of the spin physics involved; the basic mechanisms of most of the experiments to be discussed have been treated fairly extensively in the literature.[1-3]

In the next section, a brief summary will be made of the various 2D NMR techniques that have been developed in recent years. Before proceeding to deal with the bewildering variety of pulse sequences and data manipulation methods available, it is instructive to consider a simple example of a double Fourier transform NMR experiment. One of the earliest techniques proposed was that of Müller, Kumar and Ernst,[4] which allows multiplet structure to be separated from chemical shifts in proton-coupled carbon-13 NMR. The pulse sequence for this experiment, illustrated in Figure 1, consists simply of a 90° carbon-13 pulse followed by a delay t_1, after which wideband proton decoupling is turned on and a proton-decoupled free induction decay recorded. Since no decoupling is used during t_1, the resonance positions in f_1 in the resultant 2D spectrum are those expected for the proton-coupled conventional spectrum, and the positions in f_2 are the chemical shift frequencies found in the normal decoupled spectrum.

Figure 2 shows a Müller-Kumar-Ernst 2D spectrum for 1,3-butanediol, illustrating the separation of overlapping multiplet structure which the 2D experiment makes possible. Cross-sections through the 2D spectrum at appropriate f_2 values will give a proton-coupled carbon-13 multiplet for each distinct carbon site. Although more effective methods for untangling overlapping spectra will be described shortly, the utility of such 2D NMR techniques, which take the signals of the conventional spectrum and disperse them in two frequency dimensions, was made clear very early.[4]

The experiment of Figure 1 relies on the use of double resonance to cause a change in resonant frequencies between t_1 and t_2, one

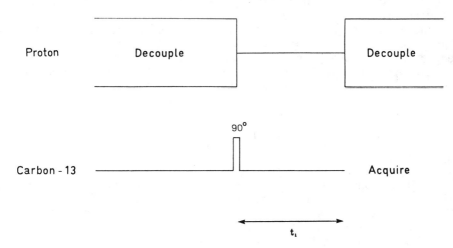

Figure 1. Pulse sequence for the experiment of Müller, Kumar and Ernst for the separation of overlapping proton-coupled carbon-13 multiplets.[4]

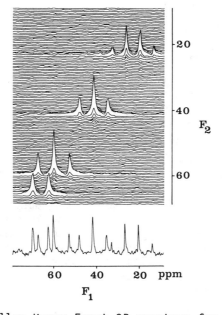

Figure 2. Top, Müller-Kumar-Ernst 2D spectrum for 1,3-butanediol; bottom, the normal proton-coupled carbon-13 spectrum for this system. Both spectra were recorded on a CFT-20 spectrometer operating at 20 MHz for carbon-13.

example of the sort of nonlinear behavior which sets NMR apart from most other forms of spectroscopy. All 2D NMR experiments rely on exciting a nonlinear response from the nuclear spins, usually by exploiting a mixing of spin states by 90° or 180° pulses or the changes in resonance frequencies and intensities brought about by double resonance. In the next section, some of the variations that have been proposed on these basic themes will be described.

PULSE SEQUENCES

The core of a two-dimensional NMR experiment is the sequence of radiofrequency pulses used to obtain the matrix of time-domain signals $S(t_1,t_2)$, since it is this which governs the signal distribution in the final spectrum. In the preceding section, one of the earliest high-resolution 2D NMR experiments was used to exemplify the basic features of two-dimensional methods. The details of the experimental methods and data manipulations used are deferred to the next section; this section sets out to make a brief summary of the many different types of 2D spectrum that may be produced, and the many and various pulse sequences used to generate them.

No attempt is made here to review the rapidly expanding literature on NMR imaging, or zeugmatography, although many imaging experiments make use of double and even triple Fourier transforms. Attention is concentrated here on high-resolution NMR methods; most of the solid-state applications of 2D NMR were made early in the development of the field and are dealt with in reference 1, which reviews the literature up to July, 1978, although more recently there has been a revival of interest.[5-7] Progress since the summer of 1978 has been in three main directions: the development of techniques for studying multiple quantum coherences, the refinement and extension of the the early experiments for enhancing resolving power and for correlating spectra, and the application of 2D NMR methods to the resolution and assignment of particularly difficult spectra.

Most high-resolution 2D NMR methods can be divided into one of three classes. The first, involving the separation of multiplet structure from chemical shifts, consists mainly of methods relying on the generation of spin echoes to suppress chemical shift effects during t_1. These methods may be thought of as descendants of the one-dimensional J spectroscopy[8] or spin-echo spectroscopy[9] experiments. The second class of techniques involves the correlation of spectra, usually autocorrelation in homonuclear experiments and cross-correlation in heteronuclear. These experiments may be used to map out coupling patterns in proton spectra, or to make indirect measurements of the signals of other nuclei in heteronuclear experiments. The third and most diffuse class is that of experiments designed to measure NMR parameters not directly obtainable from a simple conventional spectrum, particularly those associated with multiple quantum coherences.

All of the experiments to be described share the same basic features. A pulse sequence containing a variable delay t_1 is used to build up a matrix of free induction decays $S(t_1,t_2)$, which describes the NMR signal as a function of the delay t_1 during excitation of the spin system and of the time t_2 after excitation. This time-domain matrix is then weighted and Fourier transformed, once with respect to t_1 and once with respect to t_2. The f_2 signal dispersion, resulting from free precession, is the same as that in a normal spectrum; the f_1 signal dispersion depends on the behavior of the nuclear magnetizations during t_1. The purpose of the pulse sequence is to cause each of the resonances observed during t_2 to have its phase and intensity modulated as a function of t_1 by some useful frequency, often a chemical shift or coupling constant. In discussing the various pulse sequences, then, the central task is to analyze the t_1 dependence of the signals which they bring about.

An attempt has been made in Table 1 to summarize the publications to date in the main 2D NMR techniques. The rapid development of such methods has led to a considerable proliferation of nomenclature, as the first column of Table 1 bears witness. Where one paper describes several distinct experiments it is referred to separately under each head, but only publications reporting significant advances in technique or new applications are included. In subsequent paragraphs, the techniques mentioned in Table 1 are discussed in a little more detail, the classification introduced being further subdivided into homonuclear (usually proton) and heteronuclear (usually carbon-13 and irradiate proton) experiments.

Homonuclear J Spectroscopy

The relatively small range of proton chemical shifts, and the extensive proton-proton scalar coupling found in most systems, place a considerable premium on the ability to distinguish multiplet splittings from chemical shift differences. One of the most effective ways of distinguishing chemical shifts from scalar couplings is to examine the modulation of spin echoes; indeed, the existence of scalar couplings was first inferred from such experiments.[10,11] The mechanism of formation of spin echoes, and their modulation by J coupling, has been dealt with extensively elsewhere.[8-13]

Two-dimensional homonuclear J spectroscopy uses a simple Carr-Purcell type A pulse sequence[14] to generate a spin echo with its maximum at time t_1:

$$90° - t_1/2 - 180° - t_1/2 - \text{acquire } S(t_2) \qquad [1]$$

The free induction decay acquired is thus the second half of a spin echo, the amplitude of which is modulated as a function of t_1 by scalar couplings. In weakly coupled systems, then, the effect of the 180° pulse is to suppress the effects of chemical shifts (and indeed

Table 1. Two-dimensional NMR techniques.

Name	Nucleus Observed	Heteronucleus Pulsed	f_1 Parameters	f_2 Parameters	Techniques	Applications
Homonuclear 2D J Spectroscopy	H C		J_{HH} J_{CC}	δ_H, J_{HH} δ_C, J_{CC}	14	1,2,14-19,23-34 21,22
Heteronuclear 2D J Spectroscopy; 2D Spin Echo Spectroscopy	C	H	J_{CH}	δ_C (J_{CH})	35,41	1,2,20,35-48
2D Correlated NMR Homonuclear Shift Correlation; Jeener Two-pulse Expt.	H		δ_H, J_{HH}	δ_H, J_{HH}	3,50,52	3,49-52
Chemical Shift Correlation Heteronuclear 2D Correlated NMR	C	H	δ_H, J_{HH} (J_{CH})	δ_C (J_{CH})	58-63	1,2,30,58-65
	H	C	δ_C, J_{CH}	δ_H, J_{CH}, J_{HH}	57	57
	31P	H	δ_H, J_{HH}, J_{PH}	δ_P, J_{PP}, J_{PH}	53,54	53,54
	31P	H	J_{PH}, J_{HH}	δ_P, J_{PP}, J_{PH}	68	68
	H	15N	δ_N (J_{NH})	δ_H, J_{HH}, J_{NH}	55	55
	11B	H	δ_H, J_{BH}, J_{HH}	δ_H, J_{HH}, J_{NH}	56	56

Müller-Kumar-Ernst Expt.	C	H	δ_C, J_{CH}	δ_C	4	4
	C	H	J_{CH}	δ_C	69	69
2D T_1, T_2 Expt.	C		$1/T_1, 1/T_2$	δ_C	70	70
Homonuclear Multiple Quantum 2D NMR	H		MQT's	δ_H, J_{HH}	71-73,75-79	71-73,75-79
Heteronuclear Multiple Quantum 2D NMR	H	C	MQT's	δ_H, J_{HH}	74	74
	C	H	MQT's	δ_C	74	74
2D Magnetization Exchange	H	H	δ_H, J_{HH}	δ_H, J_{HH}	80,81	80,81

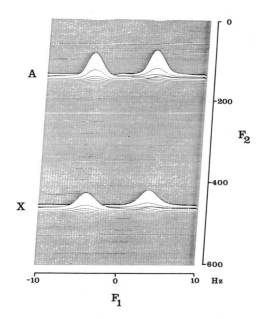

Figure 3. Proton 2D J spectrum for the AX subsystem of thiouracil, signal distribution in f_2 showing the usual two doublets, and in f_1 showing separated doublets for the A and the X protons.

of static field inhomogeneity) during t_1, although in strongly coupled systems some modulations dependent on shifts reappear. Thus, a homonuclear AX spin system, with chemical shifts δ_A and δ_X and coupling constant J, will give rise to a 2D J spectrum with signals at frequency coordinates $(-J/2, \delta_A-J/2)$, $(J/2, \delta_A+J/2)$, $(-J/2, \delta_X-J/2)$ and $(J/2, \delta_X+J/2)$. This effects a partial separation of multiplet structure from chemical shifts, as can be seen from the experimental spectrum of Figure 3. The further data manipulations which can complete this separation will be discussed in the next section.

Homonuclear 2D J spectroscopy was first proposed in a succinct communication by Aue, Karhan and Ernst,[14] followed by two further notes[15,16] on the application of the technique to the proton spectra of biomolecules. Several theoretical analyses of the effects of strong proton-proton coupling on 2D J spectra have been produced,[17-19] reference 19 including some experimental data for simple spin systems. Such spectra can be analyzed iteratively if necessary, in order to extract accurate values for coupling constants.[19,20] Homonuclear 2D J spectroscopy can also be used for the study of carbon-13 - carbon-13 scalar couplings,[21,22] although dynamic range and sensitivity considerations make this difficult in unenriched compounds. Variations on the basic experiment have included the introduction of homonuclear double resonance[23,24] and solvent nulling by the WEFT method.[25] Recent work has included analyses of proton carbohydrate, nucleotide and steroid spectra[26-32] and of small molecules in liquid crystals;[33] previously unpublished spectra have also appeared in several reviews.[1,2,34]

Heteronuclear J Spectroscopy

The principle of two-dimensional J spectroscopy may easily be extended to the study of heteronuclear scalar couplings, notably those between protons and carbon-13. A simple spin-echo pulse sequence will however not normally generate echoes modulated by heteronuclear couplings; since heteronuclei are not affected by the 180° pulse (unless there is strong coupling), the heteronuclear splittings are refocussed in the same way as chemical shifts. In order to reintroduce echo modulation, it is sufficient either to ensure that all coupled nuclei experience a 180° pulse at the midpoint of t_1, or to apply wideband decoupling during one half of t_1 only. Thus, for the measurement of carbon-13 - proton 2D J spectra, the two pulse sequences would be:

$$90°_C - t_1/2 - 180°_C, 180°_H - t_1/2 - \text{acquire } ^{13}C \ S(t_2) \qquad [2]$$

and

$$90°_C - t_1/2 \ (\text{no dec.}) - 180°_C - t_1/2 \ (^1H \text{ dec.}) - \text{acq. } S(t_2) \qquad [3]$$

These two sequences are known as the "proton flip" method and the "gated decoupler" method[35] respectively.

Once again, only scalar couplings affect the t_1 modulation of signals in weakly coupled spin systems, so that only multiplet structure appears in f_1. Since decoupling suppresses the heteronuclear scalar couplings for one half of t_1 in sequence 3, the "gated decoupler" method gives rise to spectra in which the multiplet splittings in f_1 appear to be halved. Although this degrades the limiting resolving power obtainable for a given line width, this type of pulse sequence does have one advantage over those of type [2], as will emerge shortly.

A further degree of freedom is available in heteronuclear (as opposed to homonuclear) 2D J spectroscopy, since it is possible to suppress heteronuclear couplings during the acquisition of the free induction decay $S(t_2)$ with the aid of wideband proton decoupling. If this feature is added to pulse sequences [2] and [3], the resulting spectra will have pure heteronuclear coupling structure in f_1, and pure carbon-13 chemical shifts in f_2. Since the main aim of 2D J spectroscopy is to separate multiplet structure from chemical shifts, proton decoupling is almost always used during acquisition of free induction decays.

Heteronuclear 2D J spectroscopy was first demonstrated on the proton-coupled carbon-13 quartet of iodomethane;[36] spectra of this type appeared in a bewildering variety of guises in the subsequent months, as heteronuclear J spectroscopy proved an ideal vehicle for the development of experimental methods in two-dimensional NMR.[37,38] The principal motivation for early work on J spectroscopy was the suppression of static field inhomogeneity effects in f_1, in order to allow very high resolution studies of proton-coupling fine structure in carbon-13 spectra.

Measurement of heteronuclear 2D J spectra for pyridine,[20,35,39] which has strong coupling among its protons, illustrated the important difference between the pulse sequences 2 and 3. Where strong coupling is present, the "gated decoupler" method always gives rise to a 2D spectrum in which the f_1 multiplets are isomorphous with those obtainable from conventional spectra, which can lack symmetry about the chemical shift frequency, as in the C-2 multiplet of pyridine. The "proton flip" pulse sequence 2 on the other hand always gives rise to multiplets that are symmetrical about $f_1 = 0$, and the detailed analysis of such spectra is complicated by the need to take into account the mixing of spin states caused by the 180° pulses. Indeed, the theoretical analysis of the "proton flip" experiment is exactly analogous to that of homonuclear J spectroscopy.[18-20]

The choice between "gated decoupler" and "proton flip" pulse sequences for a given application depends on the relative importance attached to resolution and to ease of analysis, since the former method gives conventional multiplet patterns, while the latter gives slightly distorted multiplets but can yield slightly higher resolution. Since the difference in effective obtainable resolution is usually rather less than the factor of two that might be expected, and the "proton flip" method can suffer from artifacts due to imperfect proton pulses, most spectra are recorded using the "gated decoupler" technique. Several other sequences may be used for the investigation of other types of echo modulation, and are included in a survey of heteronuclear 2D J spectroscopy methods by Ernst and coworkers.[40,41] The effects of off-resonance proton decoupling on carbon-13 echo modulation have been investigated by Müller[42] and by Henrichs et al.[43] Apart from its use for measuring high-resolution multiplet structure, 2D J spectroscopy may also be used for unravelling overlapping proton-coupled carbon-13 spectra, as in an early study of the proton-coupled spectrum of cholesterol,[44] and in a more recent measurement of one-bond coupling constants in the spectra of oligosaccharides.[45] A number of experimental techniques of general utility in two-dimensional NMR were developed using 2D J spectroscopy, including the use of phase cycling for artifact suppression,[38] and the production of phase-sensitive 2D spectra.[46-48]

Homonuclear Correlation

One of the first steps in a conventional assault on a complex assignment problem in proton NMR is to run a series of homonuclear double resonance experiments. Not only can this establish which multiplets have a mutual J coupling, but it can reveal hidden resonances obscured by overlap, through their couplings to less crowded parts of the spectrum. Since a separate experiment is needed for each multiplet, this approach can be quite time-consuming. Similar information can however be obtained more efficiently by 2D NMR methods; indeed, this was the purpose of the first 2D NMR experiment, proposed by Jeener in 1971.

Jeener's unpublished suggestion, which lay dormant for several years until taken up and enlarged upon by Ernst and coworkers,[3,49] was to perform a double Fourier transformation on data collected using the following pulse sequence:

$$90°_H - t_1 - 90°_H - \text{acquire } S(t_2) \qquad [4]$$

This pulse sequence, perhaps simultaneously the simplest of the early 2D NMR sequences and yet the most daunting to analyze, exploits the mixing of spin states by the second 90° pulse to transfer transverse magnetization between multiplets.

Three classes of response appear in 2D spectra obtained using the sequence 4: axial peaks, diagonal peaks, and cross peaks.[3] The first class, attributable to longitudinal magnetization remaining after the t_1 period, appear with normal f_2 frequencies along the $f_1 = 0$ axis, and are comparatively uninteresting. The diagonal signals lie on or near the diagonal, $f_1 = f_2$, and are due to transverse magnetizations which remain associated with the same multiplet before and after the second 90° or "mixing" pulse. Again these are of secondary importance, although the fine structure of these responses may carry useful information. The principal interest lies in the cross peaks, which are due to transverse magnetizations transferred via scalar couplings from one multiplet to another by the mixing action of the second 90° pulse. The existence of a cross peak correlating two multiplets thus indicates the presence of a mutual scalar coupling. Since the amount of magnetization transferred depends (among other things) on the quantity $\sin(\pi J t_1)$, where J is the mutual coupling, experimental conditions can be chosen so as to reveal large couplings only, or to include signals due to long range couplings.

The full analysis of the Jeener two-pulse experiment was presented in a seminal paper by Aue, Bartholdi and Ernst,[3] although the first experimental spectra appeared slightly earlier.[49] Unfortunately, the density matrix methods needed for the analysis rapidly become unwieldy with systems of more than two or three spins. This, together with the need for large data matrices to hold all the information produced by the experiment, has until recently tended to discourage its use. Although very few spectra obtained using the basic Jeener method have been published,[3,49] a recent modification of the experiment has attracted considerable interest.

The "spin echo correlated spectroscopy" technique retains the basic two-pulse sequence, but delays acquisition of a free induction decay until the maximum of the Hahn echo:[50,51]

$$90°_H - t_1/2 - 90°_H - t_1/2 - \text{acquire } S(t_2) \qquad [5]$$

In this way, the f_1 dimension is made to reflect chemical shift differences, and the cross peaks are emphasized at the expense of the diagonal responses. Where there are no couplings between multiplets at opposite ends of the spectrum, the measurement of chemical shift differences rather than the shifts themselves can allow smaller f_1 spectral widths and hence smaller data matrices to be used. The relative phases of the two 90° pulses must be varied in a cyclic manner during time-averaging, in order to allow the signs of shift differences to be discriminated and to suppress axial peaks. Unfortunately, it is rarely possible to reduce the f_1 spectral width significantly, making the data matrices needed the same size as those for the original experiment, and in spectra where natural line widths predominate the extra $t_1/2$ delay can cause considerable sensitivity losses. A compromise technique, which returns to the Jeener pulse sequence but uses a phase cycling scheme related to that of reference 50 to allow optimal use of data storage, has recently been proposed;[52] this modification will be discussed after the next section.

Heteronuclear Correlation

The improvement in resolving power brought about by using 2D J spectroscopy in proton NMR increases the limiting size of spin system for which detailed analysis is possible by a useful factor, perhaps about three or four. The improvements in information capacity, in the number of signals that can be present in a spectrum and yet still be resolved individually, are much greater for homonuclear correlation experiments. The most spectacular gains are however to be made in heteronuclear correlation methods; a 2D spectrum correlating proton and carbon-13 chemical shifts, for example, will be about 250 x 10 parts per million, as opposed to 10 ppm square for a homonuclear correlation spectrum. As a very rough guide to the relative resolving powers and information capacities, it is perhaps worth considering the relative areas covered by different 2D NMR experiments.

At 200 MHz, a typical homonuclear J spectrum will have an area in frequency space of some 5×10^4 Hz^2, a homonuclear correlation spectrum 4×10^6 Hz^2, a heteronuclear J spectrum 5×10^6 Hz^2, and a heteronuclear correlation spectrum 3×10^7 Hz^2.

The basic aim of heteronuclear correlation 2D NMR is, as in the homonuclear experiment, to indicate which nuclei are coupled to one another. In heteronuclear NMR the correlation information is doubly useful since it provides chemical shifts for two different nuclear species in a single experiment. In the discussion which follows, the nuclei concerned will be assumed to be protons and carbon-13, although similar experiments have been reported for $^{31}P^{-1}H$,[53,54] $^{15}N^{-1}H$[55] and $^{11}B^{-1}H$[56] correlation.

Heteronuclear correlation techniques exhibit considerable variety, as pulses affecting only one nuclear species at a time allow the two spectra being correlated to be manipulated independently. The simplest experiment is to extend the basic Jeener two-pulse sequence to two nuclei:

$$90°_H - t_1 - 90°_H, 90°_C - \text{acquire } ^{13}C \ S(t_2) \quad [6]$$

Here proton magnetization generated by the first 90° pulse is transferred to carbon-13 nuclei by the $90°_H, 90°_C$ pulse pair, and is recorded as a carbon-13 free induction decay. Magnetization transfer in such experiments can be viewed either as a direct transfer of coherence from one part of a spin system to another, brought about by the mixing action of the pair of 90° pulses, or as arising from differential perturbation of the populations of the carbon-13 spin states brought about by the two proton 90° pulses. The mechanism of this step has been discussed at length elsewhere.[1,53,54,57-60] The critical point is that the amount of magnetization transferred to carbon-13 depends on the phase and amplitude of the proton magnetization at the end of the t_1 period, so that t_1 signal modulation reflects proton resonance frequencies, and t_2 modulation carbon-13 frequencies.

Such experiments may be performed either by transferring proton signals to carbon-13 and observing a carbon-13 free induction decay, or by transferring in the opposite direction and monitoring proton signals. The latter method was the first demonstrated,[57] and has the higher basic sensitivity by virtue of the proton magnetogyric ratio, but suffers from a number of disadvantages. The main problems are the need to suppress signals from the 99% of protons not bound to carbon-13, the need for large numbers of t_1 samples to digitize adequately the wide carbon-13 spectrum, and the presence of proton-proton scalar coupling. For these reasons, almost all other heteronuclear $^{13}C^{-1}H$ experiments reported have observed carbon-13 signals.

The main interest in heteronuclear correlation methods lies in the indirect measurement of correlated chemical shifts; the fine multiplet structure which the basic experiment of sequence [6] produces can usually be obtained more easily from other sources. The ideal experiment, from the point of view of assigning spectra and characterizing spin systems, would be one in which a 2D spectrum was produced with one signal for each directly bonded carbon-13 - proton pair in a molecule, the coordinates of the signal being the carbon-13 and proton chemical shift frequencies. Something very close to this can be achieved if the pulse sequence [6] is extended to include a carbon-13 pulse at the midpoint of t_1 and proton decoupling during t_2, in order to suppress heteronuclear scalar couplings. This leaves only the comparatively small proton-proton couplings, which normally remain unresolved in practical correlation 2D experiments. An effective pulse sequence for this experiment is then[58,61,62]

$$90°_H - t_1/2 - 180°_C - t_1/2 - \Delta_1 - 90°_H, 90°_C - \Delta_2$$

$$- \text{acquisition } (^1H \text{ dec.}) \; S(t_2) \qquad [7]$$

The delays Δ_1 and Δ_2 are chosen to be of the order of $1/2J_{CH}$ and $1/3J_{CH}$ respectively, in order to prevent the mutual cancellation of antiphase signals. If a phase-sensitive display of the entire spectrum is desired, then it is useful to insert $180°_C, 180°_H$ pulse pairs into the midpoints of the Δ_1 and Δ_2 delays, but such a display mode is rarely helpful.

The first communication describing heteronuclear correlation 2D NMR[57] emphasized the potential sensitivity improvements for low magnetogyric ratio nuclei, and the connectivity information made available. Subsequent publications have laid more stress on the indirect measurement of correlated chemical shifts.[1,30,58-64] Reference 58 describes the two most effective ways of obtaining net transfer of transverse magnetization between different nuclear species: the "pulse interrupted free precession" of sequence [7], and Hartmann-Hahn coherence transfer in the rotating frame. For experimental reasons, the first of these is usually to be preferred, although a detailed analysis of the rotating frame contact method has been published[63] and very good quality spectra shown.[58,63]

The pulse sequence [7], which will be returned to in section 4, has been extensively used in carbohydrate NMR.[30,60,64] With minor modification, it may also be used for the indirect measurement of proton spin-lattice relaxation rates via polarization transfer to carbon-13,[65] although this is very time-consuming compared to the analogous one-dimensional experiment.[66,67] Heteronuclear correlation methods have also been used for correlating phosphorus and proton spectra in cellular phosphates,[53,54] for the detection of nitrogen-15 signals with enhanced sensitivity,[55] and for correlating boron-11 and proton signals in carboranes.[56] Recently an elegant modification has been proposed in which a proton refocussing pulse during the t_1 period of sequence [6] is used to suppress proton chemical shifts during t_1, so that the resultant 2D spectrum correlates ^{31}P chemical shifts with proton-proton multiplet structure. This corresponds to the measurement of a separate proton J spectrum for each ^{31}P (or in the corresponding experiment, ^{13}C) site.[68]

Other Techniques

A number of the early 2D NMR experiments fall outside the classification used in the preceding pages, including the experiment used to introduce 2D spectra in section 1.[4] This latter method was overtaken by J spectroscopy techniques, but has recently been revived by Muller[69] in slightly different form. J spectroscopic methods require the use of 90° and 180° pulses, whereas improved sensitivity can sometimes be obtained from experiments which perturb the spin system less, by using smaller flip angle pulses. The need for large numbers of t_1 samples in the original method[4] can be done away with by application of a cyclic shift to f_1 data in the final 2D spectrum. Although the gains in sensitivity are not great, the resultant technique can be a simple and effective way of disentangling overlapping proton coupled carbon-13 spectra. Another early experiment, for the simultaneous measurement of approximate T_1 and T_2 values,[70] has not found widespread application, but affords a nice illustration of the use of phase shifts to discriminate between different orders of coherence, a principle important in multiple quantum NMR.

Multiple quantum coherences are not directly observable in magnetic resonance experiments, but their properties may be inferred from 2D NMR experiments in which a mixing pulse is used to transfer multiple quantum to single quantum coherence at the end of the t_1 period. Different orders of multiple quantum transition may be distinguished either by their differing sensitivity to changes in the radiofrequency phase of pulses[71] or to magnetic field gradients.[72,73] Most experiments reported have involved homonuclear coherences, but heteronuclear multiple quantum coherences may also be studied.[74] Double resonance may be used to increase the information content of multiple quantum 2D spectra,[75] but the greatest interest attaches to the use of relaxation properties to fill in missing elements of the Redfield relaxation matrix.[76-79]

A recent experiment that seems likely to assume considerable importance is unusual in not making use of coherence transfer, but rather using t_1 modulation to "label" longitudinal magnetizations involved in chemical exchange or cross-relaxation.[80,81] The basic pulse sequence has the form:

90° - t_1 - 90° - τ - 90° - acquire $S(t_2)$ [8]

with the effects of transverse magnetization during the "exchange" period τ being suppressed either by suitable phase cycling of pulses or by a magnetic field gradient "homospoil" pulse. The initial 90° - t_1 - 90° pulse pair results in a consinusoidal modulation of the longitudinal magnetizations according to their resonance frequencies. Chemical exchange or cross-relaxation during the period τ results in some of this modulated magnetization being transferred to other resonances, thus generating cross peaks in the resulting 2D spectrum. In homonuclear coupled spin systems, zero quantum coherences might possibly lead to some spurious responses, and there is an obvious ambiguity between chemical exchange and relaxation processes, but these should cause few problems in practice. The use of this experiment to map out cross-relaxation in complex proton spectra is particularly appealing, since the identification of such "through-space" interactions nicely complements the use of homonuclear correlation 2D NMR to identify "through-bond" scalar interactions.

EXPERIMENTAL METHODS

The task of an NMR spectrometer system in performing a two-dimensional experiment may be divided into three parts. The first is the acquisition of data, which requires the generation of accurately timed sequences of pulses followed by the digitization of the resultant free induction decays. The second task is the processing of this matrix of time-domain data to yield a two-dimensional spectrum, and the third and final stage is the presentation of the data in a form suitable for analysis. No commercial spectrometer is ideally suited to this type of experiment, but almost all spectrometers have the basic hardware required. The implementation of 2D NMR techniques on a given instrument usually reduces to the development of the necessary computer software for carrying out the steps outlined above; these will now be discussed in a little more detail.

Data acquisition

The pulse sequences involved in 2D NMR vary in complexity from the simple two-pulse Jeener[3] and J spectroscopy[14] sequences to those involving numbers of pulses at different frequencies with gated wide-band modulation, such as that for chemical shift correlation.[62] The basic requirement remains the same: the ability to generate accurately timed pulses with reproducible widths, amplitudes, phases and frequencies. Since the most rapid time scale in high resolution NMR is usually that of the chemical shift, the very rapid pulse sequences associated with solid state methods are not usually necessary, and delays of a few microseconds between pulses can be tolerated.

Although static field inhomogeneity is often less important in two-dimensional[82] than in conventional NMR, radiofrequency field inhomogeneity assumes rather greater significance. Inhomogeneities in B_1 field (observed nucleus) and B_2 field (decoupled heteronucleus,

Two-Dimensional Fourier Transform NMR Spectroscopy

if any), generally give rise only to small phase and intensity anomalies in conventional NMR. In 2D NMR, on the other hand, areas of the sample subjected to imperfect 90° or 180° pulses can give rise to spurious signals appearing at frequencies quite different from those expected. These artifacts can cause considerable problems, particularly in spectra with large dynamic range. Typical examples are the so-called "ghost" and "phantom" signals seen in 2D J spectroscopy,[38] which may be suppressed by suitable cycling of the radiofrequency phases of the pulses.

One of the most stringent requirements is that of stability. Since 2D methods rely on mapping out the t_1 dependence of signals in experiments often spread over several hours, stability of the static field and reproducibility of pulses are extremely important. Since any change in amplitude from one time-averaged free induction decay to the next is interpreted as a modulation, random changes in signal phase, amplitude or frequency give rise to random signals in the f_1 dimension of the resultant 2D spectrum. This gives rise to the characteristic ridges of so-called "t_1-noise" which disfigured many early 2D spectra. Such effects are much less important on superconducting than on iron magnet systems, and only obtrude in high field spectrometers when very strong signals are present. One interesting way of further reducing "t_1-noise" would be to extract some unmodulated reference signal (for example, TMS) from each averaged free induction decay, and then use this to correct the phase and amplitude of the rest of the decay, thus "phase-locking" the modulated signals to the reference. This would require very good signal-to-noise ratio for the reference signal, but would greatly reduce problems associated with radiofrequency and static field instability.

Once generated, the free induction decays are detected, filtered, digitized and stored just as in a normal FT NMR experiment. The only unusual requirement is that a large number of decays must be stored simultaneously, making the use of a magnetic disk necessary. Although hard disk systems have normally been used in 2D NMR to date, the demands on speed and capacity are not incompatible with floppy disk hardware.

Data processing

With a complete set of free induction decays stored on magnetic disk, the next step is to perform any weighting required, followed by a double Fourier transformation. The most satisfactory solution would be to use an array processor to perform all these manipulations in the computer memory, but as such hardware is not yet generally available, a piecemeal approach has to be adopted. Since data acquisition rarely monopolizes the computer's attention, free induction decays are usually weighted and transformed as soon as time-averaging for a particular t_1 value is complete. Thus, when data acquisition is over, the data are available on disk in the form of a series of f_2 spectra stored sequentially for increasing values of t_1.

To perform the second transformation efficiently, it is now necessary to transpose this data matrix, so that blocks of data mapping out the t_1 dependence of signals are stored on the disk in order of f_2 value. These t_1 domain decays are normally referred to as "interferograms," to distinguish them from free induction decays, which are recorded in real time rather than being built up from successive experiments. Each interferogram in turn may then be read back into the computer memory, weighted, transformed and returned to disk; if relatively small blocks of data are involved, it may be more efficient to read in several decays at once. The disk interactive matrix transposition program is usually the principal extra software needed to adapt a normal FT spectrometer control program to 2D NMR.

As in conventional NMR, it is customary to use some weighting and apodization of time domain data prior to transformation, although some extra points must be taken into consideration in choosing a suitable weighting function. The most common weighting functions in use are exponential, Gaussian and convolution difference, used either singly or in combination. Digitization is usually at a premium in 2D NMR, both because of computer storage limitations and because experiment time increases with improving t_1 digitization. As a result, poor digitization often prevents accurate representation of line shapes, and in unfavorable cases signal loss may result. This can be prevented if suitable weighting is used, for example exponential weighting with a time constant equal either to the decay constant of the signal (matched filtering) or to the length of the time domain signal, whichever is the lesser. At the other extreme, if lines are well digitized in a crowded spectrum, it may be useful to effect a Lorentzian to Gaussian conversion using biexponential weighting, in order to reduce the tendency of the wide "skirts" of two-dimensional Lorentzians to overlap and interfere with other signals. Some experimental spectra using a variety of different weighting functions are discussed in section 4.

After double Fourier transformation, it usually remains only to present the 2D spectrum in a form suitable for interpretation. In some cases, however, an extra data processing step can be of use, the most significant example being 2D J spectroscopy. The double Fourier transformation of the modulated spin echoes gives rise to a 2D spectrum in which the f_2 signal dispersion reflects the normal mix of chemical shifts and multiplet splittings, while in f_1 only scalar couplings appear for weak coupling. Each multiplet then lies along a line with gradient $df_2/df_1 = 1$. Thus, to approach the desirable goal of a proton 2D spectrum in which only chemical shifts appear in f_2 and only multiplet structure in f_1, all that is necessary is to "tilt" the entire frequency domain data matrix through 45° in frequency space:[14-16,29,31]

Two-Dimensional Fourier Transform NMR Spectroscopy

$$S(f_1, f_2') = S(f_1, f_2-f_1) \quad [9]$$

Although the separation of shifts and couplings is still not complete where strong coupling is present, such "tilted" spectra are usually considerably easier to interpret than the raw 2D spectra.

Data presentation

Given a matrix of complex data points digitizing a two-dimensional spectrum $S(f_1, f_2)$, in which signals are distributed according to their NMR parameters as outlined in the preceding section, two questions remain. First, what is the nature of the line shape associated with each signal; and second, how can the positions and intensities of these signals best be determined? In one-dimensional NMR, the answers are relatively simple: the line shape is a complex Lorentzian, and a simple plot of the real part of the spectrum against frequency is a very effective way of extracting the required information. Matters are less simple in two dimensions; if the line shape produced by a phase modulated signal decaying exponentially with time is a complex Lorentzian, then the two-dimensional line shape produced by a signal phase modulated and decaying exponentially with respect to two independent times will be the product of two complex Lorentzians. Thus, if Δ_1 and Δ_2 are the offsets from resonance in f_1 and f_2 of a signal decaying with time constants τ_1 and τ_2, then its two-dimensional line shape in frequency space will be:

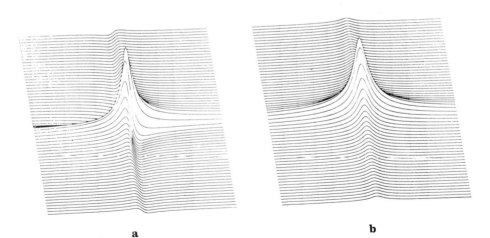

Figure 4. The real part (a) and the modulus (b) of the "phasetwist" line shape characteristic of a signal phase modulated in both t_1 and t_2.

$$S(f_1,f_2) = \left(\frac{\tau_1 - i\Delta_1\tau_1^2}{1 + \Delta_1^2\tau_1^2}\right) \cdot \left(\frac{\tau_2 - i\Delta_2\tau_2^2}{1 + \Delta_2^2\tau_2^2}\right) \qquad [10]$$

The real part of equation 10 has been plotted out in Figure 4a; as is immediately obvious, this line shape lacks the convenient property, found in the real part of a one-dimensional Lorentzian, of being positive for all frequencies. The negative-going parts of the two-dimensional line shape make for severe problems in plotting out 2D spectra, since overlapping signals suffer from destructive interference and negative signals tend to be obscured by positive. The apparent rotation of phase as the line shape of Figure 4a is traversed has given rise to the name "phasetwist."[37] Because of these problems, it has become customary to plot spectra in absolute value mode, plotting the modulus of Equation 10, which leads to the broadened line shape of Figure 4b. Unfortunately, the broad "tails" of Figure 4b tend to interfere with neighboring resonances, and often lead to distorted peak shapes and positions.[83]

Although no general solution to the problem of the "phasetwist" line shape has yet been presented, in certain cases the antisymmetric part of the real term of Equation 10 can be cancelled out by adding 2D spectra which have the same signal distribution, but the opposite sense of phase rotation in their line shapes. This is particularly simple for spectra that are symmetrical about $f_1 = 0$, since a simple mirror image provides the necessary partner:

$$S'(f_1,f_2) = S(f_1,f_2) + S(-f_1,f_2) \qquad [11]$$

Where there is no symmetry about $f_1 = 0$, it may still be possible to find the required conjugate spectrum if a second experiment can be performed in which the pulse sequence is modified to reverse the apparent sense of the t_1 modulation.[46,47]

Once a full 2D spectrum has been obtained in a form suitable for plotting, the two most popular plot modes are the stacked trace plot (used throughout this chapter), with its three-dimensional effect, and the intensity contour plot (see for example reference 1). Neither method is entirely satisfactory, as trace plots can allow small peaks to be hidden by larger, and contour plots have difficulty representing noisy spectra and dealing with the steeply rising flanks of most two-dimensional line shapes.

The problems discussed in the preceding paragraphs only obtrude when entire 2D spectra are to be recorded, which is not generally necessary. The significant information content of a spectrum can usually best be summarized by a series of cross-sections through f_1, which can be plotted out individually.[84] Indeed, since the f_2 distribution of signals is usually known to the experimenter from the conventional spectrum, only the traces that are known to bear signals need be examined. Since all resonances centered on a particular f_2 value will have similar line shapes, it is usually possible to adjust the phase of a single cross-section to pure absorption mode and hence to plot out individual traces in phase-sensitive form.[20,39] Any "tails" of lines from other f_2 values will however still appear out of phase, making it desirable to cancel the antisymmetric part of the phasetwist to yield a "double absorption" line shape whenever possible.[46,47]

So far the methods of data presentation discussed have concentrated on summarizing the information content of 2D spectra on paper, taking full advantage of the dispersion of signals into two frequency domains. It can however be useful to sacrifice some of this resolving power in order to condense a 2D spectrum into a one-dimensional representation. This can be achieved by "integral projection," integrating a 2D spectrum along some direction in order to project all signals onto a given axis. The two trivial projections are along the f_1 and f_2 axes, to produce pure f_2 and pure f_1 one-dimensional spectra, respectively. A little algebra suffices to show that the resultant spectra are identical to the Fourier transforms of the time-domain signals $S(0,t_2)$ and $S(t_1,0)$, these being the conventional spectrum excited by the appropriate pulse sequence with $t_1 = 0$, and the analog of the classical J spectrum.[8,9] These two relations are special cases of a general theorem relating integral projections and cross-sections.[84]

Projections onto the f_2 axis are of little interest since the same information can be obtained from a conventional experiment, while projections of complete 2D spectra onto f_1 suffer from poor sensitivity. In the time domain this reflects the loss of all signal energy from times greater than $t_2 = 0$, while in the frequency domain it corresponds to the integration of all the noise, as well as all the signals, onto the f_1 axis. Since, however, signals are constrained to appear at the f_2 frequencies of the conventional spectrum, only traces at these frequencies need be integrated, other cross-sections through the 2D spectrum carrying only noise. The process of double Fourier transformation followed by integration of selected cross-sections may be replaced by the more efficient "tailored detection" method of time domain filtering, in order to obtain f_1 projections with optimum sensitivity.[85] The subject of signal-to-noise ratio in 2D NMR has been dealt with in detail by Ernst and coworkers.[84,86]

The most widely used form of integral projection is not however onto the f_1 or f_2 axis, but onto the axis $f_1 = f_2$. For weakly coupled spin systems, homonuclear 2D J spectroscopy gives rise to muliplets which lie along lines with gradient $df_2/df_1 = 1$.[15] Thus, a projection at 45° in frequency space should give rise to a spectrum devoid of homonuclear couplings; such a "proton decoupled proton" spectrum would be an extremely useful assignment tool. Unfortunately, an ideal projection of this type results in zero signal intensity. Application of the cross-section-projection theorem to this case shows that the desired projection is equivalent to the Fourier transform of the time domain cross-section with $t_1 = -t_2$. Since data are not recorded for times t_1, t_2 less than zero, this constrains the "45° projection" to be zero; in frequency domain terms, this corresponds to the fact that the integral of a 45° section through a "phasetwist" line shape is zero.

It is thus not possible to obtain a phase-sensitive 45° projection of a normal homonuclear 2D J spectrum, and instead it is usual to project the absolute value or the power spectrum. This gives rise to distorted intensities and peak positions, but nevertheless is an extremely useful aid to assignment. The subject of phase-sensitive proton spectra without multiplet splittings is not however entirely closed, despite the clear message of the projection-cross-section theorem, and will be returned to later. In the interim, some of the many data acquisition and handling methods discussed above will be illustrated with some experimental spectra.

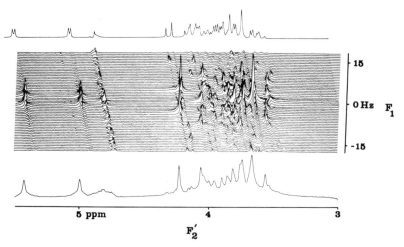

Figure 5. Top, normal proton spectrum of raffinose (O-α-D-galactopyranosyl-(1→6)-α-D-glucopyranosyl-(1→2)-β-D-fructofuranoside) in D_2O); middle, tilted 2D J spectrum displayed as a series of traces across f_2'; bottom, integral projection of the absolute value of the tilted 2D spectrum onto the f_2' axis, showing partially suppressed multiplet structure. All spectra were obtained on an XL-200 spectrometer, using continuous saturation to suppress the HDO solvent signal. The 2D spectrum employed 256 x 64 complex data points.

APPLICATIONS

In this section, a typical spectrum from each of the four main classes outlined above will be presented and discussed. The spectra were chosen for didactic reasons, rather than to illustrate the performance of the methods at full stretch; consequently all of the techniques dealt with here can be applied profitably to considerably more complex systems.

Homonuclear J spectroscopy

A simple AX proton 2D J spectrum was chosen as Figure 3 to illustrate the basic features of a 2D spectrum in the introduction to this chapter. Figures 5 and 6 show the results of applying two-dimensional J spectroscopy to a practical problem, the 200 MHz proton spectrum of a D_2O solution of the trisaccharide raffinose. Figure 5 shows the tilted 2D J spectrum, plotted as a series of traces through f_2', together with the conventional spectrum and a "proton decoupled proton" spectrum obtained by 45° projection of the absolute value 2D spectrum. The multiplet structure for each chemical shift can be seen a little more clearly if the spectrum is re-plotted as a series of f_1 traces, as in Figure 6. Clearly separated multiplets are obtained, in marked contrast with the severe overlap problems in the normal spectrum. Note that certain areas of the tilted 2D spectrum are set to zero, since these correspond to regions of frequency space that lay outside the original experimental data.

The usual 2D J spectroscopy pulse sequence of Figure 7 was used to collect the time domain data, with the relative phases of the two pulses and of the receiver reference being cycled on successive transients to suppress artifacts due to pulse imperfections.[38] The phase cycling sequence used was:

Transient no.	1	2	3	4	...	
90° pulse phase	0	0	0	0	...	
180° pulse phase	0	270	180	90	...	
Receiver phase	0	90	180	270	...	[12]

In order to minimize interference between adjacent signals, the same biexponential weighting was used in both time domains, to effect a Lorentzian to Gaussian transformation. As usual, the 2D spectra of Figures 5 and 6 are plotted in the absolute value mode. For accurate measurements of coupling constants, phase-sensitive cross-sections through the tilted 2D spectrum should be examined,[31,83] although it should be emphasized that such cross-sections still suffer from the basic problems of interference between out-of-phase signals from different f_2' values.

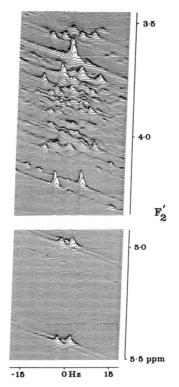

Figure 6. The 2D J spectrum of Figure 6, re-plotted as a series of traces across f_1 in order to display the multiplet structure associated with each chemical shift value in f_2' more clearly.

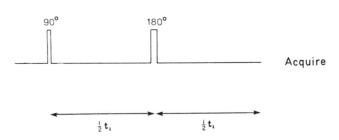

Figure 7. Pulse sequence for homonuclear 2D J spectroscopy, used in acquiring data for Figures 5 and 6.

Heteronuclear J spectroscopy

The two principal uses of heteronuclear J spectroscopy discussed earlier were first, for the measurement of proton coupled carbon-13 multiplets with line widths approaching the natural limit of $1/\pi T_2$; and second, for the untangling of overlapping multiplets in crowded spectra. Figure 8 shows an example of the latter application, being a carbon-13 2D J spectrum at 50.3 MHz for raffinose in D_2O. Similar spectra at 67.8 MHz have been used for the measurement of one-bond couplings in this system.[60] Time domain data for the spectrum of Figure 8 required about 4 hr for acquisition, in contrast to the experiment of Figures 5 and 6 which lasted less than 5 min.

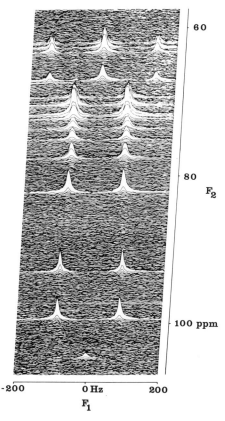

Figure 8. Heteronuclear carbon-13 2D J spectrum for raffinose in D_2O, measured using the broadband 10 mm probe of an XL-200 spectrometer, operating at 50.3 MHz for carbon-13. A data matrix of 512 x 64 complex points was used. Note that the frequency scale for f_1 has been doubled to show the true splittings.

The gated decoupler pulse sequence of Figure 9 was employed, with the phase cycling scheme [12] used for artifact suppression. Exponential weighting was used in t_2 with a time constant equal to the maximum value of t_2 for which data were acquired; since signal-to-noise ratio and digitization were more than adequate, no weighting was used in t_1. As is often the case with large molecules that have short relaxation times, there is little evidence of long range proton-carbon-13 couplings, although these can reappear if higher temperatures and more dilute samples are used. Had fine structure been resolvable, optimum use of the t_1 data would have meant zero-filling the 64 t_1 samples up to 128 or more prior to the second Fourier transformation.

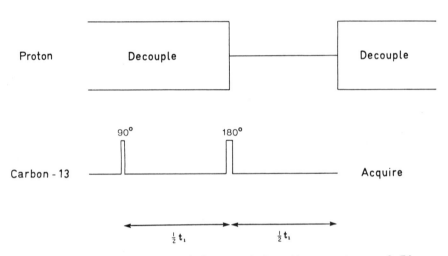

Figure 9. Pulse sequence used in acquiring the spectrum of Figure 8; this "gated decoupler" method for heteronuclear 2D J spectroscopy results in the multiplet splittings in f_1 appearing to be halved.

Homonuclear correlation

The basic Jeener two-pulse experiment [3] generates amplitude modulation as a function of t_1, yielding 2D spectra symmetrical about $f_1 = 0$. As a result, positive and negative f_1 frequencies are indistinguishable, making it necessary to place the transmitter frequency at one end of the proton spectrum and forgo the advantages of quadrature detection. To regain these advantages, it is necessary to convert amplitude into phase modulation, allowing positive and negative frequencies to be distinguished in both frequency dimensions.

In the "spin echo correlated spectroscopy" experiment, Nagayama et al.[50] use phase cycling to select one only of the two counter-rotating components into which the t_1 amplitude modulation may be decomposed, specifically that component which refocusses to give a Hahn echo. When static field inhomogeneity dominates line widths, this method gives good sensitivity, discriminates against diagonal peaks and suppresses axial peaks. When line widths are dominated by transverse relaxation, as is usually the case in complex proton spectra, it can be more effective to revert to the original Jeener pulse sequence [4] using phase cycling to select the other component. This leads to the following sequence of phases:

Transient no.	1	2	3	4	···	
1st 90° pulse phase	0	0	0	0	···	
2nd 90° pulse phase	0	90	180	270	···	
Receiver phase	0	0	0	0	···	[13]

As it stands, the extended Jeener method has the advantage of minimizing the size of data matrix needed, and of giving good sensitivity for large molecules, but gives rise to substantial diagonal peaks which clutter up the spectrum with uninformative signals. The phase properties of diagonal and cross peaks may be exploited to discriminate against the former, since convolution difference weighting will have much less effect on the cross peaks, which have antiphase fine structure and hence are rapidly attenuated by line broadening.

As Figure 10 shows, the pulse sequence of Figure 11 with the phase cycling scheme [13] can give 2D spectra that are quite clean and easy to interpret. The spectrum of Figure 10, for an aqueous solution of the cyclic peptide antibiotic viomycin, summarizes all the coupling and connectivity information that would be obtained from a complete series of homonuclear double resonance experiments. A similar spectrum is discussed in more detail in reference 52.

Heteronuclear correlation

The basic chemical shift correlation pulse sequence [7] outlined earlier leads, as does the Jeener experiment discussed in the preceding section, to amplitude modulation as a function of t_1, and hence to the loss of sign information for f_1 frequencies. Most shift correlation experiments in the literature have therefore placed the proton transmitter frequency to one side of the proton spectrum, using either a single phase transformation for f_1 or examining one side of the full 2D spectrum only. This wastage of data storage and proton transmitter power may be avoided if the phase cycling sequence [13] is adapted for the heteronuclear experiment, restoring the discrimination of positive and negative f_1 frequencies. No extra

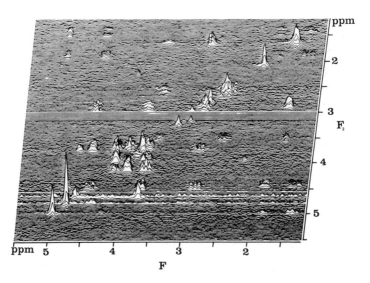

Figure 10. Homonuclear correlation 2D spectrum for viomycin in D_2O, obtained using the Jeener two-pulse sequence with the added phase cycling scheme [13] to allow frequency discrimination in both dimensions, recorded on an XL-200 using a data matrix of 256x256 complex points. Solvent saturation was again used to suppress the residual HDO signal. Spurious signals near f_2 = 3.1 and 4.5 ppm have been removed in order not to obscure cross peaks.

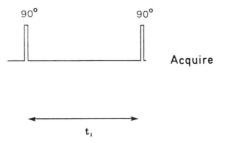

Figure 11. Two-pulse sequence used in acquiring the data of Figure 10.

data manipulation is required, so this phase cycling method avoids some of the extra complexity of the hypercomplex double Fourier transformation method proposed by Müller and Ernst.[63]

Figure 12 shows the complete 2D shift correlation spectrum obtained for a solution of menthol in deuteriobenzene. The f_1 digitization used was rather better than is usual, in order to show some of the proton-proton splittings that can be resolved. Although these multiplet patterns can be useful for assignment purposes, they should be interpreted with caution.[54] Note the different axial and equatorial proton chemical shifts observed for the ring methylene groups. Small spurious reponses occur due to polarization transfer via long-range, rather than one-bond couplings; see for example the small signal between the axial and equatorial responses for the methylene group at 45 ppm in f_2. These extra signals rarely cause problems, but can be further reduced by choosing shorter delays Δ_1 and Δ_2.

The pulse sequence used is illustrated in Figure 13; the phase cycling sequence ran as follows:

Transient number	1	2	3	4	5	6	7	8	...	
^{13}C 180° phase	0	0	0	0	180	180	180	180	...	
1st ^1H 90° phase	0	0	0	0	0	0	0	0	...	
2nd ^1H 90° phase	0	90	180	270	0	90	180	270	...	
^{13}C 90° phase	0	90	180	270	0	90	180	270	...	
Receiver phase	0	0	0	0	0	0	0	0	...	[14]

Spurious signals due to imperfect carbon-13 180° pulses are suppressed; if receiver phase imbalance causes zero frequency artifacts, these may be removed by adding receiver phase cycling, extending the sequence to include 32 different sets of phases.

Exponential weighting was used in both t_1 and t_2, the time constants being set to the maximum values of t_1 and t_2 respectively. The delays Δ_1 and Δ_2 used were 3.5 and 2 msec respectively, giving good polarization transfer for all the CH groups in menthol. These delays may be manipulated in order to emphasize or suppress signals of a given multiplicity or coupling constant, although the chemical shift dispersion of carbon-13 is usually sufficient to make further efforts of this sort to improve resolving power unnecessary.

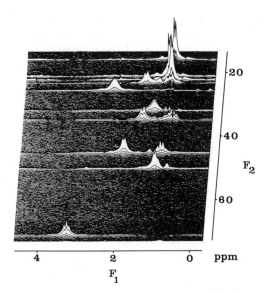

Figure 12. Heteronuclear chemical shift correlation 2D spectrum displaying the proton (f_1) and carbon-13 (f_2) shifts of a solution of menthol in deuteriobenzene, obtained by observing carbon-13 signals at 50.3 MHz in an XL-200 instrument. The data matrix used contained 256 x 256 complex points.

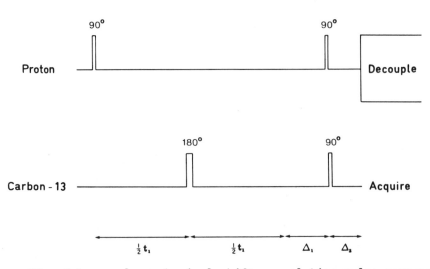

Figure 13. Heteronuclear chemical shift correlation pulse sequence used with the phase cycling scheme [14] in acquiring the spectrum of Figure 12.

Chemical shift correlation is a good example of a 2D NMR experiment in which it is not generally necessary or useful to plot out a complete 2D spectrum such as Figure 12. The most effective use of available computer storage and time is usually to extract signal-bearing cross-sections from the matrix $S(t_1,f_2)$, and dispense entirely with a second Fourier transformation for the rest of the data. The significant interferograms thus obtained can be zero-filled and plotted in phase sensitive mode, allowing proton shifts to be measured with quite good precision. Examples of this mode of data recording can be found in references 30, 60 and 64.

DISCUSSION

In the preceding pages, an attempt has been made to describe progress so far in the field of double Fourier transform NMR, and to illustrate some of the ways in which these new techniques can widen the range of chemical and biochemical systems amenable to study by NMR. Indeed, experiments such as the homonuclear and heteronuclear correlation methods can offer distinct advantages over conventional methods, not just at the frontiers of NMR applications, but also in the routine assignment of new spectra. The technical problems of implementing 2D NMR techniques on commercial instruments are now being greatly reduced as manufacturers introduce suitable software packages; these are now available for most modern high field spectrometers. The complexity of the data manipulations involved in double Fourier transform NMR need not impinge on the instrument operator, since these may be automated in just the same way as the data processing in a conventional FT NMR experiment, making routine 2D NMR spectra quite practicable.

A useful by-product of interest in 2D NMR has been the added impetus given both to the development of improved theoretical methods, and to the development of new techniques in one-dimensional NMR. Some of these new methods make use of the idea of an "interferogram" to produce one-dimensional spectra that are equivalent to integral projections of 2D spectra, but have properties not obtainable by conventional means. Examples include the "J-scaling" experiment for the meausrement of proton coupled carbon-13 spectra with uniformly scaled multiplet splittings,[85,87] and an interesting technique described by Bax and co-workers[88] for obtaining phase-sensitive proton spectra with suppressed scalar coupling structure. Although the method is not completely general, it does provide scope for further developments, which may yet lead to the desirable goal of a fully "proton decoupled proton" spectrum. Other new methods make use of pulse sequences adapted from 2D NMR to excite free induction decays in normal one-dimensional FT experiments, particularly successful examples being the INEPT experiment for enhancing the sensitivity of low magnetogyric ratio nuclei[66,67,89] and the INADEQUATE experiment[90,91] for studying natural abundance carbon-13 - carbon-13 scalar couplings.

Although two-dimensional NMR is still in its infancy, it seems probable that it will be turned to increasingly in the more demanding areas of NMR applications. Already problems are being tackled which would be difficult or impossible to solve by more conventional means; see for example references 23,30,60,64, and 79. If present promise is fulfilled, the coming of age of 2D NMR during the next few years should add a powerful array of weapons to the armory of chemical and biochemical NMR.

ACKNOWLEDGMENTS

The work described in this chapter owes a great debt to the continuing advice and encouragement of Dr. Ray Freeman, in whose laboratory all of the experimental spectra were obtained. I am also indebted to past and present colleagues, including Prof. L. D. Hall, Dr. G. Bodenhausen, Dr. D. L. Turner, Prof. J. B. Grutzner, and A. Bax, for many helpful discussions. Technical assistance from Dr. S. Smallcombe, who provided much of the extra software necessary to produce the experimental spectra of Figures 3,5,6,8,10, and 12, and from T. Mareci, who produced the experimental spectrum of Figure 3, is also gratefully acknowledged; Figure 10 is the work of A. Bax.

REFERENCES

1. Freeman, R., and Morris, G. A. 1979, Bull. Magn. Reson. 1, 5.
2. Freeman, R. 1980, Proc. Roy. Soc. A373, 149.
3. Aue, W. P., Bartholdi, E., and Ernst, R. R. 1976, J. Chem. Phys. 64, 2229.
4. Müller, L., Kumar, A., and Ernst, R. R. 1975, J. Chem. Phys. 63, 5490.
5. Höhener, A., Müller, L., and Ernst, R. R. 1979, Mol. Phys. 38, 909.
6. Bodenhausen, G., Stark, R. E., Ruben, D. J., and Griffin, R. G. 1979, Chem. Phys. Lett. 67, 424.
7. Polak, M., Highe, A. J., and Vaughan, R. W., 1980, J. Magn. Reson. 37, 357.
8. Freeman, R., and Hill, H. D. W. 1971, J. Chem. Phys. 54, 301.
9. Vold, R. L., and Chan. S. O. 1970, J. Chem. Phys. 53, 447.
10. Gutowsky, H. S., McCall, D. W., and Slichter, C. P. 1951, Phys. Rev. 84, 589.
11. Hahn, E. L., and Maxwell, D. E. 1951, Phys. Rev, 84, 1246.
12. Abragam. A. 1961, The Principles of Nuclear Magnetism, Oxford University Press, Oxford.
13. Freeman, R., and Hill, H. D. W. 1975, in Dynamic Nuclear Magnetic Resonance, ed. Jackman, L. M. and Cotton, F. A., Academic Press, New York.
14. Aue, W. P., Karhan, J., and Ernst, R. R. 1976, J. Chem. Phys. 64, 4226.

15. Nagayama, K., Wüthrich, K., Bachmann, P., and Ernst, R. R. 1977, Naturwiss. 64, 581.
16. Nagayama, K., Wüthrich, K., Bachmann, P., and Ernst, R. R. 1977, Biochem. Biophys. Res. Commun. 78, 99.
17. Kumar, A., and Khetrapal, C. L. 1978, J. Magn. Reson. 30, 137.
18. Kumar, A. 1978, J. Magn. Reson. 30, 227.
19. Bodenhausen, G., Freeman, R., Morris, G. A., and Turner, D. L. 1978, J. Magn. Reson. 31, 75.
20. Bodenhausen, G., Freeman, R., Morris, G. A., and Turner, D. L. 1977, J. Magn. Reson. 28, 17.
21. Niedermeyer, R., and Freeman, R. 1978, J. Magn. Reson. 30, 617.
22. Barfield, M., Brown, S. E., Canada, E. D., Jr., Ledford, N. D., Marshall, J. L., Walter, S. R., and Yakali, E. 1980, J. Amer. Chem. Soc. 102, 3355.
23. Nagayama, K., Bachmann, P., Ernst, R. R., and Wüthrich, K. 1979, Biochem. Biophys. Res. Commun. 86, 218.
24. Nagayama, K. 1979, J. Chem. Phys. 71, 4404.
25. Hall, L. D., and Sukumar, S. 1979, Carbohydr. Res. 74, C1-C4.
26. Hall, L. D., Sukumar, S., and Sullivan, G. R. 1979, J. Chem. Soc. Chem. Commun., 292.
27. Hall, L. D., and Sukumar, S. 1979, J. Amer. Chem. Soc. 101, 3120.
28. Everett, J. R., Hughes, D. W., Bain, A. D., and Bell, R. A. 1979, J. Amer. Chem. Soc. 101, 6776.
29. Hall, L. D., Morris, G. A., and Sukumar, S. 1979, Carbohydr. Res. 76, C7-C9.
30. Hall, L. D., Morris, G. A., and Sukumar, S. 1980, J. Amer. Chem. Soc. 102, 1745.
31. Hall, L. D., and Sukumar, S. 1980, J. Magn. Reson. 38, 555.
32. Hall, L. D., Sanders, J. K. M., and Sukumar, S. 1980, J. Chem. Soc. Chem. Commun., 366.
33. Khetrapal, C. L., Kumar, A., Kunwar, A. C., Mathias, P. C., and Ramanathan, K. V. 1980, J. Magn. Reson. 37, 349.
34. Rabenstein, D. L. 1978, Anal. Chem. 50, 1265A.
35. Bodenhausen, G., Freeman, R., Niedermeyer, R., and Turner, D. L. 1976, J. Magn. Reson. 24, 291.
36. Bodenhausen, G., Freeman, R., and Turner, D. L. 1976, J. Chem. Phys. 65, 839.
37. Bodenhausen, G., Freeman, R., Niedermeyer, R., and Turner, D. L. 1977, J. Magn. Reson. 26, 133.
38. Bodenhausen, G., Freeman, R., and Turner, D. L. 1977, J. Magn. Reson. 27, 511.
39. Freeman, R., Morris, G. A., and Turner, D. L. 1977, J. Magn. Reson. 26, 373.
40. Kumar, A., Aue, W. P., Bachmann, P., Karhan, J., Müller, L. and Ernst, R. R. 1976, Proc. XIXth Congress Ampère, Heidelberg.
41. Müller, L., Kumar, A., and Ernst, R. R. 1977, J. Magn. Reson. 25, 383.
42. Müller, L. 1980, J. Magn. Reson. 38, 79.

43. Zumbulyadis, N., Henrichs, P. M., and Schwartz, L. J. 1977, J. Chem. Phys. 67, 1780.
44. Turner, D. L., and Freeman, R. 1978, J. Magn. Reson. 29, 587.
45. Hall, L. D., and Morris, G. A. 1980, Carbohydr. Res. 82, 175.
46. Bachmann, P., Aue, W. P., Muller, L., and Ernst, R. R. 1977, J. Magn. Reson. 28, 29.
47. Freeman, R., Kempsell, S. P., and Levitt, M. H. 1979, J. Magn. Reson. 34, 663.
48. Levitt, M. H., and Freeman, R. 1979, J. Magn. Reson. 34, 675.
49. Ernst, R. R. 1975, Chimia 29, 179.
50. Nagayama, K., Wüthrich, K., and Ernst, R. R. 1979, Biochem. Biophys. Res. Commun. 90, 305.
51. Bain, A. D., Bell, R. A., Everett, J. R., and Hughes, D. W. 1980, J. Chem. Soc. Chem. Commun., 256.
52. Bax, A., Freeman, R., and Morris, G. A. 1981, J. Magn. Reson. 42, 164.
53. Bolton, P. H., and Bodenhausen, G. 1979, J. Amer. Chem. Soc. 101, 1080.
54. Bodenhausen, G., and Bolton, P. H. 1980, J. Magn. Reson. 39, 399.
55. Bodenhausen, G., and Ruben, D. J. 1980, Chem. Phys. Lett. 69, 185.
56. Finster, D. C., Hutton, W. C., and Grimes, R. N. 1980, J. Amer. Chem. Soc. 102, 400.
57. Maudsley, A. A., and Ernst, R. R. 1977, Chem. Phys. Lett. 50, 368.
58. Maudsley, A. A., Muller, L., and Ernst, R. R. 1977, J. Magn. Reson. 28, 463.
59. Bodenhausen, G., and Freeman, R. 1977, J. Magn. Reson. 28, 471.
60. Morris, G. A., and Hall, L. D. 1981, J. Amer. Chem. Soc. (in press).
61. Bodenhausen, G., and Freeman, R. 1978, J. Amer. Chem. Soc. 100, 320.
62. Freeman, R., and Morris, G. A. 1978, J. Chem. Soc. Chem. Commun., 684.
63. Müller, L., and Ernst. R. R. 1979, Mol. Phys, 38, 963.
64. Morris, G. A., and Hall, L. D. 1981, Can. J. Chem. (in press).
65. Avent, A. G., and Freeman, R. 1980, J. Magn. Reson. 39, 169.
66. Morris, G. A. 1980, J. Amer. Chem. Soc. 102, 428.
67. Morris, G. A. 1980, J. Magn. Reson. 41, 185.
68. Bodenhausen, G. 1980, J. Magn. Reson. 39, 175.
69. Müller, L. 1979, J. Magn. Reson. 36, 301.
70. Bodenhausen, G., and Freeman, R. 1977, J. Magn. Reson. 28, 303.
71. Wokaun, A., and Ernst, R. R. 1977, Chem. Phys. Lett. 52, 407.
72. Maudsley, A. A., Wokaun, A., and Ernst, R. R. 1978, Chem. Phys. Lett. 55, 9.
73. Bax, A., DeJong, P., Mehlkopf, A. F., and Smidt, J. 1980, Chem. Phys. Lett. 69, 567.

74. Müller, L. 1979, J. Amer. Chem. Soc. 101, 4481.
75. Wokaun, A., and Ernst, R. R. 1979, Mol. Phys. 38, 1579.
76. Wokaun, A., and Ernst, R. R. 1978, Mol. Phys. 36, 317.
77. Bodenhausen, G., Szeverenyi, N. M., Vold, R. L., and Vold, R. R. 1978, J. Amer. Chem. Soc. 100, 6265.
78. Bodenhausen, G., Vold, R. L., and Vold, R. R. 1980, J. Magn. Reson. 37, 93.
79. Vold, R. L., Vold. R. R., Poupko, R., and Bodenhausen, G. 1980, J. Magn. Reson. 38, 141.
80. Meier, B. H., and Ernst, R. R. 1979, J. Amer. Chem. Soc. 101, 6441.
81. Jeener, J., Meier, B. H., Bachmann, P., and Ernst, R. R. 1979, J. Chem. Phys. 71, 4546.
82. Bodenhausen, G., Kempsell, S. P., Freeman, R., and Hill, H. D. W. 1979, J. Magn. Reson. 35, 337.
83. Turner, D. L. 1980, J. Magn. Reson. 39, 391.
84. Nagayama, K., Bachmann, P., Wüthrich, K., and Ernst, R. R. 1978, J. Magn. Reson. 31, 133.
85. Morris, G. A., and Freeman, R. 1978, J. Amer. Chem. Soc. 100, 6763.
86. Aue, W. P., Bachmann, P., Wokaun, A., and Ernst, R. R. 1978, J. Magn. Reson. 29, 523.
87. Freeman, R., and Morris, G. A. 1978, J. Magn. Reson. 29, 173.
88. Bax, A., Mehlkopf, A. F., and Smidt, J. 1979, J. Magn. Reson. 35, 167.
89. Morris, G. A., and Freeman, R. 1979, J. Amer. Chem. Soc. 101, 760.
90. Bax, A., and Freeman, R. 1980, J. Amer. Chem. Soc. 102, 4849.
91. Bax, A., Freeman, R., and Kempsell, S. P. 1980, J. Magn. Reson. 41, 349.

ENDOR SPECTROSCOPY BY FOURIER TRANSFORMATION OF THE ELECTRON SPIN ECHO ENVELOPE

W. B. Mims

Bell Laboratories
Murray Hill, NJ 07974

ELECTRON SPIN ECHO DECAY ENVELOPE

The envelope of electron spin echoes[1-3] is the function obtained by plotting the amplitude of the spin echo signal against the time between the echo-generating microwave pulses (see Figure 1). In this function, it is often possible to observe a periodic "modulation," which can be shown to arise from superhyperfine structure (shfs) in the resonance line. The echo decay envelope is modulated by the "ENDOR" frequencies, and it is possible, by recording the echo envelope and then Fourier transforming the result, to obtain an ENDOR spectrum. We discuss here some of the advantages and also some of the problems of this form of "echo envelope spectroscopy."

Figure 1. Diagram showing how the two-pulse echo envelope may be obtained by repeating the electron spin echo cycle with gradually increasing values of τ. The "modulation" of the envelope is due to electron-nuclear coupling and is caused by interference between allowed and semi-forbidden microwave transitions (see e.g. Figure 2). The periods observed in the echo envelope correspond to the nuclear shfs, or ENDOR frequencies. (Reproduced with permission from Biochemistry 15, 3863 (1976).)

Two-pulse echo

The physical origins of the effect can be illustrated from Figure 2, which shows the energy level scheme for an I = 1 nucleus, such as ^{14}N, coupled to an S = 1/2 electron spin. In the case considered here (i.e., the electron-nuclear interaction, the nuclear Zeeman interaction, and the nuclear quadrupole interaction, all of the same order), microwaves can induce both allowed and semi-forbidden transitions between states in the M_S = 1/2 manifold (α) and the M_S = - 1/2 manifold (β). Simultaneous excitation of both kinds of transitions by the echo generating microwave pulses gives rise to interference effects, which manifest themselves as variations in the echo amplitude and thus cause the modulation of the echo envelope. Where a number of nuclei are coupled to the same electron spin, the level scheme becomes more complicated, but it is possible to factor out contributions due to coupling with each nucleus in the overall modulation pattern. If $V_{mod}(I_1, I_2, \cdots I_n)$ is the modulation function due to coupling with n nuclei, then

$$V_{mod}(I_1, I_2, \cdots, I_n) = V_{mod}(I_1) \times V_{mod}(I_2) \times \cdots \times V_{mod}(I_n); \qquad [1]$$

i.e., the overall modulation function is the product of the modulation functions due to each nucleus considered separately.

For a two-pulse electron spin echo sequence such as that shown in Figure 1, the function modulating the echo envelope has the general form

$$V_{mod} = X_0 + \sum_{i,j}^{i \neq j}{}' X_{ij}^{(\alpha)} \cos(\omega_{ij}^{(\alpha)} \tau) + \sum_{k,n}^{k \neq n}{}' X_{kn}^{(\beta)} \cos(\omega_{kn}^{(\beta)} \tau)$$

$$+ \sum_{i,j}^{i \neq j}{}' \sum_{k,n}^{k \neq n}{}' X_{ij,kn}^{(\alpha,\beta)} \cos((\omega_{ij}^{(\alpha)} + \omega_{kn}^{(\beta)})\tau) + \cos((\omega_{ij}^{(\alpha)} - \omega_{kn}^{(\beta)})\tau) \qquad [2]$$

in which the $\omega_{ij}^{(\alpha)}$ and $\omega_{kn}^{(\beta)}$ are radian frequencies corresponding to shfs intervals in the α and β manifolds, and τ is the pulse separation. [The notation $\sum'$ is used to indicate that any pair of indices i,j or k,n is to be used only once in the sum.[4,5]] There are certain relationships between the coefficients, X, which determine the amplitudes of the various frequency components. Thus, all of the X coefficients (including X_0) sum to unity, which is the value of V_{mod} at $\tau = 0$. Also,

ENDOR by Fourier Transformation of the Electron Spin Echo Envelope

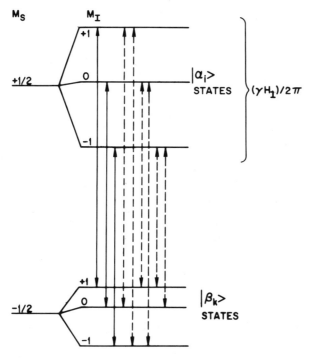

Figure 2. Energy level scheme for an I = 1 nucleus (e.g., ^{14}N) weakly coupled to an S = 1/2 electron spin. For the case shown here, the nuclear quadrupolar interaction and the nuclear Zeeman interaction are of the same order. Interference between allowed transitions (———) and semi-forbidden transitions (------) gives rise to the nuclear modulation effect in the envelope of electron spin echoes. This interference can only occur, however, if the microwave magnetic field H_1 is large enough to excite both kinds of transitions simultaneously. The quantity $\gamma H_1/2\pi$ should be at least equal to the spread in the nuclear shfs levels.

$$\sum_{i,j}^{i \neq j} {}' \chi_{ij}^{(\alpha)} + \sum_{k,n}^{k \neq n} {}' \chi_{kn}^{(\beta)} = - \sum_{i,j}^{i \neq j} {}' \sum_{k,n}^{k \neq n} {}' \chi_{ij,kn}^{(\alpha,\beta)} \qquad [3]$$

For a nucleus with I = 1/2, there are only two shfs levels in the α and β manifolds, and Equations 2 and 3 reduce to

$$V_{mod} = (1 - \frac{k}{2}) + \frac{k}{2} [\cos(\omega_\alpha \tau) + \cos(\omega_\beta \tau)$$

$$- \frac{1}{2} \cos((\omega_\alpha + \omega_\beta)\tau) - \frac{1}{2} \cos((\omega_\alpha - \omega_\beta)\tau)] \qquad [4]$$

in which ω_α and ω_β are the radian frequencies for the two intervals concerned. In this case, the coefficients X reduce to a single modulation depth parameter k that depends on the ratio between "allowedness" and "forbiddenness" for the microwave transitions connecting the α and β manifolds. In the more general case, the X coefficients are calculable from a matrix M which specifies the allowedness and forbiddenness of all the transitions connecting the α and β manifolds. M is given by the product $M_\alpha^\pm M_\beta$, in which M_α and M_β are unitary matrices describing the nuclear substate mixing in the α and β manifolds. If, for example, a representation is chosen so as to be diagonal for the shfs states in the α manifold, then $M = M_\beta$, where M_β is the matrix whose elements specify the re-mixing of nuclear substates caused by changing the electron spin state from $M_s = 1/2$ to $M_s = -1/2$.

The source of the nuclear modulation effect is an interference effect between allowed and semi-forbidden microwave transitions. This must be remembered whenever echo envelope spectroscopy is considered as a substitute for ENDOR. It is essential for M to contain off-diagonal elements in order that there should be non-zero values for the coefficients $X\binom{\alpha}{ij}$, $X\binom{\beta}{kn}$, $X\binom{\alpha,\beta}{ij,kn}$ (see Ref. 5, Eq. 4-6). If the nuclear substate compositions are the same in both manifolds, $M_\alpha = M_\beta$, $M = I$ (the unit matrix), and transitions are either wholly allowed or wholly forbidden. Only one transition can connect each α state with each β state, the X coefficients are all zero, and there is no interference effect or envelope modulation. This happens when the electron-nuclear coupling term is very small compared with other nuclear terms in the spin Hamiltonian--for example, when the local field due to the electron is small compared to the Zeeman field, or when the nuclear quadrupole splitting is very large. However, it should be noted that the same situation is also encountered in some instances in which the electron-nuclear coupling is much larger than the other terms in the nuclear spin Hamiltonian, as is often the case for a coupling term of the form, A I·S Quantization in both the α and β manifolds is then determined almost entirely by the A I·S term and the states (appropriately ordered) have the same composition in both. Here again, the microwave transitions are either highly allowed or highly forbidden, and no significant modulation effect can be seen.

A less fundamental reason for failing to observe a modulation effect may be that the microwave field amplitude H_1 is too small to excite both the allowed and semi-forbidden transitions for a given paramagnetic center. One or the other of the transitions is then effectively off-resonance. The remedy here is to design the experimental apparatus so that $\gamma H_1/2\pi$ (where γ is the magnetogyric ratio for the electron) is at least equal to the frequency spread of the shfs levels. For example, a linearly polarized microwave field with amplitude 10 G corresponds to a circularly polarized component $H_1 = 7$ G and a factor $\gamma H_1/2\pi \cong 20$ MHz. This is more than adequate to generate the proton modulation pattern for weakly coupled hydrogen nuclei in a typical X-band experiment (operating frequency $f \cong 9.4$ GHz, Zeeman field $H_0 \cong 3000$ G, proton shfs splittings $\cong 13$ MHz). A microwave field of $H_1 = 7$ G will turn a g = 2 electron spin by 120° in 17 nsec and would normally be available in a spectrometer designed to work with 20 nsec microwave pulses. An interesting figure showing the disappearance of proton modulation as H_1 is reduced for a sample containing both ^1H and ^2H is given by Kevan et al.[6] (Figure 5 in ref. 6).

Failure to observe a modulation effect (provided that it is not due merely to an inadequate apparatus) can sometimes provide a limited amount of useful chemical information about the structure of a chemical complex. Either the electron-nuclear coupling corresponding to the nucleus in question is very weak, indicating that the nucleus does not belong to the coordination sphere of the atom on which the electron is located, or the coupling is very strong and might with some effort be detected as a fine structure in the EPR spectrum. There are, however, certain other shortcomings of the two-pulse echo envelope spectroscopy when used as a replacement for ENDOR. The modulation pattern contains sum and difference frequencies in addition to the desired ENDOR frequencies. These will clutter up the spectrum and make it hard to resolve individual components when several different nuclei are involved. The shfs lines will, moreover, be lifetime-broadened by the phase memory decay which is characteristic of the two-pulse electron spin echo experiment. In materials of interest the phase memory time tends to be quite short. For example, in biological or other organic materials it is $\cong 2$ μsec, this limit being set by local magnetic field disturbances caused by the flip-flopping of protons in the environment of the electron spin.

Three-pulse echo

Fortunately, both of these problems can be solved by performing three-pulse echo experiments (Figure 3) in which the stimulated echo (SE) amplitude is used to plot the envelope. The mechanism underlying stimulated echo generation is as follows. The first two pulses impose a cosine-shaped toothed pattern of pitch $\Delta f = 1/\tau$ on an initially smooth resonance line. This toothed pattern is then subsequently detected by the method of free induction spectroscopy, i.e., by applying an additional pulse (pulse III). The resulting Fourier transform of the toothed pattern consists of a single pulse (SE) offset from pulse III by an interval τ.

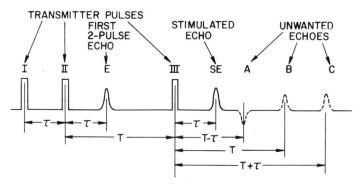

Figure 3. Transmitter pulse sequence and spin echo signals observed in a stimulated echo experiment. In order to obtain the envelope function, τ is set to a fixed value and T is slowly increased. Superhyperfine frequencies appear as terms of the form $\cos(\omega(T+\tau))$. The unwanted echoes C, B, A are two-pulse echoes generated by combinations of pulse III with pulse I, with pulse II, or with the first two-pulse echo E. Overlap of echoes B, A, with the stimulated echo, SE, occurs when $T = \tau$ and $T = 2\tau$ and may cause glitches to appear in the echo envelope function. (Reproduced with permission from the Journal of Biological Chemistry.)

The manner in which pulses I and II give rise to a toothed pattern may be understood by considering an idealized spin echo experiment performed on an EPR resonance line narrow compared to H_1. All three pulses would then be 90° pulses, and for spins that were exactly on resonance, the first two 90° pulses would sum their effects to produce a spin inversion. Spins $\pm \Delta f/2$ off resonance would execute a 180° precession relative to the phase of the microwave field in the time τ between pulses I and II, and would therefore be returned to their initial orientation. Spins $\pm \Delta f$ off resonance would execute a 360° precession relative to the phase of the microwave field during the time τ and would be inverted, like those on exact resonance, and so on. The important feature to note here is that the pattern imposed by pulses I and II exists as a z-component of magnetization between pulses II and III and is not subject to phase memory decay until it is converted into a precessing magnetization by pulse III. The toothed pattern fills in slowly by cross-relaxation mechanisms, thus leading to a decay in stimulated echo amplitude with increasing time T, but decay is slow compared to the phase memory decay in a two-pulse echo experiment, and it usually affords ample time for the observation of envelope modulation effects.

ENDOR by Fourier Transformation of the Electron Spin Echo Envelope

The function modulating the stimulated echo envelope is given by

$$V_{mod}(\tau,\tau') = X_0 + (1/2) \sum_{i,j}{}^{'\,i \neq j} X_{ij}^{(\alpha)} [\cos(\omega_{ij}^{(\alpha)}\tau) + \cos(\omega_{ij}^{(\alpha)}\tau')]$$

$$+ (1/2) \sum_{k,n}{}^{'\,k \neq n} X_{kn}^{(\beta)} [\cos(\omega_{kn}^{(\beta)}\tau) + \cos(\omega_{kn}^{(\beta)}\tau')]t$$

$$+ \sum_{i,j}{}^{'\,i \neq j} \sum_{k,n}{}^{'\,k \neq n} X_{ij,kn}^{(\alpha,\beta)} [\cos(\omega_{ij}^{(\alpha)}\tau') \cos(\omega_{kn}^{(\beta)}\tau)$$

$$+ \cos(\omega_{kn}^{(\beta)}\tau') \cos(\omega_{ij}^{(\alpha)}\tau)] \quad [5]$$

in which $\tau' = T + \tau$. When $T \to 0$, i.e., when $\tau = \tau'$, Equation 5 reduces to the two-pulse modulating function of Equation 2. (In the idealized narrow-line case, the 90° pulses II and III would combine to form a single 180° pulse.) For an $I = 1/2$ nucleus, the modulating function can be written in the form

$$V_{mod}(\tau,\tau') = 1 - (1/4)\, k\, [(1 - \cos(\omega_\alpha \tau))(1 - \cos(\omega_\beta \tau'))$$

$$+ (1 - \cos(\omega_\alpha \tau'))(1 - \cos(\omega_\beta \tau))] \quad [6]$$

The formula for the modulating function in the case where n nuclei are coupled to the same electron is[7]

$$V_{mod}^{(I_1,I_2,\cdots,I_n)} = (1/2) \prod_{i=1}^{n} V_i(\tau,\tau') + (1/2) \prod_{i=1}^{n} V_i(\tau',\tau) \quad [7]$$

in which the functions V_i are obtained by writing V_{mod} in Eqs. 5 and 6 as the sum of two terms

$$V_{mod}(\tau,\tau') = V(\tau,\tau') + V(\tau',\tau) \quad [8]$$

Thus, in the case of Equation 6,

$$V(\tau,\tau') = \frac{1}{2} - \frac{1}{4} k \, (1 - \cos(\omega_\alpha \tau))(1 - \cos(\omega_\beta \tau')) \qquad [9]$$

The frequency suppression effect

It can be seen from Equations 5 and 6 that, if one measures the stimulated echo envelope by setting τ to a fixed value and varying T (i.e., varying $\tau' = T + \tau$), there will be no sum and difference frequencies to contend with. (The same is true if one sets τ' to a constant value and varies τ, but this is generally less convenient in practice.) The sum and difference terms of Equation 2 have, indeed, not disappeared, since Equation 5 reduces to Equation 2 in the limit that T = 0, but they appear in Equation 5 in a different form. The amplitudes of frequency components in the envelope belonging to the α manifold are now partly dependent on the values of frequencies in the β manifold and on the setting chosen for τ. Thus, in the I = 1/2 case (Equation 6), if τ is set equal to a whole number of cycles of the frequency ω_β, the factor $(1 - \cos(\omega_\beta \tau)$ is zero, and V_{mod} reduces to $[1 - (1/4)k(1 - \cos(\omega_\alpha \tau))(1 - \cos(\omega_\beta \tau'))]$. Since τ' is the experimental time variable, only the ω_β oscillation appears in the envelope and the ω_α oscillation is suppressed.

In the more general case (Equation 5), frequency suppression is not complete, but it is often still observable, and it can be experimentally useful as a means of assigning shfs transitions to the correct manifold. The frequency suppression effect is illustrated in Figure 4, which shows stimulated echo envelopes obtained for a frozen solution sample of the blue copper protein stellacyanin at a series of τ settings. The modulation is due to coupling between Cu(II) and a ^{14}N (I = 1) nucleus in an imidazole ligand belonging to a histidine residue in the protein.[8] Two frequencies are prominent in the envelope, one at 4 MHz associated with the transition between the outermost levels in the α manifold (Figure 2), and one at 1.47 MHz associated with the same transition in the β manifold. The corresponding two periods in the envelope are 250 nsec and 680 nsec. It will be noted that when τ = 224, 296 nsec (which is close to one period of the 4 MHz component), the 680 nsec modulation becomes appreciably weaker, and when τ = 516 nsec, the 680 nsec period is practically unobservable. A very elegant demonstration of this effect in a single crystal sample has been given by Merks and de Beer.[9] It will be noted incidentally, that none of the envelope tracings in Figure 4 show any proton modulation, although there are numerous weakly coupled hydrogen nuclei in the ligands and in the surrounding protein. This modulation, which would occur at $\cong$ 13.5 MHz, has been eliminated from the data by selecting τ values which are approximate multiples of the proton precession period ($\cong$ 74 nsec), thereby using the frequency suppression effect to simplify the form of the envelopes.

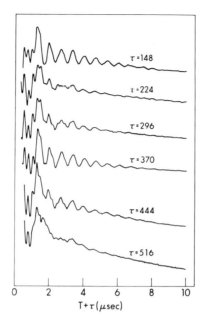

Figure 4. Stimulated echo envelope recordings for a frozen solution sample of the blue copper protein stellacyanin at a field $H_0 = 3165$ Gauss and a frequency 9.175 GHz. The modulation pattern is due to the remote ^{14}N nucleus in a histidyl ligand which coordinates the Cu(II) ions. These results are taken at times τ which illustrate the "frequency suppression effect" (see text).

We see than that three of the problems likely to be encountered when using echo envelope spectroscopy in place of ENDOR can either be discounted or disposed of. Failure to observe modulation is, assuming adequacy of the electron spin echo equipment, a possible source of information in itself. Lifetime broadening due to phase memory decay, and unwanted complexity of the modulation pattern due to sum and difference terms can both be avoided by performing experiments in the stimulated echo mode. The remaining problem is less tractable and has not yet been solved in a wholly satisfactory manner. This is the problem of transforming the echo envelope to yield a frequency spectrum.

FOURIER TRANSFORMATION OF THE INCOMPLETE TIME WAVEFORM

From the form of the theoretical expressions (Equations 2 and 5) it would seem that the shfs frequencies might be obtained quite straightforwardly by making a Fourier cosine transform. Some minor problems could arise at the very low frequency end on account of phase memory decay or in cases where the modulation is shallow, but these could easily be dealt with by subtracting out the d.c. and low frequency components before the data are transformed. A polynomial of the form $P(\tau') = a_0 + \Sigma a_i \tau'$ containing three or four terms with coefficients chosen to give a rough fit to the data has been found useful for this purpose.[10] The main difficulty, however, is that the time waveform is incomplete at the low τ (or τ') end. In the two-pulse case, the missing portion of the data corresponds to the time during which the microwave transmitter signal in pulse II decays to the thermal noise level. Decay is over a range $\cong$ 140 dBs and requires $\cong$ 32 cavity decay time constants. In the stimulated echo case, the problem is apparently worse, since, in Figure 3, T cannot be less than zero and τ' cannot therefore be less than the preset value τ. In actual fact the $\cos(\omega\tau')$ terms continue their evolution when T is made negative by moving pulse III backward in time across pulse II. (The experiment remains a stimulated echo experiment with the roles of τ and τ' interchanged.) However, the envelope is broken by "glitches" occurring at $T = 0$, $T = -(1/2)\tau$, and $T = -(1/3)\tau$, where two pulse echoes overlap the stimulated echo, making this region of little use. Similar glitches also occur in the "positive T" range at $T = 2\tau$, and $T = 3\tau$, but they tend to be weaker and are more easily excised from the envelope.

Window smoothing

There are two possible approaches to the problem of transforming the incomplete time waveform. One is to smooth over the discontinuity at the start of the data by means of a window function; the other is to attempt to reconstruct the missing portion. The choice between the two depends largely on the nature of the sample, and on how much of the waveform is missing. The window function approach has been explored by Merks and de Beer,[11] who start out with stimulated echo envelopes obtained from a single crystal sample of Cs_2ZnCl_4 doped with Cu(II). The modulation pattern here is due to ^{133}Cs and continues for $\cong$ 300 μsec. Data is multiplied by the "cosine bell" window function

$$h(\tau') = 1, \qquad \tau'_i + \alpha(\tau'_f - \tau'_i) \leq \tau' \leq \tau'_f - \alpha(\tau'_f - \tau'_i)$$

$$h(\tau') = \frac{1}{2}\left[1 - \cos\left\{\frac{\pi(\tau'-\tau'_i)}{\alpha(\tau'_f - \tau'_i)}\right\}\right], \qquad \tau'_i \leq \tau' \leq \tau'_i + \alpha(\tau'_f - \tau'_i)$$

$$h(\tau') = \frac{1}{2}\left[1 - \cos\left\{\frac{\pi(\tau'-\tau'_f)}{\alpha(\tau'_f - \tau'_i)}\right\}\right], \qquad \tau'_f - \alpha(\tau'_f - \tau'_i) \leq \tau' \leq \tau'_f$$

$h(\tau') = 0$ otherwise, [10]

in which τ_i' and τ_f' are the beginning and end points of the data. The window function rises smoothly to unity over a central region covering a fraction, $1 - 2\alpha$, of the recorded data, and falls smoothly at the end. (The windowing operation is usually unnecessary at the τ_f' end, since the recorded modulation amplitudes decay smoothly to zero. It was required in certain of the examples discussed in Ref. 11 because of the use of a procedure to correct for phase memory decay.)

The advantage of multiplying the data by a smooth window function can be appreciated from the following considerations. The recorded waveform may be thought of as the ideal waveform with a short section deleted at the start. The resulting FT is given by the correct spectrum minus a spectrum obtained by Fourier transforming the deleted portion. The latter is an artifact. If we make no attempt to deal with the problem and simply treat the waveform during the dead time from $\tau' = 0$ to $\tau' = \tau_i'$ as nonexistent, the missing portion consists of the true waveform multiplied by a step function

$$h'(\tau') = 1, \qquad 0 \leq \tau' \leq \tau_i$$
$$h'(\tau') = 0, \qquad \tau' > \tau_i \qquad [11]$$

The artifact is therefore given by the convolution of the true spectrum, including whatever d.c. and low frequency portions have not yet been removed by pre-processing, with the FT of Equation 11, i.e., with $\sin(\omega\tau_i)/(\omega\tau_i)$. This artifact will introduce subsidiary peaks offset from zero by Δf, and lying Δf to either side of the genuine lines, where $\Delta f = 0.75\ \tau_i'$, $1.75\ \tau_i'$, $\cdots$, etc. (allowing for the fact that the artifact consists of a subtraction from the true spectrum). The worst effects here are due to the sharpness in the fall of the step function $h'(\tau')$, and they can be minimized by making the data rise gently from zero at the start. Thus, if a cosine bell window function is used, the artifact consists of the true spectrum convoluted with the FT of the complement of $h(\tau')$ in Equation 10. In this case, the artifact may integrate up to a quantity that is actually larger than it would be if no window at all were used, but it lacks prominent peaks and side lobes, and is therefore less likely to generate spurious spectral lines. By applying window functions to recorded stimulated echo envelope data, Merks and de Beer have been able to obtain clear and well-resolved shfs spectra for ^{133}Cs in single crystals of Cs_2ZnCl_4:Cu(II)[11] and also for ^{14}N in single crystals of $Bi_2Mg_3(NO_3)_{12}24H_2O$:Gd(III).[9] They have avoided making a phase correction by taking the Fourier modulus transform, i.e., the square root of the Fourier power transform, when deriving the spectra.

These methods succeed less well with data obtained for frozen solution samples. The shfs lines tend to be much broader here, and the modulation pattern decays in times < 10 μsec (see, e.g., Figure 4), thus making the deleted portion of the data a more significant fraction of the whole. The use of the window function [10] in conjunction with the curves in Figure 4 would seriously reduce the amount of time-domain information available and could perhaps eliminate some components from the spectrum (e.g., the 4 MHz, 250 nsec component in Figure 4). It is, moreover, inadvisable to sacrifice phase information when processing data of this kind, since the broad shfs line may only be represented by one or two oscillations. For example, the two maxima at $\cong$ 1.5 μsec and $\cong$ 3.0 μsec that can be seen in the τ = 516 tracing of Figure 4 afford clear evidence for the presence of a broad shfs line at $\cong$ 0.7 MHz. This inference might indeed be made from any of the tracings in Figure 4, since they all contain a maximum at $\cong$ 1.5 μsec. However, if the time origin were to be disregarded, and if the data were to be processed by using the Fourier modulus transform, this component of the shfs spectrum would appear to be much weaker and might be hard to identify with certainty.

Reconstruction of the waveform

In a case such as this, there is much to be gained by reconstructing the waveform in the dead time interval, provided that this can be done without generating spectral artifacts worse than those generated by using alternative procedures. A certain amount of general knowledge concerning the form of the modulation function can be drawn on when making the reconstruction. Thus, the envelope consists of a fairly small number of damped cosine terms and rises to a maximum at τ' = 0. Also, there is often some additional experimental information, such as two-pulse data or stimulated echo data at other τ settings, which suggests the correct form of the function in the dead time region. Making use of this information, one can sketch in a tentative reconstruction line, digitize the completed curve, and perform a Fourier cosine transform.

There remains, of course, the question as to whether all of the spectral features appearing in the transform of a reconstructed curve are implicit in the data, or whether certain of them have been introduced by the reconstruction procedure. This might be tested by redrawing the line on a different day, or by soliciting the aid of an impartial colleague. However, a computer procedure offers an easier means for checking and refining the completed time waveform.

The procedure is illustrated in Figures 5 and 6. Figure 5a consists of the stimulated echo envelope for a bis-imidazole complex of heme-a in frozen solution.[12] Small glitches, occurring at τ' = 300 nsec and 450 nsec due to the momentary overlap of the stimulated echo and two pulse echoes shown as "unwanted echoes" in Figure 3, have been excised. The tentative "zero-order" reconstruction

ENDOR by Fourier Transformation of the Electron Spin Echo Envelope

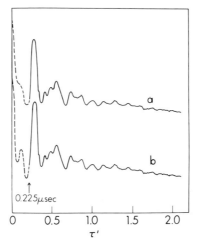

Figure 5. (a) Stimulated echo envelope recording for a frozen solution of the low spin Fe(III) compound, bis-imidazole heme a in H_2O:glycerol. In the figure, $\tau' = T + \tau$. The broken line at the start is a tentative reconstruction of the missing portion of the envelope, made in order to facilitate Fourier cosine transformation. (b) Envelope as in (a), but with computer refinement of the reconstruction as described in the text and shown in Figure 6.

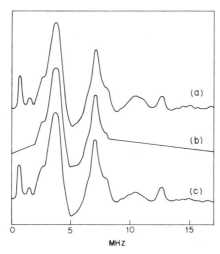

Figure 6. (a) Fourier cosine transform of the time waveform in Figure 5a. (b) Spectrum as in 6a with deletion of all but the two most prominent features. This function is transformed back into the time domain, and the initial portion of it (from $\tau' = 0$ to $\tau' = 0.225\,\mu sec$) is used to replace the broken line reconstruction in Figure 5a. (c) Fourier cosine transform of the revised time waveform (Figure 5b).

between $\tau' = 0$ and $\tau' = 0.225$ nsec, where data recording commenced, is shown as a broken line. The cosine transform of 5a is shown in Figure 6a.

The next step is to identify the most prominent peaks in the spectrum and compute a new reconstructed section based on these frequencies only, deleting the rest of the spectrum as a potential artifact. The simplified spectrum obtained by making these deletions is shown in Figure 6b. It is transformed back into the time domain, and the initial portion of it, between $\tau' = 0$ and $\tau' = 0.225$ μsec, is joined smoothly (by adding a small constant amount) to the recorded data. This revised time waveform (Figure 5b) is once again transformed into the frequency domain (Figure 6c). The refinement procedure described here could be repeated any number of times. In the illustration, once was enough to produce a spectrum that did not change significantly in subsequent iterations.

From a comparison of Figures 6a and 6c, it can be seen that little was altered as a result of the computer refinement procedure. The two weak low-frequency lines at $\cong 0.7$ and $\cong 1.5$ MHz continued to show up in the spectrum, although they were deliberately excluded from the simplified spectrum (Figure 6b) used in the reconstruction step. These two lines can therefore be presumed to be genuine and not a product of the reconstruction itself. However, the same is not true of a broad spectral feature at $\cong 10.5$ MHz, which is significantly weaker in Figure 6c than in Figure 6a, suggesting that this might be an artifact due to imperfect reconstruction of the time waveform.

The procedure illustrated above is obviously a very rudimentary one and might be improved in a number of ways. Deletion of spectral components by straight line sections in order to obtain the simplified spectrum to be used in the reconstruction step is an excessively crude expedient and can lead to difficulties if the straight line sections do not lie close to the zero abscissa. A considerable amount of experimentation, based on different kinds of recorded data, is still needed in order to arrive at the procedure best suited to the derivation of ENDOR data from electron spin echo envelope measurements.

ADVANTAGES OF FT OF ELECTRON SPIN ECHO ENVELOPE

Alternative to ENDOR

In the foregoing discussion, echo envelope spectroscopy has been presented as an alternative to ENDOR. The method cannot, as pointed out earlier, be used unless there is a significant degree of branching between allowed and semiforbidden transtions. The field of application is therefore a restricted one. However, it should be noted that, in certain circumstances, the echo envelope method may prove more successful than ENDOR in detecting shfs lines. The data shown in Figure 4 provide an example. The modulation patterns in the Figure are due to the low-frequency shfs intervals associated with the coupling between Cu(II) and the remote nitrogen nucleus of an imidazole ligand. These patterns, or similar ones, are readily seen for a number of Cu(II)-containing proteins,[8,13,14] but the corresponding shfs lines are difficult to detect by ENDOR.

The reason here is that the ^{14}N nucleus has a small magnetic moment, which is hard to drive by r.f. fields, the difficulty being compounded in the case of frozen solution samples by line broadening due to the ^{14}N nuclear quadrupolar coupling. Echo envelope modulation effects are, on the other hand, quite easy to see for nuclei with small moments. For the simple case of an $I = 1/2$ nucleus weakly coupled to an $S = 1/2$ electron, it can be shown that the modulation depth is independent of the nuclear moment and depends only on the ratio between the Zeeman field and the local field at the nucleus due to the electron.[15] Breadth of the shfs line is, moreover, not a serious obstacle to detection, provided that at least one modulation cycle can be seen in the echo envelope.

Quantitation of shfs intensities

Another important advantage of echo envelope spectroscopy is that it can be used for quantitative measurements of line intensities in the shfs spectrum, because the depth of modulation is exactly calculable from the spin Hamiltonian parameters and is independent of apparatus settings (provided that the microwave magnetic field amplitude H_1 is not too small to excite the band of microwave transitions concerned). Shimizu et al.[16] have demonstrated the use of echo envelope methods in the study of chemical complex formation. In this work, the increase in depth of the ^{31}P modulation pattern in a frozen aqueous solution of Nd(III) and adenosine triphosphate (ATP) was observed as a function of ATP concentration. The ^{31}P pattern reached a maximum depth at an ATP:Nd(III) ratio of 3:1, this being the same as the ratio required to produce a maximal effect in the optical spectrum. Deuterium can also be used as a ligand marker, as is shown by the experiments of Kevan and co-workers on the coordination of solvated electrons and radicals.[3] In these experiments, the data were not transformed into the frequency domain, since the modulation patterns were fairly simple in form and could be recognized in the time waveform. A reliable Fourier transformation procedure would, however, extend the scope of the method by making it possible to isolate the shfs lines of interest, and to plot titration curves analogous to the curves commonly plotted by using other physical methods of detection.

SUMMARY

Superhyperfine spectra due to electron-nuclear coupling are often more easily obtained by recording and transforming the electron spin echo decay envelope than by performing double resonance experiments. Electron spin echo measurements have the additional advantage that the magnitude of the effect, i.e., a modulation of the echo envelope by the superhyperfine frequencies, depends only on the coupling parameters and not on instrumental settings. It is also possible to correlate superhyperfine transitions belonging to each of the two electron spin states. The main practical problem is that the echo envelope function cannot be recorded in its entirety owing to the existence of an observational dead time following the microwave transmitter pulses. The dead time can be reduced to $\cong$ 100 nsec ($\cong$ 1 1/2 proton periods in an X-band experiment) by careful attention to microwave circuit design. This missing portion of the envelope can also be reconstructed by computer methods to yield satisfactory Fourier transform spectra.

REFERENCES

1. Mims, W. B. 1972, in "Electron Paramagnetic Resonance," ed. S. Geschwind, Plenum, New York, Chapter 4.
2. Salikhov, K. M., Semenov, A. G., and Tsvetkov, Yu. D. 1976, "Electron Spin Echoes and Their Applications," Nauka, Novosibirsk.
3. Kevan, L. 1979, in "Time Domain Electron Spin Resonance," ed. L. Kevan and R. N. Schwartz, Wiley-Interscience, New York.
4. Mims, W. B. 1972, Phys. Rev. B5, 2409.
5. Mims, W. B. 1972, Phys. Rev. B6, 3543.
6. Kevan, L., Bowman, M. K., Narayana, P. A., Boeckman, R. K., Yudanov, V. F., and Tsvetkov, Yu. D. 1975, J. Chem. Phys. 63, 409.
7. Dikanov, S. A., Yudanov, V. F., and Tsvetkov, Yu. D. 1979, J. Magn. Reson. 34, 631 (1979).
8. Mims, W. B., and Peisach, J. 1979, J. Biol. Chem. 254, 4321.
9. Merks, R. J. P., and de Beer, R. 1979, J. Phys. Chem. 83, 3319.
10. Blumberg, W. E., Mims, W. B., and Zuckerman, D. 1973, Rev. Sci. Instrum. 44, 546.
11. Merks, R. J. P., and de Beer, R. 1980, J. Magn. Reson. 37, 305.
12. Peisach, J., and Mims, W. B. (unpublished data).
13. Kosman, D. J., Peisach, J., and Mims, W. B. 1980, Biochemistry 19, 1304.
14. Mondovi, B., Graziani, M. T., Mims, W. B., Oltzik, R., and Peisach, J. 1977, Biochemistry 16, 4198.
15. See, e.g., Equations 21, 22 in Reference 3, p. 293. If A and B are less than ω_I, the depth parameter $k = (B/\omega_I)^2$, which is proportional to $(g\beta/H_0 r^3)^2$.
16. Shimizu, T., Mims, W. B., Peisach, J., and Davis, J. L. 1979, J. Chem. Phys. 70, 2249.

ADVANCES IN FT-NMR METHODOLOGY FOR PARAMAGNETIC SOLUTIONS: DETECTION

OF QUADRUPOLAR NUCLEI IN COMPLEX FREE RADICALS AND BIOLOGICAL SAMPLES

N. S. Dalal

Chemistry Department
West Virginia University
Morgantown, WV 26506

INTRODUCTION

This chapter reviews our recent studies aimed at developing the FT-NMR method as a "beat-relaxation" technique which permits the detection of fast-relaxing spins such as quadrupolar nuclei in paramagnetic solutions. The technique is termed "beat relaxation" since the FT-NMR method takes advantage of the short spin-lattice relaxation times ($<10^{-11}$ sec) whereas fast relaxation renders other paramagnetic resonance techniques unsuitable for detecting such species. Our primary focus will be on measuring small ($\leq$ 1 Gauss) hyperfine (hf) couplings from quadrupolar nuclei in large organic radicals in solution where such measurements have been feasible for the first time. The significance of the FT-NMR advances may be gauged from the fact that measurements of small ^{14}N couplings in free radicals had previously not been possible despite the use of EPR, conventional NMR, electron-nuclear double-resonance (ENDOR),[1,2] electron-electron double-resonance (ELDOR),[1,2] and electron-nuclear triple resonance.[3]

Since this chapter deals mainly with NMR, we will assume familiarity with EPR, ENDOR, ELDOR and triple resonance. For background in these techniques, the two recent books[1,2] on this subject may be consulted. An excellent description of ENDOR via spin-echo techniques has been presented by W. B. Mims in this book. The chief aim in our undertaking is to demonstrate that the fast electron-nuclear relaxation processes that make other electron resonance techniques unsuitable for measuring hf couplings can be harnessed to yield the same data via FT-NMR techniques. Illustrative examples will be chosen from our recent studies of large organic free radicals. Finally, the extension of the FT-NMR method to paramagnetic biological systems is pointed out, with the conclusion that even there the method holds considerable promise.

COMPARISON OF EPR, ENDOR, ELDOR, AND FT-NMR

EPR is the most common technique for measuring hf couplings. For large radicals, however, difficulties could arise in the analysis of EPR spectra, either from too many spectral lines or from the inherently low resolution due to a superposition of signals or from the line-broadening caused by electronic spin-lattice relaxation. Although ENDOR and ELDOR[1,2] can aid in the analysis of EPR spectra, these techniques have not been useful in cases where some of the splittings are due to quadrupolar nuclei possessing small magnetic moments and situated at sites of low spin density (i.e., have small hf coupling constants). A typical example is the small ^{14}N hf couplings in large organic free radicals containing -NO$_2$ groups on the aromatic rings, such as 2,2-diphenyl-1-picrylhydrazyl (DPPH) and picryl-N-aminocarbazyl (PAC), whose structures are shown in Figure 1.

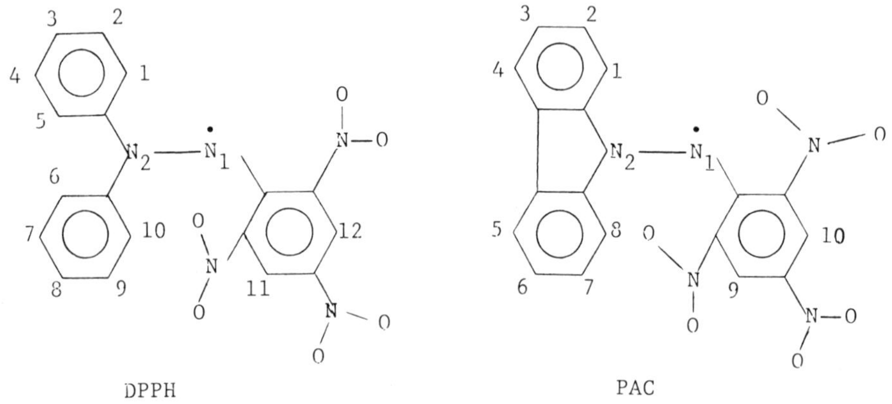

Figure 1. Structures of DPPH and PAC.

As an example, the EPR spectrum of DPPH, one of the first radicals investigated by EPR,[4-6] is not yet understood, despite its extensive study by EPR,[4-16] proton NMR,[15,17-22] ^{14}N broad-line NMR,[23] ENDOR,[14,22,24,25] ELDOR,[13] and theoretical calculations.[16,26] It seems appropriate here to summarize the salient features of these studies, since they will serve as a critique of the complementary nature and limitations of the above-mentioned techniques.

A typical high-resolution EPR spectrum of DPPH, shown in Figure 2, will also serve to illustrate the difficulties encountered in analyzing the EPR spectrum of a large radical. Only at best 150, out of potentially 10^5 lines, are resolved. Thus, simulation of such an EPR spectrum with a computer will necessarily have a high degree of ambiguity in the (assumed) set of hf couplings. However, part of the

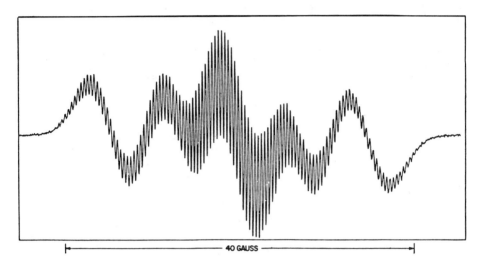

Figure 2. A high-resolution EPR spectrum of DPPH.

ambiguity could be removed with the help of double-resonance experiments. The largest coupling constants, from the central nitrogens (N_1 and N_2)[7,10], were measured accurately with ELDOR by Hyde et al.[13] The resolution of ELDOR did not permit measurements of hf couplings from the twelve ring protons. Dalal et al.[14] subsequently used ENDOR to measure hf couplings from all 12 protons and these measurements were in essential agreement with previous less accurate NMR results. More recently Biehl et al.[22] have reported detailed EPR, NMR, ENDOR, and electron-nuclear triple resonance studies of DPPH. That work confirmed Hyde's ELDOR measurement of the N_1 and N_2 couplings and much of our own ENDOR results[14,24,25] for the protons, except for an interchange of couplings for the ortho- and para-phenyl protons. A major disappointment was that despite signal enhancement with triple resonance, no satisfactory signals were observed for the -NO_2 nitrogens.

We believe that the failure to detect multiple electron resonance signals from the -NO_2 nitrogens is related to the low γ_N, fast relaxation, and the small hf coupling for ^{14}N. As has been shown phenomenologically be Allendoerfer and Maki[27] and more rigorously by Freed,[28] the ENDOR enhancement in solution, I_{ENDOR}, is given approximately by

$$I_{ENDOR} \propto \nu_E^{3/2} [T_2 A^2/(T_2^2 A^2 + 0.063)] \qquad [1]$$

$$\nu_E = (\gamma_N H_o \pm A/2) \qquad [2]$$

Here ν_E is the ENDOR resonance frequency, A is the hf coupling, T_2 is a phenomenological electronic spin-spin relaxation time, and H_0 is the Zeeman field. It can thus be seen that $I_{ENDOR} \to 0$ as the spin-spin relaxation time $T_2 \to 0$, as is expected to be the case with quadrupolar nuclei. Similarly $I_{ENDOR} \to 0$ as $A \to 0$. Furthermore, for a given set of isotopes, the quadrupolar members have significantly smaller γ_N than those with spin 1/2. The proton-deuteron set provides a particularly good example: $\gamma_H/\gamma_D = 6.514$. Thus for a small A, $\nu_E \cong \gamma_N H_0$, contributing to a further reduction in ENDOR intensity for the quadrupolar nucleus. All these factors combined together appear to be largely responsible for the failure of ENDOR measurement of small hf coupling from any quadrupolar nucleus in solution.

Here it is worthwhile to present a concise summary of the advantages and disadvantages of the various electron-nuclear resonance techniques and Table 1 contains this information.

Although a large portion of the contents of Table 1 has been discussed above, some further clarifying comments seem desirable. An important advantage of NMR over EPR, ENDOR and ELDOR is that whereas all the techniques may determine the magnitude of a hyperfine coupling constant, only NMR can yield the absolute sign. This conclusion has its origin in the fact that in EPR, ENDOR and ELDOR, the multiplet structure is (normally) symmetric with respect to center (i.e., position of zero coupling constant), but in NMR the contact shift is proportional to the coupling constant itself; hence its direction relative to the position for the zero coupling constant yields the absolute sign directly. Another big advantage for NMR is the simplicity of the spectra. For a paramagnetic molecule with, say, 10 coupling constants, the number of lines will be 2^{10} for EPR, $2 \times 10 = 20$ for ENDOR, 20 or more for ELDOR, but only 10 for NMR. This advantage becomes more significant, the more complex the molecule. Finally, the relaxation processes dominate the intensity of an ENDOR and ELDOR signal, thereby making it unsuitable for determining the number of nuclei contributing to the signal. On the other hand, the fast relaxation processes help sharpen the NMR signals. The NMR intensity is thus directly proportional to the number of contributing nuclei, a major analytical advantage over the other techniques.

THEORY OF NMR OF QUADRUPOLAR NUCLEI IN FREE RADICAL SOLUTIONS

Theoretical aspects of NMR of free radicals containing spin-1/2 nuclei have been thoroughly discussed in the literature.[29] Extension of these studies to free radicals containing quadrupolar nuclei (i.e., nuclei with nuclear spin ≥ 1) is relatively straightforward, especially in the solution phase with which we are mainly concerned here. The quadrupolar interaction, however, does lead to fast relaxation and hence broadens the NMR signals. We make a further assumption that the electronic Zeeman term, $g\beta H$, is much

Table 1. Comparison of various techniques of studying paramagnetic molecules.

FEATURE	EPR	NMR	ENDOR* & TRIPLE	ELDOR
Sensitivity	$\gtrsim 10^{-5}$ M (~10^{12} spins)	$\gtrsim 10^{-2}$ M $> 10^{16}$	$\gtrsim 10^{-4}$ M ~10^{13}-10^{14}	$\gtrsim 10^{-4}$ M 10^{13}-10^{14}
Resolution	Poor 1-10 MHz	High (~kHz)	High (~kHz)	Poor ~10 MHz
Identification of nuclei directly	No	Yes	Yes	No
Overlapping from spectra of several species present simultaneously	Yes, severe	Yes	No	No
Relative sign determination	Seldom	Yes	Yes	Not usually
Absolute sign determination	No	Yes	Seldom	No
Intensity proportional to number of nuclei	Yes, if resolved	Yes, generally	Seldom done but possible	Seldom investigated
Study of relaxation phenomena	Not very convenient	Indirect	Indirect	Direct
Time scale	~10^{10} Hz	10^6 Hz	10^8-10^{10} Hz	10^9 Hz
Range of relaxation times	$> 10^{-9}$ s	$\lesssim 10^{-11}$ s	$\gtrsim 10^{-7}$ s	$> 10^{-9}$ s
Measurement of small (1 Gauss) couplings	Not possible for complex species	Yes, via FT-NMR	Usually difficult	No
Spectral complexity for N coupling constants	2^N	N lines	2N lines	$\geq 2N$ lines

* Some of these requirements are less stringent for spin-echo ENDOR, but such experiments have not yet been reported for solutions (see W. B. Mims chapter in this volume).

larger than the hyperfine and quadrupolar terms. To derive the contact shift, we begin from the spin Hamiltonian, H, for the nuclear spin, I, coupled to an electronic spin $S = 1/2$, in which β is the Bohr magneton, H is the applied Zeeman field,

$$H = \beta \, \bar{\bar{H}} \cdot \bar{\bar{g}} \cdot \bar{S} - \hbar \gamma_N \, \bar{H} \cdot \bar{I} + \bar{I} \cdot \bar{\bar{A}} \cdot \bar{S} + \bar{I} \cdot \bar{\bar{Q}} \cdot \bar{I} \qquad [3]$$

$\bar{\bar{A}}$ the total (isotropic + dipolar) hyperfine coupling, and $\bar{\bar{Q}}$ the quadrupole tensor. The rapid tumbling of the molecules due to Brownian motion averages the anisotropic parts of the $\bar{\bar{g}}$ and the $\bar{\bar{A}}$ tensor, leaving behind their isotropic parts, g and A, respectively. The quadrupolar term averages to zero since it has no isotropic part, thereby yielding

$$H = g\beta H \bar{S}_z - \hbar \gamma_N H \bar{I}_z + A \, \bar{I} \cdot \bar{S} \qquad [4]$$

The isotropic shift is given by

$$(\Delta H/H)^{iso} = (A/\hbar \gamma_N H) \langle S_z \rangle \qquad [5]$$

where

$$\langle S_z \rangle = Tr(\rho S_z)/Tr(\rho) \qquad [6]$$

and ρ is the density matrix for the spin system. For a system in thermal equilibrium

$$\rho = e^{-H/(kT)} = 1 - H/(kT) \cong 1 - g\beta H S_z/(kT) \qquad [7]$$

in the high temperature approximation ($kT \gg$ eigenvalues of H), and where $g\beta H \gg (\hbar \gamma_N H + A/\hbar)$, a valid assumption for free radicals.

$$\langle S_z \rangle = \frac{Tr(\rho S_z)}{Tr(\rho)} = \frac{Tr(S_z - (g\beta H/kT)S_z^2)}{Tr(\rho)} \qquad [8]$$

Evaluation[30] of the traces leads to

$$\langle S_z \rangle = - \frac{g\beta H S(S+1)}{3kT} \qquad [9]$$

and

$$(\Delta H/H)^{iso} = \frac{g\beta S(S+1)A}{3\hbar \gamma_N kT} \qquad [10]$$

Thus, in solutions the expression for contact shift for quadrupolar nuclei is the same as that for non-quadrupolar ($I = 1/2$) nuclei. However, the quadrupolar interaction does contribute to nuclear relaxation; the expression for this contribution is given by[31-33]

$$\frac{1}{T_{1Q}} = \frac{1}{T_{2Q}}$$

$$= \frac{3}{40} f(I) \left(\frac{e^2qQ}{\hbar}\right)^2 \left(1 + \frac{\eta'^2}{3}\right)\tau_r \qquad [11]$$

in which $f(I) = (2I + 3)/(I^2(2I - 1))$, $e^2qQ/\hbar$ is the quadrupole coupling constant, $\eta' = (V_x - V_y)/V_z$ is the asymmetry parameter where V_x, V_y, V_z are the principal components of the electric field gradient tensor, and τ_r is the correlation time. The above equation has been derived[31] for the rapid motion limit, $\omega^2\tau_r^2 \ll 1$, and under the assumption that the time dependence of nuclear magnetization follows a simple exponential decay with time constant τ_r. It may be noted that τ_r can be obtained from the celebrated Einstein-Debye relationship

$$\tau_r = \frac{4\pi r^3 \eta}{3kT} \qquad [12]$$

in which η is the viscosity, r the molecular radius, k the Boltzmann constant, and T the absolute temperature. This relationship is important in that it provides a fairly good idea of the temperature dependence of the relaxation times for a given molecule, and can thus be utilized to advantage for choosing experimental conditions which would favor the "fast" relaxation criterion for the success of NMR studies.

An especially useful result obtainable from the T_2-data in solutions is the determination of the quadrupole coupling constants, since the observed total NMR line width $(2/T_2)$ can be attributed to three intramolecular interactions

$$(1/T_2) = (1/T_2)_{FC} + (1/T_2)_D + (1/T_2)_Q \qquad [13]$$

in which $(1/T_2)_{FC}$ is the Fermi-contact term, $(1/T_2)_D$ is the dipolar term, and $(1/T_2)_Q$ is the quadrupolar term. The $(1/T_2)_Q$ term is given by Equation 11, and the other two are given by[31-33]

$$(1/T_2)_{FC} = \frac{1}{3} A^2 \gamma_e^2 S(S+1)\left[\tau_e + \frac{\tau_e}{\omega_e^2 \tau_e^2}\right] \qquad [14]$$

$$(1/T_2)_D = \frac{1}{15} \frac{\rho^2 \gamma_e^2 \gamma_N^2 \hbar^2}{r^6} S(S+1)\left[7\tau_d + \frac{13\tau_d}{1 + \omega_e^2 \tau_d^2}\right] \qquad [15]$$

Here τ_e is the correlation time for the electron exchange, τ_d is the dipolar correlation time defined by $(1/\tau_d) = (1/\tau_e) + (1/\tau_r)$, in which τ_r is the rotational correlation time and P is the fraction of the electron which interacts with the nucleus. It should be noted that the Fermi contact term varies as A^2 while the dipolar term varies as A^2 only when the dominant dipolar interaction is between the nucleus and the spin at the adjacent carbon. We shall see later how Equations 11, 14 and 15 can be used to estimate quadrupolar couplings in solutions of free radicals.

EXPERIMENTAL RESULTS

As discussed in the introduction, the free radical DPPH was taken to be the model for the FT-NMR studies. Small hf couplings from ^{14}N-nitrogroup nitrogens and from deuterium substituted at various positions on the rings have been measured.

^{14}N NMR studies of DPPH

In order to demonstrate the advantages of the FT-NMR method, we shall first discuss the application of continuous-wave (cw) NMR studies of DPPH reported earlier by Dalal et al.,[23] which also marked the beginning of ^{14}N NMR studies of quadrupolar nuclei in free radicals.

The ^{14}N NMR signals were recorded as first derivatives with a Varian V-4210 rf unit and a magnetic field sweep. The magnetic field was modulated at 84 Hz and lock-in detection was employed. The radiofrequency was 4.335 MHz, stabilized by a frequency synthesizer.

Saturated solutions of commercial DPPH in CH_2Cl_2, CH_2Br_2, and tetrahydrofuran were examined. The sharpest ^{14}N signals were obtained with the CH_2Cl_2 solutions and these were used for further studies of the $-NO_2$ ^{14}N contact shifts. The shifts were measured relative to the $-NO_2$ resonances in CH_2Cl_2 solutions of the diamagnetic precursor, diphenyl picryl hydrazine. At 298 K, this shift was determined as +375 (+ 20) parts per million. As a check, some spectra were obtained at higher temperatures in a sealed tube. At +71°C, for example, the shift was found to be +339 ppm. The addition of the free radical solvent, di-t-butyl nitroxide, in an attempt to obtain narrower lines resulted in only a marginal decrease in the observed line width. The inverse dependence of ΔH^{iso} on T was verified through variable-temperature measurements. For example, the shifts of +375 ppm at 298 K and of +339 ppm at 344 k, as cited above, conform to the above relationship to within 4%. This indicates that, within experimental accuracy, the shift is dominated by the isotropic hyperfine interaction (cf. Figure 5, below). The measured shift corresponds to an hf coupling of -0.42 + 0.05 G at 298 K. This coupling must correspond to the smallest splitting observed in the high-resolution EPR spectrum of DPPH.

Advances in FT-NMR Methodology for Paramagnetic Solutions

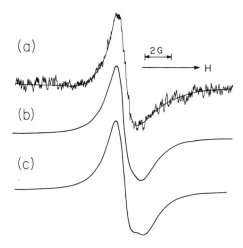

Figure 3. (a) The observed cw ^{14}N NMR signal from $-NO_2$ groups of DPPH in CH_2Cl_2.
(b) Computer-simulated spectrum using $A^N_{ortho} = -0.46$ and $A^N_{para} = -0.38$ G, corresponding to the solution I in the text.
(c) Simulated spectrum using solution II, $A^N_{ortho} = -0.39$ and $A^N_{para} = -0.48$ G.

The observed line shape, presented in Figure 3a, is asymmetric and suggests that the signal is composed of more than one resonance line, i.e., that the ^{14}N hf couplings from the three $-NO_2$ groups might be slightly different. To examine this point more carefully, we wrote a computer program to analyze the experimental spectrum in terms of two overlapping Lorentzian derivatives by a non-linear least squares fitting procedure. Line intensities of 1:2 were assumed, and both the positions and widths of the two lines were fitted. It turned out to be impossible to decide, on the basis of the minimum sum of squares, whether A^N_{ortho} or A^N_{para} is greater, for the following solutions have the same minimum sum of squares:

Solution	Intensity	Line Width (Hz)	Shift (ppm)	A^N(gauss)
I	1	370	+350	-0.38
	2	630	+427	-0.46
II	1	530	+450	-0.48
	2	440	+360	-0.39

The simulated spectra using both sets are shown in Figure 3. Of course, this example demonstrates the limitation of conventional cw NMR techniques; if the spectrum in Figure 3a had better signal-to-noise, then it would have been possible to make the proper assignment. As a matter of fact, in the original paper,[23] solution I was preferred over solution II, based on several theoretical calculations[16,17,26] of the spin density distribution in DPPH.

Figure 4 shows our recent measurements[34] of the ^{14}N FT-NMR of similar samples. Sample preparation was similar to that described earlier, except that we added the free radical, di-t-butyl nitroxide in an attempt to obtain narrower lines.[29] The NMR measurements were made with a Bruker pulse spectrometer operating at a frequency of 13.004 MHz.

For comparison, Figure 4e shows the cw spectrum corresponding to Figure 3a. It may be noted that for the same scan time, the FT-NMR had about two orders of magnitude better signal-to-noise ratio. It was thus possible to carry out a detailed temperature dependence study of the signal line shapes with a view toward detecting possible conformational changes and molecular motion. Several interesting results follow from Figure 4. The observation of two well-resolved signals in Figures 4b to 4d shows that in the investigated range

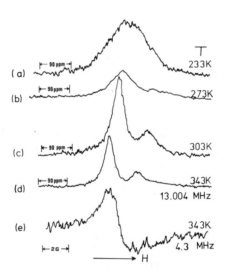

Figure 4. ^{14}N NMR spectra of DPPH.
(a)-(d) FT spectra at various temperatures.
(e) cw spectrum corresponding to Figure 3a.

(230-343 K), the A^N (-NO$_2$) couplings fall into two sets. A comparison of the relative signal intensities immediately reveals that the set corresponding to the smaller A^N contains two magnetically equivalent nitrogens, while the larger coupling contains only one nitrogen. This observation in itself is insufficient to assign the signals to specific -NO$_2$ groups, since crystal structure[35] and molecular packing models of DPPH show[24,25] that the molecule contains no twofold or higher rotation axis nor any mirror plane. However, the temperature dependence of the NMR line shapes and a comparison with earlier proton ENDOR studies of DPPH[14,22,24,25] yield information on molecular motion in this molecule and an important clue to the signal assignment. Definitive evidence for hindered rotation of the picryl ring about the N_1-C (picryl) bond comes from ENDOR studies,[22,25] especially on the deuterium-labeled DPPH.[22] At about 250 K and lower, two well-resolved signals were observed for the (two) picryl protons, but these signals coalesced at about 300K and above. The temperature dependence of the line shape was interpreted using modified Bloch equations. The observed rates above 250 K are $> 10^6$ sec^{-1}. This rate is fast on the NMR time scale, which is defined by the resolvable difference between contact-shifted lines from two species, typically $\sim 10^4$-10^5 Hz. Independent evidence for the hindered motion of the picryl group in the DPPH precursor (1,1-diphenyl-2-picrylhydrazine line) has been obtained by proton NMR;[36] again the rate was found to be fast on the NMR time scale. This fast hindered motion of the picryl ring about the N_1-C bond provides an explanation for the observed two (rather than three) ^{14}N NMR signals despite the lack of symmetry in DPPH. In addition, the presence of fast motion already at 250 K implies that if the line of higher intensity belonged to one ortho and one para nitrogen, then a further increase in temperature would lead to a broadening and eventual coalescence of the two signals. If, however, the higher intensity NMR signal corresponds to the two ortho nitrogens, then an increase in temperature above 250 K would lead either to further narrowing or no change in the spectra. Figures 4a to 4d show the latter to be indeed the case, hence establishing that the higher intensity signal must be assigned to the two ortho-NO$_2$ nitrogens. Figure 5 shows the temperature dependence of the observed shifts. The plots are essentially linear, establishing that the dominant mechanism for the observed shift is indeed the isotropic hyperfine interaction, as mentioned earlier. Accurate values of A^N_{ortho} and A^N_{para} were obtained by a least squares fitting of the observed temperature dependent contact shifts. The plots yielded A^N_{ortho} = -0.38 $\pm$ 0.02 and A^N_{para} = 0.48 $\pm$ 0.03 Gauss. These values have been collated in Table 2, which also contains earlier A^N estimates reported in the literature.

An unexpected result was that the assignment of $A^N_{para} > A^N_{ortho}$ is in conflict with current theoretical models for estimating ^{14}N hyperfine couplings, as discussed earlier.[16,23] The most detailed results on the spin density distribution on large DPPH-like radicals have been reported by Gubanov and Chirkov,[16] using the unrestricted Hartree-Fock procedure. The spin density is found to be -0.0051 on the ortho -NO_2 nitrogens and -0.0050 on the para -NO_2 nitrogen. The reliability of the reported spin densities was strongly suggested by a very good agreement between the calculated and observed values[23] for the proton hyperfine couplings in DPPH. Assuming that this also applies to spin densities on the hetero atoms, the calculated spin densities were converted to A^N_{ortho} and A^N_{para} by using the three-parameter Karplus-Fraenkel formula[37] and the relationship[38] $A^N = 48\ \rho_N$. The former approach yielded $A^N_{ortho} = -1.56$ and $A^N_{para} = -1.14$ G, and the latter gave $A^N_{ortho} = -0.27$ and $A^N_{para} = -0.25$ G. Thus, both approaches predict that A^N_{ortho} is larger than A^N_{para}, indicating the inadequacy of current theoretical models of ^{14}N hf couplings.

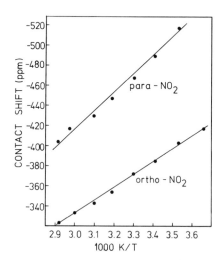

Figure 5. Temperature dependence of ^{14}N contact shifts in DPPH.[34]

Table 2. ^{14}N hf couplings (in Gauss) for the -NO$_2$ groups in DPPH.

Technique	A^N_{ortho}	A^N_{para}	Reference
Computer simulation of EPR spectra	0.754	0.377	12
Fourier transform analysis of EPR	0.38	0.33	15
Combined EPR and proton ENDOR	1 ± 0.3	0.7 ± 0.3	14
Theoretical values using Hartree-Fock spin densities and			
(a) Karplus-Fraenkel approach	-1.56	-1.14	16,23,37
(b) Nanda et al. approach	-0.27	-0.25	16,23,38
CW-NMR at 4.3 MHz and computer simulation	-0.46 ± 0.02 -0.39 ± 0.02	-0.38 ± 0.02 -0.48 ± 0.02	23 23
FT-NMR at 13.3 MHz	-0.38 ± 0.02	-0.48 ± 0.03	34

Discrepancy was also found in the results for molecular motion, as seen from ^{14}N NMR and proton ENDOR. Figure 6a shows the temperature dependence of the observed line widths of the ortho -NO$_2$ signal. The sharp decrease in line width with increase in temperature is indicative of the rotational narrowing, as observed in proton ENDOR studies.[22,25] The observed line width could be interpreted in the usual way by employing modified Bloch equations.[39] In the "fast" motion regime, the line broadening, $\delta\omega$, caused by motional effects is given by

$$\delta\omega \equiv \delta\omega_T - \delta\omega_0 = \frac{(\Delta W)^2}{8P} \qquad [16]$$

in which $\delta\omega_T$ is the observed line width at temperature T, $\delta\omega_0$ is the line width in the absence of motion, P is the probability per unit time of the molecular reorientation, and ΔW is the separation (in rad/sec) of the two signals being averaged. Equation 16 can be used to evaluate P if ΔW is known. In the present case, the motion is too fast over the entire temperature range (230-350 K) over which reliable line shape studies were feasible. Thus it has not been possible to measure ΔW. Measurements below 230 K were not made, because it was noted that at lower temperatures, the solubility of DPPH in

CH_2Cl_2 becomes low enough to deposit some powder at the bottom of the sample tube. This change in concentration led to additional signal broadening and hence undesirable complication in the line width analysis.

However, Equation 16 can be used to measure E_a, the effective barrier to the picryl ring rotation. Since $P = P_0 \exp(-E_a/kT)$, Equation 16 may be written as

$$\ln(\delta\omega) = \ln[(\Delta W)^2/(8P_0)] + E_a/(kT) \qquad [17]$$

Thus, if motional effects are the dominant cause of line broadening, then the plot of $\ln(\delta\omega)$ vs. $1/T$ should be linear, with slope E_a/k. One can then compare this value with that from ENDOR.[22] In particular, if E_a is the same for NMR and ENDOR, then the ENDOR data can be utilized for deducing the "rigid limit" couplings for the ^{14}N nuclei.

Figure 6b shows a plot of measured $\ln(\delta\omega)$ against $1/T$. The values are the widths at half height. The plot is fairly linear, thereby suggesting that the temperature dependence of the observed line width is indeed due to hindered rotation of the picryl ring. A least-squares fitting of the plot yielded $E_a = 6.3 \pm 1$ kcal/mole. This value is in excellent agreement with that determined by Weil and coworkers[36] for the picryl ring motion in the parent hydrazine when account is taken of the stabilization due to the hydrogen-bonding

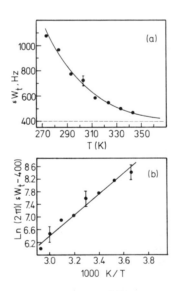

Figure 6. Temperature dependence of ^{14}N FT-NMR line width. (a) (a) Line width versus absolute temperature. (b) Arrhenius-type plot of log(line width) versus 1/(temperature).

between an NO_2 oxygen and the N-H (hydrazine) proton. These authors deduced an activation energy of ~12.5 kcal/mole for the picryl-ring rotation. The stabilization energy from the above-mentioned hydrogen-bonding is ~5 kcal/mole, thus leaving 7.5 $\pm$ 1 kcal/mole as E_a for the picryl ring motion in the absence of hydrogen bonding. Furthermore, our value of 6.3 $\pm$ 1 kcal/mole agrees with the value of 5 $\pm$ 0.5 kcal/mole determined from the temperature dependence of ENDOR signals from picryl protons.[22] Based on this agreement, we can assume that the data for the reorientation rate $K = 1/P$, and determine the difference, ΔW, between the A^N_{ortho}-values in the "rigid-lattice" limit. ENDOR data show that for the picryl ring motion, $P_O = 10^{12}$ sec^{-1}. Using this value of P_O in Equation 4, we deduce that $|\Delta W| \cong 1$ G. This result shows that in the "frozen motion" limit, the two A^N_{ortho} couplings are $\cong -0.0 \pm 0.2$ and -1.0 ± 0.2 Gauss. Comparison of these values with the EPR studies on several substituted nitrobenzene anions[40] suggests that the larger coupling should be assigned to the nitro group which is less sterically hindered, i.e., which moves further out of the planes of the phenyl rings.

Just as ENDOR studies[14,22,24,25] help the analysis of NMR work, the FT-NMR work provides valuable complementary information. First, the NMR work extends the temperature range covered by ENDOR. Second, the motional narrowing observed with $-NO_2$ groups presents definitive evidence for the picryl group motion. On the other hand, the motional effects observed with picryl-protons do not necessarily need the rotation of the whole picryl ring. For example, fast $-NO_2$ reorientations alone can lead to the inequivalence of the picryl protons via transient intramolecular >N-O···H-C hydrogen bonding, as suggested in NMR studies of the DPPH precursor hydrazine.[36] However, the inequivalence of $-NO_2$ nitrogens implies a definitive evidence for hindered motion of the picryl ring.

The line width of the para-NO_2 group also shows a comparable temperature variation, implying that this group also reorients in and out of the plane of the picryl ring in concert with the ortho-NO_2 groups. This observation provides direct support for the earlier proton NMR[36] studies, in which a cooperative motion of ortho and para $-NO_2$ groups was anticipated.

Deuteron NMR studies

The advantages of FT-NMR over ENDOR and triple resonance have been demonstrated also for deuterium NMR. Again, DPPH is a good example since, despite serious attempts,[14,24,25] ENDOR enhancements were not observed for deuterons in DPPH. The deuteron results were important for the assignment of ENDOR signals to various protons. Thus recently Biehl et al.[22] have used a Bruker B-KR 3225 FT-NMR spectrometer for measuring contact shifts from deuterons in selectively deuterated DPPH samples. To obtain narrow signals, it was necessary to spin the samples and lock the magnetic field. Sample solvent signals were too broad for an efficient lock, so the protons in tetramethylsilane contained in a coaxial tube provided the lock signal. Lines obtained by this method were at least an order of magnitude narrower than spectral lines without this procedure. At T = 295 K, the deuterium hf coupling, A^D, can be related to the contact shift according to

$$A^D = -2.0458 \times 10^{-3} \left(\frac{\Delta H}{H}\right) \qquad [18]$$

in which A^D is in gauss and $\Delta H/H$ is in parts per million.

Biehl et al.[22] noted, however, that the proton hf couplings measured by ENDOR differed significantly (5-10%) from the corresponding values deduced via deuteron NMR, using the relationship,

$$A_H = (\gamma_H/\gamma_D)A_D = 6.514\ A_D \qquad [19]$$

It was suggested that the cause of this discrepancy may be either the use of different solvents in the ENDOR and NMR studies, or the inadequacy of Equation 19, since it ignores any possible change in the structure of the radical upon replacement of hydrogen by deuterium. Solvent effects on the hf couplings are, of course, well known,[39] and there is some evidence even for the isotope effect on the magnitudes of the hf coupling constants.[41]

While attempting to provide an explanation for this anomaly, we noted that the deuteron results were consistently higher than those obtained via the proton (ENDOR) measurements. Since the proton studies were carried out at lower temperatures, we suspected that motional effects might be responsible for the discrepancy. With a view toward verifying these convictions, and in order to study the corresponding motion quantitatively, we have recently carried out a detailed variable-temperature deuteron FT-NMR study of specifically deuterium-labeled DPPH. Although the details of these studies will be reported elsewhere,[42] some illustrative results are summarized below. We believe that this is the first such study of motional effects for an organic radical, using deuterium NMR.

Figure 7 shows the temperature dependence of the deuterium NMR signals from DPPH-(ortho d_4), which designates DPPH molecules labeled with deuterium only at positions 1, 5, 6, and 10 (see Figure 1).

In contrast to the failure of ENDOR for detecting these deuterium couplings even approximately, the FT-NMR yields not only the magnitudes and the signs of these hf couplings, but also the small changes in these couplings due to the temperature variation. Based on the proton ENDOR results, the signal corresponding to the larger contact shift can be assigned to both of the ortho-deuterons in the phenyl ring which is twisted out of the plane of the picryl ring. The observation of only two signals from all four ortho-deuterons of the phenyl rings must result from "fast" hindered rotation of the two phenyl rings in DPPH about the $C-N_2$ bond, thus making the two ortho-deuterons on a given ring magnetically equivalent on the NMR time scale (Table 1). This result corroborates the proton ENDOR results, where such a motion was detected. It can also be seen from Figure 7 that with the increase in temperature, the two NMR signals start to broaden, then coalesce to one broad peak, and finally this peak

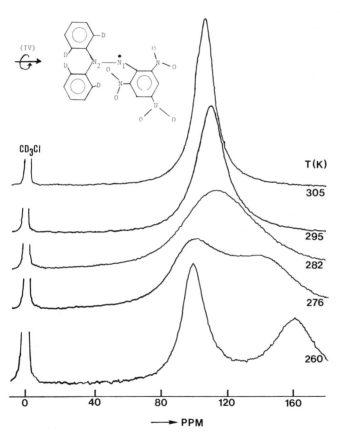

Figure 7. Temperature dependence of deuterium NMR spectra of DPPH-(ortho d_4).

starts to narrow down as the temperature is further increased. These observations show directly that the two phenyl rings execute hindered rotation about the N_2-N_1 bond at a rate that is faster than the line spacing between the two signals in the "frozen motion" temperature range. The observed motion corresponds to process IV as defined by Biehl et al.[22] and indicated in Figure 7. Analysis of the observed line shape using modified Bloch equations yielded a barrier height of 13 ±2 kcal/mole for this rotation of the phenyl ring about the N_2-N_1 bond. It is, again, noted that ENDOR/triple resonance had failed to detect this motion.

Another conclusion derived from the results of Figure 7 was that, as anticipated, the deuteron hf couplings themselves exhibited significant temperature dependence. Moreover, at low temperatures the magnitudes of the deuteron couplings increase monotonically and sufficiently to account for the difference between the proton (ENDOR) and deuteron (FT-NMR) results. On the basis of these results, it is clear that for studying these small deuterium hf couplings also (as for ^{14}N), the FT-NMR is more powerful than ENDOR/ELDOR/triple resonance techniques.

Deuteron shifts have also been measured for the negative ions of biphenyl-d_{10},[43] naphthalene,[44] and triphenyl benzene.[45] A striking result was the determination of deuteron quadrupole coupling via the comparison of the line widths, as mentioned earlier. The line widths of the deuterium peaks in naphthalene-d_8 did not vary linearly with A^2, as is expected from Equations 17 and 18. However, the variation in line widths was accounted for[44] by considering the contribution of the dipolar interaction. If one uses the proton line width to estimate $(1/T_2)_{FC}$ and $(1/T_2)_D$, one can estimate $(1/T_2)_Q$. The only unknown parameter is then τ_r, the rotational correlation time, which can be estimated from the widths of lines from the parent diamagnetic molecule. Then one can use Equation 11 to deduce quadrupolar coupling in the radical. This procedure was in fact used for quadrupolar couplings for the α and β deuterons in the naphthalene-d_8 negative ion.[44]

Alkali metal couplings

Another impressive application of the NMR technique is the study of alkali metal ions in aromatic anion alkali metal cation radicals. Here also, the ENDOR and triple resonance techniques were useful only for measuring fairly large couplings (1 Gauss).[46,47] On the other hand, deBoer and coworkers[44,45,48] have taken advantage of the NMR technique to measure very small (1 Gauss) hf couplings for ^6Li, ^7Li, ^{23}Na, ^{39}K, ^{85}Rb, ^{87}Rb, and ^{133}Cs. As mentioned earlier, an important advantage of NMR is that the signals provide the absolute sign of the hf couplings. The sign information is then used to deduce whether the interaction between the unpaired electron and the metal ion is direct or indirect (via spin-polarization). These NMR studies represent a significant step in understanding the structure and dynamics of ion-pair formation, and the method has proved to be an important complement to EPR, ENDOR, and ELDOR techniques.

CONCLUSIONS AND POSSIBLE FUTURE DEVELOPMENTS

The major theme of the present understanding has been to demonstrate that the FT-NMR technique, coupled with the use of spin-relaxers, constitutes a unique method for studying quadrupolar nuclei in paramagnetic solutions. Whereas the fast spin-lattice relaxation processes hamper signal enhancement in ENDOR, ELDOR, and triple resonance techniques, these same processes help to improve the signal-to-noise ratio in the NMR method. A general discussion of the advantages and disadvantages of EPR, ENDOR, ELDOR, triple resonance and FT-NMR techniques has been summarized in Table 1. The importance of the NMR method can be gauged from the fact that only with NMR has it been recently possible to measure small ($\leq$ 1 Gauss) hf couplings. In fact, the results obtained indicate that current theoretical models are not sufficiently accurate to account for the measured nitrogen couplings, although the same models adequately explain the proton hf couplings.[14] The nitrogen hf coupling can thus provide more sensitive probes for studying the electronic structure of complex molecules, and FT-NMR appears to be the only practical method for such measurements.

The NMR method has also been used successfully in measuring deuterium hf couplings as well as in estimating deuterium quadrupole couplings in solutions. Important data on both the signs and magnitudes of hf couplings have been obtained for ion-pairs of alkali metal ions with aromatic radical anions. It must be noted that such results were not obtainable with other techniques.

In conclusion, we mention three areas where further work could be quite profitable. The first is the extension of these measurements to other large organic radicals such as PAC, whose structure is shown in Figure 1. In particular, studies in solid powders should be attempted, since the fast relaxation processes are expected to lead to averaging of the anisotropic dipolar and quadrupolar interactions. To our knowledge, such experiments have not yet been reported. Second, detailed measurements of quadrupolar relaxation processes are lacking. For example, relaxation studies of ^{14}N nuclei have not yet been reported for any free radical. Comparative studies of ^{14}N and other nuclei could lead to an estimate of quadrupole couplings for ^{14}N. Third, there are several nitrogen-containing biological systems, such as paramagnetic porphyrins, where FT-NMR would yield valuable information. In general, such molecules have been studied by ENDOR. Because of fast relaxation processes, such ENDOR measurements have been carried out[49] at liquid helium temperatures. As discussed in this chapter, such relaxation can only help in making NMR signals easier to obtain, thus making possible the _in vivo_ detection of small hf couplings. The corresponding hf power needed for in vivo ENDOR studies with present instrumentation is too high to be practical.

ACKNOWLEDGMENTS: The author wishes to thank Drs. J. A. Ripmeester and A. H. Reddoch (National Research Council of Canada) and R. I. Walter (University of Illinois) for permission to quote some unpublished work on FT-NMR of DPPH,[34,42] and Najma Dalal for a critical reading of the manuscript.

REFERENCES

1. Kevan, L., and Kispert, L. D. 1976, Electron Spin Double Resonance Spectroscopy, John Wiley & Sons, New York.
2. Freed, J. H., and Dorio, M. M. 1979, Multiple Electron Resonance Spectroscopy, Plenum Press, New York.
3. (a) Freed, J. H. 1969, J. Chem. Phys. 50, 2271.
   (b) Dalal, N. S., and McDowell, C. A. 1970, Chem. Phys. Lett. 6, 17.
   (c) For a recent review, see K. Mobius and R. Biehl in ref. 2, p. 475.
4. Holden, A. N., Kittel, C., Merrit, F. R., and Yager, W. A. 1949, Phys. Rev. 75, 1614.
5. Townes, C. H., and Turkevich, J. 1950, Phys. Rev, 77, 147.
6. Hutchison, C. A., Pastor, R. C., and Kowalsky, A. G. 1952, J. Chem. Phys. 20, 534.
7. Deal, R. M., and Koski, W. S. 1959, J. Chem. Phys. 31, 1138.
8. Deguchi, Y. 1960, J. Chem. Phys. 32, 1584.
9. Holmberg, R. W., Livingston, R., and Smith, W. T. 1960, J. Chem. Phys. 33, 541.
10. Chen, M. M., Sane, K. V., Walter, R. I., and Weil, J. A. 1961, J. Amer. Chem. Soc. 65, 713.
11. Lord, N. W., and Blinder, S. M. 1961, J. Chem. Phys. 34, 1693.
12. Haniotis, Z., and Gunthard, Hs. H. 1968, Helv. Chim. Acta. 51, 561.
13. Hyde, J. S., Sneed, R. C., and Rist, G. H. 1969, J. Chem. Phys. 51, 1404. In addition to the pioneering ELDOR work, this paper presents an excellent summary of magnetic resonance studies of DPPH up to 1969.
14. Dalal, N. S., Kennedy, D. E., and McDowell, C. A. 1973, J. Chem. Phys. 59, 3403.
15. Gubanov, V. A., Koryakov, V. I., and Chirkov, A. K. 1973, J. Magn. Reson. 11, 326.
16. Gubanov, V. A., and Chirkov, A. K. 1973, Acta Phys. Pol. A, 43, 361.
17. Gutowsky, H. S., Kusumoto, H., Brown, T. H., and Anderson, D. H. 1959, J. Chem. Phys. 30, 860; 1960, 33, 720.
18. Anderson, M. E., Pake, G. E., and Tuttle, T. R. 1960, J. Chem. Phys. 33, 1581.

19. Sagdeev, R. Z., Molin, Yu. N., Koryakov, V. I., Chirkov, A. K., and Matevosyan, R. O. 1972, Org. Magn. Reson. 4, 365.
20. Verlinden, R., Grobet, P., and Van Gerven, L. 1974, Chem. Phys. Lett. 27, 535.
21. Yoshioka, T., Ohya-Nishiguchi, H., and Deguchi, Y. 1974, Bull. Chem. Soc. Japan 47, 430.
22. Biehl, R., Mobius, K., O'Conner, S. E., Walter, R. I., and Zimmerman, H. 1979, J. Phys. Chem. 83, 3449.
23. Dalal, N. S., Ripmeester, J. A., and Reddoch, A. H. 1978, J. Magn. Reson. 31, 471.
24. Dalal, N. S., Kennedy, D. E., and McDowell, C. A. 1974, J. Chem. Phys. 61, 1689.
25. Dalal, N. S., Kennedy, D. E., and McDowell, C. A. 1975, Chem. Phys. Lett. 30, 186.
26. Walter, R. I. 1966, J. Amer. Chem. Soc. 88, 1930.
27. Allendoerfer, R. D., and Maki, A. H. 1970, J. Magn. Reson. 3, 396.
28. See, for example, the excellent summary of this work by J. H. Freed in reference 2, p. 73.
29. See, for example, W. Kreilick 1973, in NMR of Paramagnetic Molecules, Principles and Applications, ed. G. N. La Mar, W. deW. Horrocks, Jr., and R. H. Holm, Academic Press, p. 595.
30. See, for example, J. P. Jesson in reference 29, p. 47.
31. Shimizu, H. 1964, J. Chem. Phys. 40, 754.
32. Huntress, W. T., Jr. 1963, J. Chem. Phys. 48, 3524.
33. Wallach, D., and Huntress, W. T., Jr. 1969, J. Chem. Phys. 50, 1219.
34. Dalal, N. S., Ripmeester, J. A., and Reddoch, A. H., paper in preparation.
35. Williams, D. E. 1966, J. Amer. Chem. Soc. 88, 5665.
36. Heidberg, J., Weil, J. A., Janusanis, G. A., and Anderson, J. K. 1964, J. Chem. Phys. 41, 1033.
37. Rieger, P. H., and Fraenkel, G. K. 1963, J. Chem. Phys. 39, 609.
38. Nanda, D. N., Subramanium, J., and Narsimhan, P. T. 1971, Theor. Chim. Acta, 22, 369.
39. See, for example, A. Carrington and A. D. McLachlan 1967, Introduction to Magnetic Resonance, Harper & Row, New York.
40. Geske, D. H., Ragle, J. L., Bambenek, M. A., and Balch, A. L. 1964, J. Amer. Chem. Soc. 86, 987.
41. Lawler, R. G. and Fraenkel, G. K. 1968, J. Chem. Phys. 49, 1126; Lawler, R. G., Bolton, J. R., Karplus, M., and Fraenkel, G. K. 1967, J. Chem. Phys. 47, 2149.
42. Dalal, N. S., Ripmeester J., and Walter, R. I., paper in preparation.
43. Canters, G. W., Hendriks, B. M. P., and deBoer, E. 1970, J. Chem. Phys. 53, 445.
44. Hendriks, B. M. P., Canters, G. W., Corvaja, C., deBoer, J. W. M., and deBoer, E. 1971, Mol. Phys. 20, 193.
45. van Broekhoven, J. A. M., Hendriks, B. M. P., and deBoer, E. 1971, J. Chem. Phys. 54, 1988.

46. van Willigen, H., Plato, M., Biehl, R., Dinse, K. P., and Mobius, K. 1973, Mol. Phys. 26, 793.
47. Atherton, N. M., and Day, B. 1973, J. Chem. Soc. 69, 1801.
48. Canters, G. W., deBoer, E., Hendriks, B. M. P., and van Willigen, H. 1969, Chem. Phys. Lett. 1, 627; Canters, G. W., deBoer, E., Hendriks, B. M. P., and Klaasen, A. 1969, Colloque Ampere XV.
49. See, for example, C. P. Scholes in reference 2, p. 297.

FOURIER TRANSFORM μSR

Jess H. Brewer

Department of Physics
University of British Columbia
Vancouver, B.C. V6T 2A6 CANADA

Donald G. Fleming

Department of Chemistry
University of British Columbia
Vancouver, B.C. V6T 1Y6 CANADA

Paul W. Percival
Department of Chemistry
Simon Fraser University
Burnaby, B.C. V5A 1S6 CANADA

INTRODUCTION

The technique of muon spin rotation (μSR) is described, with examples of its application in the fields of chemistry and solid state physics. It is shown how the raw experimental data contains information about the evolution of the spin polarization of muons stopped in matter. Fourier transformation provides a means of extracting the precession frequencies characteristic of various muonic species. Some manipulation of the raw data is essential to ensure accurate representation of the frequency information, and further techniques are often used to improve the final spectrum. These are discussed, and some examples are given of their effects. This is followed by descriptions of specific applications of Fourier transform μSR in the study of the light hydrogen isotope muonium (Mu = μ^+e^-), muonium-substituted free radicals, and paramagnetic states of the μ^+ in solids.

Muons and μSR

The muon (μ) and its decay via the weak interaction provided one of the first demonstrations of the failure of nature to pass "through

the looking glass" of inversion symmetry.[1] The details of these
phenomena continue to enchant particle physicists,[2] but for the
purpose of this article the muon is regarded as an infinitely dilute
probe of matter whose parity-violating decay gives the experimenter
easy access to the interactions of its spin.[3-5] In this role, its
two most important properties are its mass ($m = 206.7\ m_e = M/9$) and
its charge. Muons come in two charge states, μ^+ and μ^-; the corresponding interactions of positive and negative muons in matter are
completely different. The μ^- has long and properly been regarded as
a "heavy electron," since it readily takes a role like that of an
orbital electron in its slowing down process, forming a muonic atom.[6]
The μ^+, on the other hand, shuns the positive nuclear charge and
behaves in matter as if it were a very light proton. This article
will be concerned exclusively with the positive muon, although Fourier Transform techniques analogous to those described herein are very
useful in studies with negative muons.[7] One manifestation of the
similarity between the μ^+ and the proton is the formation of the
muonium atom (μ^+e^-), whose chemical symbol is usually written Mu.
The Mu atom is simply and truly a light isotope of hydrogen, but a
very light one. The ratio of 27 between the mass of the heaviest
isotope (tritium) and that of the lightest (Mu) is unprecedented. In
a sense, the muon, being a structureless Dirac particle (unlike the
proton), is an ideal electromagnetic probe of matter. It has extensive interdisciplinary applications in the fields of atomic and solid
state physics, radiation chemistry, and chemical reaction dynamics.
In all these areas, Fourier transforms play a major role.

Techniques for monitoring the muon's behaviour in matter are
known generically as "Muon Spin Rotation" since the vast majority of
experiments in the field have relied upon the precession of the
muon's magnetic moment in a transverse magnetic field. Recent
reviews can be found in Refs. 3-5. The acronym "μSR" was coined to
suggest obvious analogies with the more established techniques of NMR
and ESR. However, in recent years longitudinal and zero magnetic
field techniques[8] have become increasingly important, and true
resonance techniques[9] are now being developed, so that the meaning
of "μSR" needs to be continually expanded to include "Relaxation,"
"Resonance," or simply "Research." Nevertheless, even in zero
magnetic field oscillatory behaviour can be observed, so that Fourier
transform techniques are still very useful; e.g., Mu atoms can interact with a nonuniform local electric field in a crystal to yield
complicated hyperfine spectra.[10]

The μSR method

The starting point for muon experiments is the production of
pions (π) in a nuclear reaction initiated by energetic protons
(e.g., $^9Be + p \longrightarrow {}^{10}Be + \pi^+$). Threshold energies for pion production are > 150 MeV but the cross sections peak in the "intermedi-

ate energy" range of 400-1000 MeV, necessitating the use of larger accelerators. In recent years, the demand for high intensity proton beans (> 100 μA) in this energy range has led to the creation of three "meson factories:" LAMPF (Clinton P. Anderson Meson Physics Facility, Los Alamos, New Mexico, U.S.A.: 800 MeV), SIN (Swiss Institute for Nuclear Research, Villigen, Switzerland: 600 MeV), and TRIUMF (Meson Facility of the Universities of British Columbia, Victoria, Alberta, and Simon Fraser University, Vancouver, B.C., Canada: 520 MeV).

The positive pion decays with a mean lifetime of 26 ns to give a muon and a neutrino:

$$\pi^+ \to \mu^+ + \nu_\mu \qquad [1]$$

Since the pion has no spin and the neutrino has negative helicity (spin opposite to its momentum), linear and angular momentum conservation force the muon also to have negative helicity in the pion's rest frame. The muon momentum in the pion's rest frame is uniquely 29.8 MeV/c. Thus, muons emitted from the decay of positive pions in the outer skin of the production target are 100% polarized, nearly monochromatic, and have very high stopping density. These "surface muons" have become very useful in μ^+SR research in the last few years.[3,4,11] Judicious selection of the momentum of muons from the in-flight decay of fast pions can also be used to produce a muon beam of high polarization (80%) and momentum in the typical range 50-125 MeV/c. In either case, the resultant muon beam is degraded and stopped in the target of interest. Some properties of positive muons are compared with those of protons in Table 1.

While in the sample, the muon decays, emitting a positron (which is detected in the experiments), a neutrino, and an antineutrino:

$$\mu^+ \to e^+ + \nu_e + \overline{\nu}_\mu \qquad [2]$$

Table 1. Properties of Positive Muons.

Property	μ^+	H$^+$
Mass (MeV)	106.7	938.3
Spin	1/2	1/2
Lifetime	2.1974 μs	Stable
Magnetic Moment	3.18 μ_N	μ_N
Magnetogyric Ratio	0.01355 MHz/gauss	0.004258 MHz/gauss
Atoms Formed	μ^+e^- (Muonium)	H (Hydrogen)
Ionization Potential of Atom	13.539 eV	13.595 eV
Radius of Atom	0.532 Å	0.529 Å

The mean lifetime of the μ^+ is 2.1974 μs. The angular distribution of decay positrons is anisotropic. This can be understood most easily for the case where the neutrino (ν_e) and antineutrino ($\bar{\nu}_\mu$) are emitted exactly opposite the positron, which then exits with the maximum possible energy ($m_\mu/2$ = 52.3 MeV). In this case, since ν_e and $\bar{\nu}_\mu$ have opposite helicities, angular momentum conservation forces the positron (which at high energies acts like an antineutrino and must have positive helicity) to exit along the original muon spin direction. The e$^+$ emission probability is in general proportional to 1 + a cosθ:

$$dN_e/d\Omega \cong (1 + a\cos\theta) \quad\quad [3]$$

in which θ is the angle between the muon spin and positron momentum directions. The asymmetry coefficient, a, is a function of positron energy, with a spectrum-weighted average value of 1/3. As a result, the time variation of positron detection probability in a given direction reflects the evolution of the muon spin polarization.[3-5]

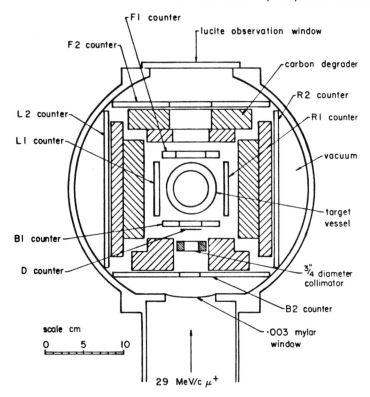

Figure 1. Picture of apparatus showing four e$^+$ counter telescopes. The target vessel is often a cryostat for low temperature studies. The counters shown are set up for "surface muons" and have holes in the forward pair to allow the incident beam to enter.

Fourier Transform µSR

A typical experimental setup is shown schematically in Figure 1. An incident "surface muon" triggers the "D" counter (a thin plastic scintillator) and (inevitably) stops in the target. For higher energy "conventional muons" the "B" counters would not have holes in them and the incoming muon would trigger B2, D and B1 but (if it stopped in the target as desired) not F1. Conversely, a "Forward" decay e+ triggers counters F1 and F2 but not D, B1 or B2. Thus the logic "signature" for a stopping "surface" muon is "μ_{stop}" = D·$\overline{F1}$, that for a stopping "conventional" muon is "μ_{stop}" = B2·D·B1·($\overline{F1+F2}$), and for a Forward positron, "e_F" = ($\overline{B1+B2+D}$)·F1·F2. Several e+ counter telescopes are normally used, with analogous logic used to form the "e" trigger for each. Figure 1, for instance, shows Forward (F) and Backward (B), Left (L) and Right (R) telescopes. The "μ_{stop}" and "e" triggers form the "start" and "stop" inputs for a time-digitizing system (a "clock") with typical time resolution of 1 ns; the digitized time interval and associated routing information (e.g., to tell which "e" telescope fired) are sent to an on-line minicomputer which calculates memory addresses and increments histogram "bins" to form time distributions such as those shown in Figure 2. Additional logical data are transmitted to the computer to reject imperfect events; e.g., only unequivocal single μ^+ stop/correlated e+ emission events are retained.[12]

As noted earlier, the classic μ^+SR experiment is carried out in a transverse magnetic field, in which case the μ-e decay pattern precesses in the field at the muon Larmor frequency,

$$\nu_\mu (\text{MHz}) = 13.55 \, B \, (\text{kG}) \qquad [4]$$

so that in Equation 3, $\theta = (2\pi\nu_\mu t + \phi)$, in which ϕ represents the initial phase angle between the muon spin and the positron telescope axis. (For instance, $\phi_B \approx 0°$ and $\phi_L \approx 270°$ for the B and L telescopes shown in Figure 1). Typical experimental spectra for muons precessing in a copper target in a transverse field of about 100 G are shown in Figure 2; the difference in phase ϕ is evident. Such data are easily fitted to an expression of the form

$$N(t) = N_0 \, [B + \exp(-t/\tau_\mu) |1 + S(t)|] \qquad [5]$$

in which τ_μ is the mean muon lifetime, B is a time-independent background term (usually small), N_0 is a simple scale factor, and S(t) is the "µSR Signal" containing information about the magnitude, precession and relaxation of the muon polarization. S(t) is the analogue of the NMR free induction decay signal observed following a $\pi/2$ pulse. In practice it is usually simple to fit and remove the background, normalization, and muon decay factors, leaving the signal of interest,

$$S(t) = [(N(t)/N_0) - B]\exp(+t/\tau_\mu) - 1 \qquad [6]$$

as shown in Figure 3. This procedure will be discussed in more detail in the next section.

The simplest version of $S(t)$ is the relaxing single-frequency precession in Figure 3:

$$S(t) = A_0 \, G_x(t) \, \cos(2\pi \nu_\mu t + \phi) \qquad [7]$$

in which A_0 is the empirical initial asymmetry coefficient and $G_x(t)$ is an envelope function describing the transverse relaxation. (The initial direction of the muon polarization is taken to be x, with z the direction of the applied magnetic field.) The initial amplitude A_0 includes the beam polarization, the asymmetry coefficient, a, from Equation 3 and various geometrical factors.

Even for this simplest case, there is considerable variety due to the flexibility of the transverse relaxation function $G_x(t)$, whose

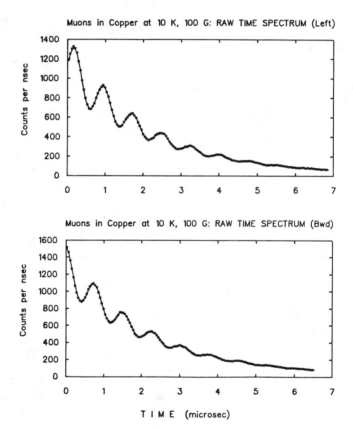

Figure 2. Raw μ^+SR time spectra for μ^+ in Cu at 10 K and 100 Gauss: (Top) "L" telescope; (Bottom) "B" telescope.

Fourier transform is in face the line shape one would observe directly in an actual resonance experiment. The width of the line corresponds to $1/T_2$, where T_2 is the characteristic transverse relaxation time. In many instances $G_x(t)$ can be written as a simple exponential relaxation $G_x(t) = \exp(-t/T_2)$, corresponding to a Lorentzian line shape. However, more complicated functional forms (the next simplest being Gaussian) are often used to fit the data, particularly in studies of μ^+ diffusion and trapping in solids.[3,13]

So far only single-frequency spectra have been considered. In these cases the Fourier transform exhibits only one peak, and is really needed only for an initial guess at the frequency. A chi-squared-minimization fit then gives the various parameters of Equations 5 and 7 and their respective errors. In general, however, one might observe a number of precession frequencies (and relaxation functions) corresponding to the various different environments in which the μ^+ might find itself. In such cases the "signal" can be represented by a sum of precession terms,

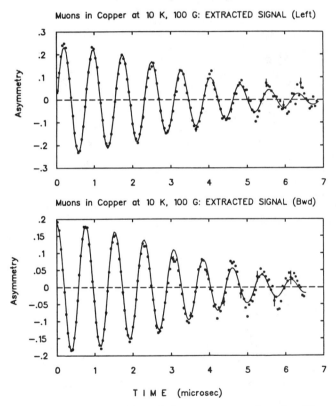

Figure 3. Extracted μSR "Signals" corresponding to Figure 2 (Top) and (Bottom).

$$S(t) = \sum_i A_i \, G_i(t) \, \cos(2\pi\nu_i t + \phi_i) \qquad [8]$$

in which A_i is the amplitude of the ith component, ν_i is its frequency, ϕ_i its initial phase, and $G_i(t)$ its relaxation function. An example of such a multi-frequency signal is shown in Figure 4; the Fourier power spectrum of the same data is shown in Figure 5. In these cases, χ^2-fitting is posisble only (if at all) after exhaustive Fourier analysis, and most often the Fourier transform spectra constitute the final results. We now turn to a more detailed description of the processes by which Fourier spectra are extracted from μSR data.

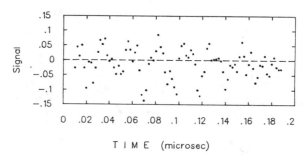

Figure 4. μSR time signal from Si at 85 K in 75 G. Only the first 200 ns are shown, but the signals visible here are still evident out to about 2 μs.

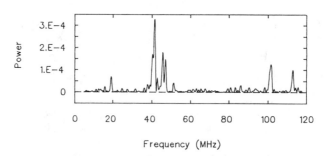

Figure 5. μSR power spectrum from a Fourier transform of the first 1.0 μs of the time signal shown in Figure 4.

FOURIER TRANSFORMING μSR DATA

Preparation of the "μSR signal" from raw data

The "raw" time spectrum obtained in a μSR experiment is not yet in a form suitable for input to a Fourier transform, as explained in the previous Section. The original time spectrum, except for a time-independent background fraction B, is weighted by a normalization factor whose initial value is N_0, but which decays exponentially with the muon lifetime τ_μ. Although these uninteresting factors are easy to remove, their effect on the statistical scatter of the data is a serious problem which can be ameliorated in a variety of ways, none of which will be appropriate in every situation. Considerable controversy surrounds this topic in the μSR community, and one purpose of this article is to define the problem clearly and outline various approaches to it. [For instance, Martoff and Rosenblum[14] have justly lamented the ambiguity of results reported in terms of Fourier transforms in which the data preparation, time range, and filtering procedures are unspecified. They recommend a standardized treatment in which the raw time spectrum itself is transformed; this is, however, tantamount to prescribing one particular filtering function for all μSR data, as will be explained in the next Section. Such standardization might well be useful, but one must still be aware of the alternatives that are being rejected.]

Recalling Equations 5-8, the information of interest is contained in $S(t)$, the "μSR Signal." The empirical maximum asymmetry A_0, corresponding to the precession amplitude actually measured in the same apparatus for a case where no depolarization occurs after the muon enters the target (e.g., for an aluminum or CCl_4 target), usually has a value of the order of 1/3. A_0 includes the beam polarization, the average positron anisotropy coefficient, and various geometrical factors, none of which are usually of interest; in general it is also a function of target density and magnetic field, but with careful calibration it can be divided out of the μSR signal, leaving the "normalized" signal,

$$F(t) = S(t)/A_0 \qquad [9]$$

This represents the time-dependence of the projection of the muon polarization on the axis of the positron detectors, effectively now for muons that are initially 100% polarized. This is the quantity provided by most theories, and is therefore what one wishes most to learn from the experiment. However, A_0 can be divided out of the Fourier transform of the data as easily as from the signal, so it usually suffices to transform $S(t)$.

The first step in extraction of the μSR signal from the raw time spectrum is removal of the constant background B, which arises through noise in the counters (dark current in the phototubes, background radiation, light leaks, etc.) and imperfect electronics. By introducing an electronic delay in the "μ_{stop}" pulse to the clock, one can shift the histogram so that the early portion corresponds to negative times (before the muon arrives), and thus contains <u>only</u> the background. (This is a standard procedure employed in PAC and other time-differential techniques.) This is preferable to measuring B at long times, but in any case it is assumed that B is truly constant throughout the spectrum. Alternatively, it is possible to obtain an estimate of B and N by means of a simple least squares algorithm such as

$$B = (N_0 \Delta)^{-1} \mid n \left(\sum_i \frac{E_i^2}{N_i} \right) - \left(\sum_i \frac{E_i}{N_i} \right) \left(\sum_j E_j \right) \mid \qquad [10a]$$

$$N_0 = \Delta^{-1} \mid \left(\sum_i E_i \right) \left(\sum_j \frac{1}{N_j} \right) - n \sum_i \frac{E_i}{N_i} \mid \qquad [10b]$$

in which
$$\Delta = \left(\sum_i \frac{1}{N_i} \right) \left(\sum_j \frac{E_j^2}{N_j} \right) - \left(\sum_i \frac{E_i}{N_i} \right)^2$$

and
$$E_i = \exp\left(\frac{-t_i}{\tau_\mu}\right), \quad i = 1, 2, 3, \ldots, n.$$

(N_i is the number of counts in the ith bin.)

This algorithm yields B and N_0 with adequate precision as long as S(t) oscillates rapidly or has small-amplitude oscillations.

An alternative approach is often used when two positron telescopes are arranged on opposite sides of the target. Suppose, for example, that the counters are on either side of the target, out of the beam. The corresponding histograms would normally be labelled L (left) and R (right). Let us further suppose that by luck or design one can obtain (for unpolarized muons) the same counting rate in each telescope, and that the background can be subtracted by the "negative-time" method described above, leaving L'(t) and R'(t), the "background-free" time spectra. In this case, the μSR signal can easily be obtained:

$$S(t) = [L'(t) - R'(t)]/[L'(t) + R'(t)] \qquad [11]$$

For each histogram bin thus defined, a statistical counting error can be assigned as described below.

Fourier Transform μSR

In general, of course, one telescope will intercept positrons with higher efficiency, and a relative normalization factor must be included:

$$S(t) = [L'(t) - fR'(t)]/[L'(t) + fR'(t)] \quad [12]$$

in which f can be obtained simply as the ratio of the total counts in the R' and L' histograms (following background subtraction). This method of extracting S(t) is simple and convenient, accurate enough for input to a Fourier transform, and has the advantage of combining two data sets into one with reduced noise.

Effects of statistical noise

Once the μSR signal S(t) has been extracted from the raw time spectrum by one of the above methods, one is in principle ready to Fourier transform. However, before proceeding it is wise to consider the "statistical noise" across the spectrum. Since μSR is a "counting" technique, Poisson statistics apply: The standard deviation for each histogram channel is (for channels with at least a few dozen counts) the square root of the number of counts in that channel. Since muon decay gives less counts at long times, the signal-to-noise ratio decreases with time, roughly as $N_0^{1/2} e^{-t/2\tau_\mu}$. When muon decay is divided out to give the time spectrum S(t), the noise at long times is effectively magnified exponentially. Figure 6 shows this effect in the signal from a time spectrum with a typical number of counts. To achieve constant noise across the spectrum S(t) can be multiplied by $\exp(-t/2\tau_\mu)$:

$$S'(t) = [|(N(t) - B)/N_0|\exp(t/\tau_\mu) - 1] \exp(-t/2\tau_\mu), \quad [13]$$

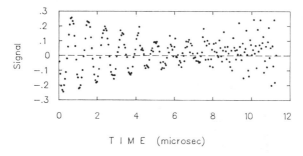

Figure 6. Typical μSR Signal, showing the exponential increase of "statistical noise" at late times.

so that the noise $\delta S'$ is given by

$$\delta S' = (\partial S'/\partial N)\,\delta N = N_0^{-1/2}$$

This procedure is nothing more than application of a digital filter with an exponential time constant equal to twice the muon lifetime. Figure 7 shows the effect of this transformation on the data of Figure 6. Shorter time constants for the exponential filter may be appropriate for short-lived signals, in order to discriminate against noise at later times; however, broadening of the peaks in the Fourier spectrum is an unavoidable and undesirable consequence.

Alternatively, the time spectrum may be cut short and zero-filling used to preserve the resolution of the frequency spectrum. Of course, this is really tantamount to apodizing with a step function, and unless the signal has already decayed to the noise level by the cutoff time, it will naturally introduce side-lobes about each peak. This spectral distortion can be alleviated by use of apodization functions of "intermediate smoothness," i.e., in between the discontinuous extreme of the sharp cutoff and the gentle extreme of the exponential decay. A typical example would be an envelope function in the shape of a quarter-cycle of a sine function, so phased that it reaches zero just at the end of the desired time region. (For further discussion of these and related points see Ref. 14-16).

One must not lose sight of the fact that the troublesome noise is the result of finite counting statistics; in principle, any desired signal-to-noise ratio can be achieved anywhere in the time spectrum (even at late times) simply by taking more data. This can hardly be regarded as a practical solution, of course; the largest number of counts ever taken in one histogram, to these authors' knowledge, was about 10^8, which sets a practical lower limit of about 0.5% on the statistical scatter in a 1-ns bin at early times,

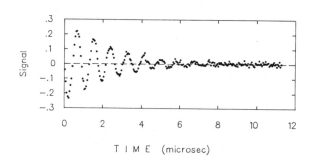

Figure 7. Effect of multiplying the signal of Figure 6 by an exponential apodizing filter function $\exp(-t/2\tau_\mu)$.

and more at later times. However, it is important to keep in mind that the problem of statistical noise is a strong function of the data-taking conditions, bin size, etc., and that measures which are necessary in some circumstances are superfluous in others.

A more detailed treatment of the effects of statistical noise on the Fourier spectrum will be given below.

The Fourier Transform Algorithm

The Fourier spectrum is obtained from the final time spectrum by application of a standard discrete fast Fourier transform algorithm:

$$a_r = \sum_{k=0}^{N-1} x_k W^{rk}, \quad r = 0,1,2,\ldots,N-1 \quad [14]$$

where

$$W = \exp(-2\pi i/N), \quad [15]$$

x_k is the kth complex time domain point, and a_r is the rth complex frequency point.

An example of a theoretical x array would be

$$x_j = \frac{A_0}{2} \exp\left| i\phi + \frac{i\pi j}{N}(k_0 + i\alpha) \right| \quad [16]$$

where

$$j = \frac{2Nt}{T}, \quad k_0 = \nu_0 T, \quad \text{and} \quad \alpha = \lambda T/2\pi, \quad [17]$$

corresponding to the simple exponentially-damped single-frequency signal

$$S(t) = A_0 e^{-\lambda t} \cos(2\pi\nu_0 t + \phi), \quad 0<t<T \quad [18]$$

where T is the time interval being transformed. However, μSR histograms are usually Fourier analyzed separately, so that only the real data is available for the x_k array, in which case the full Fourier spectrum is contained in the complex array a_r, $r=0,1,2,\ldots N/2$. (More will be said later about complex transforms.)

Most commonly the power spectrum

$$|a_r|^2 = [\text{Re}(a_r)]^2 + [\text{Im}(a_r)]^2 \qquad [19]$$

is displayed, since all the frequency information is available without the complications of phase. Recent applications, however, entail the extraction of more than just frequency information from the Fourier spectrum. Since all noise in the power spectrum gives positive amplitude it does not average out in fitting signal peaks. For such purposes it is better to use either the real or imaginary part of a_r alone. Provided the time spectrum has been extended to at least twice its original length by zero-filling, both halves of the complex frequency spectrum contain all the available information.[16] In principle, the real (cosine) part is associated with an absorption line shape and the imaginary (sine) part with a dispersion line shape; in practice the real and imaginary parts of a_r contain mixtures of absorption and dispersion lineshapes. The major cause of this mixing is the disparity between the nominal start of the time spectrum and the true zero of time. Although it is possible to determine the latter accurately, practical considerations in the electronic logic setup result in distortions at early times. If the time spectrum used in the transform starts at time t_1, then the corrected spectra are given by

$$F_1(\nu_r) = \cos\phi_1 \, \text{Re}[a_r] + \sin\phi_1 \, \text{Im}[a_r] \qquad [20a]$$

$$F_2(\nu_r) = \sin\phi_1 \, \text{Re}[a_r] - \cos\phi_1 \, \text{Im}[a_r] \qquad [20b]$$

where the phase correction is first order in frequency ($\phi_1 = \nu_r t_1$). Zero-order phase errors can also occur; for example, if the orientation of the positron telescope is not exactly parallel or perpendicular to the initial muon spin polarization. The correction can be computed by requiring that the area under a selected peak be a maximum--i.e., by forcing a pure absorption shape on that peak. The phase shift is given by

$$\phi_0 = \arctan[\sum F_2(\nu_r)/\sum F_1(\nu_r)] \qquad [21]$$

where the summation covers that portion of the frequency spectrum containing the selected peak.

Noise Peaks and "Confidence Level Spectroscopy"

As mentioned earlier, the population of any bin of the experimental histogram has a statistical uncertainty describable by a Poisson distribution. This is manifest in the discrete Fourier transform as "noise peaks." Here we will briefly calculate the expected noise spectrum for a μSR histogram with no signal [$S(t) = 0$], following the procedure outlined by B. D. Patterson.[17] (See also Ref. 14 and 15).

The first approximation that is made for this description is that B = 0; that is, that there is no time-independent background of events unrelated to muon decay. Since experimental values of B are usually less than a few percent, this is a tolerable approximation, especially for early times. The raw histogram bins can therefore be represented as

$$\overline{x_j} = N_0 \exp(-j/j_\mu) \qquad [22]$$

where $j_\mu = \tau_\mu/\Delta T$, ΔT being the bin width.

The number of time bins is taken to be N. The "null signal" to be transformed will thus have the form

$$s_j = (\overline{x_j}/N_0)\exp(j/j_\mu) - 1 \qquad [23]$$

and its Fourier transform will have an average power

$$\langle p \rangle = (1/N^2) \sum_{j=0}^{N-1} (\overline{x_j}/N_0^2) \exp(2j/j_\mu) \qquad [24a]$$

or $$\langle p \rangle = (1/N^2 N_0) \sum_{j=0}^{N-1} \exp(j/j_\mu) \qquad [24b]$$

It is easy to show that this gives a signal-to-noise ratio proportional to A^2/N_0 for small asymmetries A, which is what we are trying to detect when we are worrying about signal-to-noise ratios. The distribution of amplitudes of noise peaks can also be calculated.[17] For the power spectrum $p_k = |a_k|^2$ of pure noise we obtain

$$P(p_k) = \exp(-p_k/\langle p \rangle) \qquad [25]$$

where $\langle p \rangle$ is given by Equations 24. The second approximation is to sum over x_j instead of $\overline{x_j}$. Then, using the value of N_0 extracted by fitting or by some algorithm such as Equations 10, we can numerically extract $\langle p \rangle$ and thus predict the distribution of noise peaks. Figure 8 shows that the distribution of noise peak power in a real μSR spectrum is actually exponential, as predicted.

Given the distribution of noise peaks, one can define the probability that a peak of a given size p is "significant" rather than just a noise peak. This is called the "confidence level" C(p) and is given by

$$C(p) = (p/\langle p \rangle)^{N/2} \tag{26}$$

where $\langle p \rangle$ is the average power and $N/2$ is the number of frequency bins. The entire frequency spectrum can be converted into a "confidence level spectrum" by means of this algorithm; this sometimes provides a very nice way of exhibiting the peaks of interest, as illustrated in Figure 9. In principle, the average power $\langle p \rangle$ can be calculated from the original time histogram bins; ideally this method should be used. However, this occasionally gives an under- or overestimate of $\langle p \rangle$, perhaps because of nonzero B; in these cases, one can take advantage of the fact that most frequencies have only noise peaks to use a simple empirical method for extracting $\langle p \rangle$: first the straight average of the measured power is calculated, then a second average is taken, excluding bins with significant peaks (i.e., more than, say, 3 times the initially-calculated $\langle p \rangle$). This "brute-force" empirical approach is inelegant, but reliable; experience has shown that it gives consistent and reasonable confidence levels to known frequency peaks. (This algorithm is used in generating Figure 9). Another advantage of the empirical approach is that it can be used with a wide variety of apodization and filtering without modification. Of course, in such cases one has long since abandoned any possibility of making a rigorously precise quantitative interpretation of the results; but (as mentioned earlier) this is not really the forte of Fourier transform methods.

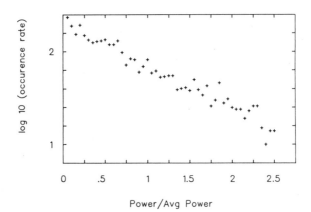

Figure 8. Distribution of Fourier power peaks from statistical noise in an actual μSR spectrum. Muons in Si at 5 K, 40 G.

Complex Fourier Transforms

Most μSR experimental assemblies include more than one positron telescope, so histograms are accumulated for positrons emitted in different directions. (See Figure 1.) Although the histograms are usually analyzed separately, it can be advantageous to transform them pairwise or even four at a time (when there are two pairs of opposite telescopes in orthogonal directions). The two time spectra resulting from a pair of mutually perpendicular telescopes can be used as the real and imaginary parts of the data array input to a complex Fourier transform. This is the same as the "quadrature detection" method well-known in NMR.[18] One advantage is the doubling of the "Nyquist frequency" ν_{max}, the highest frequency that can be represented without aliasing (folding back). For a single histogram with time resolution δt, $\nu_{max} = (2\delta t)^{-1}$, but for quadrature detection $\nu_{max} = (\delta t)^{-1}$, since the ensemble muon polarization is sampled twice for each Larmor precession cycle. A further feature of quadrature detection is the additional information provided by the phase relation of the two signals. It is possible to distinguish between precession frequencies of opposite sign, as will be demonstrated later in the discussion of radical spectra.

To ensure a faithful transform, free of artifacts, is impossible; to do as well as one can, certain conditions must be met: The two time spectra x_k and y_k must be orthogonal,

$$\sum_k x_k y_k = 0 \qquad [27]$$

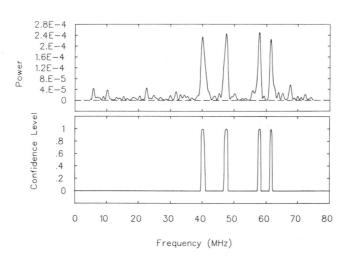

Figure 9. Fourier power spectrum and corresponding "confidence level spectrum" for muons in Si at 5 K, 40 G, showing statistically significant peaks.

and their powers must be equal,

$$\sum_k x_k^2 = \sum_k y_k^2 \qquad [28]$$

In principle it is possible to satisfy the orthogonality condition by rigorous attention to the geometrical placement of the counter telescopes. The second condition is more difficult, since it depends not only on geometrical effects (the solid angles sampled by the telescopes) but also on counter efficiency and the amount of material the positrons must penetrate to be counted. The last factor is particularly sensitive, since higher-energy positrons are more correlated with the muon spin and exhibit a higher asymmetry.[2,3] In practice one can make some empirical corrections such as[19]

$$x_k' = \beta(x_k - \alpha y_k) \qquad [29a]$$

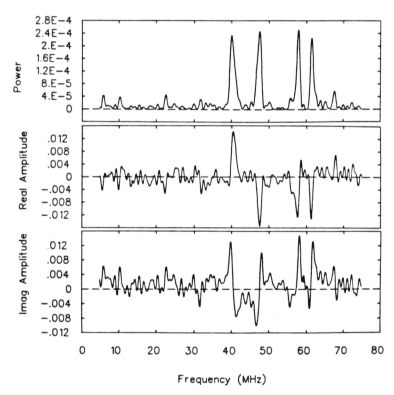

Figure 10. Complex Fourier transform information. Muons in Si at 5 K, 40 G. Top: Power spectrum from data taken with two orthogonal pairs of positron telescopes. Middle: Real part of the same spectrum. Bottom: Imaginary part of the same spectrum.

where

$$\alpha = \sum_k x_k y_k / \sum_k y_k^2 \qquad [29b]$$

and

$$\beta = [\sum_k y_k^2 / \sum_k (x_k - \alpha y_k)^2]^{1/2} \qquad [29c]$$

Typical values of these correction factors are $\alpha = 0.095$ and $\beta = 1.032$ (P. W. Percival, unpublished results from S.I.N.). However, in the end a finite systematic uncertainty is always introduced by this procedure; this is balanced by the statistical and analytical advantages.

Examples of the sorts of analysis one can make of data taken very carefully and treated judiciously are illustrated in Figure 10.

Why Fit in Time Space and When?

When the spectral content of a μSR signal is unknown, or when a signal of a certain type is expected but its frequency is not known, one has no choice but to begin the analysis with a Fourier transform. Even after the initial identification of signals, Fourier transform μSR spectroscopy is the most economical and convenient way to characterize data with numerous frequencies, examples of which will be mentioned in the following section. However, it is usually a mistake to try to extract final results from frequency spetra. This is a somewhat contentious statement, especially in view of the fact

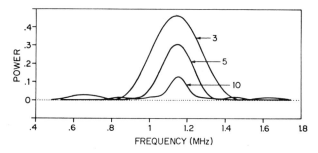

Figure 11. Effect of different time ranges (3μs, 5μs, and 10μs) on the Fourier transform of a relaxing μSR signal. No filter used.

that each person, regardless of awareness of the one-to-one correspondence between time space and frequency space,* will have a preference for one or the other as being more "real" or "fundamental." De gustibus, of course, non disputandum est; however, μSR data is in most cases collected in time space, and the statistical uncertainties are simple, unambiguous, and model-independent in time space, whereas formulae for translating statistical uncertainties into frequency space are complicated and sensitive to the treatment given the data before and during the Fourier transform.

It is almost always true, therefore, that a parameter of interest (frequency, phase, amplitude, relaxation time, or shape of a relaxation function) can be extracted with a well defined uncertanity from the original time histogram with more confidence than from a derived frequency spectrum. Several examples can be offered to illustrate this point.

First consider the case of a single frequency with large amplitude and good statistics, whose relaxation function is desired. A Fourier transform yields a single line whose center defines the precession frequency and whose width defines the relaxation rate. There are other factors, however, which also affect the width and shape of the line: the number of bins of data trasformed, the apodization or filter used, if any, and subtle effects like "ringing" if no filter is used. In principle these effects can be accounted for, but in practice this approach is unreliable. Figure 11 shows the unfiltered signal for different time ranges from the same μSR time spectrum. Figure 12 shows the same effect using an exponential filter $\exp(-t/2\tau_\mu)$. In each case a tendency can be seen, but the fluctuations are very problematic. Fitting the same data in time space by χ^2-minimization yields the parameters of interest, with uncertainties, and provides a concise comparison of the quality of fits to different relaxation functions (e.g., exponential or Gaussian).

---

*The one-to-one correspondence between time and frequency spectra can lead to misunderstandings with regard to "zero-filling" for example, which can be used to decrease the frequency bin size and thus increase the apparent resolution in frequency space. The argument that "no information can be gained from such manipulations" is based on the above correspondence principle. But the Fourier transform is a model-independent analysis in which every frequency bin is presumed to represent an independent "signal;" this is not at all what we have in mind when we interpret the data looking for a few discrete "peaks." Thus our interpretation of the Fourier transform is normally a very model-dependent distortion, consistent with the zero-filling technique for acquiring better resolution with which to judge the position and shape of the isolated "line".

Fourier Transform μSR

Another example is the very weak signal. It is often claimed that a signal with small amplitude at a frequency high enough to have many periods in the region of interest can only be detected by Fourier transform. This is certainly true for the initial step of locating the signal; but again, once the approximate frequency is found, a "sweep" through fits in the time domain using different (fixed) values of the frequency yields the clearest and most useful results. An example of a very small signal is analyzed in this way in Figure 13. Again the results can be interpreted more rigorously from the time-domain data than from the Fourier transform. However, the necessity of a Fourier transform as input to this treatment must not be overlooked.

Finally, consider a μSR signal containing a number of small-amplitude frequencies; this is the classic case in which a Fourier transform is considered the final word. However, one would like to interpret the sizes and widths of the various peaks in terms of the fractions of the muons in the corresponding states and their relaxation times. This is tedious and expensive in any case, but fitting in the time domain is still effective. This is because several signals can be fitted simultaneously, and if the remaining peaks are well-separated from those being fitted, their "spoiling" effect on χ^2 is essentially independent of the χ^2 minimization in process. Then, once several frequencies have been fitted satisfactorily, they can be subtracted from the overall signal and the next two or three fitted in succession. The only danger in this procedure is when one signal

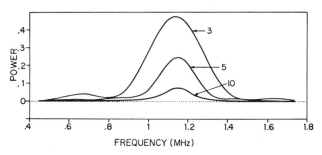

Figure 12. Effect of time range on Fourier transform of a relaxing signal. Same as Figure 11 except filtered with an exponential apodization function $\exp(-t/2\tau_\mu)$.

is large enough to substantially affect the counting statistics over its period--a condition which violates the definition of this type of spectrum, given above. Consistency checks can also be made by starting over and fitting/removing frequencies in a different order. This procedure is tedious and expensive, but it has been used in several circumstances to provide the maximum possible detail from the data available.

With all these qualifications, it is still accurate to say that Fourier transform methods are indispensible to μSR, especially as an on-line tool. Some representative examples of this use are given in the next Section.

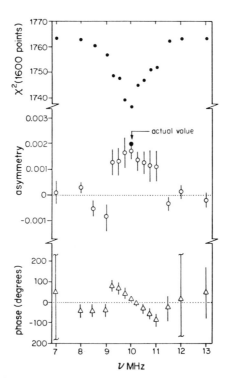

Figure 13. χ^2-minimization frequency analysis of a "generated" μSR signal with 0.2% asymmetry at 10 MHz. For each point the frequency is held fixed while all other parameters are free to vary. Note the simultaneous effects in χ^2, A, and ϕ.

Fourier Transform μSR

APPLICATIONS

Muonium in a Tranverse Magnetic Field

The simplest and most common paramagnetic environment in which muons are found is the muonium (Mu) atom itself, μ^+e^-.[3,4,20] In the gas phase, Mu is formed via an epithermal charge-exchange process, from which muons emerge either as "free" μ^+ (most likely bound in molecular ions) or as Mu atoms, with kinetic energies of $\lesssim$ 20 eV, depending upon the moderator gas. Subsequent elastic and inelastic collisions complete the energy-loss process, leaving the Mu atoms with kinetic energy $\sim k_B T$.[21] The same mechanism may operate in some condensed media (on a much shorter time scale) but the situation is less clear than in gases, and radiation "track" or "spur" effects may play a major role.[22] In solids, the μ^+ may thermalize at particular interstitial sites in which the $\mu^+ - e^-$ hyperfine interaction is considerably modified from the "vacuum" value observed in gases, particularly in semiconductors (recall Figures 5 and 10) but also, as we shall see below, in insulating crystals such as quartz. In organic liquids, the Mu atoms may add to double-bond systems, forming muonic radicals.[23] In each of these cases, characteristic frequencies are most easily revealed through Fourier transform techniques.

The spin Hamiltonian for muonium in a magnetic field is given by

$$H = (g_e \mu_B \vec{S} - g_\mu \mu_\mu \vec{I}) \cdot \vec{B} + A \vec{S} \cdot \vec{I} \quad [30]$$

where g_e and g_μ are the electron and muon g-factors, μ_B and μ_μ are the Bohr magneton and the muon magneton, $\vec{B}$ is the external field, $\vec{S}$ in the electron spin, $\vec{I}$ is the muon spin, and A is the strength of the Fermi contact interaction, proportional to the electron density at the μ^+:

$$A = \frac{8\pi}{3} g_e \mu_B g_\mu \mu_\mu |\psi(0)|^2 = h \nu_0 \quad [31]$$

where h is Planck's constant and ν_0 is the hyperfine frequency.

The eigenvectors of this Hamiltonian (for the H atom) have been familiar for about 50 years now; the energy eigenvalues are shown in frequency units in the Breit-Rabi diagram of Figure 14. The hyperfine frequency of Mu in vacuum is ν_0 = 4463.302 35(52) MHz,[24] which can be compared with the corresponding value for the H atom, $\nu_0(H)$ = 1420.405751(1) MHz, to give almost exactly the ratio predicted by Equation 31 when $g_\mu \mu_\mu$ is replaced by $g_\mu \mu_\mu$ for the proton. In fact, a more detailed calculation, combined with the experimental values of g_μ and μ_μ, provides the best test of quantum electrodynamics at present,[24] thanks to the structurelessness of the μ^+.

There are four allowed ($\Delta M = \pm 1$) transition frequencies in Figure 1: ν_{12}, ν_{23}, ν_{34} and ν_{14}. The muon, initially 100% polarized in the x-direction, captures an unpolarized electron to form a mixed ensemble whose time dependence in a tranverse magnetic field (along z) has the form[3,4]

$$P_\mu(t) = 1/4[(1 + \delta)(e^{i\nu_{12}t} + e^{-i\nu_{34}t})$$
$$+ (1 - \delta)(e^{i\nu_{23}t} + e^{i\nu_{14}t})] \qquad [32]$$

where
$$\delta = x/\sqrt{1 + x^2}$$

$x = H/H_0$ and $H_0 = 1585$ G, the "contact field" of the μ^+ on the electron. (Compare 503 G for the H atom.) Here the real part of $P_\mu(t)$ is the polarization component in the x-direction and the imaginary part is the polarization component in the y-direction perpendicular to both x and z. If there is initially no polarization

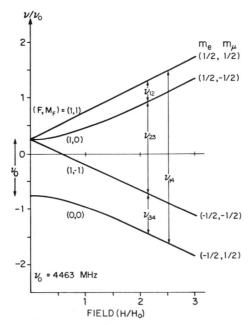

Figure 14. Breit-Rabi energy-level diagram for muonium. The contact field $H_0 = 1585$ G for muonium.

along the field direction, none will develop. In moderate magnetic fields (< 300 G) $\delta \ll 1$ but the frequencies ν_{34} and ν_{14} are comparable to ν_0 and thus not resolvable with state-of-the-art μSR time resolution (> 0.3 ns). Hence only two frequencies, ν_{12} and ν_{23}, are directly observable. The difference between these two frequencies can, however, be analyzed to extract ν_0. For low fields (< 100 G),

$$\nu_0 = \nu_{Mu}^2/\Omega \qquad [33]$$

where

$$\nu_{Mu} = (\nu_{12} + \nu_{23})/2 \qquad [34]$$

and

$$\Omega = (\nu_{23} - \nu_{12})/2 \qquad [35]$$

The average frequency, ν_{Mu}, is the precession frequency of "triplet" Mu,

$$\nu_{Mu}(MHz) = 1.39 \, B(G) \qquad [36]$$

and the "beat frequency" Ω is initially proportional to B^2.

This "two-frequency precession" is often seen in inert media at low field, usually accompanied by a "quasi-free" μ^+ precession signal from the muons that stop in diamagnetic environments. If no further signals are present, the data can usually be fitted directly by χ^2-minimization to yield all the relevant parameters, including ν_μ and ν_0; in such simple cases, Fourier transforming is valuable mainly for quick "on-line" checks of the data and providing initial guesses for the fitting program.

In a very weak fields (B < 20 G), $\nu_{12} \cong \nu_{23} \cong \nu_{Mu}$ and $\Omega \cong 0$; in this case only a single muonium frequency is observed, making χ^2 fitting even simpler. Low fields are therefore generally applied to studies of the reaction of Mu atoms in gases and liquids.[4,5,25]

Muonic Radicals

A molecule containing a muon and one unpaired electron exhibits a quadratic Zeeman splitting similar to that of muonium, but modified to account for the different muon-electron hyperfine coupling. In general, however, additional magnetic nuclei are present in muonic radicals. In particular, most organic radicals contain many protons. Although the muon-nuclear coupling is negligible ($|J|<10^2$ Hz), coupling between the additional nuclei and the unpaired electron ($0<|A|<10^9$ Hz) has a significant effect. By analogy with Equation

30, the Hamiltonian for rapidly tumbling radicals in liquids is

$$H/h = \nu_e S_z - \nu_\mu I_z^\mu - \sum_k \nu_k I_z^k + A_\mu \vec{S}\cdot\vec{I}$$

$$+ \sum_k A_k \vec{S}\cdot\vec{I}^k \qquad [37]$$

where ν_e, ν_μ and ν_k are the Larmor precession frequencies for the electron, the muon and the k'th nucleus, and A_μ and A_k are the relevant hyperfine coupling constants. The theoretical and numerical calculation of the eigenfunctions and eigenvectors appropriate for a variety of cases of different magnetic fields and couplings has been reviewed by Roduner.[26] Practical applications employ the high field limit almost exclusively. In this case the terms involving I^k may be treated as a perturbation to the case of muonium. Only two of the four muonium-like transitions have significant intensity in the high field limit ($\nu_e \gg A_\mu, A_k$):

$$\nu_{12} = \tfrac{1}{2}[A_\mu - |(\nu_e + \nu_\mu)^2 + A_\mu^2|^{1/2} + \nu_e - \nu_\mu]$$

$$\longrightarrow \tfrac{1}{2} A_\mu - \nu_\mu \qquad [38]$$

$$\nu_{43} = \tfrac{1}{2}[-A_\mu - |(\nu_e + \nu_\mu)^2 + A_\mu^2|^{1/2} + \nu_e - \nu_\mu]$$

$$\longrightarrow -\tfrac{1}{2} A_\mu - \nu_\mu \qquad [39]$$

That only these two frequencies are resolvable at high fields can be easily understood from Equation 32 and Figure 14. In the case of a radical, the contact field $B_o(R)$ is typically 0.1 of that in normal Mu, so that δ rapidly approaches 1 even in moderate fields (> 1.5 kG). Consequently only the first term of Equation 32 survives, giving the observable frequencies ν_{12} and ν_{43}.

Each set of equivalent nuclei contributes a first order correction

$$\Delta = A_k M_k S^2 \qquad [40]$$

where M_k is the z-component of the total angular momentum for the set, and

$$S^2 = \frac{1}{2}[1 - (\nu_e + \nu_\mu)/|(\nu_e + \nu_\mu)^2 + A_\mu^2|^{1/2}] \qquad [41]$$

To summarize, the μSR Fourier spectrum of a radical in the high magnetic field limit is comprised of two frequencies placed symmetrically about the muon precession frequency ($-\nu_\mu$, according to the sign convention used in writing the above Hamiltonian). As the field is reduced, additional splitting occurs and the two frequencies (or multiplets) are no longer quite symmetrical about $-\nu_\mu$. Nevertheless, to first order the difference between the two frequencies (or multiplets) is still equal to A_μ. Examples of spectra at different fields are given in Figure 15. The muon nuclear Larmor frequency (D) signals the presence of muons substituted in diamagnetic molecules. Such a diamagnetic fraction is found in almost all liquids, but little is known about its nature and source since the chemical shifts and line splittings so useful in NMR are

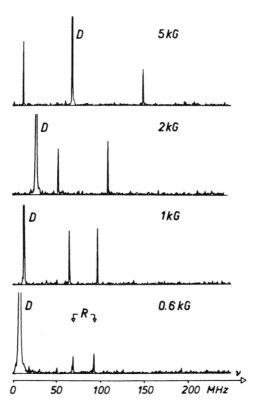

Figure 15. μSR Fourier power spectra at several magnetic fields for the muonic radical 2,3-dimethyl-2-butene. (From Ref. 23.)

unresolvable in μSR. By contrast, valuable information is provided by the radical spectrum, in the form of frequencies, splittings, linewidths, amplitudes and phases. Evidence clearly supports assignment of the frequencies R in Figure 15 to the radical $(CH_3)_2\overset{\bullet}{C}C(CH_3)_2Mu$ formed by addition of thermalized muonium atoms to 2,3-dimethyl-2-butene. All muonic radicals observed to date are formed in analogous rections of unsaturated organic compounds.

The spectra of Figure 15 were obtained by real Fourier transformation, so only absolute values of the frequencies are obtained:

$$\tfrac{1}{2} | A_\mu - \nu_\mu | \text{ and } \tfrac{1}{2} | A_\mu + \nu_\mu |: \qquad [42a]$$

$$| A_\mu | > 2 | \nu_\mu |. \qquad [42b]$$

That the two radical precessions have opposite senses can be demonstrated by complex Fourier transformation of orthogonal time spectra. An example is given in Table 2. The precession observed at 25.8 MHz in the real transform appears at 357.3 MHz in the complex transform. Given the periodic nature of the discrete Fourier transformation, the larger frequency is the "wraparound" of a negative precession frequency of 25.8 MHz (i.e., $\nu_{max} - |\nu|$). Thus, it can be seen that the true difference between the two frequencies $|A_\mu| = (135.4 + 25.8)$MHz = 161.2 MHz, and that the two frequencies are indeed close to being equidistant from the diamagnetic precession. (See Equation 37).

Since complex transformation distinguishes between precession frequencies of opposite sign, it is tempting to think that the absolute sign of A_μ can be determined. In fact, the sign ambiguity

Table 2 Muon precession frequencies in 4.4M 2,3-dimethyl-2-butene in cyclohexane. 1024 bins of 2.61 ns width were Fourier analyzed using an exponential filter with an 0.5 μs time constant.

Transform Type	Observed Frequencies (MHz)				ν_{max} (MHz)
real	25.8	54.3	135.4	--	191.5
complex	--	54.3	135.4	357.3[+]	383.1

[+] $357.3 \equiv 357.3 - \nu_{max} = -25.8$ MHz

remains since there is no practical means of telling which observed frequency corresponds to ν_{12} and which to ν_{43}. (In principle, assignment is possible on the basis of second-order shifts in the multiplets; in practice, the necessary resolution is unattainable.) In favourable cases, the sign of A_μ may be found from zero field μSR spectra.[27]

For the majority of applications, determination of the absolute value of A_μ is enough. It is related to the distribution of unpaired spin density in the radical, and can be interpreted in a manner analogous to proton coupling constants as measured by ESR. As an illustration consider the muon coupling constants found for samples of thiophene and furan at room temperature. By analogy with H atom chemistry, the radicals are probably

A_μ = 339 MHz   and A_μ = 379 MHz

The higher A_μ for furanyl is consistent with a greater spin density on the 3-carbon, which would arise through the reduced delocalization in the oxygen-containing heterocycle of furan compared with the sulphur in thiophene. Differences in molecular geometry may also have an effect.

Trends in A_μ may be used in the assignment of radical structures in cases where more than one radical can be formed from a single parent molecule. A particularly thorough study has been made of the radicals that ensue form a large family of methyl substituted butadienes.[28] Information on molecular geometry is gained from the temperature dependence of A_μ. Knowledge of the qualitative dependence is often enough to determine the equilibrium conformation, whereas quantitative data leads to details of the internal dynamics.[26].

In general, the amplitudes and phases of the two radical precession frequencies differ. To observe radicals at all, muon spin polarization must be transferred from muonium to the radical. The degree of transfer depends on reaction rates and precession frequencies, which in turn depend on the magnetic field. This field of study is still in its infancy, however, so the reader is referred elsewhere for some preliminary results.[29]

In the absence of extraneous spin relaxation effects, the decay of a radical signal is attributed to chemical reaction. Since the muon in a radical is usually at a site remote from the reaction center, it functions merely as a tracer. This is potentially a very valuable application of μSR to the field of chemical kinetics, since it is very difficult to accurately determine radical rate constants by other means. A pilot study has demonstrated the feasibility of such a determination, using radical line widths from Fourier transform spectra to measure decay rates as functions of the concentration of an added substrate.[26]

Paramagnetic States of the Positive Muon in Solids

The study of μ^+ thermalization and the formation of muonium and muonium-like states in solids, a topic dating back to the very first μSR work,[30] continues to be a major branch of μSR research. Recently, in fact, it has been one of the fastest-growing fields in μSR, perhaps due to an accelerated interest in hydrogen-like impurity states in semiconductors. Despite the fact that hydrogen is one of the main impurities in the two most important semiconductors, Si and Ge, it has never been directly seen in an EPR study, and hence has really only been characterized in these elements through its analog isotope, muonium. The early data obtained in Si in 100G transverse fields already held the promise of a wealth of information.[31] As can be seen from Figure 16, the characteristic two muonium frequencies, ν_{12} and ν_{23}, are split considerably further apart in Si than in fused quartz. Hence, from Equations 33-35 we see that the beat frequency Ω = 10 MHz, and so $(\nu_0)_{Si}$ = 0.45$(\nu_0)_{vac}$ = 2012 MHz. This means that the contact e^- density at the μ^+ site in Si is considerably reduced from its vacuum value in quartz, which presents an interesting theoretical problem.[32] The field, temperature and orientation dependence of this so-called "normal" Mu has now been studied in considerable detail, revealing its description in terms of the isotropic Hamiltonian of Equation 30.

Even more interesting of late is so-called "amomalous" muonium (Mu*), which produces the frequencies clearly observable in Figure 16 at 15, 40 and 48 MHz, and in Figure 5 at similar frequencies. Detailed studies have revealed that Mu* is describable in terms of an anisotropic but axially symmetric Hamiltonian

Fourier Transform µSR

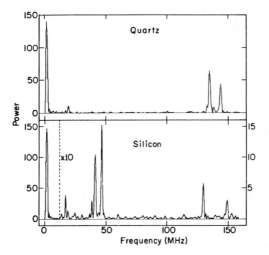

Figure 16 Top: Fourier power spectrum from fused quartz at 100 G; Bottom: similar spectrum from Si at 77 K and 100 G. The two rightmost signals in each spectrum are the muonium frequencies. (From Ref. 31.)

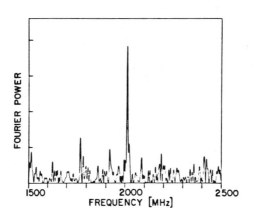

Figure 17 Mu hyperfine oscillation observed directly in silicon. (From Ref. 37.)

$$H = (g_e \mu_B \vec{S} - g_\mu \mu_\mu \vec{I}) \cdot \vec{B}$$
$$+ h A_\perp (S_x I_x + S_y I_y) + h A_\parallel S_z I_z \qquad [43]$$

where the z axis is along any of the four [111] axes of the crystal lattice.[33] The magnitudes of $A_\perp$ and $A_\parallel$ are 92.1 and 17.1 MHz, respectively, compared to $A_\perp = A_\parallel = \nu_0 = 2012$ MHz for normal isotropic Mu. The nature of the stopping site for the μ^+ is not fully understood[34] but it is clear that the interaction is enormously anisotropic. Similar states have also been seen in Ge[35] and diamond,[36] and the search is on in different seimconductors as well, promising an exciting future.

One of the most interesting applications of μSR studies is in zero magnetic field, a situation completely precluded in NMR. Recent zero-field studies of the free μ^+ in longitudinal fields have revealed new and unique information on spin glass and other systems[8] but here we concentrate on muonium and muonium-like states in zero magnetic field. In this case, precession is not observed in the classical sense of the word but rather a modulation of the muon polarization with time. This can be most easily understood in the case of muonium itself in terms of the isotropic Hamiltonian of Equation 30. As noted above, muonium is formed via the "capture" of an electron from the stopping medium. Since the μ^+ is longitudinally polarized[1-3] (α_μ) but the captured e$^-$ is not (α_e or β_e), muonium forms initially in two spin states, defined by $|A\rangle = |\alpha_\mu \alpha_e\rangle = |1\ 1\rangle$ and $|B\rangle = |\alpha_\mu \beta_e\rangle = 1/\sqrt{2} \ ||1\ 0\rangle + |0\ 0\rangle|$ where in weak magnetic fields the total spin states $|F\ M_F\rangle$ are nearly energy eigenstates. In zero or weak longitudinal fields, $|A\rangle$ is an eigenstate but $|B\rangle$ is not, oscillating instead between $|B\rangle = |\alpha_\mu \beta_e\rangle$ and $|B'\rangle = |\beta_\mu \alpha_e\rangle = 1/\sqrt{2} \ ||1\ 0\rangle - |0\ 0\rangle|$ at the hyperfine frequency $\nu_0 = 4463$ MHz. This modulation is in principle observable as an (F,M) = (0,0) $\longrightarrow$ (1,0) transition using zero degree (forward/backward) detectors, but with current time resolutions of typically 1 ns it is not quite feasible to measure the vacuum hyperfine frequency directly. Thus Mu in zero field normally appears effectively depolarized by a factor of two (a measurement of the difference in counting rates between 0° and 180° yields an asymmetry 50% of the maximum value corresponding to the $|\alpha_\mu \alpha_e\rangle$ state, but no observable modulation). In the semiconductors Si and Ge, however, as noted above, Mu has an isotropic hyperfine coupling about half that of the vacuum value, and recent zero field experiments have succeeded in measuring the hyperfine frequency directly as a modulation of the μ^+ polarization. This is beautifully illustrated in the Fourier transform spectrum of Figure 17, which shows a single frequency at (2011.8 + 0.5) MHz.[37] The time resolution in this experiment was 350 ps; 1000 data bins were transformed corresponding to a time window of 500 ns.

Fourier Transform μSR

In zero field, the Hamiltonian of Equation 43 predicts three frequencies, since the $|F = 1\ M = 0\rangle$ substate is shifted in energy with respect to the degenerate case depicted in the Breit-Rabi diagram of Figure 14. These frequencies are given by $A_\perp$ and $1/2\ |A_\perp + A_\parallel|$ and have been seen in Si,[33] Ge[35] and most recently in diamond.[36] A sample Fourier transform spectrum for diamond in zero field is given in Figure 18, where the frequencies ν_1, ν_2 and ν_3 are defined by $A_\parallel = \nu_2 - \nu_1 = 168 \pm 0.4$ MHz and $A_\perp = \nu_1 + \nu_2 = \nu_3 = 392.5 \pm 0.4$ MHz. Normal Mu has also been seen in diamond[36] with an isotropic hyperfine coupling $\nu_0 = 3790$ MHz, which is 83% of the vacuum value, far larger than the values found in either Si or Ge. This is in itself an interesting result with consequences for the model interpretation of Mu in solids; for instance, it is inconsistant with the prediction of Ref. 32. It is also interesting to note the high field limit of Equation 30 ($B \gg B_0$, the effective magnetic field of the μ^+ at the electron; e.g., $B_0 \sim 35G$ for Mu* in Si). In this case, only the frequencies ν_{12} and ν_{34} (in the notation of Figure 14) are observable, a situation formally identical with the case of muonic radicals discussed above.

As a final example of Fourier transform μSR in solids, we will mention recent results obtained in quartz crystals. The study of μ^+ and Mu in quartz has long been of interest in μSR studies.[3,38] Two-frequency Mu precession is routinely observed in fused quartz in magnetic fields ~ 50 G (Figure 16), yielding the

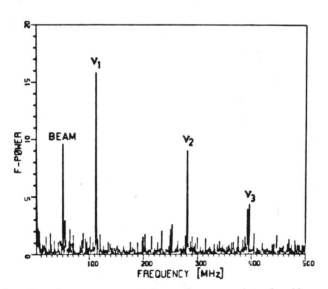

Figure 18. Fourier power spectrum for muonium in diamond at zero magnetic field. (From Ref. 36.)

"vacuum" value ν_0 = 4463 MHz dicussed above. In very weak magnetic fields (10 G), only a single precession frequency is observed in fused quartz (since $\Omega \to 0$), giving an amplitude A_{Mu} in fused quartz corresponding to ~ 60% Mu formation, still the largest fraction seen in any solid other than the solid phases of noble gases.[39] The first studies in quartz crystals were carried out in weak fields (6.3 G) in an attempt to see if there was any difference in A_{Mu} for Mu formation in D and L quartz.[40] These studies showed (not unexpectedly) that the difference was consistent with zero but in the process revealed an unexpected splitting of the Mu precession frequency (the quadratic Zeeman splitting Ω should only have been 0.0047 MHz, not resolvable).[41] The time spectrum from Ref. 41 is shown in Figure 19; "beats" are clearly apparent in the data. The splitting $\Delta \nu$ between these frequencies is markedly dependent on the orientation angle θ of the axis of the quartz crystal with respect to the applied magnetic field. Figure 20 presents room temperature Fourier power specra taken at orientations of 0° and 90°, which provide respectively the maximum and minimum splittings of 0.8 and -0.4 MHz, and at 55°, where the splitting disappears. In fact, the angular dependence follows a simple $(3\cos^2\theta - 1)$ dependence as shown in Figure 21. This led to an initial interpretation of this phenomena in terms of the intrinsic electric quadrupole moment of the F = 1 Mu atom.[41] Such an interpretation is understandable in terms of the axially symmetric Hamiltonian of Equation 43 describing the Mu* states in semiconductors, as discussed already.

The main difference between Mu in single-crystal quartz near room temperature and Mu* in Si (for example) is in the magnitudes of $A_\parallel$ and $A_\perp$; in Si $\langle A \rangle = (A_\perp + A_\parallel)$ is 2% of the vacuum value ν_0 and the anisotropy $(A_\perp - A_\parallel)$ is -69% of $\langle A \rangle$; whereas in quartz, $\langle A \rangle$ has essentially the vacuum value and the anisotropy is only 0.00018 $\langle A \rangle$. (Recent precise measurements[42] have shown that $\langle A \rangle$ = 4509+3 MHz, which is 1.03% bigger than the vacuum value.) Since the magnitude of the anisotropy is so tiny in quartz, Equation 43 can be solved within

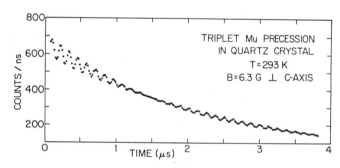

Figure 19. μSR time spectrum for muonium in a quartz crystal at 6.3 G and room temperature.

Fourier Transform μSR

a subspace of just "triplet" muonium, which means that only two frequencies should be seen in weak magnetic fields, as observed (Figures 19 and 20). This result can also be seen immediately from the F = 1 part of the Breit-Rabi diagram of Figure 14, recognizing that the zero field M = 0 sublevel should be shifted up relative

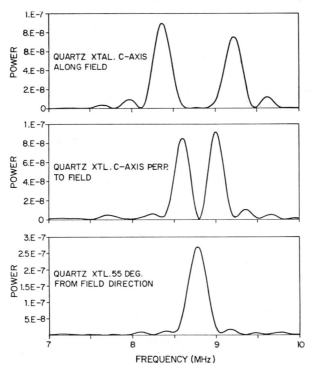

Figure 20. Fourier power spectra for muonium signals from a single crystal of quartz at 4.0 G and room temperature, with the c-axis of the crystal at 0°, 90° and 55° to the applied magnetic field.

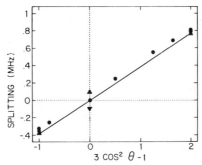

Figure 21. Orientation dependence of the splitting shown in Figure 20.

sublevel should be shifted up relative to M = 1 because of the anisotropic interaction. This picture also predicts only one frequency in zero field, which has also been observed. With the crystalline c axis perpendicular to the μ^+ spin, a single "quadrupole oscillation" frequency of 0.412±0.004 MHz has been measured,[10,41] which vanishes when the $\hat{c}$ axis is parallel to the μ^+ spin, in accord with detailed predictions of a simple model.[41]

More recent low temperature measurements[10] have, however, revealed severe shortcomings in this simple model. Data obtained at 6K are shown in Figure 22 which presents both time (left) and frequency (right) spectra as a function of crystal orientation. Three frequencies are observed: 1.7 ± 0.1, 6.1 ± 0.1 and 7.9 ± 0.1 MHz, the amplitudes of which depend markedly on the orientation. Figure 22 again illustrates the utility of Fourier transform μSR. In the middle spectrum three frequencies are clearly seen, although this may not be obvious from the corresponding time spectrum. The observation of three frequencies at low temperature in zero field means that the uniaxial Hamiltonian of Equation 43 no longer applies and one must consider a general anisotropic Hamiltonian which in the case of zero field can be written in the form:

$$H/h = A_{11}S_xI_x + A_{22}S_yI_y + A_{33}S_zI_z \quad [44]$$

where A_{11}, A_{22} and A_{33} are the three principal values of a 3x3 (F = 1) matrix decribing the hyperfine coupling. Such a Hamiltonian produces three frequencies (as in the case of zero field Mu* in Si) which can be defined by $\nu_{12} = (1/2)|A_{11} - A_{22}|$, $\nu_{23} = (1/2)|A_{22} - A_{33}|$ and $\nu_{13} = (1/2)|A_{11} - A_{33}|$, where it should be noted that $\nu_{12} + \nu_{23} = \nu_{13}$, as observed experimentally.

These measurements are of course interesting in their own right but also important in the sense that they establish for the first time the close parallel between Mu and H atoms in solids. There has been a considerable amount of EPR work done in quartz;[43] in particular, the H site and principal values of the hyperfine matrix in low temperature X-irradiated quartz have been determined. Scaling these values for muonium by multiplying by a factor of 3.18 (the ratio of muon to proton g factors) predicts zero-field oscillation frequencies for Mu which are about 1.4 times higher than the observed values but the same for all three frequencies. This difference can be understood in terms of the distortion of Mu in the lattice, which is expected to be less than that of H because of the smaller mass of the μ^+ and hence relatively larger zero point motion (i.e., it pushes the neighbouring atoms further away). Such an interpretation is consistent with the previously mentioned observation that ν_0 for Mu in quartz is 1.03% larger than the vacuum value[42]; whereas for H it is larger by 2.23%.[43]

Fourier Transform μSR

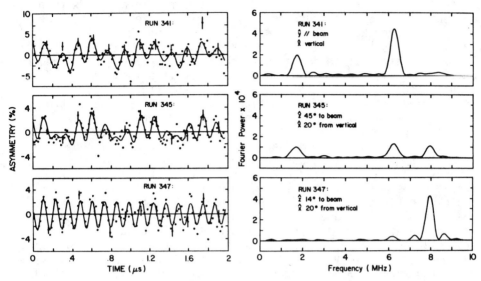

Figure 22. (Left) Time spectra and (Right) corresponding frequency spectra for three different orientations of a quartz crystal at 6 K and zero magnetic field.

SUMMARY

Although the basic features of the μSR technique have been known for about 25 years, it has really only been in the last 5 years with the advent of the world's meson factories (LAMPF, TRIUMF and SIN) that the techniques have been routinely applied to a range of studies of muonium chemistry and reaction kinetics in gases and liquids, radiation chemistry, magnetism and hyperfine fields, quantum diffusion in solids and paramagnetic states in solids and liquids. These subjects are well represented in the Proceedings of two recent Topical Meetings on Muon Spin Rotation (the first in Rorschach, Switzerland, in 1978, and the second in Vancouver, Canada, in 1980), to which we have made frequent reference in this article. The future success of the field will depend in large measure upon the willingness of chemists and atomic and condensed matter physicists to venture away from "traditional" experimental areas and embark on research in μSR.

In those cases in which the muon experiences a transverse magnetic field at its stopping site, the muon spin precesses with a characteristic Larmor frequency; in general, there will be several different frequencies corresponding to different environments for the muon. In these cases, one has classic Muon Spin Rotation (μSR) and Fourier transformation is an invaluable guide towards understanding a complicated histogram. Even in the case of zero magnetic field studies, fast modulation of the muon spin polarization can easily be seen through Fourier transformation. As an illustration of the technique we have focussed on muonic radicals and on paramagnetic states in solids, particularly zero field studies in quartz crystals. Fourier transformation provides a measure of the muon frequencies, which in some cases (radicals) may be the only information of interest. A sophisticated on-line FT routine thus "immediately" provides the essential results of some experiments. More generally, though, the frequency is only one of many parameters of interest. The initial amplitude and relaxation rate of the signal, as well as the initial phase of its precession, can all be roughly determined from Fourier transform spectra and so an immediate on-line analysis can again provide a valuable estimate of desired information. However, the answers are not unambiguous, being affected by counting statistics, time range, apodization and other filtering techniques; in particular, if reliable <u>errors</u> for the parameters of interest are also sought, these can only be provided by fitting the the time spectrum itself. In this case, FT provides an invaluable "guess" for initial values and thus saves many hours (and dollars) of computer time.

In this article we have dealt only with the positive muon (μ^+SR) although negative muon (μ^-SR) experiments are also of interest. In μ^-SR, in fact, signal amplitudes are much weaker (a combination of μ^- lifetimes and residual polarization) than in μ^+SR and FT often provides an invaluable measure as to the existence or nonexistence of a signal. No doubt, as the awareness of the μSR community of the strengths (and limitations) of FT grows to a level comparable to that in FT-NMR, FT-IR, etc., it will be used more and more routinely to reliably extract more of the parameters of interest in a μSR experiment, perhaps eventually even supplanting the need for concomitant fitting of the data in time space.

ACKNOWLEDGEMENTS

We are indebted to Dr. Bruce Patterson, now returned to Switzerland, who helped develop Fourier transform software at TRIUMF from 1978 to 1979, and to whose work we have referred liberally (see Ref. 17) in this article. Had Bruce remained in Vancouver he would certainly have been a coauthor.

The TRIUMF research described herein was performed with the financial assistance of the Natural Sciences and Engineering Research Council of Canada.

REFERENCES

1. Garwin, R. L., Lederman, L. M. and Weinrich, M. 1957, Phys. Rev. 105, 1415.
2. See for instance "Muon Physics,"1975, ed. V. W. Hughes and C. S. Wu, Academic Press, Vol. II.
3. Brewer, J. H., Crowe, K. M., Gygax, F. N., and Schenck, A., in "Muon Physics," 1975, Ed. V. W. Hughes and C. S. Wu, Academic Press, Vol. III; Brewer, J. H., and Crowe, K. M., 1978, Ann. Revs. Nucl. Part. Sci., 28, 239; Schneck, A., in "Nuclear & Particle Physics at Intermediate Energies," 1975, ed. J. B. Warren, Plenum Press, New York, 159.
4. Fleming, D. G., Garner, D. M., Vaz, L. C., Walker, D. C., Brewer, J. H., and Crowe, K. M., 1979, Adv. in Chem. Series 175, 279.
5. Percival, P. W., 1979, Radiochimica Acta, 26, 1.
6. Wilets, L. and Wu, C. S., 1969 Ann. Rev. Nucl. Sci. 19, 527; Kim. Y.N., 1971 "Mesic Atoms and Nuclear Structure," North Holland, Amsterdam; Hufner, J. and Scheck, F. 1975 in "Muon Physics," ed. V. W. Hughes and C. S. Wu, Academic Press, Vol. I.
7. See for instance Yamazaki, T., Hayano, R. S., Kuno, K., Imazato, J., Nagamine, K., Kohn, S. E., and Huang, C. Y., 1979 Phys. Rev. Lett., 42, 1241; Nagamiya, S., Nagamine, K., Hashimoto, O. and Yamazaki, T., 1975 ibid, 35, 308.
8. Uemura, Y. J., Yamazaki, T., Hayano, R. S., Nakai, R. and Huang, C. Y., 1980 Phys. Rev. Lett. 45, 853; Hayano, R. S., Uemura, Y. J., Imazato, J., Nishida, N., Yamazaki, T. and Kubo, R., 1979 Phys. Rev. B20, 850.
9. Nagamine, K. N., 1981 Proc. of 2nd Int. Topical Mtg. on Muon Spin Rotation, Hyp. Int. 8, 787.
10. Brewer, J. H., Spencer, D. P., Fleming, D. G., and Coope, J. A. R., 1981, Proc. of 2nd Int. Topcial Mtg. on Muon Spin Rotation, Hyp. Int. 8, 405; Brewer, J. H., Fleming, D. G. and Spencer, D. P., 1981, Nuclear and Electron Resonance Spectroscopies "Applied to Materials Science", eds. Kaufmann and Shenoy, Elsevier, North Holland, p. 487.
11. Brewer, J. H., in Proc. of 2nd Int. Topical Mtg. on Muon Spin Rotation, 1981, Hyp. Int. 8, 831.
12. Garner, D. M., Ph.D. Thesis, Department of Chemistry, University of B.C., Aug. 1979 (unpublished).
13. Seeger, A., 1978, "Hydrogen in Metals", Springer-Verlag Topics in Applied Physics, Vol. 28, eds. Alefeld, G. and Volke, J., p. 349; Hintermann, A., Schenck, A., Schilling, H. and Hartmann, O., 1981, O. Proc. of 2nd Int. Topical Mtg. on Muon Spin Rotation, Hyp. Int. 8, 539.

14. Martoff, C. J., and Rosenblum, Proc. of 2nd Int. Topical Mtg. on Muon Spin Rotation, 1981, Hyp. Int. $\underline{8}$, 805.
15. Marshall, A.G. 1980 in "Phys. Meth. of Mod. Chem. Anal." Vol. III, ed. T. Kuwana, Academic Pres, New York; Andre, J. C., Vincent, L. M., O'Connor, D. and Ware, W. R., 1979 J. Phys. Chem. $\underline{83}$, 2285; Bloomfield, P., 1976, <u>Fourier Analysis of Time Series: An Introduction</u>, Wiley, New York.
16. Bartholdi, E. and Ernst, R.R. 1973 J. Magn. Res. $\underline{11}$ 9.
17. Patterson, B.D., Ph.D. Thesis, Department of Physics, Univ. of Calif., Berkeley, (Lawrence Berkeley Lab. Report No. LBL-3817) (1975) unpublished.
18. Redfield, A. G. 1976 in "NMR, Basic Principles and Progress, Vol. 13: Introductory Essays," ed. M. M. Pinter, Springer-Verlag, Berlin, 137.
19. S. I. Parks and R. B. Johannesen 1976 J. Magn. Res. $\underline{22}$, 265.
20. Hughes, V. W. 1966 Ann. Rev. Nucl. Sci. $\underline{16}$, 445; Hughes, V. W., McColm, D. W., Ziock, K. and Prepost, R. 1970 Phys. Rev. 1A, 595.
21. Fleming, D. G., Mikula, R. J. and Garner, D. M., Proc. of 2nd Int. Topical Mtg. on Muon Spin Rotation, 1981, Hyp. Int. $\underline{8}$, 307; Mikula, R. J., Garner, D. M., Fleming, D. G., Marshall, G. M., and Brewer, J.H., 1979 Hyp. Int. $\underline{6}$, 379; Stambaugh, R. D., 1974, Ph.D. Thesis, Dept. of Physics, Yale Univ., unpublished; Stambaugh, R. D., Casperson, D. E., Crane, T. W., Hughes, V. W., Kaspar, H. F., Souder, P., Thompson, P. A., Orth, H., zuPutlitz, G., and Denison,. A. B., 1974, Phys. Rev. Letts. $\underline{33}$, 568.
22. Percival, P. W., Roduner, E. and Fischer, H., 1978, Chem. Phys. $\underline{32}$, 353; Walker, D. C., Jean Y. C., and Fleming, D. G., 1980, J. Chem. Phys. $\underline{72}$, 2902; Percival, P. W., Proc. of 2nd Int. Topical Mtg. on Muon Spin Rotation, 1981, Hyp. Int. $\underline{8}$, 315; Walker, D. C., Proc. of 2nd Int. Topical Mtg. on Muon Spin Rotation, 1981, Hyp. Int. $\underline{8}$, 329.
23. Roduner, E., Percival, P. W., Fleming, D. G., Hochmann, J. and Fischer, H. 1978 Chem. Phys. Lett. $\underline{57}$, 1.
24. Casperson, D. E., Crane, T. W., Denison, A.B., Egan, P. O., Hughes, V. W., Mariam, F. G., Orth, H., Reist, H. W., Souder, P. A., Stambaugh, R. D., Thompson, P. A. and Zu Putlitz, G. 1977 Phys. Rev. Lett. $\underline{38}$, 956.
25. Fleming, D. G., Mikula, R. J. and Garner, D. M. 1980 J. Chem. Phys. $\underline{73}$, 2751; Jean, Y. C., Fleming, D. G., Ng, B.W. and Walker, D. C. 1978 Chem. Phys. Lett. $\underline{60}$, 125.
26. Roduner, E., 1980 in "Exotic Atoms '79", ed. K. M. Crowe, J. Duclos, G. Fiorentini and G. Torelli, Plenum Press; Roduner, E. Proc. of 2nd Int. Topical Mtg. on Muon Spin Rotation, 1981, Hyp. Int. $\underline{8}$, 561.
27. Roduner, E. and Fischer, H. 1981 Chem. Phys., $\underline{54}$, 261.
28. Roduner, E., Strub, W., Burkhard, P., Hochmann, J. Percival, P. W.and Fischer, H. 1981 Chem. Phys., in press.

29. Percival, P. W. and Hochmann, J. 1979 Hyperfine Int. 6, 42.
30. Feher, G., Prepost, R. and Sachs, A. M. 1960 Phys. Rev. Lett. 5, 515; Eisenstein, B., Prepost, R. and Sachs, A. M. 1966 Phys. Rev. 142, 217.
31. Brewer, J. H., Crowe, K. M., Gygax, F. N., Johnson, R. F., Patterson, B. D., Fleming, D. G. and Schenck, A. 1973 Phys. Rev. Lett. 31, 143.
32. Wang, J. S. Y. and Kittel, C. 1973 Phys. Rev. B7, 713; Altarelli, M. and Hsu, Y. W. 1979 Phys. Rev. Lett. 43, 1346.
33. Meier, P. F. in "Exotic Atoms '79:, ed. K. M. Crowe, J. Duclos, G. Fiorentini and G. Torelli, Plenum Press (1980); Patterson, B. D., Hintermann, A., Kundig, W., Meier, P. F., Waldner, F., Graf, H., Recknagel, E., Weidinger, A. and Wichert, Th. 1978 Phys. Rev. Lett. 40, 1347.
34. Clawson, C. W., Crowe, K. M., Rosenblum, S. S. and Brewer, J. H., 1981, Hyp. Int. 8, 397; Clawson, C. W. Haller, E. E., Crowe, K. M., Rosenblum, S. S., and Brewer, J. H., ibid, p. 417.
35. Holzschuh, E., Graf, H., Recknagel, E., Weidinger, A., Wichert, Th. and Meier, P. F. 1979 Phys. Rev. B20, 4391.
36. Holzschuh, E., Estreicher, S., Kundig, W., Meier, P. F., Patterson, B.D., Appel, H., Sellshop, J. P. F., and Stemmet, M. private communication; Brewer, J. H., Fleming, D. G. and Spencer, D. P., unpublished data from TRIUMF.
37. Holzschuh, E., Kundig, W. and Patterson, B.D., Proc. of 2nd Int. Topical Mtg. on Muon Spin Rotation, 1981, Hyp. Int. 8, 819.
38. Myasishcheva, G.G., Obukhov, Yu. V., and Firsov, V. G. 1967 Zh. Eksp. Teor. Fiz. 53, 451 [Sov. Phys. JETP 26 (1968) 298.]
39. Kiefl, R. F., Warren, J. B., Marshall, G. M., Oram, C. J. and Clawson, C. W., 1981, J. Chem. Phys. 74, 308.
40. Spencer, D. P., Fleming, D. G., Brewer, J. H. and Mikula, R. J. 1979 Proc. of Symp. on Origins of Optical Activity in Nature, Vancouver, June 1979, ed. D. C. Walker, Elsevier, 87.
41. Brewer, J. H., Beder, D. S., and Spencer, D. P. 1979 Phys. Rev. Lett. 42, 808; Beder, D. S. 1980 Phys. Rev. B21, 3861.
42. Brown, J. A., Dodds, S.A., Estle, T. L., Heffner, R. H., Leon, M. and Vanderwater, D. A., 1980 Solid State Comm. 33, 613.
43. Weil, J. A., 1981 Proc. of 2nd Int. Topical Mtg. on Muon Spin Rotation, Hyp. Int. 8, 371; Weil, J. A. 1974 J. Magn. Res. 15, 594; Rinneberg, H. and Weil, J. A. 1971 J. Chem. Phys. 56, 2019.

FOURIER TRANSFORM INFRARED SPECTROMETRY

James A. de Haseth
Department of Chemistry
The University of Alabama
University, Alabama 35486

INTRODUCTION

Fourier Transform Infrared (FT-IR) Spectrometry has evolved during the last quarter century from a limited and specialized technology to a widely accepted and powerful tool. Although FT-IR was originally the province of astronomers it quickly found application in high-resolution spectroscopy and analytical chemistry. This chapter will cover the basic aspects of Fourier Transform Infrared Spectrometry as well as explore some recent applications of the technique. More in-depth treatments of the subject may be found elsewhere.[1,2]

The Michelson Interferometer

The basic component of most Fourier Transform Infrared spectrometers is the Michelson interferometer. This is not the only interferometer used in FT-IR, but it is employed more often than other designs. A treatment of many other interferometer designs is available.[3] The Michelson interferometer in a Fourier Transform Infrared spectrometer replaces the monochromator in a dispersive instrument, although the functions cannot be correlated. A monochomator divides a continuous bandwidth into its component frequencies, whereas an interferometer produces interference patterns of the bandwidth in a precise and regulated manner. It should be noted that this type of interferometer is not restricted to the infrared region and its use can be extended to the visible and millimeter regions of the electromagnetic spectrum.

A schematic of a Michelson interferometer as used in its most common configuration is shown in Figure 1. The radiation paths have been offset for clarity. It is preferable to consider the case in which the incoming radiation is monochromatic before the general polychromatic case is presented. The incoming radiation of intensity $I(\bar{\nu}_1)$ first strikes the beamsplitter which reflects some of the radiation and ideally transmits the remainder. (An ideal beamsplitter will not absorb or scatter any of the radiation.) If the frequency of the monochromatic radiation is $\bar{\nu}_1$ wavenumbers, we may consider the reflectance of the beamsplitter to be $R\bar{\nu}_1$ and the transmittance to be $T\bar{\nu}_1$, at frequency $\bar{\nu}_1$. Both the reflected and transmitted beams, $R\bar{\nu}_1$ and $T\bar{\nu}_1$, are directed towards mutually perpendicular plane mirrors which return the beams to the beamsplitter. As can be seen from Figure 1, a portion of the radiation equal to $R_{\bar{\nu}_1}^2 + T_{\bar{\nu}_1}^2$ is returned to the source and a portion equal to $2R\bar{\nu}_1 T\bar{\nu}_1$ is emitted perpendicular to the source. The magnitude of the two portions is determined by the relative distances of the two mirrors from the beamsplitter.

For monochromatic radiation the radiation from both arms of the Michelson interferometer will be in phase at the beamsplitter if the two mirrors are equidistant from the beamsplitter. In practice one mirror is fixed, the other is variable in its distance from the beamsplitter. Using the conventional symbol, δ, for the optical path difference between the two mirrors, constructive interference of the two beams from both arms will occur when they are in phase, or when

$$\delta = \frac{n}{2\bar{\nu}_1} \qquad [1]$$

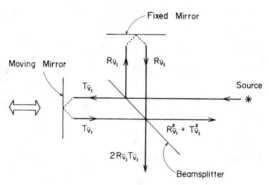

Figure 1. Optical layout of a Michelson interferometer. The beams in the two arms of the interferometer have been offset for clarity.

where n is an even integer. Destructive interference occurs when n is an odd integer. The radiation passed by the interferometer and directed perpendicular to the source, i.e. to the detector, has an intensity or energy proportional to the optical path difference. This intensity, $I'(\delta)_{Det}$, can be shown to be equal to:

$$I'(\delta)_{Det} = 2R_{\bar{\nu}_1}T_{\bar{\nu}_1}I(\bar{\nu}_1)(1 + \cos 2\pi\bar{\nu}_1\delta) \qquad [2]$$

As energy must be conserved, the remainder of the total source intensity $I(\nu_1)$ is returned to the source. The intensity of this radiation is:

$$I'(\delta)_{Src} = (R_{\bar{\nu}_1} + T_{\bar{\nu}_1})I(\bar{\nu}_1) - 2R_{\bar{\nu}_1}T_{\bar{\nu}_1}I(\bar{\nu}_1)\cos 2\pi\bar{\nu}_1\delta \qquad [3]$$

As noted above an ideal beamsplitter does not absorb or scatter radiation, thus:

$$R_{\bar{\nu}_1} + T_{\bar{\nu}_1} = 1 \qquad [4]$$

Using this relationship it can be verified easily that

$$I'(\delta)_{Det} + I'(\delta)_{Src} = I(\bar{\nu}_1) \qquad [5]$$

It is interesting to note that a Michelson interferometer is most efficient when $T_{\bar{\nu}_1} = R_{\bar{\nu}_1} = 0.50$. The most intense signal is received by the detector when the radiation interferes constructively, i.e. when n is an even integer in Equation 1, and then the intensity at detector is $1/2\ I(\bar{\nu}_1)$. This shows that a Michelson interferometer can have a maximum efficiency of only 50 percent.

In the configuration shown in Figure 1 only the beam $I'(\delta)_{Det}$ is measured. As can be seen from Equation 2, there are two parts to $I'(\delta)_{Det}$. One part is constant which is equal to $2R_{\bar{\nu}_1}T_{\bar{\nu}_1}I(\bar{\nu}_1)$, and may be considered a DC component. The other part, $2R_{\bar{\nu}_1}T_{\bar{\nu}_1}I(\bar{\nu}_1) \cos 2\pi\bar{\nu}_1\delta$, is modulated or an AC component. The composite of the AC and DC parts is called the interference record, whereas the AC component alone is called the interferogram. Only the interferogram is recorded; thus, Equation 2 becomes:

$$I(\delta)_{Det} = 2R_{\bar{\nu}_1}T_{\bar{\nu}_1}I(\bar{\nu}_1)\cos 2\pi\bar{\nu}_1\delta \qquad [6]$$

If we let

$$Q(\bar{\nu}_1) = 2R_{\bar{\nu}_1}T_{\bar{\nu}_1} \qquad [7]$$

then Equation 6 becomes

$$I(\delta)_{Det} = Q(\bar{\nu}_1)I(\bar{\nu}_1)\cos 2\pi\bar{\nu}_1\delta \qquad [8]$$

Equation 8 describes the situation when the source is monochromatic, yet FT-IR is most useful when the source is polychromatic. By integrating over all possible values of $\bar{\nu}$ we obtain:

$$I(\delta)_{Det} = \int_{-\infty}^{\infty} Q(\nu)I(\bar{\nu})\cos 2\pi\bar{\nu}\delta\, d\bar{\nu} \qquad [9]$$

Infrared detectors do not measure the energy of the photons, but rather the power (watts) of the radiation striking the detector. The detector has an AC-coupled amplifier which converts the detector response to a voltage, and thus it will have a response function $R(\bar{\nu})$. The detector output $E(\delta)_{Det}$ is equal to:

$$E(\delta)_{Det} = R(\bar{\nu})I(\delta)_{Det} \text{ (volts)} \qquad [10]$$

From Equation 9

$$E(\delta)_{Det} = \int_{-\infty}^{\infty} B(\bar{\nu})\cos 2\pi\bar{\nu}\delta\, d\bar{\nu} \qquad [11]$$

where

$$B(\bar{\nu}) = Q(\bar{\nu})R(\bar{\nu})I(\bar{\nu})$$

Equation 11 indicates that $E(\delta)_{Det}$ is the cosine (or real) Fourier transform of $B(\bar{\nu})$. Unfortunately, this is rarely the case due to the practical constraints of the apparatus. A frequency-dependent phase error, θ_ν, is often present and affects the phase angle $2\pi\bar{\nu}\delta$. The source of this phase error can be misalignment of the interferometer, dispersion by the beamsplitter, errors in data acquisition, or caused by electronic filters in the detector amplifier. If the phase error is considered Equation 11 becomes:

$$E(\delta)_{Det} = \int_{-\infty}^{\infty} B(\bar{\nu})\cos(2\pi\bar{\nu}\delta + \theta_{\bar{\nu}})\, d\bar{\nu} \qquad [12]$$

This is equivalent to

$$E(\delta)_{Det} = \int_{-\infty}^{\infty} B(\bar{\nu})e^{2\pi i\bar{\nu}\delta}\, d\bar{\nu} \qquad [13]$$

where $i = \sqrt{-1}$. Equation 13 states that the interferogram, that is, the electrical response at the detector $E(\delta)_{Det}$, is the complex inverse Fourier transform of $B(\bar{\nu})$. $B(\bar{\nu})$ is the spectrum of the polychromatic source. This spectrum can be calculated from the measured interferogram by performing the complex Fourier transform of Equation 13, hence,

Fourier Transform Infrared Spectrometry

$$B(\bar{\nu}) = \int_{-\infty}^{\infty} E(\delta)_{Det} e^{-2\pi i \bar{\nu} \delta} d\delta \qquad [14]$$

An example of an interferogram is given in Figure 2. Although this interferogram is nearly symmetrical a small amount of phase error is present. The short leading side of the signal is used to calculate the phase error and correct the spectrum. The methods of FT-software are discussed elsewhere in this book.

SPECTRAL RESOLUTION AND APODIZATION

To obtain an accurate representation of the spectrum $B(\bar{\nu})$, the interferogram $E(\delta)_{Det}$ must be sampled for all possible values of δ, according to Equation 14. In practice it is not possible to collect all values of δ, that is, the maximum optical path difference (or retardation) must be limted to some practical value, L. If interferogram data are collected in the interval of optical retardation extending from -L to +L, then Equation 14 becomes

$$B'(\bar{\nu}) = \int_{-L}^{L} E(\delta)_{Det} e^{-2\pi i \bar{\nu} \delta} d\delta \qquad [15]$$

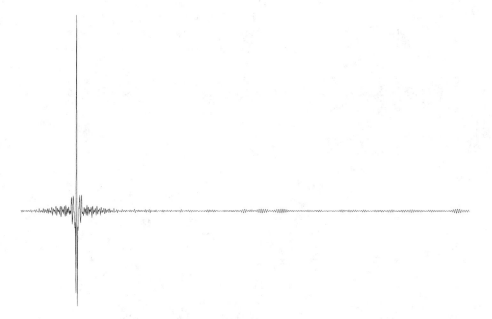

Figure 2. A typical interferogram of the mid-infrared region.

The spectrum $B(\bar{\nu})$ is affected by this change, consequently it must be renamed $B'(\bar{\nu})$. This is equivalent to multiplying $E(\delta)_{Det}$ over the original limits by a truncation function $U(\delta)$ so that

$$B'(\bar{\nu}) = \int_{-\infty}^{\infty} U(\delta) E(\delta)_{Det}\, e^{-2\pi i \bar{\nu} \delta} d\delta \qquad [16]$$

where

$$U(\delta) = \begin{cases} 1. & |\delta| < L \\ 0. & |\delta| > L \end{cases} \qquad [17]$$

Equation 16 states that the spectrum $B'(\bar{\nu})$ is the complex Fourier transform of the product of $E(\delta)_{Det}$ and $U(\delta)$. By recalling that the product of two functions in one domain (distance, δ) is the convolution of the functions in the other domain (spectral, $\bar{\nu}$) then

$$B'(\bar{\nu}) = B(\bar{\nu}) * S(\bar{\nu}) \qquad [18]$$

where $S(\nu)$ is the Fourier transform of $U(\delta)$. Upon taking the Fourier transform of $U(\delta)$, the rectangular truncation function, one obtains:

$$S(\bar{\nu}) = \int_{-L}^{L} e^{-2\pi i \bar{\nu} \delta} d\delta = 2L\, \text{sinc}(2\pi \bar{\nu} L) \qquad [19]$$

$S(\bar{\nu})$ is known as the instrument line shape function (ILS) and is illustrated in Figure 3. The effect of convolving the spectrum $B(\bar{\nu})$

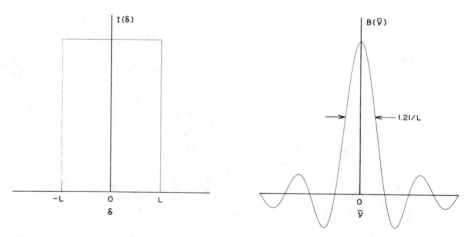

Figure 3. A schematic representation of a rectangular truncation function and its Fourier transform, the sinc function.

Fourier Transform Infrared Spectrometry

with $S(\bar{\nu})$ is to cause all the absorption or emission bands to modify their profiles from the normal Gaussian or Voigt functions.

The problem with using a rectangular function to truncate the interferogram is the production of the ringing or "feet" around the infrared absorption or emission band. This ringing produces an error in the spectrum; it affects the absolute absorption intensity of the band itself, and the feet may interfere with the absorption intensity of any closely spaced band.

The ILS not only distorts the true bandshape in terms of the "feet", but the magnitude of L determines the width of the ILS. It can be shown that the full width at half height (FWHH) for a sinc function is 1.21/L. Consequently as L becomes greater, the ILS becomes narrower and the resolution of the spectrum increases, but is of course limited to the inherent sample resolution.

As might be expected the rectangular truncation function may not be desirable because it produces the feet around bands which include regions of negative signal. A method by which the effect of the feet can be lessened is by apodization (from the Greek, meaning "without feet"). If the rectangular truncation function is replaced by a triangular function, as shown in Figure 4, the resulting line shape is equivalent to a $sinc^2$ function. In this case there is no negative region in the bandshape, and the relative amplitudes of the feet have been reduced by a factor of four. Although these benefits are readily apparent, the use of a triangular apodization function has

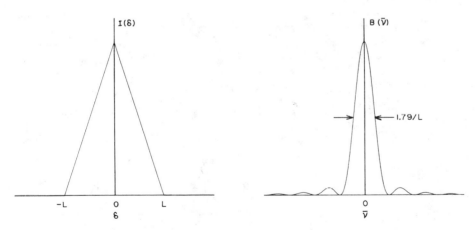

Figure 4. A triangular apodization function and its Fourier transform, the $sinc^2$ function.

other effects. The FWHH is increased to 1.79/L, which is considerably wider than the ILS.

As was stated above the apodization functions affect the resolution of the spectrum. The measurement of resolution is rather arbitrary, and various criteria have been used. For example, resolution is often measured by the ability to separate two closely-spaced bands. In some doctrines resolution is not complete unless the signal baseline is recorded between two bands. This is a rather stringent criterion. Another that is often accepted is the Rayleigh criterion. The Rayleigh criterion states that two bands are resolved if the first zero value of one band is at the peak of the other band. This also assumes the magnitudes of the two bands are nearly identical. When the two bands are recorded according to this criterion, the signal between the two bands is approximately eighty percent of the full band height. The Rayleigh criterion is illustrated in Figure 5 for two bands with a sinc^2 function profile. As can be seen the presence of two bands can be easily recognized.

When a rectangular truncation function is used on the interferogram, the inherent resolution of the spectrum can be shown to be

$$\text{Resolution (ILS)} = \frac{1}{2L} \qquad [20]$$

On the other hand, it can be shown that if the Rayleigh criterion is obeyed and a triangular apodization function is used the resolution becomes

$$\text{Resolution (triangular)} = \frac{1}{L} \qquad [21]$$

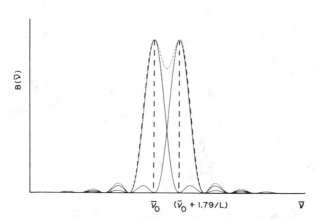

Figure 5. The Rayleigh criterion for resolution of two sinc^2 functions of equal amplitude.

Fourier Transform Infrared Spectrometry

By comparing Equations 20 and 21, the resolution decreases by a factor of two in going from rectangular truncation to triangular apodization.

Other apodization functions are possible, such as algebraic and trigonometric functions. It has been shown that no matter what apodization function is used, there is a trade-off between FWHH and the relative amplitudes of the feet.[4] As a result there is no such thing as a perfect apodization function. Line-widths cannot be kept narrow and ringing eliminated simultaneously, hence compromises must be made.

FT-IR DATA COLLECTION

Thus far descriptions of the basic optical components and mathematical theory have been given, yet nothing has been mentioned as to how the interferogram is acquired. Figure 2 is a single-scan interferogram from an FT-IR and to all appearances is an analog signal. As shown in Equation 13 the interferogram is indeed continuous. Due to practical constraints it is not possible to store a continuous interferogram in a digital computer: the storage allocations on even the largest systems would be quickly exceeded, hence the interferogram must be sampled at discrete intervals. The size of the interval must be carefully selected so as not to introduce errors into the computed spectrum. The choice of sampling interval is made according to the Nyquist criterion.

It is known that to avoid the phenomenon of spectral folding or aliasing, one must sample at a rate of at least twice the maximum frequency in the bandwidth being measured. This is a criterion of Nyquist's Sampling Theorem. In the mid-infrared region for example, the region extends from approximately 4500 to 400 wavenumbers. In this case, it is necessary to modulate and record the signal from the inteferometer at a frequency of at least 9000 wavenumbers. It is not always convenient to have a sampling interval exactly twice the maximum frequency of the highest frequency in the bandwidth. As a result, reference is generally made to a stable, very high frequency, such as a single-mode He:Ne laser.

A single-mode He:Ne laser has an emission at 632.8 nm (visible, red) which corresponds to a frequency of $\sim 15,800$ wavenumbers. The reference He:Ne laser beam is coupled to the moving mirror of the interferometer by either its own beamsplitter in parallel with the infrared beamsplitter, or directly through the infrared optics. The laser radiation is detected separately from the infrared radiation and is recorded as a cosine wave, as indicated by Equation 6. According to the Nyquist sampling criterion, to correctly sample the He:Ne signal data would have to be collected at twice its frequency, or at ~31,600 wavenumbers. If the AC component of the laser radiation is measured, a frequency of 31,600 wavenumbers is achieved by sampling the signal every time the cosine wave equals zero. Consequently if we sample the laser interferogram at every second zero value (or zero crossing), sampling is made at a frequency of ~15,800 wavenumbers and the maximum frequency that can be faithfully recorded according to the Nyquist sampling criterion is one-half that frequency, or ~7,900 wavenumbers. Using every second zero crossing is called undersampling by a factor of two, or alternatively, using an undersampling ratio of two. This ratio is most often used in mid-infrared spectroscopy as it will conveniently cover the bandwidth 4500 to 400 wavenumbers.

Higher undersampling ratios may be used. For example, a ratio of four would cover the range 3950 to 0 wavenumbers. This may be convenient if the upper limit of the bandwidth can be restricted to below 3950 wavenumbers. Of course, because of spectral aliasing, the bandwidths, 7900 to 3950, 11850 to 7900 and 15800 to 11850 wavenumbers will all be sampled accurately at the same time. Usually only one of the potential bandwidths is wanted; hence, the appropriate optical and electronic filters must be used to reject the unwanted bandwidths. By the use of optical and electronic filters high undersampling ratios may be used to collect interferograms of spectra of limited bandwidth. This is especially useful for high-resolution studies where large interferograms are collected [Resolution $\propto L^{-1}$, see Equations 20 and 21]. If high undersampling ratios are used, interferogram storage and data transformation time can be minimized.

It is possible to apply the rapid-scanning Michelson interferometer to far-infrared studies (below 400 wavenumbers). Although the conversion of most commercial rapid-scanning FT-IR spectrometers to the far-infrared region is a simple task (vide infra), many far-infrared spectroscopists employ slow-scanning interferometers, with which the data are collected over a relatively long time period for each sampling interval. These spectrometers often use the fringes of a Moire grating to provide the sampling interval. Such devices are not sufficiently accurate to be used by rapid-scanning interferometers, yet they are adequate for the relaxed tolerances of slow-scanning instruments at long wavelengths.

At the high frequency end of the mid-infrared region is the near-infrared. This can be adequately sampled with the He:Ne laser system at the appropriate undersampling ratio and by the use of the correct filters. Rapid-scanning Michelson interferometers have been extended to the visible and ultraviolet regions of the electromagnetic spectrum.[5-7] As was discussed above the He:Ne laser can be used to sample frequencies accurately up to and including 15,800 wavenumbers, if an undersampling ratio of 1 is used (every zero

Fourier Transform Infrared Spectrometry

crossing). By chopping the He:Ne cosine signal into smaller segments, that is by sampling the maxima and minima of the signal and half-way between zero and the maxima or minima as well as at the zero crossings, the bandwidth can be extended to 63,200 wavenumbers (or as high as to 158 nm).

INSTRUMENTATION

The function of the interferometer in a Fourier transform infrared spectrometer has been presented. An FT-IR spectrometer optical layout is now described and information is provided for each element of a typical spectrometer design. A schematic of a typical FT-IR optical design is given in Figure 6.

Source and Associated Optics

In the mid-infrared region the source is generally a filament in which a resistive wire is wrapped around a ceramic element. Another

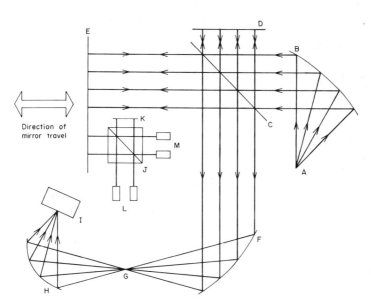

Figure 6. Schematic optical layout of FT-IR spectrometer. A = Source, B = Off-axis paraboloid, C = Beamsplitter, D = Fixed mirror, E = Movable mirror, F = Off-axis paraboloid, G = Sample point, H = Off-axis ellipsoid, I = Infrared detector, J = Visible beamsplitter, K = Fixed mirror, L = He:Ne and white light sources, M = He:Ne and white light detectors.

common source is a globar which is a silicon carbide rod electrically heated to about 1300°C. The source is depicted at point A in the schematic in Figure 6. The source radiation is collected by a collimating mirror B.

In the above treatment of FT-IR the source has been considered to radiate from a point. If the collimator is perfect, the rays entering the interferometer are parallel. This was assumed in the treatment given above. In actual fact the source is not of point dimensions but subtends a solid angle which we shall call Ω. In this case the collimator is unable to produce a parallel ray pattern; hence, some rays oblique to the parallel path will have a larger retardation than the on-axis rays. If the source solid angle becomes too great the path differences between the off-axis and on-axis rays may be considerable and resolution will be lost. As the retardation becomes larger each sampling interval has a higher probability of having information about more than one path difference. It can be shown[8] that the resolution is dependent upon the solid angle subtended by the source by the relationship

$$\text{Resolution} = \bar{\nu}'\Omega/2\pi \qquad [22]$$

where $\bar{\nu}'$ is the frequency (in wavenumbers) of the band whose resolution is being measured. As the resolution is frequency dependent, a maximum allowable solid angle for the source may be calculated by using the maximum frequency in the spectrum to be measured. Although the size of the source is fixed, the solid angle it subtends may be reduced by the use of an aperture. Many optics designs incorporate

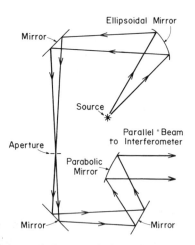

Figure 7. Source optics with provision for aperture. (As used in Digilab[R] FTS Systems).

a focal point before a final collimation mirror so that apertures may be inserted for high-resolution studies. A schematic of such a source optical design is presented in Figure 7. In the schematic in Figure 6 no such allowance is made, yet if an aperture is put at the sampling point G it has the same effect as if it were placed before the interferometer. The function of the aperture is somewhat akin to slit width in a dispersive spectrometer.

The mid-infrared sources can be used in the far-infrared region down to frequencies of approximately 100 wavenumbers. Below this region a high-pressure mercury vapor lamp can be used, provided the ultraviolet and visible frequencies are removed by the suitable optical filter. This filter is usually carbon impregnated polyethylene. In the near-infrared region a tungsten filament lamp functions well as the source.

Beamsplitter

The beamsplitter is labeled C in Figure 6. The beamsplitter is usually placed at 45° between the two mutually perpendicular mirrors in the interferometer. Beamsplitters must have low absorption properties and high reflectance and transmittance. The beamsplitter material has a high refractive index in order to affect these criteria. For the mid-infrared region the actual beamsplitter material is either germanium (refractive index, $n \cong 4.0$) or silicon ($n \cong 3.5$). As these beamsplitters are by necessity thin films, the beamsplitter material must be deposited onto a suitable substrate, as the material itself is too brittle to have any durability as a self-supporting film. The suitability of the substrate is determined by its bandpass and its refractive index as well as the ability to polish and coat the substrate. The refractive index of the substrate should be as low as possible to minimize reflections from the surface and the substrate should be polished to one wavelength flatness of the shortest wavelength to be measured. These criteria are satisfied by salts such as NaCl, CsI and KBr. Potassium bromide is the common choice as it has a lower frequency cutoff than NaCl and is less deliquescent. Even though CsI has a lower frequency cutoff than KBr, it is too plastic to retain a high degree of flatness, hence the high frequency limit is considerably lower than that of potassium bromide. Because the beamsplitter is on a substrate which affects the pathlength in only one arm of the interferometer, a compensator plate of the substrate material is placed in the other arm of the interferometer. This will lessen the phase error. Near-infrared beamsplitters have either a calcium fluoride or quartz substrate with a ferric oxide beamsplitting film.

Far-infrared spectrometers usually use a self-supporting film of polyethylene terphthalate, which has the trade name Mylar (or Melinex in Europe). The Mylar sheet has sufficient mechanical strength not to require a substrate. The thickness of the sheet determines the wavenumber range passed by the beamsplitter, but these sheets are usually between 50 and 5 μm thick. A Mylar beamsplitter has a relatively low refractive index (n ≈ 1.8) and the efficiency of the beamsplitter is highly dependent upon the incident angle of the radiation. Highest efficiencies are found when the incident angle is greater than 75°.[9] Other beamsplitters for the far-infrared are wire meshes; whereas these have not been as widely used as the Mylar beamsplitters, at least one manufacturer supplies a wire grid beamsplitter.

Interferometer Mirrors and Drive

The fixed and moving mirrors of the Michelson interferometer are labelled D and E, respectively, in Figure 6. The alignment of the plane mirrors must be stringently controlled to maintain the integrity of the interferometer. If one mirror tilts slightly so as to perturb the alignment, the beam that is reflected from that mirror will no longer coincide with the beam that has been reflected from the other mirror. That is, if mirror E tilted some small angle $d\theta$, the beams transmitted by the beamsplitter would strike E at an angle of incidence equal to $d\theta$ and have an equal angle of reflection. The reflected beams would not be coincident with the corresponding beams from D at the beamsplitter. Beams from different areas of E will travel different distances, possibly sufficiently different so that they may not all interfere constructively or destructively with the beams from D. In other words, the detector will record several different retardations at a single sampling interval. This effect is similar to that of an extended source (vide supra) and will lead to a degradation of spectral resolution. Of course, as the retardation becomes larger, and thus the resolution greater, the tolerances for $d\theta$ become smaller. One way to alleviate the problem is by the use of retroreflectors, which will always produce parallel input and output beams.[10]

Various mirror drive systems have been used. For far-infrared spectrometers where the wavelengths are long and the tolerance for mirror tilt is large, simple drives such as lubricated pistons or screw drives have been used. In the mid-infrared most moving mirrors are supported on air bearings that are moved rapidly by a solenoid. Scan rates vary from about one millimeter per second to about 6 cm per second. These air bearing drives can maintain optical alignment to produce spectral resolutions of approximately 0.02 wavenumbers. Traditionally, step-scan interferometers have been used for higher resolutions with which the moving mirror has been stopped at each sampling position and realigned. The signal is averaged over a relatively long time period at each position. Recently, one commercial supplier of FT-IR spectrometers has started providing systems in which the drive is rapidly driven by pulling a carriage through a hollow tube. The carriage is supported by roller

Fourier Transform Infrared Spectrometry   401

bearings and the moving mirror allowed to tilt. The fixed mirror is
tilted in unison with moving mirror by employing a dynamic alignment
system. Using this system resolutions as high as 0.002 wavenumbers
in the mid-infrared have been demonstrated.[11]

Sampling Optics

The output of the interferometer is collected and focused at the
sampling position. The focus is usually 3 to 12 mm in diameter, but
this value often is not determined by the focusing mirror (Figure
6F), but by the solid angle subtended by the detector. Commercial
FT-IR systems have beam diameters of about 2.5 to 7.5 cm as the beam
emerges from the interferometer. The sampling optics do not produce
a point focus, which is preferred as the sample will cover a finite
area. Figure 6 illustrates a single-beam spectrometer design, yet
instruments are available that have a "double-beam" option. Unlike
many dispersive systems an FT-IR beam is not chopped and ratioed
through sample and reference cells. In a typical FT-IR experiment
the sample and reference are always separated in time, i.e. one is
collected, then the other; and, they are sometimes separated in
space. The sampling optics (labelled F, G, and H in Figure 6) may be
replaced by two parallel sets for the double beam option. Either
beam can be selected at the operator's discretion. One set holds the
reference, the other the sample. An example of such an optical
set-up is given in Figure 8. In this set-up, the two flipper mirrors
are coupled so that the beam is alternatively focused at the two
sampling points.

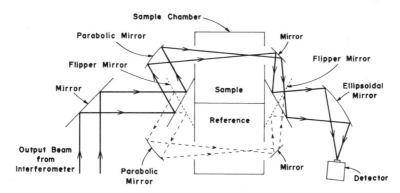

Figure 8. Double-beam optics for sample and reference spectra
           collection. (Digilab[R] FTS Systems.)

In practice if a sample and reference are collected in a single experiment, usually signal-averaging is alternated for a short time between the sample and reference to compensate for long term drifts in the source or detector and associated electronics.

An interesting aspect of FT-IR is that it makes little difference where the sample is placed in the system. Equal results are obtained most often by placing the sample before the interferometer. (For example, the source and detector in Figure 6, A and I respectively, could be interchanged.) Scattered radiation is not a problem as it is in dispersive instruments, because scattered radiation will be constant and only change the DC component of the interference record. It is often preferable to place the sample in one position or the other. Emission measurements require that the sample be the source. Absorption studies of high temperature samples, such as vapors emitted by molten salt cells, should have the samples placed after the interferometer so that infrared radiation emitted by the sample is not modulated.

Detector

The type of detector used in an FT-IR spectrometer is highly dependent upon the bandwidth (i.e. the spectral frequencies), the modulation rate of the interferometer, and the intensity of the radiant flux. Several types of detectors are used in the infrared regions: photoconductive, photovoltaic, bolometers, pyroelectric and Golay cells. A detailed discussion of detectors may be found elsewhere.[12] In general, the photovoltaic and photoconductive detectors can be used in the near- and mid-infrared regions as rapid response, high sensitivity detectors. Usually the bandwidths are limited and will not cover the total range passed by the beamsplitter. Examples of such detectors are given in Table I. As can be seen from the Table, these detectors are cooled. The value $D_{\lambda_0}^{**}$ corresponds to the sensitivity of the detector and may be considered a figure of merit associated with the detector, where λ_0 is the long wavelength limit of the detector. This value is independent of detector field-of-view and has the units cm $H_z^{1/2}$ ster$^{1/2}$/watt. The larger the value of $D_{\lambda_0}^{**}$, the more sensitive the detector.

Other detectors that are useful in the near- and mid-infrared regions are bolometers and pyroelectric detectors. Both these detectors have very large bandwidths and can operate at room temperature; however, they have long response times compared to the photodetectors and they have low $D_{\lambda_0}^{**}$'s. Pyroelectric detectors are useful in the far-infrared region with rapid-scanning spectrometers whereas Golay cell detectors are often used with slow scanning far-infrared interferometers. These cells are modulated at or below 20 Hz.

Table 1. Infrared Detector Characteristics*

Name	Approximate Range (wavenumbers)	Operating Temperature (K)	$D^{**}_{\lambda_o}$ (cm Hz$^{1/2}$Sr$^{1/2}$/watt)
Triglycine sulfate (pyroelectric)	10-5000	300	2×10^9
Golay cell	10-1000	300	6×10^9
Mercury Cadmium Telluride (photoconductive)	700-3800	77	3.1×10^{10}
Indium Antimonide (photovoltaic)	1850-6600	77	6×10^{10}
Thermistor (bolometer)	5000-10000	300	2×10^8

* This table is not exhaustive and only a few of all possible detectors are presented.

At first inspection it may appear desirable to use the most sensitive detector available, assuming the bandwidth and modulation spectrometer criteria can be met. This may not necessarily be the case because the spectrometer is constrained by the dynamic range of the data acquisition system. If a broad band source is used, the interferogram will have a large value at the centerburst (approximately the point of zero retardation, i.e. both interferometer mirrors equidistant from the beamsplitter), assuming the phase error is small. The wings of the interferogram will have small values relative to the centerburst (see Figure 2). The detector responds to the radiant flux by producing an analog signal which is digitized for the data system. Analog-to-digital converters (ADC's) are usually 12 to 16 bits. A 15-bit ADC, for example, has a full dynamic range of about 30,000. If the signal-to-noise ratio (SNR) of an interferogram being measured is greater than 30,000 the noise level will be limited by the ADC, rather than by the inherent noise level in the signal or detector. This SNR is not unreasonable for an interferogram, especially for absorbance measurements.

The interferogram signals are signal-averaged by coadding the digitized interferogram values. In order to avoid adding noise to the signal-averaged data, it is often necessary to use computer word lengths of at least 27 bits.[13]

Sampling and Trigger Optics

It was explained above that data must be sampled at a rate at least twice the frequency of the highest frequency to be recorded. A convenient way to do this is measure the mirror movement, that is, optical path difference, by the fringes from a He:Ne laser beam passed through the interferometer. The measurement using the laser beam can be accomplished in two basic ways, directly through the beamsplitter or remotely. As the beamsplitter material is often opaque to visible He:Ne radiation, remote sensing is used as shown in Figure 6. A laser source is situated at L and directed to a visible radiation beamsplitter J which directs the radiation to the fixed mirror K and to the moving mirror E. As E is scanned a detector at M records the He:Ne cosine wave and this signal can be used to measure the optical retardation.

An alternative method is to use a beamsplitter that transmits both infrared and visible radiation, such as Sb_2S_3 on a KBr substrate. The only beamsplitter in the system would be that at position C in Figure 6. The He:Ne beam could be passed through an edge of the main beamsplitter and recorded physically adjacent to the infrared radiation. Otherwise, small holes could be cut in mirrors B and F so that a laser source could be put behind B and a visible detector behind F. The laser beam is then coincident with the infrared beam and sampling errors due to mirror tilt are minimized. For

Fourier Transform Infrared Spectrometry

this alternate method, the beamsplitter does not have to be transparent to visible radiation, only a small area need be.

In very high resolution spectrometers, such as the dynamic alignment spectrometer mentioned above, three laser beams are run in parallel to the infrared beam. These three laser beams are positioned at the apices of an equilateral triangle. Laser fringe measurements are taken on all three beams and any error, (that is, if all the zero crossings are not simultaneous) is corrected by tilting the fixed mirror. By this method, both mirror tilt and sampling frequency measurement are accomplished.

It is evident there are two sources at position L and two detectors at M in Figure 6. The second source is a broad-band white light source and it has its detector at M. The function of the white light is to produce an interferogram centerburst, offset to the infrared interferogram centerburst. The white light centerburst, arranged so that it occurs in the scan before the infrared centerburst, serves as a trigger for the commencement of data collection. As the moving mirror is scanned, the white light interferogram is monitored. When the intensity of the interferogram reaches a preset threshold (in the centerburst) data collection of the infrared signal is begun on the next laser fringe. This is a reliable method to assure that data collection is started at the same retardation each scan, making coaddition for signal averaging possible. The white light signal can be remote as shown, or run through an infrared beamsplitter that will also transmit visible radiation. This case is analogous to the He:Ne system.

Computer System

Commercial FT-IR spectrometers have dedicated computer systems to collect, store, transform and output the data. These computer systems range from limited-task microcomputers to versatile, powerful minicomputers. Computer systems are general to all Fourier transform techniques and will not be discussed in this chapter.

FT-IR ADVANTAGES

At this point the FT-IR experiment may appear to be a rather difficult way to produce an infrared spectrum. Clearly, there must be some advantages to the technique to justify its existence. There are three distinct facets of FT-IR which make FT-IR superior to conventional dispersive infrared spectroscopy.

Fellgett Advantage

The Fellgett or multiplex advantage deals with the fact that a Fourier transform spectrometer records data from the entire spectral region throughout the experiment. This is quite different to the case with a dispersive spectrometer, as the grating or prism instrument only measures a narrow bandwidth at any time. The measurement bandwidth of the dispersive spectrometer is regulated by the instrument's exit slit. This difference has important effects on the acquisition of data.

It is well-known that the signal-to-noise ratio (SNR) in many data acquisition systems can be improved by signal-averaging. This is accomplished by either sampling the signal for a long period of time so that each datum is more accurately measured, or by rapidly measuring the entire signal repeatedly and coadding the results. In spectroscopy account must be taken of the number of spectral elements, N, that are being collected. The number of spectral elements is simply the bandwidth divided by the resolution. In a grating or prism spectrometer, when the noise is random and independent of signal strength, the SNR should decrease with sampling time, t, but should also depend upon N, thus

$$SNR_D \propto (t/N)^{1/2} \qquad [23]$$

where the subscript D refers to a dispersive instrument. An interferometer obeys different rules because of the Fellgett advantage. If the noise is random and independent of signal strength, the SNR for the interferometer (SNR_I) decreases with sampling time, but is independent of N. As stated above, the entire bandwidth is being sampled at all times, thus N plays no role in the signal-to-noise ratio, and

$$SNR_I \propto t^{1/2} \qquad [24]$$

Taking the ratio of the two SNR's one has:

$$\frac{SNR_I}{SNR_D} = N^{1/2} \qquad [25]$$

That is, in a fixed time an interferometer system will produce a spectrum with a signal-to-noise ratio greater than that of a dispersive system by the square root of the number of spectral elements. In a high resolution system this can be significant, on the order of 10 to 10^3. In the infrared region the noise is generally independent of signal strength so the advantage holds.

The multiplex advantage is not valid in the visible region of the spectrum as the noise is signal strength dependent. The difference arises from the fact that optical detectors are better than the infrared detectors, hence the signal is source limited in the visible region. It can be shown that SNR_D and SNR_I are equal in the visible region, hence there is no multiplex advantage for an interferometer system.

Jacquinot Advantage

An interferometer has no slits, so theoretically there is no attenuation of the source intensity by the spectrometer. This is a basis of the Jacquinot advantage (also called the throughput or étendue advantage.) Jacquinot stated that the brightness of an image is equal to the brightness of the source in a lossless optical system. Therefore, a slitless Michelson interferometer assures that all the information (radiant flux) will reach the detector at all times, dependent upon the interference pattern provided by the interferometer. This is not so in a dispersive instrument because the radiant flux is attenuated by the spectrometer exit slit. It can be shown that the throughput for a Michelson interferometer is considerably greater than that of a prism or grating instrument. Throughputs for a Fourier Transform Infrared instrument can be in excess of a factor of 100 greater than those of dispersive instruments.

Unlike the Fellgett advantage, throughput is not dependent upon the wavelength or detector used in the instrument. For this reason interferometers enjoy the ability to record and detect spectra of weakly emitting or weakly absorbing samples in either the optical or infrared regions of the spectrum. This property has been heavily exploited in the field of FT-IR.

Connes Accuracy

A third advantageous property of interferometers can be found in the inherent spectroscopic accuracy of the technique. The use of a He:Ne laser to measure the optical retardation has been discussed above. If this laser is a single-mode laser, and thus monochromatic, the optical retardation can be measured to a high degree of accuracy. If the instrument is correctly manufactured data will be recorded in a precise manner. Accurate and precise measurement in the interferogram yields accurate frequency assignment to the spectral values. With most interferometers it is possible to assign spectral features to better than 0.01 wavenumbers of the true position.

APPLICATIONS OF FT-IR

Fourier Transform Infrared Spectrometry has been applied to a vast array of chemical problems. The majority of these problems are beyond the scope of this review, consequently, this chapter will be limited to those techniques that have been developed or greatly advanced within the last few years. Several recent reviews have appeared that deal with the established techniques and the reader is referred to these reviews for further information.[14-16]

Infrared spectroscopy often is applied to absorption studies of organic and inorganic materials. In this area Fourier Transform Infrared Spectrometry often has little to offer over and above conventional dispersive spectroscopy, especially in the area of solid and liquid or condensed phase samples. FT-IR excels in the area of high resolution spectrometry, therefore it is well applied to low pressure gaseous samples or matrix isolated samples where the inherent sample resolution may be exploited. Fourier Transform Infrared Spectroscopy may be very useful for absorption studies of liquid and solid samples if the samples are very minute, where the throughput advantage can be used to obtain a suitable spectrum. FT-IR is employed often in quality assurance applications. In quality assurance applications it is necessary to record spectra as rapidly as possible, as there may be a great number of samples for analysis in a short time period. FT-IR has the ability to record high signal-to-noise ratio spectra rapidly. This is due to the Fellgett or multiplex advantage. As was discussed above, a Fourier Transform Infrared Spectrometer will produce a much higher signal-to-noise ratio spectrum than will a dispersive spectrometer if the collection conditions are the same for both instruments. Conversely, if a fixed signal-to-noise ratio is sought in a spectrum then the FT-IR spectrometer can produce this spectrum faster than an equivalent dispersive spectrometer.

Those areas which have received great attention in Fourier Transform Infrared Spectrometry recently have been the chromatographic infrared spectrometry hybrid systems. This includes gas chromatography-infrared spectrometry and liquid chromatography-infrared spectrometry. Two other areas of current interest are photoacoustic-infrared spectroscopy and the use of infrared spectroscopy to determine impurities in semiconductor materials.

GC/FT-IR

The potential to interface a gas chromatograph to a rapid-scanning infrared interferometer was recognized shortly after the interferometers became available. The earliest studies used low resolution interferometers.[17,19] This work was quickly expanded to laser-referenced interferometers capable of high-resolution studies.[19,20] The goal of this work was to collect the infrared spectrum of gas chromatographic effluents as they eluted from the column. Commercial FT-IR systems accomplished this by trapping eluates in a small volume gas cell and signal-averaging a number of scans to provide a good spectrum. This necessitated the shut-down of the chromatographic process and the ensuing chromatographic degradation; or, the chromatographic process was allowed to continue, but subsequent peaks to the one in the cell were discarded. Only a fraction of all the chromatographic eluates in a single run could be adequately investigated.

Spectroscopic acquisition of dynamic GC eluates was realized in 1976 by Azarraga.[21] This is accomplished by passing the chromatographic effluent through an internally gold coated glass tube that is coincident with the infrared beam. This tube or "light-pipe" has dimensions of approximately 0.7 to 6.0 mm internal diameter and 5 to 125 cm overall length. The manufacture of the light-pipe is a straightforward procedure and can be carried out in the laboratory.[22] Using this method Azarraga was able to reduce the detection limits of GC eluates to a few hundred nanograms when capillary WCOT columns were used. With this technique both packed and capillary GC systems could be investigated. The volume of the light-pipe must be optimized to provide the best signal from the GC eluate. The optimum light-pipe parameters have been established by Griffiths.[23]

The mode of data collection for GC/FT-IR spectra has been to scan the effluent in the light-pipe at the rate of at least one scan per second. As capillary GC peaks can be quite narrow, signal-averaging is not always performed and each individual interferogram is stored. Although the resolution of the spectrum is low (often 8 cm^{-1}) a single GC run can produce massive amounts of stored data. Methods whereby the data handling could be streamlined to locate interferograms of eluates were sought. Unlike gas chromatography/mass spectrometry no direct analog of the total ion current exists in GC/FT-IR. Two basic methods of locating eluate interferograms were devised. One used the Gram-Schmidt vector orthogonalization algorithm to reconstruct the chromatogram directly from the interferogram data;[24] the other transformed a "short" interferogram to calculate the absorbance within various user-defined spectral windows.[25] The use of such algorithms permitted the spectroscopist to selectively store data on the system disk so as not to exceed the data storage capabilities of the instrument. These software developments led to the development of user-oriented packages that include GC reconstruction, data reduction, Fourier transforms, and spectral search systems.[26]

Instrumental development in GC/FT-IR has been toward the improvement of spectral detection limits. Griffiths has utilized dual-beam FT-IR.[27] Matching light-pipes are put in both beams of the system, one containing the eluate, the other a reference gas.[28]

A block diagram of the optical layout is reproduced in Figure 9. The advantage to such a technique is that the detector measures the optical subtraction signal of both beams of the interferometer. In this situation a high-sensitivity infrared detector can be use to full advantage without exceeding the dynamic range of the detector. Griffiths was able to record spectra of GC eluates of 1 ng sample size[29] (see Figure 10). Commercial systems in a single-beam FT-IR configuration are now available where the detection limits are 5 to 50 ng per eluate from a capillary column.[30] An example of a spectrum from this system is given in Figure 11.

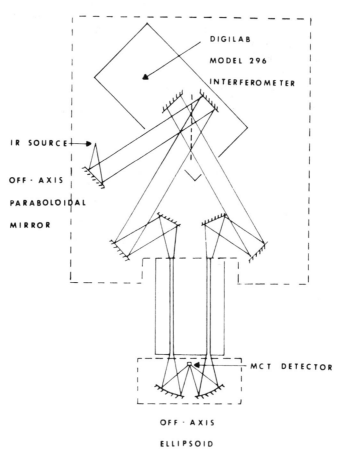

Figure 9. Dual-beam spectrometer optical layout as used for GC-IR measurements. (Reproduced from reference 28, by permission of the American Chemical Society, copyright 1978).

Other approaches to the improvement of the GC/FT-IR technique have been directed to the chromatography. One such study has been the employment of two-dimensional chromatography to separate unresolved chromatographic peaks.[31] The system trapped the peaks of interest in an activated-charcoal cold trap, then rechromatographed the peaks on a different column. No loss of spectral sensitivity was found. By another method, very low sensitivity eluates have been chromatographically separated and each eluate frozen onto a cryogenic mirror.[32] This allowed the spectroscopist to improve sensitivity by signal-averaging the trapped eluate. Of course, the dynamic advantages of GC/FT-IR were lost in favor of spectral sensitivity.

LC/FT-IR

The logical extension of GC/FT-IR is to apply that existing technology to liquid chromatographic systems. This major difference between the two technologies is the mobile phase, which is totally transparent in GC but is never transparent in LC system. There is always a high solvent:solute ratio in liquid chromatographic effluents, often on the order of at least 1000:1. As a consequence there are two contradicting criteria: the cell pathlength must be increased to improve the absorbance of the solute, but the pathlength must be minimized to decrease the absorbance of the solvent.

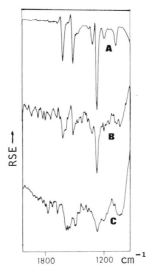

Figure 10. GC/FT-IR spectra of anisole, collected on the dual-beam spectrometer. A comparison is made between three solute quantities A = 100ng, B = 5ng, C = 1ng. (Reproduced from reference 28, by permission of the American Chemical Society, copyright 1978).

An early attempt to produce on-line LC/FT-IR spectra used a 0.030-mm pathlength flow-through cell connected to a liquid chromatograph.[33] Spectra were collected every 0.5 sec. in an on-the-fly recycling mode. The resulting spectra at first appearances were solely of the solvent, but when the pure solvent spectrum was subtracted, low signal-to-noise ratio spectra of the solutes were evident. A spectrum of 2 μg of p-nitrophenol with a signal-to-noise ratio of 10:1 was shown. It was calculated the original mixture had a solvent:solute ratio of 2500:1, the solvent being chloroform. This detection limit is not adequate for all LC systems and overall is not very practical.

Even though many solvents have regions which exhibit some high transmittance, they do not transmit at the 100 percent level. Those bands that are present in the solvent systems absorb very strongly. If a low concentration solute is present, and it is a strong infrared absorber, its spectrum may be accurately recorded in those regions that are free of solvent bands. In those regions where the solvent

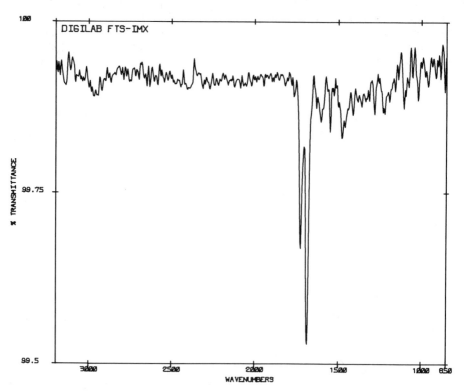

Figure 11. A capillary GC/FT-IR spectrum of 20 ng of caffeine. (Spectrum courtesy of D. Kuehl, Digilab; to be submitted for publication).

Fourier Transform Infrared Spectrometry

absorbs very strongly, the spectrum of the solute will not be accurately recorded as the dynamic range of the detector at those frequencies will be too low. In other words, very high signal-to-noise ratios will be required to distinguish the solute features. This is not possible with single-scan dynamic acquisition systems. In those cases where a polar solvent is used, such as in aqueous phase chromatographic systems, much of the bandwidth is obliterated by solvent and the solute is difficult to detect.

A more recent attempt at LC/FT-IR was made by Vidrine and Mattson[34] who coupled a gel permeation chromatograph to an FT-IR spectrometer. As some instrumental advances had been made in spectrometer hardware since the work of Kizer et al.[33], primarily in the area of detection, the sensitivities of the LC/FT-IR work was improved. The cells were of flow-through design with pathlengths varying from 200 μm to 11.2 mm. When tetrahydrofuran was used as the solvent, detection limits of 100 ng were achieved, but spectra could only be recorded in regions of good solvent transparency. Very few flow-through cell system papers have been published, but exclusion chromatography is often used because of the relatively low solvent: solute ratios and the ability to use nonpolar solvents.

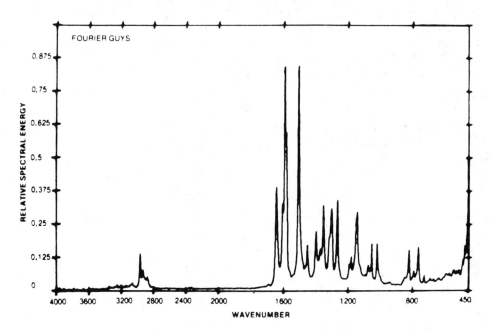

Figure 12. Spectrum of Indophenol Blue dye by solvent elimination LC/FT-IR method. 1 μg of Indophenol Blue produced this spectrum. (Reproduced from reference 36, by permission of Preston Publications, Inc., copyright 1979).

Another method to obtain LC/FT-IR spectra is by solvent elimination. This has been accomplished by Kuehl and Griffiths[36,37] by concentrating the LC effluent onto sintered KCl powder, then evaporating the solvent by heating and inert gas flow. The spectra are measured by diffuse reflectance spectroscopy.[28] These workers were able to separate stereoisomers by adsorption chromatography and record the spectra. Figure 12 reproduces the spectrum of Indophenol Blue dye obtained from the LC/FT-IR technique. The sensitivities extended to the submicrogram sample sizes. The primary drawback of such a system is that only organic, highly-volatile solvents could be used. The system could not function if the solute were evaporated with the solvent, or if the solvent dissolved the KCl powder.

Griffiths has discussed the possibility of using dual-beam FT-IR for liquid chromatography.[39] This in part would reduce the dynamic range problem by putting the LC effluent in one beam and a solvent reference in the other. The dynamic range problem would not be totally eliminated, however, as those regions of high transmittance will still exhibit a high signal throughput. Other regions may be essentially opaque, beyond the range of the ADC. Further problems would occur with gradient elution systems. It would be very difficult to assure that the solvent make-up is identical in both beams. For isocratic systems the reference cell could be fixed and would not require flow-through capability. The feasibility of the dual-beam apparatus for low concentration solutions has been demonstrated.[40] Spectra of 0.001% v/v anisole in CCl_4 and 5 ppm t-butyl methacrylate in CCl_4 have been shown, but up to 10,000 scans of the interferometer were required.

FT-IR-PAS

Fourier Transform Infrared Photoacoustic Spectroscopy (FT-IR-PAS) has been developed recently. In this technique the standard

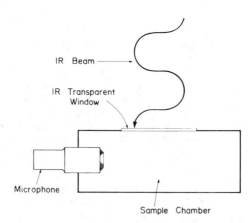

Figure 13. Schematic of FT-IR-PAS sample cell.

Fourier Transform Infrared Spectrometry

infrared detector, usually a deuterated triglycine sulfate detector, is replaced by a microphone. A simple schematic of a typical sample cell is shown in Figure 13. Both the Fellgett and Jacquinot advantages play important roles in this spectroscopy. The throughput or Jacquinot advantage is necessary so that sufficient radiant flux will reach the sample to be converted into heat by the sample. This in turn can be detected as pressure waves by the detector-microphone. Conventional spectrometers generally have too weak a radiant flux to permit photoacoustic spectra being collected in the infrared region. A Fourier Transform Infrared Spectrometer has a throughput advantage of 50 times higher than a conventional dispersive infrared spectrometer. The multiplex or Fellgett advantage also improves the photoacoustic spectroscopic technique as high resolution PA spectra can be collected. This is not to imply that infrared photoacoustic spectra cannot be collected on a dispersive instrument, as this has been demonstrated by Low and Parodi.[41] Low and Parodi were able to produce low resolution photoacoustic infrared spectra of solids using a single beam grating spectrometer.

The Michelson interferometer used in a Fourier Transform Infrared Spectrometer acts as a modulator of the infrared signal. The detector receives the modulated signal in the audio frequency range. The audio frequency is wavelength dependent and of course dependent on the scan velocity of the moving mirror. For example, if the mirror moves at the rate of 3 mm per second, the modulation frequency of the helium neon laser wavelength, 632 nm, is 5 kHertz; on the other hand, an infrared wavelength of 20 μm (or frequency of 500 wavenumbers) has a modulation of only 316 Hertz. As a result, the FT-IR-PAS spectrum is collected at a series of different modulation frequencies. It has been shown[42] that signal intensity decreases as the modulation frequency increases. As a result, the PA spectra collected on a FT-IR spectrometer tend to have more intense signals at lower frequencies. Regardless, FT-IR-PAS spectra can be corrected by various techniques to produce accurate, ratioed spectra. Figure 14 illustrates the comparison between PA spectra and a transmission spectrum of polymethylmethacrylate.

Although infrared photoacoustic spectra have been collected of gaseous samples[43,44] the primary advantage appears to lie in the area of solid and surface samples. Very little has been published thus far in the area of FT-IR-PAS, however, several publications have involved the measurement of solid samples,[46,47] the measurement of surface samples,[48] biological materials,[49] and binary mixtures.[50]

Silicon Impurity Determinations

The determination of impurities in silicon is important to the semiconductor industry. If impurity determinations can be made in silicon wafers prior to burning and etching of integrated circuits (IC's) on those wafers, the rejection rate and long-term reliability of the IC devices can be greatly improved. This is of economic benefit to the manufacturer and consumer alike. Most manufacturers produce single crystal silicon from molten silicon that is handled in quartz crystals and heated by graphite furnaces. This is known as the Czochlarski method. Oxygen and carbon find their way into the molten silicon by various pathways, but originate in the quartz crucibles and graphite heaters. These impurities change the electrical properties of the silicon and can destroy the semiconductor properties.

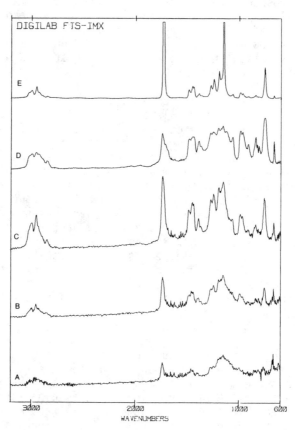

Figure 14. Spectra of plexiglass films cast on AgCl. A = PA spectrum of 0.3 µm thick film; B = PA spectrum of 0.7 µm thick film; C = PA spectrum of 3.7 µm thick film; D = PA spectrum of 13.8 µm thick film; and, E = Transmission spectrum of 13.8 µm thick film. Note saturation effects in D. (Spectra courtesy of K. Krishnan, Digilab; to be submitted for publication.)

Fourier Transform Infrared Spectrometry

There are various methods by which the impurities may be quantified, but dispersive infrared spectroscopy is often the method of choice.[51,52] This method has serious shortcomings due to inherent problems with dispersive instruments such as poor resolution and poor wavenumber repeatability in the spectral regions of interest. A further criterion is that the wafer under test must be matched with a pure wafer of identical thickness. Only recently have FT-IR spectroscopic techniques been applied to the problem and considerable success has been realized.

The carbon and oxygen content of silicon wafers can be determined at room temperature.[53] The procedure is to record the spectrum of the sample wafer and ratio it against a background (empty cell) spectrum. The resulting absorbance spectrum is stored on the spectrometer data system. A similar absorbance spectrum of a pure wafer is recorded. A spectral subtraction is performed between the two spectra and only the impurity bands of the sample wafer remain. The peak height can be correlated to impurity concentration. This process can detect oxygen and carbon down to a few parts per million. This method has been shown to be very reliable and can be automated.[54]

The impurity bands have a pronounced temperature dependence and spectral sensitivity and can be improved by lowering the temperature of the wafer to 20 K or below.[55] At least one system has been developed that is totally automated: it introduces the sample, cools it, records the spectrum and quantifies the impurities.[56] Such a system has a sensitivity of carbon detection four times that of the room temperature system. Cryogenic analysis can improve the oxygen detection limits by a factor of twenty. Analysis at cryogenic temperatures is not restricted to oxygen and carbon, but can also be applied to dopants. These include, phosphorus, boron, antimony, arsenic, aluminium, gallium and indium. Here detection limits can exceed a few parts per billion. An example of a silicon wafer spectrum is shown in Figure 15.

The strength of the Fourier transform method lies in the throughput of the spectrometer that allows very minute quantities of impurities to be detected. Furthermore, the absorbance spectral subtraction can compensate for sample and reference thickness differences. The automated cryogenic system is capable of handling a sample every few minutes.[56]

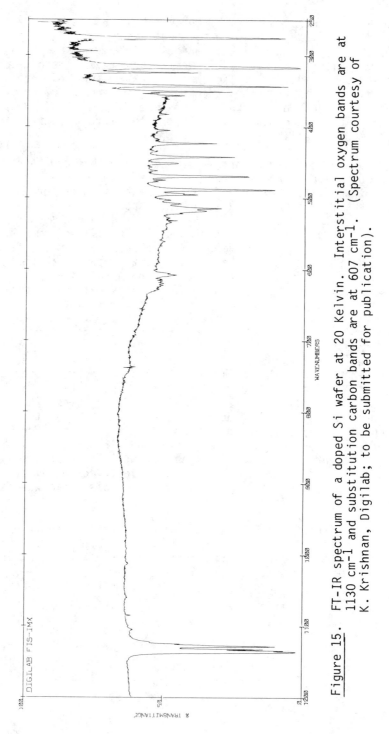

Figure 15. FT-IR spectrum of a doped Si wafer at 20 Kelvin. Interstitial oxygen bands are at 1130 cm^{-1} and substitution carbon bands are at 607 cm^{-1}. (Spectrum courtesy of K. Krishnan, Digilab; to be submitted for publication).

REFERENCES

1. Griffiths, P. R., 1975, Chemical Infrared Fourier Transform Spectroscopy, John Wiley & Sons, New York.
2. Bell, R. J., 1972, Introductory Fourier Transform Spectroscopy, Academic Press, New York.
3. Steel, W. H., 1967, Interferometry, Cambridge University Press, Cambridge.
4. Beer, R. and Norton, R. H., 1976, J. Opt. Soc. Am., 66, 259.
5. Horlick, G. and Yuen, W. K., 1975, Anal. Chem., 47, 775A.
6. Yuen, W. K. and Horlick, G., 1977, Anal. Chem., 49, 1446.
7. Horlick, G. and Yuen, W. K., 1978, Appl. Spectrosc., 32, 38.
8. Ref. 2, Chapt. 11.
9. Griffiths, P. R., 1978, "Fourier Transfrom Infrared Spectroscopy: Theory and Instrumentation" in Transform Techniques in Chemistry, Griffiths, P. R., Ed., Plenum, New York, p. 127.
10. Ref. 2, Appendix C.
11. "Specifications of Bomem Model DA3.XX Interferometric Spectrophotometers," Tech Note 18, Bomem, Inc., 910 Place Dufour, Ville de Vanier, Quebec G1M 3B1, Canada.
12. Optical and Infrared Detectors, Keys, R. J., ed., Springer-Verlag, Berlin, 1977.
13. Foskett, C. T., 1976, Appl. Spectrosc., 30, 531.
14. Griffiths, P. R., 1978, "Infrared Fourier Transform Spectrometry: Applications to Analytical Chemsitry," in Transform Techniques in Chemistry, Griffiths, P. R., Ed., Pelnum, New York, Chapt. 6.
15. Fourier Transform Infrared Spectroscopy: Application to Chemical Systems, Ferraro, J. R. and Basile, L. J., Eds., Academic Press, New York; Vol. I, 1978; Vol. II, 1979.
16. Analytical Applications of FT-IR to Molecular and Biological Systems, Durig, J. R., Ed., D. Reidel Publishing Company, Dordrect, Holland, 1980.
17. Low, M. J. D. and Freeman, S. K., 1967, Anal. Chem., 39, 194.
18. Low, M. J. D., 1968, Anal. Letters, 1, 819.
19. Low, M. J. D., Mark, H. and Goodsel, A. J., 1971, J. Paint Technol., 43 (562), 49.
20. Kizer, K. L., 1973, Am. Lab., 5 (6), 40.
21. Azarraga, L. V., 1976, Improved Sensitivity of On-the-Fly-GCIR Spectroscopy, Paper No. 334, Pittsburgh Conf. Anal. Chem. Appl. Spectrosc., Cleveland, OH.
22. Azarraga, L. V., 1980, Appl. Spectrosc., 34, 224.
23. Griffiths, P. R., 1977, Appl. Spectrosc., 31, 284.
24. de Haseth, J. A. and Isenhour, T. L., 1977, Anal. Chem., 49, 1977.
25. Coffey, P., Mattson, D. R. and Wright, J. C., 1978, Am. Lab., 10 (5), 126.
26. For example, Azarraga, L. V. and Hanna, D. A., GIFTS, Athens GC/FT-IR Software User's Guide, U.S. EPA/ERL, Athern, Ga., 1979. (Available from authors.).

27. Kuehl, D. and Griffiths, P. R., 1978, Anal. Chem., 50, 418.
28. Gomez-Taylor, M. M. and Griffiths, P. R., 1978, Anal. Chem., 50, 422.
29. Kuehl, D., Kemeny, G. J. and Griffiths, P. R., 1980, Appl. Spectrosc., 34, 222.
30. Kuehl, D., 1981, Hardware Considerations for a Capillary GC/FT-IR Interface, Paper No. 242, Pittsburgh Conf. Anal. Chem. Appl. Spectrosc., Atlantic City, NJ.
31. Azarraga, L. V. and Potter, C. A., 1981, J. High Resol. Chromatog. & Chromatog. Comm., 4, 61.
32. Reedy, G. T., Bowne, S. and Cunningham, P. T., 1979, Anal. Chem., 51, 1535.
33. Kizer, K. L., Mantz, A. W. and Bonar, L. C., 1975, Am. Lab., 7 (5), 85.
34. Vidrine, D. W. and Mattson, D. R., 1978, Appl. Spectrosc., 32, 502.
35. e.g., Brown, R. S., Hausler, D. W., Taylor, L. T. and Carter, R. C., 1981, Anal. Chem., 53, 197.
36. Kuehl, D. and Griffiths, P. R., 1979, J. Chromatogr. Sci., 17, 471.
37. Kuehl, D. T. and Griffiths, P. R., 1980, Anal. Chem., 52, 1394.
38. Fuller, M. P. and Griffiths, P. R., 1978, Anal. Chem., 50, 1906.
39. Griffiths, P. R., 1978, "Infrared Transform Specrometry: Applications to Analytical Chemistry," in Transform Techniques in Chemistry, Griffiths, P. R., Ed., Plenum, New York, p. 147.
40. Kemeny, G. J. and Griffiths, P. R., 1980, Appl. Spectrosc., 34, 95.
41. Low, M. J. D. and Parodi, G. A., 1980, Appl. Spectrosc., 34, 76.
42. Black, R. E. and Wakefield II, T., 1979, Anal. Chem., 51, 50.
43. Busse, G. and Bullemer, 1978, Infrared Phys., 18, 255.
44. Busse, G. and Bullemer, 1978, Infrared Phys., 18, 631.
45. Rockley, M. G., 1979, Chem. Phys. Lett., 68, 455.
46. Rockley, M. G., 1980, Appl. Spectrosc., 34, 405.
47. Vidrine, D. W., 1980, Appl. Spectrosc., 34, 314.
48. Rockley, M. G. and Devlin, J. P., 1980, 34, 407.
49. Rockley, M. G., Davis, D. M. and Richardson, H. H., 1980, Science, 210, 918.
50. Rockley, M. G., Davis, D. M. and Richardson, H. H., 1981, Appl. Spectrosc., 35, 185.
51. Annual Book of ASTM Standards, part 43, p. 518, Standard #F-121-76, 1977.
52. Ibid, p. 523, Standard #F-123-74.
53. Mead, D. G. and Lowry, S. R., 1980, Appl. Spectrosc., 34, 167.
54. Vidrine, D. W., 1980, Anal. Chem., 52, 92.
55. Mead, D. G., 1980, Appl. Spectrosc., 34, 171.
56. Krishnan, K., "Measurement of Impurities in Silicon Using a Conventional FT-IR System & the New Digilab SIM-100," Digilab, Inc., 237 Putnam Ave., Cambridge, Ma. 02139.

ASPECTS OF FOURIER TRANSFORM VISIBLE/UV SPECTROSCOPY

Robert J. Nordstrom

Principal Research Scientist, Physico-Chemical Systems
Battelle Columbus Laboratories
505 King Avenue
Columbus, OH   43201

INTRODUCTION

Fourier transform spectroscopy, a technique which Mertz[1] once called "a disagreeable indirect method to record a spectrum", has matured in recent years into a widely used and accepted spectroscopic technique. The rapid increase in the popularity of Fourier spectroscopy can be attributed, at least in part, to the technological advances in computer hardware and to the development of the Cooley-Tukey alogrithm[2] for computing Fourier transforms quickly. Both of these developments have helped to popularize Fourier spectroscopy and have given strong impetus to its growing appeal and commercialization by significantly reducing the time required to compute a final spectrum from the recorded interferogram.

Prior to the proliferation of commercial instrumentation, Fourier transform spectroscopy was used only by those relatively few spectroscopists who needed to overcome the difficulties of collecting spectra under conditions of very low signal levels.[3-6] The inconveniences associated with this indirect method were tolerated in order to gain and to use the inherent advantages offered by this type of instrumentation over the more direct dispersive spectroscopic methods. The present commercialization of Fourier transform instrumentation, primarily for the infrared spectral region, has brought this form of spectroscopy out of the development and prototype laboratories and into widespread use.

Today this widespread use includes a great deal of routine spectroscopic research and chemical analysis. So much so, in fact that Pierre Connes[7] has recently expressed concern that prospective spectroscopists might indeed become "over-Fourierized". He pointed out that advantages offered by the Fourier multiplex method were not needed in all applications, and that many published results might have been obtained just as well from spectra recorded with the more traditional (but therefore less glamorous) dispersive instrumentation.

The path of scientific research and technological development which has led to Fourier transform spectroscopy in its present form began with the pioneering work of A. A. Michelson.[8] The problems which motivated Michelson and the process which he made are reviewed in an article by Shankland.[9]

Rubens and Wood[10] are credited with the first publication of a recorded interferogram in 1911. From this beginning, the evolution of Fourier transform spectroscopy to its present capabilities has produced many significant publications which can be used as milestones to chart the growth of this form of spectroscopy. Several of these milestones are the publications which announced and discussed the fundamental advantages offered by the Fourier transform method over conventional spectroscopic techniques. These advantages, first discussed by Fellgett[11] and by Jacquinot[12,13] were central to the initial acceptance of the Fourier transform method and will be discussed in more detail later in this chapter. Other milestones, less central to the acceptance of the method, but nevertheles significant to its development and advancement are found in review articles,[14,15] conference proceedings,[16] collections of articles,[17,18] and text books.[19,20]

The evolution of Fourier transform spectroscopy is measured by improvements in instrumental capabilities. In the past, increases in capability and performance have included such operating parameters as resolution, speed of data collection, the number of points which can be transformed, and the spectral region which can be covered. Of particular interest in this chapter is the extension of Fourier transform spectroscopy to the visible and ultraviolet regions.

Historically, Fourier spectroscopy was developed for use in the far-infrared and mid-infrared spectral regions. Commercially available Fourier transform systems today are operating primarily in these spectral regions. The extension of Fourier transform spectroscopy to shorter wavelengths has been proposed[21,22] and demonstrated.[23-25] However, rapid commercialization of Fourier transform instrumentation for visible/UV spectroscopy is not occuring. Instead, visible/UV Fourier transform spectroscopy is being done in just a few spectroscopy laboratories with specialized, one-of-a-kind instrumentation.

To some extent, the motivation for the extension of the Fourier method into visible/UV spectroscopy is similar to the mountain climbers' reason "because it is there". However, it is important to remember the concern expressed by Connes, and to make sure that this region does not become "over-Fourierized". In some sense, the

Aspects of Visible/UV Fourier Transform Spectroscopy

capacity for the visible/UV spectral region to be "Fourierized" without becoming "over-Fourierized" is much less than in the infrared and far-infrared spectral regions. Advantages of the Fourier method which are quoted religiously by infrared spectroscopists do not necessarily apply in the visible/UV region. Furthermore instrument specifications and limits of acceptable performance are much more difficult to attain in the visible/UV region.

Nevertheless certain research goals in visible/UV spectroscopy do require the capabilities offered by the Fourier method. This chapter discusses some of the concepts concerning Fourier transform spectroscopy and the advancements which have been made in the attempt to extend Fourier spectroscopy to the visible/UV spectral region.

INSTRUMENTATION

The Michelson Interferometer

At the heart of a Fourier transform spectrometer system is the Michelson interferometer. There are many variations of this type of

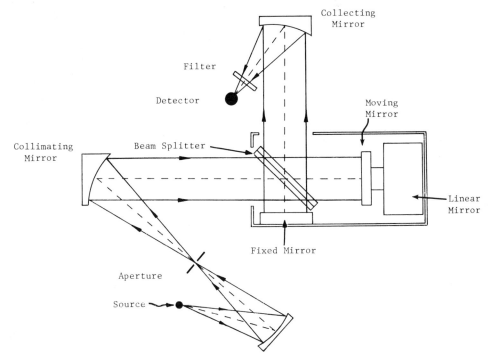

Figure 1. Basic interferometer design with source optics and detector optics.

interferometer, and Steel[26] has discussed the details and uses of many of these variations. Figure 1 shows the optical diagram of a basic Michelson interferometer. Radiation emitted by the source passes through an optical system and is collimated into the interferometer. The beamsplitter divides the radiation into two beams, ideally of equal amplitudes. Each beam is reflected back to the beamsplitter by identical mirrors. In this diagram, one mirror is fixed while the other is moveable along a track. The reflected beams recombine at the beamsplitter. Finally the radiation is collected and focused on the detector. Spectral filters are used to eliminate unwanted radiation.

The method by which an interferogram is produced has been thoroughly discussed by several authors[16,19] and therefore it will only be outlined here. Consider a point source which emits monochromatic radiation of wavelength λ. The optical system ahead of the interferometer produces plane wave radiation, and the interferometer is illuminated along a direction parallel to its optic axis. When the moveable mirror is in the position where the two paths of the interferometer are equal in length, the recombined beams will interfere constructively, giving a maximum output signal from the interferometer. If the moving mirror travels a distance $\lambda/4$, the two beams interfere destructively giving no output signal. When the moving mirror reaches the position $\lambda/2$, the round-trip path difference becomes λ and the two beams once again interfere constructively.

Figure 2 shows the interference fringes produced by the interferometer as the mirror moves along its track. The independent variable in this plot is optical path difference rather than mirror position. The use of optical path difference simplifies the discussion when more exotic interferometer designs employing multiple moving mirrors, or multipass optics are used.

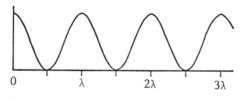

δ (OPTICAL RETARDATION)

Figure 2. Monochromatic intensity variation at the output of an interferometer as a function of optical retardation.

Aspects of Visible/UV Fourier Transform Spectroscopy

It is not difficult to show[27] that the intensity of the fringes produced by the interference of the monochromatic radiation can be expressed as

$$I(\delta,\sigma)d\sigma = \frac{I_0}{2}\{1 + \cos(2\pi\sigma\delta)\}d\sigma \qquad [1]$$

Here, δ is the optical path difference called the optical retardation in the interferometer, I_0 is the maximum intensity, and σ is the frequency of the radiation measured in cm^{-1} ($\sigma = 1/\lambda$). This is a convenient unit to use for measuring optical frequencies with Fourier spectroscopy, because it is the reciprocal of the units for measuring optical retardation. This unit for frequency simplifies the mathematical expressions and will therefore be adopted in this chapter. The increment $d\sigma$ represents a differential bandwidth which encompasses the "monochromatic" radiation.

Broadband radiation which covers a bandwidth $\Delta\sigma$ can be considered to be a composite of many spectrally non-overlapping monochromatic sources of width $d\sigma$, possibly with different intensities. It is this variation in intensity which constitutes the spectral information of the radiation. The resulting fringe pattern from this broadband radiation is the summation of each of the individual monochromatic contributors. That is,

$$I(\delta) = \int_{\Delta\sigma} I(\delta,\sigma)d\sigma = \int_{\Delta\sigma} \frac{B(\sigma)}{2}\{1 + \cos(2\pi\sigma\delta)\}d\sigma \qquad [2]$$

where $B(\sigma)$, the spectral content, has replaced I_0. At equal path differences in the interferometer, $\delta = 0$, and

$$I(0) = \int_{\Delta\sigma} B(\sigma)d\sigma \qquad [3]$$

Thus, Equation 2 can be written

$$I(\delta) = \frac{1}{2}I(0) + \frac{1}{2}\int_{\Delta\sigma} B(\sigma)\cos(2\pi\sigma\delta)d\sigma \qquad [4]$$

It is useful to define the limits of integration to be from 0 to ∞, so

$$\{I(\delta) - \frac{1}{2}I(0)\} = \frac{1}{2}\int_0^\infty B(\sigma)\cos(2\pi\sigma\delta)d\sigma \qquad [5]$$

The left-hand-side of this equation is the interferogram signal, and contains two parts. The term $I(\delta)$ is the signal which varies with the optical retardation, while the constant $1/2\ I(0)$ represents a d.c. component removed from the signal. An a.c.-coupled detector will detect this interferogram directly.

Apodization

Equation 5 describes how the interferogram is produced when radiation with spectral content $B(\sigma)$ passes through an ideal interferometer. The final goal, however, is to gain information about the spectral content $B(\sigma)$, which can be calculated from the inverse Fourier transform of the interferogram. For this development, we can write

$$B(\sigma) = \frac{4}{\pi} \int_0^\infty \left\{ I(\delta) - \frac{1}{2} I(0) \right\} \cos(2\pi\sigma\delta) d\delta \qquad [6]$$

Under ideal situations, the spectrum constructed by this method is identical to the true spectral content of the radiation. However, system imperfections, misalignments, and limitations cause the recovered spectrum to differ from the true spectrum. The quality of the final spectrum is sacrificed when these system problems are present. Some specific system problems will be examined in this chapter.

One limitation which is caused by the system hardware is the fact that the moveable mirror has a finite length of travel, giving the optical retardation a maximum value L. As a result, the integral in Equation 6 cannot be evaluated at the limit $\delta \to \infty$. Instead, an approximation $\overline{B(\sigma)}$ to the true spectrum is recovered from the integral

$$\overline{B(\sigma)} = \frac{4}{\pi} \int_0^L \left\{ I(\delta) - \frac{1}{2} I(0) \right\} \cos(2\pi\sigma\delta) d\delta \qquad [7]$$

It is useful to rewrite the limits of integration as 0 to ∞ by including a weighting function in the integrand. This is written

$$\overline{B(\sigma)} = \frac{4}{\pi} \int_0^\infty A(\delta) \left\{ I(\delta) - \frac{1}{2} I(0) \right\} \cos(2\pi\sigma\delta) d\delta \qquad [8]$$

The function $A(\delta)$ is called the apodization function.[19] Its purpose is to force the integrand to zero for values of δ greater than L. That is,

$$A(\delta) = \begin{cases} a(\delta) & 0 \leq \delta \leq L \\ 0 & \delta > L \end{cases} \qquad [9]$$

The function $a(\delta)$ is a weighting function which can be applied to the collected interferogram, but the final spectrum will of course be affected by the chosen form of $a(\delta)$.

By the convolution theorem,[28] the approximation of the true spectrum is related to the true spectrum by

$$\overline{B(\sigma)} = B(\sigma) * Q(\sigma) \qquad [10]$$

where $B(\sigma)$ is the true spectrum and $Q(\sigma)$ is the cosine transform of the apodization function $A(\delta)$. In a full treatment of the problem, $Q(\sigma)$ is written as the inverse Fourier transform of the apodization function. However for this abbreviated development, the cosine transform is all that is needed. The operation * signifies the convolution between the two functions.

The function $Q(\sigma)$ is similar to the slit function which distorts lines in spectra collected on dispersion instruments. $Q(\sigma)$ is called the instrument line shape and can be varied by changing the maximum optical retardation L or by changing the form of $a(\delta)$. Figure 3 shows several choices for the apodization function and the resulting instrument line shape for each. It can be seen that the width of the instrument line shape is proportional to 1/L. Thus, the larger the optical retardation, the narrower the spectral lines become. For the case where $a(\delta) = 1$ for all δ between 0 and L, the narrowest lines are achieved, but the side-lobes or "ringing" are most severe. When many absorption or emission lines in a spectrum are convolved with this instrument line shape, the spectrum can become difficult to interpret. Therefore, a compromise is usually reached between an apodization function $a(\delta)$ which produces narrow spectral lines and one which reduces the side-lobes.

Field-of-View

The maximum optical retardation, then, determines the ultimate resolution which is possible with a particular Fourier transform instrument. However, other design features must also be considered when deciding whether that ultimate resolution is in fact achievable. In particular, the size of the aperture through the Michelson interferometer must also be considered.

In the derivation of Equation 5, it was assumed that the interferometer was illuminated with plane wave radiation travelling parallel to the optic axis of the interferometer. In reality, however, an extended source is used and rays which are not parallel to the optic axis are present. This produces the well-known circular fringes, which are diagrammed in Figure 4 for the case where the central fringe is an intensity maximum.

As the optical retardation increases, the fringe pattern collapses toward the center. If the field-of-view through the interferometer is so large that many interference fringes are focused onto

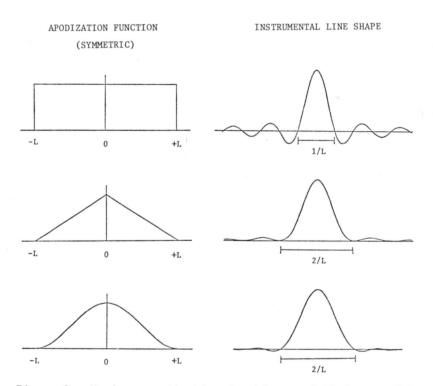

Figure 3. Various apodization functions and their resulting instrumental line profiles.

Aspects of Visible/UV Fourier Transform Spectroscopy

the detector, the resulting change in overall intensity as the optical retardation increases will be small. In order to maximize the intensity variation, or fringe contrast, it is necessary to restrict the field-of-view to only the central fringe.

This defines a maximum acceptable solid angle for the radiation passing through the interferometer. It can be shown[29] that this solid angle is

$$\Omega = \frac{\pi}{\sigma\delta} \qquad [11]$$

Since the solid angle is not adjusted for each optical frequency, or for each optical retardation, a limiting value of Ω is chosen. This limiting field-of-view is required when $\delta = L$ and $\sigma = \sigma_{max}$. Thus, the largest acceptable solid angle is

$$\Omega_0 = \frac{\pi}{L\,\sigma_{max}} \qquad [12]$$

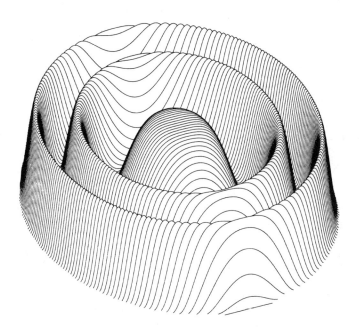

Figure 4. Circular fringe pattern caused by off-axis rays through the interferometer.

The presence of optical rays which are not parallel to the optic axis requires that Equation 5 be rewritten. The effects of the solid angle of the radiation through the interferometer have been shown to give[19]

$$I(\delta,\Omega) - \frac{1}{2}I(0,\Omega) = \int_0^\infty B(\sigma)\,\text{sinc}\,\frac{\Omega\sigma\delta}{2}\left[\cos\left(2\pi\sigma\delta - \frac{\Omega\sigma\delta}{2}\right)\right]d\sigma \qquad [13]$$

Two differences between this equation and Equation 5 are apparent. First, the integrand is multiplied by a sinc function which has its first zero at

$$\Omega = \frac{2\pi}{\sigma\delta}$$

For a given solid angle through the interferometer, each optical frequency, then, has a specific value of the optical retaradation where the sinc function is zero. If the interferometer is driven beyond this path difference, the phase of the modulation for that frequency is reversed, and energy at that frequency is removed from the spectrum rather than added to it as the optical retardation continues to increase. Therefore, in order to insure that this does not occur for any frequencies within the spectral bandwidth even at maximum path difference, the absolute maximum solid angle which can be used is

$$\Omega_1 = \frac{2\pi}{L\,\sigma_{max}} \qquad [14]$$

This value is twice that of Equation 12, and therefore includes more than the central fringe. Using the larger value of the solid angle means that the overall fringe contrast is reduced, and that at maximum optical retardation the fringe contrast of the high frequency component in the spectrum falls to zero. The sinc function acts as an apodization function which broadens the spectral lines, robbing the instrument of its designed resolution.

Using the smaller value of the solid angle will improve the fringe contrast of the highest spectral frequencies at maximum path difference and will produce higher resolution spectra for all frequencies in the bandwidth.

The second difference between Equation 13 and Equation 5 is the fact that there is now a frequency shift in the spectrum which is caused by the extended source. This frequency shift depends on the unshifted frequency σ and the solid angle Ω. The spectrum is spread to lower frequencies and the shifted frequency is given by

$$\bar{\sigma} = \sigma(1 - \frac{\Omega}{4\pi}) \qquad [15]$$

where $\bar{\sigma}$ is the mean shifted frequency. Computer software routines can be used to correct this frequency shift if the solid angle of the radiation subtended at the beam splitter is known.

In order to estimate the size of the acceptable solid angle of the radiation with which to illuminate an interferometer operating in the visible/UV spectral region, it is necessary to make a few assumptions. We will assume that Doppler-limited resolution is desired over the spectral range from about 7000 Å to 3000 Å, or in frequency, from about 14,300 cm^{-1} to 33,300 cm^{-1}. It is not likely that this entire bandwidth of 19,000 cm^{-1} would be covered in one spectrum, but it is important to know both the high frequency and low frequency limits when evaluating system performance.

For most gas-phase atoms and molecules at room temperature, the Doppler widths of spectral lines in this region fall between 0.005 cm^{-1} and 0.04 cm^{-1}. When using hot emission sources in this region (a spectroscopic problem for which visible/UV Fourier spectroscopy is well suited), the Doppler widths are larger. Thus, it is not unrealistic to consider designing a visible/UV Fourier transform spectrometer system with an ultimate resolution of 0.01 cm^{-1}. Defining the resolution conservatively to be the full width of the instrument line shape at the base line, this resolution can be achieved with a maximum optical retardation, L, of 1 meter. From Equation 12, then, the maximum allowable solid angle of the highest frequency component of the radiation through the interferometer is

$$\Omega_{max} = \frac{\pi}{L \sigma_{max}} \approx 1.05 \times 10^{-6} \text{ steradians} \qquad [16]$$

This produces a frequency shift which is on the order of 2.5×10^{-3} cm^{-1} for the high frequency component within the total bandwidth.

As mentioned earlier, it is not likely that the entire spectral region would be included in a single spectrum. Instead, optical filters are used to restrict the spectral region. In order to optimize the overall system performance, it is desirable to match the angular field-of-view to the selected spectral region. This can be done by inserting a variable aperture at a field-stop in the optical system, which provides a means for selecting the optimum field-of-view when a specific σ_{max} value is chosen by optical filtering. Such an aperture is shown in Figure 1, and is used to limit the size of the source which then restricts the off-axis rays.

The Jacquinot Advantage

Figure 5 diagrams a portion of an optical system. The increment of radiant flux dF which passes through (or is emitted by) an area dA_1 and is completely collected by the area dA_2 can be written[30]

$$dF = B \frac{(\cos\phi_1 dA_1)(\cos\phi_2 dA_2)}{r^2} \qquad [17]$$

The constant B is the brightness of the area dA_1 and is measured in watt $\cdot$ m^{-2} $\cdot$ steradian^{-1} (not just watt, or watt $\cdot$ m^{-2}). By regrouping terms in Equation 17, the radiant flux passing between the two areas is written

$$dF = B dA d\Omega \qquad [18]$$

where dA is the projected area of dA_1

$$dA = \cos\phi_1 dA_1 \qquad [19]$$

and $d\Omega$ is the solid angle subtended by dA_2

$$d\Omega = \frac{\cos\phi_2 dA_2}{r^2} \qquad [20]$$

The product $dAd\Omega$ is defined as the optical throughput or etendue between the two points of the system. The overall throughput of an optical system is equal to the smallest throughput in the system. For example, in Figure 1 three components can be identified: The source optics, the interferometer, and the detector optics. The resultant

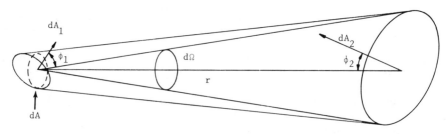

Figure 5. Optical throughput defined as dA $\cdot$ dΩ.

Aspects of Visible/UV Fourier Transform Spectroscopy

throughput of the total system is equal to the smallest throughput of the three. If the system is lossless, all component throughputs are equal.

Jacquinot[12,13] recognized that the light gathering capability of a Michelson interferometer is greater than that of a dispersive instrument operating that the same resolving power. The improvement offered by the Fourier transform spectrometer can be expressed as[19]

$$\frac{\tau_{FTS}}{\tau_D} = \frac{2\pi F}{\ell} \qquad [21]$$

where τ_{FTS} is the throughput of the Fourier transform spectrometer, τ_D is the throughput of the dispersive instrument, F is the focal length of the collimating mirror for the grating instrument, and ℓ is the slit height. It is not uncommon for this throughput ratio to reach about 200 in practical instruments. Field-widened Fourier transform spectrometers[31] can achieve even higher throughput advantages, but eventually other components begin to limit the system throughput.[32]

The improved throughput offered by the Fourier transform method increases the brightness of the image focused on the detector which in turn produces a higher signal-to-noise ratio. The signal-to-noise ratio improvement expressed by the Jacquinot advantage is proportional to $(\tau_{FTS}/\tau_D)^{1/2}$ and can be on the order of 10 to 20.

System Throughput

In Doppler-limited visible/UV spectroscopy, the interferometer is usually the component which determines the overall system throughput. The throughput of the source optics can be designed to match the throughput of the interferometer. If the beam diameter at the beam splitter of the interferometer is D (in cm), then the interferometer throughput at the high frequency limit is

$$\tau_{FTS}^{Vis/UV} = \frac{\pi D^2}{4} \Omega = (8.25 \times 10^{-7})D^2 \text{ cm}^2 \text{ steradians} \qquad (22)$$

where the value of the solid angle through the interferometer was taken from Equation 16. If the collimating component (in Figure 1 this component is a mirror) has a focal length of f, then the diameter d of the extended object at the focal point can be found by

$$\tau_{source} = \tau_{FTS}^{Vis/UV}$$

$$\left(\frac{\pi D^2}{4}\right)\left(\frac{\pi d^2}{4f^2}\right) = 8.25 \times 10^{-7} D^2 \text{ cm}^2 \text{ steradians} \qquad (23)$$

This gives an aperture diameter of

$$d = (1.16 \times 10^{-3})f \text{ cm} \qquad (24)$$

A larger diameter aperture will not provide any more usable radiation for the interferometer when attempting to achieve a spectral resolution of 0.01 cm^{-1} at 33,300 cm^{-1}. As an example, if the focal length of the collimating mirror is 50 cm, then the aperture size is required to be about 580 μm. If 1:1 optics are used to form an image of the source at the aperture, then the area of the source which is used is 0.26 mm^2.

Although silicon photodiodes (P·N or P·I·N junction) can be used as detectors to cover some of the spectral range for a visible/UV Fourier transform spectrometer system, the popular detectors are photomultipliers. Typical photomultipliers have very wide fields-of-view, usually 5 to 6 steradians. Furthermore, the minimum active detector areas are usually several square centimeters. The resulting light gathering capability for even the smallest of such detectors is measured in tens of cm^2 steradians, which is many orders of magnitude greater than the throughput of the interferometer. Therefore, the detector optics do not represent a limitation to the overall system throughput. In fact, the detector optics act more like a "photon bucket" collecting all of the radiation which exits the interferometer. Side baffles are used to shield the detector from radiation approaching from outside the field-of-view subtended by the interferometer.

Although the purpose of the detector optics is to collect all of the available radiation, there are several points to consider when looking at the detector optics. For example, short focal length optics will form a very small image on the photomultiplier cathode. Such an arrangement could be rather sensitive to any misalignments which cause the focal spot to wander on the cathode, since flying-spot scans across the face of photomultiplier cathodes show variations in photomultiplier sensitivity.[34] Furthermore electron replenishment times to the small area could reduce the speed of response of the detector. Although these variations and problems are probably not serious enough to cause major adverse effects in the spectrum, they can be at least partially eliminated by using longer focal length optics which permit more surface area of the detector to be used.

The Fellgett Advantage

Fellgett[11] recognized that the detector in a Fourier transform spectrometer instrument observes all of the spectral elements in a spectrum for the entire measurement time. This differs from the operation of a dispersive instrument where the detector observes each

Aspects of Visible/UV Fourier Transform Spectroscopy

spectral element separately for only a fraction of the total measurement time. The multiplex detection of Fourier transform spectroscopy can lead to large gains in the signal-to-noise ratio of the final spectrum under certain conditions. In order to examine the potential signal-to-noise ration improvements, it is necessary to look at the production of signal and noise in both a Fourier transform spectrometer and also a dispersive spectrometer.

Consider a spectral element of width $d\sigma$ which is contained in a total spectral bandwidth $\Delta\sigma$. The signal generated by the radiation in $d\sigma$ which is detected with a Fourier transform instrument is proportional to $B(\sigma)d\sigma T$, where $B(\sigma)$ is the spectral brightness in $d\sigma$ and T is the total observation time. The signal generated by the same spectral element which is detected by a dispersive instrument with identical transmission and detector characteristics, is proportional to $B(\sigma)\,d\sigma\,t$, where t is the time available to interrogate the single spectral element. This time is equal to $Td\sigma/\Delta\sigma$, so the signal detected by the dispersive instrument is proportional to $(B(\sigma)d\sigma)(Td\sigma/\Delta\sigma)$.

In order to determine the signal-to-noise ratio characteristics for each instrument, it is necessary to classify several sources of noise which can enter the final spectrum.

Detector Noise. This noise is generated by random processes which occur in the detector material or in the amplifier electronics. As such, it is independent of the flux of radiant energy (signal level) on the detector.

For photomultiplier tubes, the noise is caused primarily by thermionic emission from the cathode. The dark current produced by this process is proportional to the area of the photomultiplier cathode, and is typically in the range of 0.05 to 10 nA.

Since the noise is random and independent of the signal level, the noise generated output signals from the two instruments are both proportional to the square root of the observation time. For the Fourier transform spectrometer, the noise generated from the spectral element $d\sigma$ is proportional to the square root of the total measurement time $T^{1/2}$. This noise is spread uniformly over the entire spectral region by the Fourier transform process, so the contribution to the spectral noise in $d\sigma$ from a single spectral element is only $T^{1/2}(d\sigma/\Delta\sigma)$. However, there are $\Delta\sigma/d\sigma$ spectral elements which are each contributing noise everywhere in the spectrum. Since, by definition, this noise is independent of the signal level, the individual contributions are all equal, and the total noise in the spectral element $d\sigma$ is in fact proportional to $T^{1/2}$.

For the dispersion instrument on the other hand, the noise in the spectral element $d\sigma$ is proportional to $t^{1/2}$. The signal-to-noise ratios for the two instruments can be calculated, and an improvement in the registration of the spectral element $d\sigma$ by the Fourier method can be expressed as

$$\frac{SNR_{FT}}{SNR_D} \propto \left(\frac{\Delta\sigma}{d\sigma}\right)^{1/2} \tag{25}$$

Since the right-hand-side of this expression is usually much larger than unity, a substantial increase in signal-to-noise ratio can be achieved with the Fourier method when detector noise is dominant.

Photon Noise. Sometimes called source noise or photon shot noise, this noise is the result of random fluctuations in the number of photons transmitted to the detector. In infrared spectroscopy, this noise does not dominate, but in the visible/UV spectral region where photon counting is possible with high quantum-efficiency detectors, photon noise must be considered.

The signals detected separately by a Fourier transform instrument and by a dispersion instrument remain the same as stated earlier. Now, however, for photon shot noise the noise level is proportional to the square root of the signal level.[35-37] For the dispersion instrument, the signal-to-noise ratio in the spectral element $d\sigma$ is simply

$$SNR_D \propto (B(\sigma)d\sigma t)^{1/2} = \frac{B(\sigma)^{1/2}d\sigma\, T^{1/2}}{\Delta\sigma^{1/2}} \tag{26}$$

For the Fourier transform instrument the final spectral signal-to-noise ratio in $d\sigma$ is more difficult to calculate. The noise generated by the spectral element $d\sigma$ is given by $(B(\sigma)d\sigma T)^{1/2}$, but this noise is uniformly distributed over the entire spectral bandwidth. In order to estimate the total amount of noise in $d\sigma$, it is necessary to know the noise generation characteristics of all of the other spectral elements in the overall bandwidth. This will, of course, depend on the total spectral content.

One way to evaluate the shot noise contribution from the entire spectrum is

$$\text{shot noise} = (\bar{B}\, \Delta\sigma\, T)^{1/2} \tag{27}$$

where $\bar{B}$ is the average spectral brightness in the total bandwidth $\Delta\sigma$. That is,

Aspects of Visible/UV Fourier Transform Spectroscopy

$$\overline{B} = \frac{1}{\Delta\sigma} \int_{\Delta\sigma} B(\sigma) d\sigma \tag{28}$$

From this, the signal-to-noise ratio for photon noise limited Fourier transform spectroscopy is

$$SNR_{FT} \propto \frac{B(\sigma) \, T^{1/2}}{\overline{B}^{1/2}} \frac{d\sigma}{\Delta\sigma^{1/2}} \tag{29}$$

The gain in the signal-to-noise ratio which can be achieved by the Fourier transform method can be expressed as

$$\frac{SNR_{FT}}{SNR_D} \propto \left[\frac{B(\sigma)}{\overline{B}}\right]^{1/2} \tag{30}$$

As indicated by this equation, the multiplex gain at the frequency σ offered by Fourier transform spectroscopy in the presence of photon shot noise is a function of the spectral brightness at that frequency and of the overall structure in the total spectrum. This is completely different from the situation presented earlier where detector noise was assumed to dominate.

In an absorption experiment, the spectral features are contrasted against an energy continuum, and $\overline{B} \approx B(\sigma)$ throughout the spectrum. This is especially true if the spectrum contains only weakly absorbing lines, and it is approximately valid when strong absorption lines are present. Thus, for visible/UV absorption spectroscopy, the multiplex gain is essentially lost. However, the Jacquinot throughput advantage still applies.

In emission spectroscopy, $\overline{B}$ is usually very low, and in the region of emission lines $B(\sigma)$ can be much larger than $\overline{B}$. Therefore, a considerable gain in the signal-to-noise ratio can be achieved when the Fourier multiplex method is applied to emission spectroscopy. However, it should be kept in mind that the distribution of the shot noise throughout the spectrum from a single strong emission line or from groups of strong lines can interfere with the measurements of weak emission lines.

Hirschfeld[38] has suggested that the photon-noise-limited multiplex advantage expressed by Equation 30 should be called the distributed Fellgett's or multiplex gain. This nomenclature is suggestive of the fact that the gain tends to be better in certain regions of the spectrum (e.g. at peaks of emission lines) than in other regions.

Modulation Noise. This noise is generated by scintillations or system modulations, and it is directly proportional to the signal level. It can be caused by refractive index fluctuations in very long-path atmospheric measurements and in astronomical measurements, or by faulty design in laboratory instrumentation.

The multiplex gain which can be achieved by the Fourier method in the presence of modulation noise is

$$\frac{SNR_{FT}}{SNR_D} \propto \frac{B(\sigma)}{\overline{B}} \left[\frac{d\sigma}{\Delta\sigma}\right]^{1/2} \tag{31}$$

The gain is distributed as it was in the photon-noise-limited case. However, since ($d\sigma/\Delta\sigma$) is usually much less than one, the distributed multiplex gain in the presence of modulation noise is much less than one, except in very rare situations. Thus, instead of a multiplex gain, there is actually a multiplex loss. Therefore modulation noise represents a situation which should be avoided in Fourier spectroscopy.

For the three types of noise discussed above, the Fellgett advantage must be carefully evaluated. This multiplex advantage is an unquestioned benefit, for example, when detector noise dominates, as is the case in infrared Fourier transform spectroscopy. In visible/UV spectroscopy, the detector noise which is present can also be minimized with the multiplex advantage. Therefore it is not always necessary to cool photomultiplier tubes to reduce the thermionic emission for Fourier transform spectroscopy.

However, when photon noise is the dominant noise souce, the multiplex advantage can be seriously reduced or even lost in certain situations. Thus, although the Jacquinot advantage still applies, the benefit of using the Fourier transform method could be questioned.

Frequency Accuracy

In addition to the throughput and multiplex advantages which are offered by Fourier transform spectroscopy, an important third advantage is often quoted: the high degree of accuracy of the frequency scale across a broad spectral range. This frequency precision occurs because the recording of the interferogram is controlled by referencing the data collection process to an accurately known frequency standard such as lamb-dip stabilized He-Ne. Of course, certain frequency corrections should be applied to the computed spectrum in order to eliminate frequency shifts caused by the index of refraction of air and by the field-of-view as given in Equation 15.

The monochromatic radiation from the frequency standard passes through the interferometer and is intensity modulated as shown in Figure 2. A separate detector is used to measure the intensity of the reference radiation as a function of the optical retardation. This information is used to determine the position of the moving mirror and to trigger the analog-to-digital converter which samples data points at the correct optical retardation positions.

Errors in mirror position or optical retardation produce a variety of unwanted effects in the final spectrum including additional noise, degradation of the spectral resolution, and ghost or satellite specral lines. These are all similar to the effects which are caused by ruling errors on gratings in dispersive instruments.

The precision with which the optical retardation can be determined is directly related to the frequency stability of the reference source. For Doppler-limited visible/UV spectroscopy, this stability is required to be on the order of one part in 10^{11}.[22,39] Although this precision cannot be achieved with small, general purpose lasers, it can be achieved by stabilizing optical oscillators to molecular transitions as is done in lamb-dip stabilization. This stability which is required to minimize spectral distortions also provides accurate calibration for the spectral frequencies throughout the bandwidth.

Other System Advantages

Fourier spectroscopists are fond of reporting other advantages which their spectroscopic method can produce. In general, these advantages are not unique to Fourier spectroscopy, and spectroscopists using dispersive instruments can enjoy essentially the same advantages.

One such advantage is the fact that the data are digital rather than analog, and therefore computer manipulation and processing of the spectra are possible. Arithmetic operations such as spectral subtraction, background correction, and equivalent width calculations can be made very easily.

Another advantage is the fact that spectra can be collected relatively quickly. Therefore, Fourier transform spectroscopy has the ability to generate an enormous amount of information. In fact, it is difficult to make full use of all of the information displayed in a single spectrum. The usual approach in spectral analysis is to extract a few parameters such as line positions, or line strengths from a spectrum in order to calculate molecular constants. By this method, a great deal of information in the spectrum remains untapped. Shaw and colleagues[40-42] are making impressive progress at extracting more of the untapped information from the spectrum, thus making Fourier transform spectroscopy more efficient from the point of view information processing.

THREE SPECIFIC INSTRUMENTS

Several general characteristics, limitations, and advantages of visible/UV Fourier transform spectroscopy instrumentation were developed in the previous section. It was assumed that Doppler-limited resolution was a desirable feature, and that the spectral range of interest extended from about 14,300 cm^{-1} to about 33,300 cm^{-1}. The system conceptualized in Figure 1 should be considered to be generic since it represents a class of experimental designs rather than one specific experimental instrument. The source could, for example, be an emission line source such as a hollow cathode discharge, or a combustion (chemiluminescent) reaction. On the other hand, if the source has a continuous spectrum such as a black body emitter, then adsorption or reflection spectra can be obtained by proper sample placement in the beam. It is even possible to perform photoacoustic spectroscopy[43,44] by replacing the detector with a photoacoustic chamber.

These experimental modifications represent various applications of the Fourier method and all can be accommodated within the design spirit of Figure 1. Furthermore, the optical features of the Michelson interferometer shown in Figure 1 are only schematic. In the actual design and construction of interferometers which operate in the visible/UV region, several variations from this schematic representation have been implemented.

As mentioned in the introduction to this chapter, visible/UV Fourier transform spectroscopy has not yet received the vigorous commercialization thrust which has occurred for infrared Fourier transform spectroscopy. As a result, Fourier transform spectrometer systems which operate in the visible/UV spectral region are unique, one-of-a-kind instruments. In order to gain an appreciation for the diversity of design ideas which have gone into the realization of visible/UV Michelson interferometers, it is necessary to discuss several of these unique instruments in some detail. The purpose of this section is not to give the reader an in-depth analysis of these instruments. The intention is, rather, to give the reader some indication of the capabilities of the selected systems, and to guide the reader to references which discuss the instrumental capabilities in more detail.

Two visible/UV Fourier transform spectrometer instruments have been well documented in the literature. The first instrument to be

Aspects of Visible/UV Fourier Transform Spectroscopy

discussed was constructed in the Laboratoire Aime Cotton in Orsay, France.[45] The second instrument is a product of the Department of Chemistry at the University of Alberta, Canada.[46,47] Features of each of these instruments will be presented and discussed.

A third instrument will also be considered. This interferometer was constructed at the Kitt Peak National Observatory near Tucson, Arizona.[48] Unfortunately this instrument is not well documented in the literature, and most of this author's knowledge of this instrument comes from private communications with its designer, J. Brault, and from time spent actually collecting data with this instrument.

<u>Orsay, France</u>: Gerstenkorn and Luc[24,25,45] have reported the instrumental characteristics and experimental results of a visible/UV Fourier transform spectrometer constructed in 1975 at the Laboratoire Aimé Cotton. This interferometer was built along the same principles which proved successful in other interferometers that were constructed there for use in the infrared spectral region. The visible/UV interferometer at the Laboratoire Aimé Cotton is referred to as interferometer V indicating that it is a fifth generation instrument. The French support (primarily through the CNRS) and commitment for Fourier transform spectroscopy has been very impressive, and spectroscopy groups such as at the Laboratorie Aimé Cotton will continue to be at the forefront of evolutionary development in this field for many years.

A drawing of this visible/UV interferometer appears in Figure 6. One important design variation which differs from Figure 1 is the fact that the interferometer mirrors used to reflect the radiation back to the beam splitter are now cat's-eye retroreflectors. These

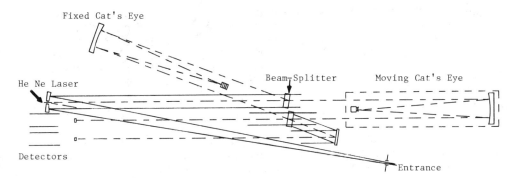

Figure 6. Optical diagram of the high-resolution, visible/UV interferometer at Aimé Cotton. Reprinted from reference 24 with permission.

reflectors significantly reduce any optical errors caused by mirror tilt.[49] The moving cat's-eye reflector in the interferometer is mounted on a carriage which is pulled along a 100 cm track (2 meter optical retardation) by a d.c. motor. A lamb-dip stabilized He-Ne laser is used to determine the location of the moving reflector as the optical retardation increases.

The carriage motion is not smooth and precise enough to provide the position accuracy required to record high resolution visible/UV spectra. Therefore provisions for fine adjustments in the position of the cat's-eye reflector relative to the moving carriage have been made. A position error signal is generated from observation of the He-Ne interference fringes, and this error signal is fed to the carriage motor through suitable electronics. Thus, the carriage is slaved to follow the reflector, and the motion of the cat's-eye is as if it were supported on an air bearing. A detailed description of the servo electronics used for precise reflector motion is given by Connes and Michel.[33] So successful is the servo "slave carriage" that Connes and Michel feel that air bearings or oil bearings in interferometers are obsolete. As with earlier French built interferometers, step-by-step scanning is used in this visible/UV instrument.

Interferograms collected with this instrument can contain up to 10^6 regularly spaced data points. If the spacing $\Delta\delta$ of the data points is $1/(2\sigma_L)$ where σ_L is the stabilized He-Ne laser frequency (about 15,798 cm^{-1}) then the 10^6 points will be collected when an optical retardation of approximately 31.64 cm is reached. The free spectral range which can be covered is governed by the density of points collected in the interferogram, as

$$\Delta\delta(2\Delta\sigma) = 1 \qquad (32)$$

where $\Delta\delta$ is the spacing between data points in the interferogram and $\Delta\sigma$ is the single-sided free spectral range. Thus, for a point spacing of $1/(2\sigma_L) = 31.64 \times 10^{-6}$ cm, the free spectral range can be extended from 15,798 cm^{-1} to 31,596 cm^{-1}. The spectral resolution which can be achieved as defined by the unapodized width at the baseline of the instrument line shape function is 0.032 cm^{-1}. The time required to record an interferogram of this kind is approximately 5 hours.

Higher resolution spectra can be recorded by restricting the spectral bandwidth with optical filters and then collecting the data points at wider spacings along the interferogram.

This instrument has been used to record both emission and absorption spectra. A particularly impressive compilation is an atlas of iodine absorption lines between 14,800 cm^{-1} and 20,000 cm^{-1}.[24] In a subsequent publication, Luc and Gerstenkorn[50] have reported that

Aspects of Visible/UV Fourier Transform Spectroscopy

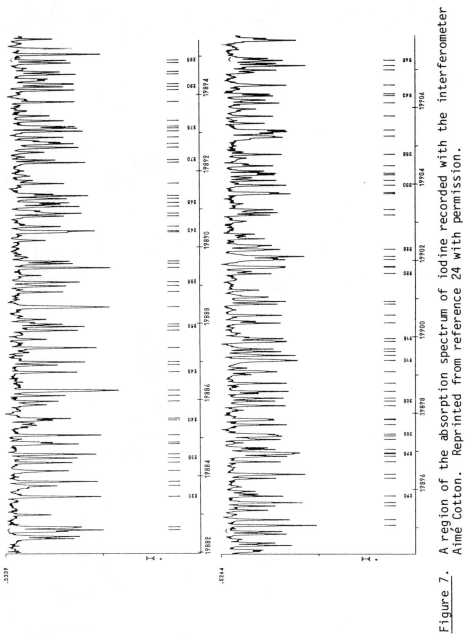

Figure 7. A region of the absorption spectrum of iodine recorded with the interferometer at Aimé Cotton. Reprinted from reference 24 with permission.

frequencies of the iodine line printed in the atlas must be corrected by subtracting 0.0056 cm^{-1} from all tabulated frequency values. Luc and Gerstenkorn believe that the Fourier transform method remains a powerful tool for both high resolution absorption spectroscopy as well as high resolution emission work. Figure 7 shows a portion of the iodine spectrum which has been collected with this instrument.

Edmonton, Alberta, Canada: A modular Michelson interferometer has been constructed at the University of Alberta by Horlick and Yuen.[46,47] As with the French instrument just discussed, this Canadian instrument represents the latest development in an evolutionary process. The interferometer consists of seven modules which are shown in Figure 8. These modules are: the mounting cube, the beam splitter assembly, the detector assembly, the white light reference system, and the He-Ne reference laser system. This modular design philosophy allows rapid interchange and realignment of optical components in order to change detectors, spectral coverage, or other system characteristics.

The mirrors in the interferometer are 4.2 cm diameter, aluminum coated flats of 1/10 λ. The moving mirror drive system consists of an electromechanically driven mirror mount supported on an air bearing. The mirror motion is continuous (rather than step-by-step) and He-Ne laser fringe referencing is used to trigger the analog-to-digital converter and to control the mirror velocity. Maximum mirror movement is about 1 cm, which gives an optical retardation of 2 cm. Time averaging of repetitive scans is used to establish a good signal-to-noise ratio in the final spectrum.

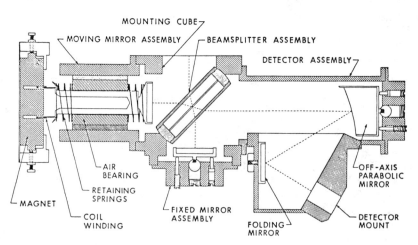

Figure 8. Optical diagram of the visible/UV interferometer at the University of Alberta. Reprinted from reference 46 with permission.

Aspects of Visible/UV Fourier Transform Spectroscopy

Detectors used by Horlick and Yuen for visible/UV spectroscopy are 1P28 or R166 (Solar blind) photomultipliers. The radiation is focused onto one of these detectors by a 167 cm focal length off-axis parabolic mirror. Horlick and Yuen[51] specifically discuss the choice of this focal length in their detector optics, giving values of image diameters for their particular instrument. Their discussion relates back to and agrees with the discussion of detector optics given in the previous section.

The optimum resolution of this instrument falls well short of the resolution capabilities of the French instrument. However, it must be pointed out that a great deal of very good visible/UV spectroscopy is being done on the lower resolution instrument.[51] Furthermore, the instrument was constructed at a component cost (optical, electronic, and mechanical, excluding machining costs and computer) of about $2,500. Therefore, with the costs in time and money for the development of another major, high-resolution visible/UV Fourier transform facility already high (and going higher), it is refreshing and even inspiring to see good spectroscopy being done with this instrument.

Flame (air-C_2H_2) emission spectra of Li, K, Rb, and Cs recorded by Horlick and Yuen are shown in Figure 9. These spectra are calculated from 512-point, double-sided interferograms. The solutions used to obtain these spectra contained about 250 ppm of each element.

Horlick and Yuen make extensive use of aliasing or folding of different spectral regions into the final spectrum. This topic is discussed in many references on Fourier transform spectroscopy (c.f. reference 19), and therefore will not be developed here. It is because of their use of aliasing that several spectra in Figure 9 have multiple frequency scales.

<u>Kitt Peak National Observatory, Arizona, U.S.A.</u>: Brault[48] has designed and constructed a high-resolution interferometer for use on the McMath Solar Telescope at Kitt Peak National Observatory. Astronomical or laboratory spectroscopy can be performed with this instrument in the visible/UV and infrared spectral regions.

The instrument is capable of a 100 cm optical retardation. However, the method by which this path difference is achieved is different from the method of single-mirror motion employed by the previous two instruments. The Kitt Peak interferometer has two cat's-eye reflectors which move in a continuous-scan, reciprocating fashion. One reflector approaches the beam splitter while the other reflector retreats from it. Figure 10 shows the details of this interferometer. Both reflectors move on oil bearings. If the beam splitter is positioned so that the zero path difference occurs when one reflector is near the front of its track while the other reflector is near the back of its track, then the maximum optical path difference which can be achieved as both reflectors move the full 25 cm is 100 cm.

It is typical, however, for the system to be used in the symmetric mode, where the zero path difference occurs when both reflectors are in the middle of their tracks. This reduces the

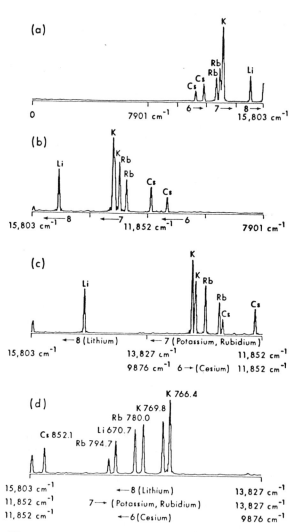

Figure 9. Flame emission spectra of Li, K, Rb, and Cs recorded with the instrument at the University of Alberta. Reprinted from reference 51 with permission.

Aspects of Visible/UV Fourier Transform Spectroscopy

achievable resolution, but simplifies other problems such as phase correction.[19] Each mirror is driven by a linear motor from a Hewlett Packard chart recorder. As with the other interferometers discussed here, a He-Ne laser is used as the position reference.

The emission spectrum of uranium in a hollow cathode discharge (neon carrier gas) has been recorded with this interferometer by Palmer et al.[52] The spectral region covered in this atlas extends from 11,000 cm^{-1} to 26,000 cm^{-1}. A part of this spectrum is shown in Figure 11.

A frequency correction was applied to the computed spectrum in order to account for the effects of the index of refraction of air and the effects of a finite aperture. The aperture correction was given as

$$\sigma' = \sigma(1 + \frac{\Omega}{4\pi}) \tag{33}$$

where σ' is the corrected frequency and σ is the measured frequency. This equation agrees with Equation 15 of the previous section, since in that equation, $\bar{\sigma}$ was the measured frequency (shifted by system performance), while σ was the true frequency. Using this frequency

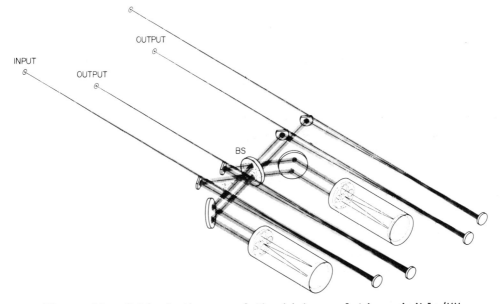

Figure 10. Optical diagram of the high-resolution visible/UV interferometer at Kitt Peak National Observatory. Reprinted with permission from J. Brault.

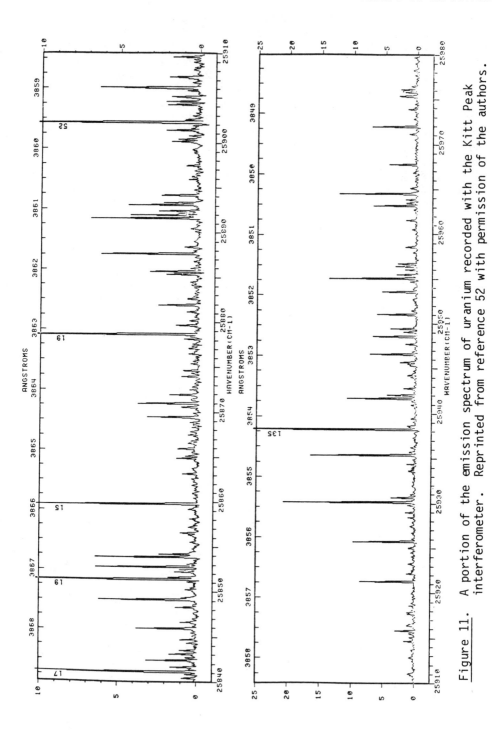

Figure 11. A portion of the emission spectrum of uranium recorded with the Kitt Peak interferometer. Reprinted from reference 52 with permission of the authors.

correction and accounting for other sources of frequency errors (such as faulty phase correction) Palmer et al determined the magnitude of random errors in the frequency positions of the emission lines to be less than 0.001 cm^{-1} while the systematic errors in frequency positions were less than 0.003 cm^{-1}.

CONCLUSIONS

There are many aspects of the mechanics of Fourier transform spectroscopy which have not been mentioned in this chapter. Problems such as digitization noise, phase correction, computer capabilities, aliasing, and sampling are better left to texts which are devoted solely to Fourier transform spectroscopy. There, adequate time can be given over several chapters to the development of these and other topics. In this brief chapter, it has been the author's intention to highlight a few of the design considerations, methodological advantages, and possible system errors which can factor into the experimenter's decision of whether or not to use the Fourier multiplex method when performing visible/UV spectroscopy.

As mentioned in the introduction to this chapter, visible/UV Fourier transform instruments are still found mainly as unique, one-of-a-kind instruments in a few spectroscopy laboratories. The research topics being pursued with these Fourier transform instruments include atomic spectrochemical measurements, atomic and molecular emission spectroscopy from hollow cathode discharges, and molecular absorption spectroscopy for accurate frequency standards and molecular constants. In each of these research efforts, the Fourier transform method has proven useful. In part, the success of this method is derived from the fundamental advantage originally stated by Jacquinot, and to some extend from the advantage stated by Fellgett.

However, there are other reasons why the Fourier method represents, or will represent in the near future, an appealing technique for several projects in visible/UV spectroscopy. The first of these reasons was stated by Fellgett in 1967:[53]

"It appears probably, moreover, that an outstanding advantage will eventually prove to be that proposed by Mertz, namely simplicity."

With only one moving component (in most interferometers) which can be controlled with a high degree of precision using standard optical methods, interferometers today are compact, simple instruments. Typing values of parameters at a computer keyboard terminal now has replaced the process of setting slitwidths, determining grating positions, and other manual operations, thus simplifying the operation of recording spectra. Furthermore, with interferometers, certain problems such as overlapping grating orders or the existence of a nonlinear frequency scale, which can occur with grating instruments, are eliminated. This simplifies later spectral interpretation.

The second reason for the appeal of Fourier transform methods is cost. While the cost for a large, visible/UV Fourier transform instrument is high, two factors should be kept in mind. First, less expensive instruments can be constructed which will provide good spectroscopic data as shown by Horlick and Yuen. Second, the true cost to be considered is the cost per spectrum. Since, in general, Fourier transform instruments can collect broad bandwidth spectra relatively quickly, many more spectra can usually be collected by this method than by conventional dispersive methods. Since spectra can be collected much more rapidly than they can be analyzed, it is not unrealistic to think that a single instrument could serve many research groups. This method of joint or shared operation keeps the instrument in a state of use which optimizes its efficiency, and spreads the cost over many users.

The largest advances in visible/UV Fourier transform spectroscopy still lie before us. Although several instruments are already in operation, and despite the fact that the visible/UV region could quickly become "over-Fourierized", there is still at this time a necessity for more instrumentation employing novel design concepts. The three instruments highlighted in this chapter operate on different principles and yet the possibilities are by no means exhausted.

Applications of visible/UV Fourier transform spectroscopy will continue to include those topics already mentioned. Additional areas of expansion will probably include amplitude spectroscopy[54,55] which has proved useful in the past by providing both absorption magnitude and phase information on transmitting samples, photoacoustic spectroscopy which can determine the composition of opaque samples, time-resolved chemical kinetic studies, and optical probing of combustion processes.

One milestone in the evolution of Fourier transform spectroscopy which will probably be reached in the next three to five years is the full development of efforts by some manufacturers to market visible/UV Fourier transform instrumentation. In fact, the process has already begun on a small scale.[56] To be sure, however, this effort even at its height will not be on the same level as present marketing efforts in infrared spectroscopy.[57] This commercialization will occur in order to provide instrumentation designed to study the problems mentioned above as well as other problems currently unforeseen. As Connes[17] pointed out, this development is a mark of maturity for the Fourier technique.

However, there will always be a few researchers who will choose not to wait for the manufacturers to design, produce, and market the desired instrumentation. Instead, they will choose to construct their own instrumentation. As a result, the community of Fourier spectroscopists will continue to include both instrument builders and commercial users as capabilities extend into the visible/UV spectral region.

REFERENCES

1. Mertz, L., 1975, Transformations in Optics, John Wiley & Sons, New York.
2. Cooley, J. W. and Tukey, J. W., 1965, Math. Comput. 19, 297.
3. Gebbie, H. A., Vanasse, G. and Strong, J., 1956, J. Opt. Soc. Am. 46, 377.
4. Gebbie, H. A., 1957, Phys. Rev. 107, 1194.
5. Connes, J. and Gush, H., 1959, J. Phys. Radium 20, 915.
6. Connes, J. and Connes, P., 1966, J. Opt. Soc. Am. 56, 896.
7. Connes, P., 1978, Appl. Opt. 17, 1318.
8. Michelson, A. A., 1891, Phil. Mag. Ser 5 31, 256.
9. Shankland, R. S., April 1974, Physics. Today.
10. Rubens, H. and Wood, R. W., 1911, Phil. Mag. 21, 249.
11. Fellgett, P., 1958, J. Phys. Radium 19, 187.
12. Jacquinot, P. and Dufour, C. J., 1948, J. Rech. C.N.R.S. 6, 91.
13. Jacquinot, P., 1960, Rep. Prog. Phys. 23, 267.
14. Loewenstein, E. W., 1966, Appl. Opt. 5, 845.
15. Bates, J. B., 1976, Sci. 191, 31.
16. Vanasse, G. A., Stair, Jr., A. T. and Baker, D. J., 1971, "Aspen International Conference on Fourier Spoectroscopy, 1970", Air Force Cambridge Research Laboratories, AFCRL-71-0019, Special Report 114.
17. 1958, J. Phys. Radium 19, Contains many papers from the 1957 C.N.R.S. Bellevue Colloquium; 1967, J. Phys. Supp. C2 28, Contains papers from the 1966 C.N.R.S. Orsay Colloquium.
18. 1978, Applied Optics 17, Contains papers from the 1977 International Conference on Fourier Transform Spectroscopy, Columbia, S.C.
19. Bell, R. J., 1972, Introductory Fourier Transform Spectroscopy, Academic Press, New York.
20. Chamberlain, J. E., 1979, The Principles of Interferometric Spectroscopy, completed, collated, and edited by G. W. Chanry and N. W. B. Stone, Wiley, Chichester and New York.
21. Jacquinot, P., 1965, Jpn. J. Appl. Phys. 4, suppl. 1, 401.
22. Filler, A. S., AFCRL 71-0019 Special Reports 114, 407.
23. Luc, P. and Gerstenkorn, S., 1972, Astron. Astrophys. 18, 209.

24. Gerstenkorn, S. and Luc, P., 1978, Atlas du Spectre d' Absorption de la Molécule d' Iodine: 14,800-20,000 cm^{-1}, Editions C.N.R.S., Paris.
25. Luc, P. and Gterstenkorn, S., 1978, Appl. Opt. $\underline{17}$, 1327.
26. Steel, W. H., 1967, Interferometry, Cambridge at the University Press.
27. Fragon, M., 1966, Optical Interferometry, Academic Press, New York.
28. Brigham, E. O., 1974, The Fast Fourier Transform, Prentice-Hall, Englewood Cliffs, N.J.
29. Loewenstein, E. V., 1971, AFCRL 71-0019 Special Reports 114, 3.
30. Heer, C. V., 1972, Statistical Mechanics, Kinetic Theory, and Stochastic Processes, Academic Pres, New York.
31. Ring, J. and Schofield, J. W., 1972, Appl Opt. $\underline{11}$, 507.
32. Hirschfeld, T., 1977, Appl. Spectrosc. $\underline{31}$, 471.
33. Connes, P. and Michel, G., 1975, Appl. Opt. $\underline{14}$, 2067.
34. An interesting view of photomultiplier tube characteristics can be found in EMI Photomultipliers, published by EMI Industrial Electronics LTD 1979.
35. Khan, F., 1959, Astrophys. J. $\underline{129}$, 518.
36. Filler, A., 1973, J. Opt. Soc. Am. $\underline{63}$, 589.
37. Chester, T., Fitzgerald, J. and Winefordner, J., 1976, Anal. Chem. $\underline{48}$, 793.
38. Hirschfeld, T., 1976, Appl. Spectrosc. $\underline{30}$, 68.
39. Sakai, H., 1971, AFCRL 71-0019 Special Reports 114, 19.
40. Chang, Y. S. and Shaw, J., 1977, App. Spectrosc. $\underline{31}$, 213.
41. Chang, Y. S. and Shaw, J., 1977, J. Quant. Spectrosc. Radiat. Transfer $\underline{18}$, 491.
42. Lin, C. L., Niple, E., Shaw, J., Uselman, W., and Calvert, J. A., 1978, J. Quant. Spectrosc. Radiat. Transfer $\underline{20}$, 581.
43. Farrow, M., Burnham, R., and Eyring, E., 1978, Appl.Phys. Lett. $\underline{33}$, 735.
44. Lloyd, L., Riseman, S., Burnham, R. and Eyring, E., 1980, Rev. Sci. Instrum. $\underline{51}$, 1488.
45. Gerstenkorn, S. and Luc, P., 1976, Nouv. Rev. Optique $\underline{7}$, 149.
46. Horlick, G. and Yuen, W. K., 1978, Appl. Spectrosc. $\underline{32}$, 38.
47. Horlick, G. and Yuen, W. K., 1975, Anal. Chem. $\underline{47}$, 775A.
48. Brault, J., 1976, J. Opt. Soc. Am. $\underline{66}$, 1081.
49. Beer, R. and Marjaniemi, D., 1966, Appl. Opt. $\underline{5}$, 1191.
50. Luc, P. and Gerstenkorn, S., 1979, Rev. Phys. App. $\underline{14}$, 791.
51. Yuen, W. K. and Horlick, G., 1977, Anal. Chem. $\underline{49}$, 1446.
52. Palmer, B., Keller, R.and Engleman, Jr., R., "An Atlas of Uranium Emission Intensities in a Hollow Cathode Discharge", Los Alamos-825/-MS, Informal Report UC-34a, July 1980.
53. Fellgett, P., 1967, J. Phys. Supp (Paris) $\underline{28}$, C2, 165.
54. Bell, E. E., 1967, J. Phys. Supp. (Paris) $\underline{28}$, C2, 165.
55. Bell, E. E., 1971, AFCRL 71-0019 Special Reports 114, 71.
56. The author is aware of the efforts of Bomem, Inc. to market an instrument which operates in the visible/UV region.
57. Dunn, S. T., 1978, Appl. Opt. $\underline{17}$, 1367.

FOURIER TRANSFORM FARADAIC ADMITTANCE MEASUREMENTS (FT-FAM):

A DESCRIPTION AND SOME APPLICATIONS

>   Donald E. Smith
>   Department of Chemistry
>   Northwestern University
>   Evanston, Illinois 60201

BACKGROUND

Small amplitude a.c. measurements always have represented a significant component in the vast array of electrochemical relaxation techniques,* which include the familiar techniques of potential step chronoamperometry, current step chronopotentiometry, conventional "d.c." polarography, linear sweep and cyclic voltammetry, normal and differential pulse polarography, square wave polarography, and a.c. polarography. Such techniques have been applied to a range of measurements from kinetic-mechanistic studies of electrode processes to highly precise and sensitive analytical applications.[1-6] The power of a.c. measurements in kinetic and analytical applications became apparent in the late 1940's and early 1950's with the pioneering work of groups led by Gerischer[7-9], Randles,[10,11] and Breyer.[12,13] The European school favored use of impedance bridge measurement methods to investigate kinetics of electrode processes, while the Australia workers introduced the concept of automatic recording of "a.c. polarograms". The latter involved the superposition of a small amplitude a.c. potential (constant amplitude of 5-20 mV and constant frequency) on the normal d.c. ramp used in conventional polarography, and recording the alternating current <u>magnitude</u> as a function of the d.c. potential. These <u>peak-shaped response</u> curves (see Figure 1) represented the first introduction of automation into a.c. measurements. The approach was used effectively for chemical analysis. Using this approach, the Australian school also discovered and interpreted the so-called tensammetric waves.[13]

---

* Any electrochemical measurement in which the observable is time-dependent. The reader is referred to the following texts and collections of monographs.[1-6]

The desirability of automation with such measurements was made apparent by early work using impedance bridges and the associated rate laws for quantitative interpretation of the measurement results. It was shown that the resistive and capacitive (in-phase and quadrature) components of the cell impedance were required over a substantial range of frequencies before conclusive quantitative interpretations could be made. These measurements had to be made in the presence and absence of the electroactive substance of interest. The background response in absence of the electroactive material allowed one to accurately measure the "nonfaradaic" solution ohmic resistance and the double-layer capacity at the electrode-solution interface. Translating this to the automated a.c. polarographic approach leads to the conclusion that the current components in-phase and quadrature to the applied a.c. potential perturbation had to be measured with and without the electroactive substance over the "substantial range of frequencies" (at least two orders-of-magnitude, preferably more). Optionally, one could measure the total current and its phase angle relative to the a.c. perturbation. The concepts of four measurements per frequency over 10-20 frequencies, while an improvement over bridge methods with regard to time required, still represented a tedious, time-consuming task, to say the least. In addition, the rather stringent requirements placed on the chemical system had to be faced. To save time the background measurements often were performed

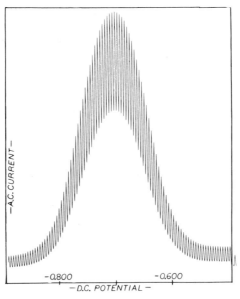

Figure 1. Classical fundamental harmonic a.c. polarogram. System: 3.0×10^3 M Cd^{2+}, 1.0 M Na_2SO_4 at the DME interface. Applied: 10 mV peak-to-peak sine wave of 320 Hz. Measured: 320 Hz current magnitude, ordinate uncalibrated, d.c. scan rate = 25 mV/min.

at just one frequency, introducing the usually valid assumption that these nonfaradaic observables behaved ideally. Once the required data were obtained, tedious calculations faced the researcher.[14,15] First one had to compute the nonfaradaic resistance and double-layer capacitance. Then one subtracted these components from the total cell response to yield the faradaic component (the current response associated with the redox process at the electrode surface), which usually is the observable of interest. In certain cases involving strong adsorption of the electroactive substance, the double layer and faradaic processes are coupled,[16,17] and thus inseparable, lead-

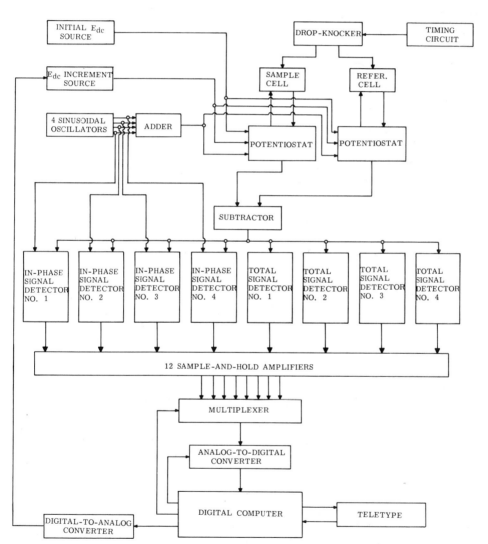

Figure 2. Schematic of original frequency multiplex instrument.

ing to even more complicated data processing.[17,18] Nevertheless, the a.c. technique was sufficiently promising that efforts to enhance measurement capability continued.

During the 1960's advances in operational amplifier and passive electronic component technology led to broadband three-electrode potentiostats with positive feedback compensation of the solution ohmic resistance, inexpensive lock-in amplifiers, low-pass and band-pass filters, analog timing circuits, signal sources and the other analog electronic components for modern a.c. polarographic instrumentation.[19-21] In the late 1960's the minicomputer and data domain converters (analog-to-digital and digital-to-analog) appeared on the scene. The advances in analog electronics were combined with this new, programmable digital data processing and storage device to produce an instrument capable of measuring two components of the a.c. polarographic response at four frequencies simultaneously, as well as the d.c. polarogram.[22] Frequency separations were performed with analog circuitry. The role of the minicomputer in this case was that of a data logger and processor. The concept of simultaneous multiple frequency measurement was shown to be effective and the term "frequency multiplexing" was introduced. A schematic of this instrument is shown in Figure 2. This device eliminated the tedious tasks of manual digitization and the "tedious calculations" referred to above. At the same time it was felt that measurements should be performed using at least 12 frequencies. Consequently, it was necessary to retune the analog signal processors two more times to obtain data at 12 frequencies. The idea of extending the analog circuitry to enable measurements at a larger number of frequencies was not pursued for several reasons, the most important being that a much more efficient digital approach had become commercially available in the form of the on-line minicomputer implemented Fast Fourier Transform (FFT) algorithm. The FFT replaces the critical tuned analog components (lock-in amplifiers) and permits measurements at very large numbers of frequencies simultaneously. A Fourier transform is equivalent to having a pair of digital phase-sensitive lock-in amplifiers at each frequency computed (excluding d.c. and the Nyquist frequency). Earlier work by Pilla[23,24] using pulse input waveforms and off-line computer computations of Laplace and Fourier transforms had established the effectiveness of this concept. Using the admittance* computation approach outlined below, the FFT enables direct measurement of in-phase and quadrature components of the cell admittance at each frequency computed by the algorithm (frequency sets computed are set by the operator within the confines of the FFT algorithm). Research into the use of on-line computerized Fourier Transform faradaic admittance* measurements (FT-FAM) began in the

---

*Cell admittance at any frequency is the cell current response divided by the amplitude of the a.c. input at that frequency. Faradaic admittance is the faradaic current response divided by the amplitude of the a.c. input.

Fourier Transform Faradaic Admittance Measurements (FT-FAM)

early 1970's. It is this writer's opinion that progress in the past decade has revolutionized instrumentation for faradaic admittance $[A_f(\omega t)]$ or impedance $[Z_f(\omega t) = A_f(\omega t)^{-1}]$ measurements, greatly enhancing the efficiency and all other aspects of this a.c. electrochemical relaxation technique.

The purpose of this chapter is to inform the reader of the current status of FT-FAM instrumentation, to feature a variety of applications which have resulted from our work, and to speculate regarding future possibilities.

An FT-FAM Instrument

The instrument described here represents a second generation FT-FAM instrument. It was developed to enable high frequency (to 125 KHz) FT-FAM measurements so that well known high frequency extrapolation procedures[16,25,26] could be used to measure and subtract the nonfaradaic solution ohmic resistance and double-layer capacity for the normal situation where the faradaic and double layer processes are not coupled. Two problems must be overcome to achieve such high frequencies. First, because of the potential control accuracy required in these measurements, such high frequencies exceed the bandpass of most potentiostats. This problem is solved by accurately measuring the applied a.c. potential waveform between the reference and working electrodes, using a broadband (100 MHz unit gain crossover frequency) unity gain voltage follower amplifier. This signal is amplified further by a similarly broadband inverting amplifier. The cell current signal is measured at the output of a broadband current follower amplifier and further amplified with the same style preamplifier used for the voltage signal. These time domain waveforms are subsequently digitized and eventually used to compute the cell admittance. By this measurement procedure it is not necessary that the potentiostat precisely control the high frequency signals. All that is required is that the potentiostat apply a significant, measurable perturbation. The second problem arises because the technique requires high speed analog-to-digital converters (ADC's) at two input channels (current and potential) and generation of the a.c. potential perturbation by a digital-to-analog converter (DAC), all done synchronously (parallel operations). No commercial laboratory minicomputer in existence can perform these parallel operations at repetition rates fast enough to achieve measurements at frequencies in the neighborhood of 100 KHz. Consequently, a high speed data buffer system was designed and constructed in-house which effected synchronous data generation and sampling at two input channels. The acronym SYDAGES was given the device in question.[27]

A schematic diagram of SYDAGES, including its connection to the minicomputer, is given in Figure 3. The unit has four external connections: two input signals transmitted through buffer amplifiers, sample-and-hold amplifiers and ADC's; one output waveform generated

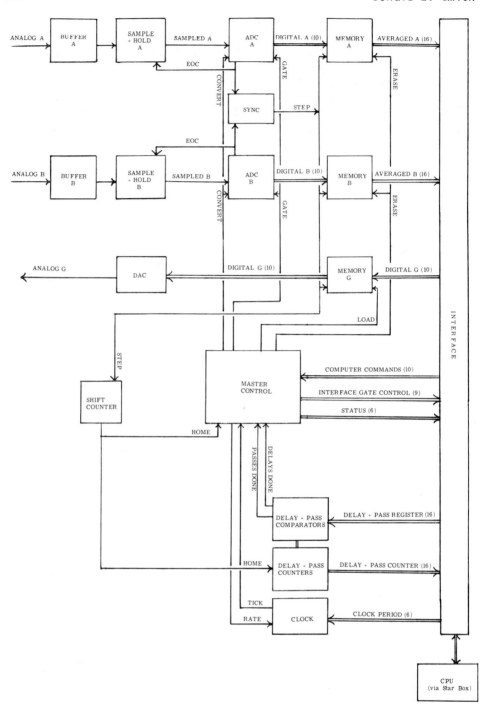

Figure 3. SYDAGES.

through the DAC; and a special data bus to the minicomputer.
Memories A, B and G are 1024-word shift register memories and, thus, are accessed sequentially. They always shift together so that synchronization is maintained. Memories A and B contain 16-bit words, while Memory G has a 10-bit word. One memory cycle involves 1024 shifts of the memories, with the shift period being controlled by the Clock Register. The Clock is a 10 MHz crystal oscillator with programmable frequency division set by the Clock Register. The Delay comparator and counter is used to control the number of a.c. waveform cycles applied before data acquisition actually begins. This is done to allow initial transients to become negligible so that the a.c. steady state approximation is validated. Typically 5-10 Delay cycles are used. The Pass comparator and counter controls the number of 1024 word Pass cycles used. Pass cycles begin immediately after the Delay cycles. During Pass cycles, output signal generation continues, but converted data are added into the A and B memories in the fashion of a signal averager. Thus, time domain averaging is made available by SYDAGES. Times used to combined Delay and Pass cycles can be quite short. For example, if the waveform period is 0.010 sec. (lowest frequency is 100 Hz.), and one performs 5 Delays and 10 Passes, the total waveform pair acquisition time is only 0.15 sec. Because the ADC's used provide 10-bit outputs (1 μsec. successive approximation conversion time), while the A and B shift register memory words are 16-bits, one can use a maximum of 2^6 or 64 Passes. Pass numbers typically used in our laboratory are 10-20. The Star Box mentioned in Figure 3 is a general input-output interface adaptor for Raytheon minicomputers. It was designed in-house.[28] The Interface operates under command of signals on the Interface Gate Control to allow data flow between the computer and the elements of SYDAGES only when the proper computer command is received. The Master Control is the SYDAGES component that interprets computer commands, monitors the status of the major control lines, and controls the interface gates, ADC's, and DAC. The Shift Counter counts the shifts and exerts the Home signal once every 1024 shifts.

Figure 4 provides a schematic of the complete FT-FAM instrumentation system. The computer (CPU) is a Raytheon Data Systems (RDS) Model 500 minicomputer with 32K of 16-bit core memory. The teletypewriter is the Teletype Model ASR-33. The 9-track magnetic tape units are Pertec tapes (Model 6860-9) with a RDS DIO controller. The discs are Diablo 1.2 Megaword units (Models 30 and 33) with a RDS DMA controller. The printer is a Centronics Model 703 and the Plotter is a Houston Complot Model DP-1. The RDS AID system provides two DAC's and a Tektronix Model 611 storage scope controller. One AID system DAC provides the d.c. waveform, while the other provides the mechanical drop-knocker pulse for experiments performed with a dropping mercury electrode (DME). The analog oscilloscope is a Hewlett-Packard Model 141B. A Datel Model 8500 d.c. voltage calibrator provides the initial voltage. The potentiostat is a standard three-electrode unit[29,30] constructed in-house using printed circuit board technology

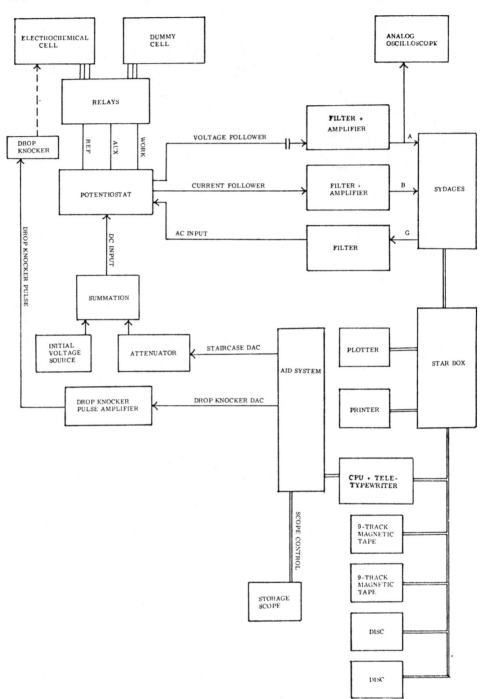

Figure 4. SYDAGES Based Instrument.

and Analog Devices Model 45J operational amplifiers. The potentiostat is attached either to the chemical or dummy cell (a precision RC circuit used for standby and calibration modes). Filters on the two SYDAGES ADC channels are 8th-order Butterworth low-pass anti-aliasing filters. The DAC channel contains an 8th-order Bessel low-pass filter to smooth the staircase character of the digitally manufactured applied a.c. waveform. A Rockland Corp. System 816 was used.

The instrumentation outlined in Figures 3 and 4 provides some truly unprecedented capabilities which are most impressive when viewed in combination, rather than individually. These capabilities include: A) a 500 KHz maximum data point acquisition (A and B memories) and output (G memory) frequency; B) Item A translates to an effective maximum cell admittance frequency of 125 KHz, a maximum faradaic admittance frequency of about 45 KHz for the in-phase component and magnitude and about 25 KHz for the quadrature component and phase angle cotangent;* C) provision for time and frequency domain averaging; D) adequate precision and accuracy for many purposes without averaging, implying vastly superior measurement quality if ensemble averaging is invoked; E) a very high maximum effective admittance spectral measurement repetition rate of 10 Hz, implying the ability to obtain, for example, 100 frequency domain replicates at a specific d.c. potential in ten seconds using a rapidly dropping mercury electrode or the use of reasonably fast d.c. scan rates in a.c. cyclic voltammetry; F) the availability of powerful numerical data enhancement procedures based on the FFT algorithm, such as FT digital filtering, interpolation, deconvolution of peak broadening effects, such as broadening due to heterogeneous charge transfer kinetics, and FFT peak narrowing by deconvolution. Examples of the advantages of each of these special capabilities will be given subsequently.

A particularly important, powerful, and novel capability of the instrument described here is the rather short 0.1 s minimum measurement interval required to define a complete admittance spectrum (Item E, above). Unfortunately, a major sacrifice has been made to realize such rapid measurement characteristics. Specifically, the ability to monitor in real time some aspect of the cell or faradaic admittance has been waived. With measurement intervals less than 5-6 seconds, the possibility of interspersing the essential data processing for some useful type of real-time cell or faradaic admittance readout does not exist with the instrumentation described. Because we normally either prefer or are forced (e.g., a.c. cyclic voltammetry) to use the advantages of short measurement intervals, our

---

*These faradaic component bandpass statements apply to optimum conditions: low resistance aqueous solvents with electroactive species concentrations $>10^{-4}$ $\underline{M}$. Lower concentrations or higher resistance solvent-electrolyte systems yield somewhat reduced bandwidth.

software has been written so that data processing is initiated following completion of the experimental measurement phase. The latter consists simply of acquiring pairs of 1024-point, 16-bit time domain waveform data arrays (input and response), and storing these arrays on a disk until the measurement is completed. The result is that measurement time can be quite short, but data processing is rather long, representing the rate determining step in obtaining the information of primary interest. The procedure of storing two 1024 point time domain arrays for each spectral measurement obviously represents rather inefficient use of disk memory. Even with the substantial storage capacity of our dual cartridge disk system, it is easily filled in experiments where ensemble averaging is desirable and/or a significant potential range is scanned with high resolution. For example, the total number of 1024-point waveform pairs (equivalent to one admittance spectrum when processed) which can be stored on the disks is 1098. For a limited experimental run involving measurements at 40 d.c. potentials over a d.c. potential range of 300 mV. (one wave) one can run (1098/40) = 27 replicates (requiring only 212 s with equipment described), after which disc memory overflow occurs. 27 replicates is not an insignificant number, but it does not provide the very high precision data (e.g., > 100 replicate results) whose use in measurement of very fast heterogeneous charge transfer has been demonstrated.[31] Similarly, if one wished to perform a more demanding a.c. cyclic experiment on a system exhibiting several waves over a potential range of 1.2 volts, using the same 7.5 mV. resolution characterizing the previous example one would require measurements at 120 points in both the forward and reverse scan (240 total points). Then only 4 replicates would be possible. It should be apparent from this and the previous paragraph that the instrument in question has some very special capabilities, but at the same time it is certainly not the final word in FT-FAM instrumentation.

An experiment is initiated with the Figure 4 instrumentation via operator-computer dialogue at the teletypewriter terminal. During this phase the operator answers computer inquires during which experimental conditions are selected. SYDAGES parameters such as clock period, delay and pass count, and G memory waveform are set, as well as non-SYDAGES parameters like the type of d.c. waveform to be used (see below), the d.c. waveform scan rate, the number and range of d.c. potentials at which frequency spectra are to be acquired and the number of frequency domain replicates to be averaged. A Start signal is given and the experiment begins. Following each waveform acquisition, the A and B memories are read by the computer and the next acquisition is initiated. The system runs automatically until all of the requested waveform acquisitions are completed. This experimental data acquisition subroutine is referred to as ACPOL1. It is used for both calibration with the dummy cell and actual electrochemical cell measurement. After the ACPOL1 phase is complete, the operator initiates the second major subroutine, ACPOL2. This subroutine runs automatically, once started by the operator. It performs the FFT on each 1024 point time domain data array generated by the experiment, computes the raw cell admittance from each input-response Fourier domain pair, and applies the frequency domain calibration function. The final major subroutine is ACPOL3, which is nothing more than a subset of interactive subroutines given in Table 1 with a brief description of each subroutine. With these routines the operator has close control of the cell admittance data processing. Full details of the instrument discussed here have been published.[27]

SOME GENERAL OBSERVATIONS

Cell Admittance Measurement Approach

Figure 5 outlines the basic approach used to compute the cell admittance. The acquired 1024-word time domain potential, $E(t)$, and current, $I(t)$, waveform arrays are Fourier transformed producing the frequency domain images, $E(\omega)$ and $I(\omega)$, respectively. $I(\omega)$ is divided by $E(\omega)$ to obtain the cell admittance. Because the frequency domain arrays contain in-phase and quadrature commponents, except for d.c. and the Nyquist frequency, they must be treated as complex numbers where the real part is the in-phase component and the imaginary part is the quadrature component. Dividing complex numbers is typically accomplished by multiplying the numerator and denominator by the complex conjugate of the denominator, as shown at the lower right of Figure 5. The denominator, $E(\omega)E^*(\omega)$, is a real number whose magnitude is equal to the square of the input <u>magnitude</u> at each frequency, and is referred to as the autopower spectrum of the input waveform.[32] The numerator, $I(\omega)E^*(\omega)$, is a complex number whose <u>magnitude</u> equals the product of the input a.c. potential and the current <u>magnitudes</u> at each frequency. It is referred to as the crosspower spectrum of the input and response waveforms.[32] The cross power spectrum contains the current-potential phase angle information at each frequency,

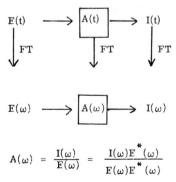

Figure 5. Computation of cell admittance.

Table 1. Interactive software subroutines available: ACPOL3

---

1. FIT1: Complex plane fits at high frequencies to obtain magnitudes of ohmic resistance, R_s, and double-layer capacitance, C_{dl}, (24-26). Operator views data on storage scope and determines interactively points to be used in the two high frequency extrapolations.

2. FITALL: Performs a FIT1 at every d.c. potential, using parameters selected by operator on FIT1.

3. RCSUB: Corrects cell admittance arrays for the non-faradaic effects; ohmic resistance and double-layer capacitance. Uses data provided by the FIT1-FITALL sequence and the formula

$$ADM_{fara.} = [(ADM_{cell})^{-1} - R_s]^{-1} - j\omega C_{dl}$$

4. BASELCOR: Corrects baseline of in-phase and/or quadrature admittance components for inaccurate RCSUB, using a least-squares quadratic fit for the baseline.

5. STORE: Performs RCSUB on all admittance data and writes corrected data on magnetic tapes.

6. SETSM: Sets parameters for FFT digital filtering.

7. SMOOTH: Performs FFT digital filtering using parameters selected with SETSM.[38]

8. LOGPLOT: From d.c. admittance (polarogram) plots $\log[(i_d - i)/i]$ vs. $E_{d.c.}$. Gives slope and zero ordinate intercept ($E_{1/2}$).

9. KIN: Corrects total admittance for effects of heterogeneous charge transfer as described.[65] Uses observed $\cot\phi$ data to calculate deconvolution function.

10. KIN2: Same as KIN, but uses $\cot\phi$ data from least-squares linear fit of $\cot\phi - \omega^{1/2}$ profile.[65]

11. KS: Calculate k_s from plot of $\cot\phi_{max}$ vs. $\omega^{1/2}$, where $\cot\phi_{max}$ is the peak $\cot\phi$ value. The equation used,

$$\text{slope} = [(2D)^{1/2}/k_s][(\alpha/1-\alpha)^{-\alpha} + (\alpha/1-\alpha)^{1-\alpha}]$$

requires prior measurement of $D^{1/2}$ and α.

12. ALPHA: Calculates α from the equation

$$[E_{d.c.}]\cot\phi_{max} = E^r_{1/2} + (RT/nF)\ln(\ /1-\ )$$

13. KSALPHA: Plots $\ln[(\cot\phi-1)(1+e^j)]$ vs. j, (y vs. x), where $j = (nF/RT)(E_{d.c.}-E^r_{1/2})$. The slope = α and the ordinate intercept is $\ln[(2\omega D)^{1/2}/k_s]$.

14. DCDIFCOF: Subtracts baseline from d.c. polarogram initially. Then by an iterative procedure determines i_d, $E_{1/2}$ and the $\log[(i_d-i)/i]$ vs. $E_{d.c.}$ plot slope. Requests parameters required for diffusion coefficient calculation and computes value, including uncertainties.

15. DCOF: Calculates diffusion coefficient using expanding sphere limiting current expression, reformulated as a quadratic in $D^{1/2}$.

16. INTERPOL: Interpolates data using FFT interpolation technique.

17. FRQDIV: Divides each point in a data array by $\omega^{1/2}$. Used only for data at one $E_{d.c.}$-value.

18. SETDATA: Performs a variety of operations on a data array, such as manually setting individual points, automatically scaling an array, etc.

19. ARAYMATH: Treats data array as complex number. Will add, subtract, multiply or divide array elements by another complex number.

20. DDMATH: Performs mathematical operations on disc files.

21. SPEAKLOC: Finds positive peak value, negative peak value, and inflection point for second harmonic a.c. polarogram.

i.e., $\phi = \cot^{-1}$ (Real part/Imaginary part). The "raw" admittance data obtained by the foregoing procedure is multiplied by a frequency domain calibration function obtained with a precision dummy cell[27] to yield the "true" cell admittance. With modern high speed analog components, the calibration function contains essentially scaling factors, except at the highest frequencies.

It should be noted that the inverse Fourier transforms of cross-power and autopower spectra are the time domain cross correlation and autocorrelation functions, respectively. Thus, another entry into FT-FAM is to record time domain cross and autocorrelation functions, Fourier transform these, and divide to yield the cell admittance or impedance. Such procedures have been applied.[33,34]

Problems with the Discrete, Finite Array Fourier Transform

Because computerized Fourier Transforms are performed on discrete, digital data arrays of finite length, two well-known problems arise. The discreteness of the data array leads to a phenomenon referred to as "aliasing" in which frequencies which are higher than one-half the data point acquisition frequency (the Nyquist frequency) appear at values which are lower than the true frequency. This effect is illustrated in Figure 6 for the case of a sine wave.

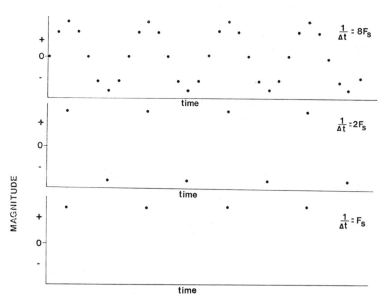

Figure 6. Example of aliasing a harmonic to a d.c. function due to inadequate sampling frequency.

Even if one knows the frequencies to be measured, as in FT-FAM measurements, one may encounter problems with noise whose frequency is higher than the Nyquist frequency. This effect is readily minimized by use of low-pass anti-aliasing filters which are part of our instrumentation (Figure 4). The finite length of the data array leads to an effect referred to as "leakage" or "bandbroadening". Figure 7A illustrates that, for a pure harmonic time domain waveform, one obtains a Fourier domain magnitude response over a range of frequencies, instead of at a single frequency. It can be shown that the discrete Fourier transform treats the finite time domain data record as if it repeated ad infinitum. Thus, for the time domain waveform array in Figure 7A, a periodic "glitch" appears because the original time domain record contains a non-integral number of cycles. This effect can be minimized by lengthening the data record, which is a general approach to minimize effects of leakage. In the special case of a periodic waveform, leakage may be eliminated entirely by making the discrete time domain data record length equal to an integral number of waveform cycles. This is illustrated in Figure 7B where exactly four cycles of the harmonic comprise the time domain record. Fourier transformation of this array yields only one non-zero point in the frequency domain magnitude spectrum. Figure 7B clearly indicates that, as far as FT data processing is concerned, a periodic signal represents the ideal situation. No other waveform type can fulfill this requirement.

Full details of the origin and effects of leakage and aliasing may be found in several texts.[35-37]

Rationale for A.C. Waveform Type Employed

Early in our work on FT-FAM, the question regarding the type of multiple frequency waveform to use for the "a.c. input" was considered. Obviously, there is a wide variety of multiple frequency waveforms. They can be encompassed by the four categories given in Table 2. The procedure outlined in Figure 5 is independent of waveform type, suggesting that any multiple frequency waveform should be applicable. The fact that multiple frequency periodic signals are best for the key data processing component, the discrete Fourier transform, might suggest that one should consider only periodic functions. However, because the faradaic admittance is nonlinear, periodic functions may be the worst case as far as the measured system is concerned. The higher harmonics and sum and difference frequencies generated by faradaic nonlinearity will precisely overlap the sought after fundamental harmonic responses, although sufficiently small amplitudes per frequency component might minimize this undersirable effect. The solution to this problem was not obvious, a priori. Consequently, we decided that the empirical approach would be the most useful and convincing.

Table 2. Major classes of multiple-frequency a.c. test signal

Waveform class	Examples
1. Complex periodic signals	Square wave, triangular wave, full-wave rectified sine wave, pseudo-random white noise, hybrids of pseudo-random white noise.
2. Almost periodic signals	Waveform typically obtained when outputs of an arbitrarily selected set of sinusoidal oscillators are added
3. Aperiodic transients	Step function, ramp function, triangular impulse, square impulse frequency sweep
4. Stochastic signals	White noise, pink noise, blue noise, flicker noise

waveforms

Special properties	Titles suggested for faradaic admittance measurement using particular waveform class
Waveform composed of set of discrete, coherently related sinusodial components, i.e., waveform is described by a conventional Fourier series.	"Faradaic admittance analysis (or a.c. polarography) in the coherent wave frequency multiplex mode"
Waveform composed of set of discrete, non-coherently related sinusoidal components, i.e., waveform is not periodic (except in limit of infinite time) even though its individual constituents are periodic.	"Faradaic admittance analysis (or a.c. polarography) in the non-coherent wave frequency multiplex mode"
Signals with a continuous, well-defined smoothly-varying phase and amplitude spectra. Bandwidth-limited in any practically-obtainable signal.	"Transient admittance analysis"
Signals with continuous spectra. Amplitude (autopower) spectrum approaches smooth distribution in long time limit. Phase spectrum randomized. Bandwidth-limited in any practicallly-obtainable signal.	"Noise response admittance analysis"

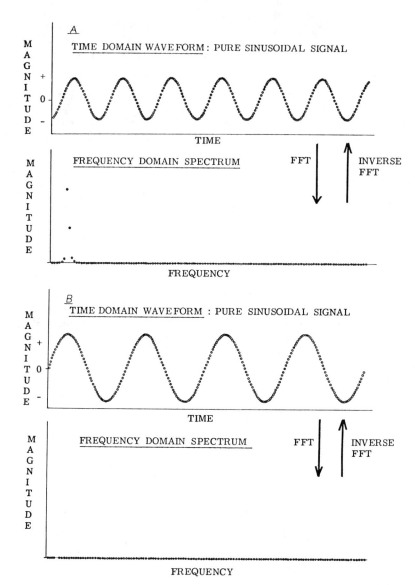

Figure 7. Illustration of leakage with sinusoidal time domain record. 7A illustrates sine wave which is inharmonic with data record, while 7B shows result when sine wave is harmonic with data record.

The model electrochemical system selected for this study was the process:[38]

$$Cr(CN)_6^{3-} + e \underset{Hg, H_2O}{\overset{1 \text{ M KCN}}{\rightleftharpoons}} Cr(CN)_6^{4-}$$

at the dropping mercury electrode (DME). It had been used previously as a model system in our laboratories. It was known to obey the rate law for the so-called quasireversible process where the electrode reaction involves mixed rate control by diffusion and the heterogeneous charge transfer step, the latter being reasonable facile (instantaneous on the d.c. polarographic time scale) but easily measured by the a.c. approach ($k_s \approx 0.30$-0.40 cm s^{-1}). The observable selected for this study was the faradaic phase angle cotangent ($\cot\phi$) versus $\omega^{1/2}$ profile at the admittance peak potential. Because this profile is linear with a slope proportional to k_s^{-1}, it provides an ideal situation for a linear least squares fit from which the slope, intercept (theoretically unity) and their standard deviations can be obtained. A typical example of a $\cot\phi$-$\omega^{1/2}$ plot is shown in Figure 8. So that statistical considerations would be reasonably valid, the average of 64 frequency domain $\cot\phi$ replicates was used for each waveform type considered. The instrument used at the time this work was performed did not permit time domain signal averaging. It did have the capability of eliminating nonfaradaic influences by analog procedures.[38] Test signals used, with brief descriptions of their characteristics, are given in Table 3. When periodic waveforms were used, the 512 word digital waveform array was synchronized to contain an integral number of waveform cycles. It is important to realize that all data were obtained with the same instrument and data processing procedures. Table 4 provides the results. The amplitudes indicated in the second column were the optimum amplitudes, except for the two cases (BLWN and almost periodic) where two amplitude entries are given, where one of the entries is optimum. The first item worth noting in Table 4 is the close agreement of the k_s values obtained. The average of the 15 values listed is 0.380 ± 0.033 cm s^{-1} (uncertainty = 1 S.D.). One also should note the close agreement between the 15 observed intercept values (1.006 ± 0.024) and the rather precise match with the theoretically predicted value of unity. This is remarkably good agreement, considering the facile nature of the heterogeneous process and the major difference in waveform types used. Some might even suggest that four different technique classes are involved. Regardless, it is clear that the Fourier transform unifies all small amplitude perturbation techniques used and confirms the generality of the calculational approach outlined in Figure 5.

Table 3. Test signal waveforms used

Waveform type	Frequency spectrum characteristics
1. Bandwidth-limited white noise (BLWN)	Continuous spectrum with random phases (flat distribution function) and amplitudes (Gaussian distribution function). Cut-off frequency at 1 kHz,
2. Combed bandwidth-limited white noise (combed BLWN)	Complex periodic signal with discrete components at frequencies = 10.07 N Hz, where N = 1,2,3,4...100. Randomized amplitudes (Gaussian distribution) and phases (flat distribution).
3. Rectangular pulse	Continuous, smoothly varying phase and amplitude spectrum. Amplitude spectrum characterized by decreasing magnitudes as frequency increases and by periodic zero-amplitude nodes.
4. Almost periodic signal	Set of discrete frequency components of identical amplitudes at 43.1, 108.7, 230.5, 382.6, 444.6 and 864.3 Hz.
5. Pseudo-random white noise (PRWN)	Complex periodic signal with discrete components of identical amplitudes at frequencies = 10.07 N Hz, where N = 1,2,3,4...100. Phases are randomized (flat distribution).
6. Phase-varying pseudo-random white noise (phase-varying PRWN)	Same as pseudo-random white noise.
7. Odd-harmonic pseudo-random white noise (odd harmonic PRWN)	Complex periodic signal with discrete components of identical amplitudes at frequencies = 10.07 $(2N-1)$ Hz where N = 1,2,3,4...50 (i.e., includes only odd harmonics of 10.07 Hz).
8. Phase varying, odd-harmonic pseudo-random white noise	Same as odd-harmonic pseudo-random white noise.
9. 15-component, odd-harmonic array	Complex periodic signal with discrete components of identical amplitudes at frequencies = 10.07 $(2N-1)$ Hz, where N = 1,2,3,4,5,7,10,13,17,21, 26,31,37,44 and 50 (components separated approximately equally in square-root-of-frequency space).
10. Phase-varying 15-component, odd harmonic array	Same as 15-component, odd harmonic array.

Relationship of signal properties on successive measurement passses (i.e., signals applied to successive mercury drops)	Signal origin
Randomized phase, amplitude, and time domain relationships on successive passes. Ensemble average of amplitude approaches smooth spectrum.	External source
Same as with bandwidth-limited white noise.	Computer-generated
All signal characteristics identical on successive passes.	Computer-generated
Amplitude spectrum invariant, time domain waveform changes on successive passes.	External source
All signal characteristics identical on successive passes.	Computer-generated
Amplitude spectrum invariant, phase spectrum and time domain waveform change on successive measurement passes. (Random number phase array rotated between each pass).	Computer-generated
Same as with pseudo-random white noise.	Computer-generated
Same as phase-varying pseudo-random white noise.	Computer-generated
Same as with pseudo-random white noise.	Computer-generated
Same as phase-varying pseudo-random white noise.	Computer-generated

Table 4. Properties of $\cot\phi-\omega^{1/2}$ data at admittance peak:

Applied a.c. potential type waveform	amplitude properties
1. BLWN[c]	2 mV/f[d]
2. BLWN	1 mV/f
3. BLWN (Gaussian window)	1 mV/f
4. Combed BLWN	1 mV/f
5. Almost periodic	5 mV/f
6. Almost periodic	2.5 mV/f
7. Unfiltered rect. pulse	+pulse, 2 mV x 2 ms
8. Filtered rect. pulse	+pulse, 2 mV x 2 ms
9. Filtered rect. pulse	-pulse, 2 mV x 2 ms
10. PRWN[f]	0.5 mV/f
11. Phase-varyng PRWN	0.5 mV/f
12. Odd-harmonic PRWN	1.5 mV/f
13. Phase-varying, odd-harmonic PRWN	1.5 mV/f
14. 15-component, odd-harmonic array	2 mV/f
15. Phase-varying, 15-component odd-harmonic array	2 mV/f

[a] Calculated from $\cot\phi-\omega^{1/2}$ slope using $\alpha = 0.59$, $\underline{n} = 1$, $\underline{T} = 298K$, $D_O = D_R = 8.20 \times 10^{-6}$ cm^{2}s^{-1}

[b] % units.

[c] BLWN = bandwidth-limited white noise.

[d] mV/f = amplitude of individual frequency components (averages in case of BLWN and combed BLWN).

Average of 64 measurement replicates using FFT

k_s/cm s^{-1} [a]	Relative slope S.D.: (σ_s) [b] = relative k_s, S.D.)	Intercept	Relative intercept S.D.: (σ_1) [b]
0.386[e]	4.70	1.012[e]	1.39
0.406	2.70	1.030	0.75
0.364	2.01	0.992	0.65
0.363	1.49	0.990	0.48
0.367	5.04	0.998	1.23
0.372	3.74	1.005	0.90
0.362	19.6	1.053	5.88
0.416	16.5	0.987	4.59
0.325	17.6	0.987	6.33
0.403	1.91	1.003	0.56
0.392	0.58	0.996	0.18
0.367	1.99	0.998	0.62
0.369	0.35	0.997	0.11
0.407	1.55	1.012	0.37
0.400	0.31	1.009	0.08

[e] Results obtained by least-squares analysis of "screened" $\cot\phi - \omega^{1/2}$ data. Screened data obtained by deleting all data points falling > 2.5 σ from original least-squares fit, and then repeating fit.

[f] PRWN = pseudo-random white noise.

S.D. = standard deviation

Table 4 does reveal that one major difference between the various waveforms employed does exist. It becomes evident when one examines measurement precision (columns 3 and 5). For both slope and intercept measurements one finds nearly a two order-of-magnitude difference in standard deviation between the best and the worst cases. These differences have been discussed and the reader is referred to the original publication for full details.[38] For present purposes it should suffice to ouline why phase-varying, odd harmonic PRWN (waveform no. 13) seemed superior (waveform no. 15 is just a special case of no. 13). First, it is periodic and, thus, ideal as far as the

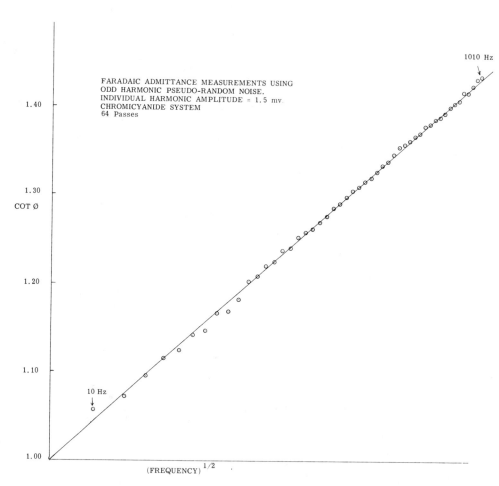

Figure 8. Faradaic admittance measurements using odd harmonic pseudo-random noise. Individual harmonic amplitude = 1.5 mV. Chromicyanide system. 64 passes.

Fourier transform operation is concerned. Second, by applying an a.c. waveform with only odd harmonics and examining only the odd harmonic responses one essentially filters out the major contributions of faradaic nonlinearity, which arise from second order terms ($\sin^2\omega t$, $\sin\omega_1 t \sin\omega_2 t$) in the series solution to the a.c. rate law,[15,39] because these higher harmonics, including sum and difference frequencies, all fall at even harmonics in the faradaic admittance spectra. However, one finds that if one uses precisely the same odd-harmonic waveform for each measurement replicate (waveform nos. 10, 12, 13), after about 35-40 replicates what appears to be random noise on the $\cot\phi$ vs. $\omega^{1/2}$ profile continues to exist, but ceases to diminish with the square root of the replicate number and eventually becomes constant ($\geq$ 55-60 replicates). It eventually was realized that this oddly behaving, apparently coherent noise component was the result of third order terms in the series solution for the a.c. response ($\sin^3\omega t$, $\sin^2\omega_1 t \sin\omega_2 t$, $\sin\omega_1 t \sin\omega_2 t \sin\omega_3 t$, etc., etc.). Such residual faradaic nonlinearity contributions do fall at odd harmonic frequencies at which the measurements were being made and, thus, produced a small faradaic nonlinearity error. What was interesting was that, because of the phase randomization characterizing this multiple frequency applied a.c. waveform, the nonlinearity error was not monotonic in the $\cot\phi$ vs. $\omega^{1/2}$ spectrum, but appeared to be random as a function of frequency, and did not change with replicate number once sufficient replicates had been run to eliminate effects of random electronic noise. Although this effect was minor, it was annoying and could be eliminated by changing the random phase array of the a.c. waveform with each frequency domain replicate. It was discovered that the simple expedient of rotating the phase array one step with each replicate was sufficient to make this residual faradaic nonlinearity effect behave like random noise. Thus, the phase varying odd harmonic waveform (no. 13) proved to be a novel hybrid a.c. waveform whose periodicity optimized the measurement system and whose particular type of phase randomization was ideal for the measured electrochemical system. Waveform 13 has been used in our FT-FAM measurements since the study just outlined was completed.

FT-FAM Versus FT-Spectroscopy

As will become apparent later, the introduction of the Fourier transform into faradaic admittance spectral measurements has yielded all the benefits for FT-FAM enjoyed by the better known Fourier transform spectroscopies (FT-IR, FT-NMR, FT-ICR, etc.). For example, the greater spectral measurement speed resulting from the use of the Fourier transform of a multiple frequency time domain signal can be converted into much higher than normal precision through use of ensemble averaging, as is evident in Table 4. Also, the measurement speed advantage may allow one to obtain spectral measurements on substituents with time varying concentrations, which cannot be done with conventional non-Fourier transform instrumentation. Various signal enhancement procedures such as FT digital filtering and interpolation become available as peripheral benefits (see below). At the same time, one major difference does exist. In the FT-spectroscopic techniques, the spectral frequencies one is dealing with usually are not precisely known and the Fourier transform is used in many cases to locate the position of the spectral lines. Leakage effects must be minimized by taking sufficiently long data records. In FT-FAM one knows and controls the frequencies to be measured, providing a distinct advantage in eliminating leakage and other undesirable effects, such as faradaic nonlinearity.

Dimensionality of FT-FAM Data

It is important to recognize that data produced by the FT-FAM procedure encompass three dimensions, response, d.c. potential (E_{dc}), and frequency (ω). Response may be the faradaic admittance (in-phase, quadrature or total) or its phase angle cotangent ($\cot\phi$). Faradaic admittance or $\cot\phi$ "polarograms" are plots of one or more of the faradaic admittance components or $\cot\phi$ versus E_{dc} at constant ω. Faradaic admittance or $\cot\phi$ spectra are plots of one or more of the faradaic admittance components or $\cot\phi$ versus $\omega^{1/2}$ at constant E_{dc}. No 3-dimensional plots will be presented in this monograph.

SOME FOURIER TRANSFORM TECHNIQUES FOR ENHANCEMENT OF ELECTROCHEMICAL DATA

A Remark

The data enhancement techniques discussed in this section are illustrated in the context of electrochemical measurements because they are employed as part of our FT-FAM data processing. However, it should be obvious that the techniques are applicable to a wide range of data types obtained with computerized instrumentation (spectroscopic, chromatographic, titrations, etc.).

FT-Digital Filtering

Figure 9 gives a simple illustration of the use of the Fourier transform in digital filtering (smoothing) of electrochemical data. A somewhat noisy digital a.c. polarogram (Figure 9A) is Fourier transformed yielding the Fourier spectrum of the original data (Figure 9B). Note that this spectrum contains large components only at the lower frequencies. Higher frequency components, which are randomly dispersed about zero, represent noise contributions. The latter are set to zero by a rectangular filtering function (Figure 9C) to produce the filtered Fourier spectrum (Figure 9D), which is inverse Fourier transformed to produce the filtered (smoothed) data

Fourier Transform Faradaic Admittance Measurements (FT-FAM)

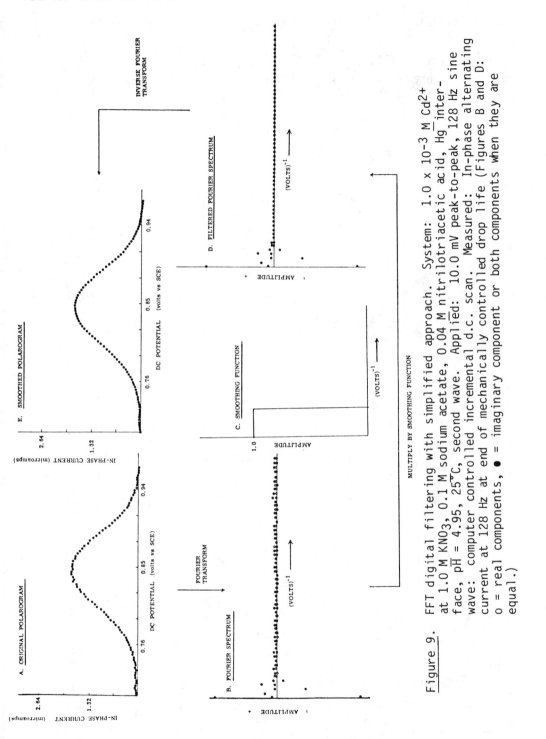

Figure 9. FFT digital filtering with simplified approach. System: 1.0×10^{-3} M Cd^{2+} at 1.0 M KNO_3, 0.1 M sodium acetate, 0.04 M nitrilotriacetic acid, Hg interface, $p\overline{H}$ = 4.95, 25°C, second wave. Applied: 10.0 mV peak-to-peak, 128 Hz sine wave: computer controlled incremental d.c. scan. Measured: In-phase alternating current at 128 Hz at end of mechanically controlled drop life (Figures B and D: o = real components, ● = imaginary component or both components when they are equal.)

(Figure 9E). Figure 10 shows an even more impressive application of the procedure outlined in Figure 9. The fact that the original data arrays in Figures 9 and 10 begin and terminate essentially at zero (+noise) is an important reason why the rectangular Fourier domain filter function worked so well. Had the data been characterized by non-zero initiation and termination (e.g., as in d.c. polarography or

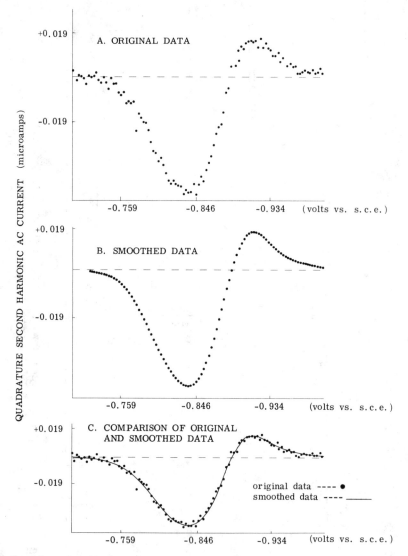

Figure 10. Conditions as in Figure 9, except applied frequency = 228 Hz and measured 456 Hz.

Fourier Transform Faradaic Admittance Measurements (FT-FAM)

normal pulse polarography where sigmoidal shapes are characteristic), the results would not be so impressive. The FT-filtered data would contain ringing as a result of this procedure.[40] It was decided that this problem could be minimized if an additional transform procedure could be applied to the original data to ensure that it would initiate and terminate at zero before the FT-filtering routine was applied. A simple rotation-translation transformation was used initially. It is illustrated in Figure 11 for a noise-free exponential decay. Here the objective was to show that the rotation translation transform followed by the FT-filtering routine of Figure 9 and the inverse rotation-translation transform would produce the original data array, whereas the simple FT-filtering routine of Figure 9 produced a badly distorted version of the original data. Subsequently it was realized that a more effective procedure would result if the original data were modified so that the initiation and termination of the data arrays were characterized by zero magnitudes and a zero first derivative. This has been accomplished by fitting a third-order polynomial to several points at the initiation and termination of the original data array and then subtracting the computed polynomial from the original data. This polynomial fit-subtract transformation and its inverse have replaced the rotation-translation procedure in our FT-filtering subroutine.

Regardless of how one handles the problem of non-zero initiation and termination, the FT digital filtering approach has the advantages of speed and ease of interpretation regarding what one is doing to the spectrum of the measured array, which do not accompany slower software digital filtering such as least-squares filtering procedures.

FT-Interpolation

Fourier transform interpolation is an effective data enhancement procedure for situations where the original digital data density is insufficient to reveal certain features of a response with satisfactory accuracy. In electrochemical measurements the special features might encompass peak magnitudes, peak potentials, half-wave potentials, peak widths at half-height, peak separations, etc. FT-interpolation has been used in FT-IR measurements.[41,42] FT-interpolation is a convenient, simple technique whereby the Fourier domain spectrum of the data array is extended by a factor of 2^n (n = positive integer) by zero filling and then inverse Fourier transformed to produce an interpolated array which contains 2^n times the number of points in the original array. This procedure assumes that the original data array is sufficient to define its Fourier domain spectrum accurately. Because data arrays which have non-zero initiation and/or termination cause problems similar to those encountered in FT-filtering, we use the polynomial fit-subtraction transform prior to Fourier transformation. After zero-filling the Fourier domain array,

482 Donald E. Smith

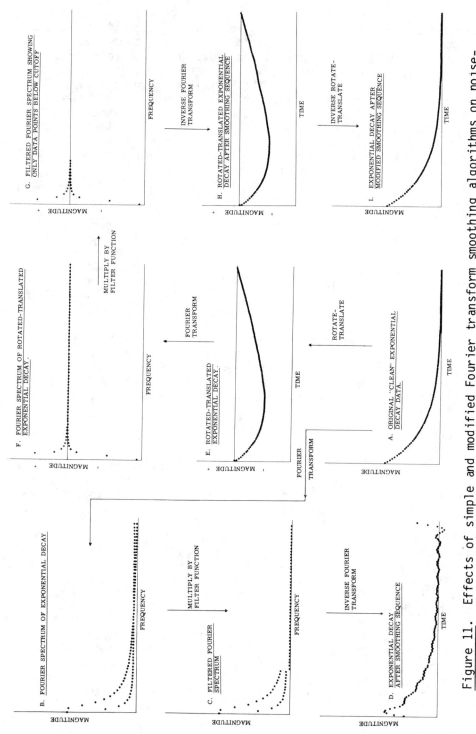

Figure 11. Effects of simple and modified Fourier transform smoothing algorithms on noise-free exponential decay.

the inverse Fourier transform followed by the inverse polynomial fit-subtraction transform yields the interpolated data array.

Figure 12 illustrates benefits obtained by interpolation of an a.c. cyclic voltammogram obtained using the FT-FAM procedure. The system involves the one-electron reduction of the tetramethylphenyl-enediamine cation (TMPD$^+$) at the platinum-0.10 $\underline{M}$ tetrabutylammonium perchlorate, acetonitrile interface. Under these conditions the electrode reaction may be characterized as quasireversible on the d.c. time scale. Such systems are predicted to exhibit unequal forward and reverse scan admittance voltammograms which include a cross-over point at a particular d.c. potential.[43] The d.c. potential of the crossover point can be used to measure an important

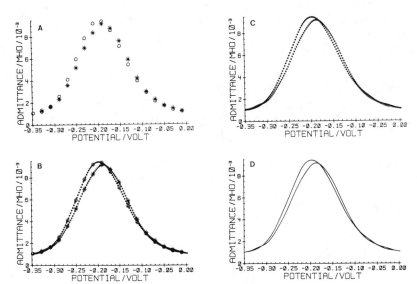

Figure 12. Fourier domain interpolation of FFT AC cyclic voltammogram. System: 8.0x10^{-4} M TMPD$^+$ at Pt-0.1 M tetrabutylammonium perchlorate, $\overline{CH}_3CN$ interface, $\overline{25}$°C. Applied: Pseudo-random, odd-harmonic a.c. waveform with 1.5 mV/frequency component, 40 components, superimposed on a staircase d.c. scan with triangular envelope whose scan rate = 150 mV/s. D.C. potentials are in volts vs. Ag/0.010 $\underline{M}$ AgClO$_4$, CH$_3$CN. Measured: faradaic admittance magnitude at 488 Hz, obtained with a single measurement pass without FFT digital filtering (39 other freq. components also measured, but not shown); (A) original data only, (B) original and interpolated (x2^3) data, (C) interpolated data points. Notation: o,* = original sampled data, forward and reverse scans, respectively; ●,x = interpolated data, forward and reverse scans.

heterogeneous charge transfer kinetic parameter, α, which may be considered as the symmetry coefficient for the electrode reaction, and often is referred to as the "charge transfer coefficient".[1-6,14,44] This parameter is dimensionless and theoretically is confined to values between zero and unity. When $\alpha \cong 0.5$, a symmetrical activation barrier is implied. The original data in Figure 12A clearly exhibit the expected crossover, but because adjacent points in the voltammogram are separated by 20 mv., one is uncertain whether the cross-over point d.c. potential estimated from Figure 12A will be sufficiently accurate ($\pm$ 2-3 mv.) to obtain a reasonably satisfactory evaluation of α. The expedient of FT-interpolation provides a simple, rapid, and precise solution to this problem. Figure 12B compares original and interpolated (x 2^3 = 8) data, Figure 12C gives interpolated data only, and Figure 12D presents the ultimate interpolation obtained by drawing straight lines between interpolated data points. Thus, by proper use of FT-interpolation and linear fitting, one can transform a discrete array into its continuous, analog image. Figure 13 gives another example of this for the d.c. cyclic voltammogram obtained as the zero frequency component of the FT-FAM spectrum of the TMPD$^+$ system. Figure 14 illustrates the benefits which can be realized by combining FT-filtering and FT-interpolation. A very noisy data array is remarkably enhanced by combining these two techniques. Although the data processing in Figure 14 was done by stepwise FT-filtering and interpolation for illustrative purposes, it

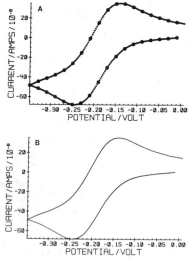

Figure 13. FD interpolation applied to a d.c. cyclic voltammogram. System, applied, and measured as in Figure 12, except d.c. component measured, (A) shows original (o) and interpolated (●) data, and (B) is a continuous curve obtained as in Figure 12D.

should be obvious that one can combine FT digital filtering and interpolation into a single operation in which the Fourier domain data are multiplied by the filtering function and zero-filled.

Peak Narrowing by FT-Deconvolution

Two well known theorems relating original data array operations and their Fourier domain images state that multiplication in the Fourier domain is equivalent to convolution in the data (time) domain, and division in the Fourier domain is equivalent to deconvolution in the data domain. The latter theorem can be used to sharpen peaks, thus, enhancing resolution of overlapping peaks. This procedure has been applied to chromatography[45] and spectroscopy.[46] It was investigated in our laboratory in the context of overlapping FT-FAM admittance peaks.[47] It can be applied to any other electrochemical technique which yields a peak shaped response, such as differential pulse polarography, square wave polarography, and semiderivative cyclic voltammetry.

Figure 15 illustrates the basic steps in the FT-deconvolution procedure, and what will occur if the peak to be sharpened is identical in all respects to the deconvolution peak. One obtains a single point located at the peak potential of the original peak. In real situations, this ultimate sharpening is impossible for a variety of reasons, including noise on the data arrays. These problems will not be discussed. What is important is what can be achieved.

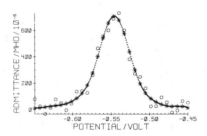

Figure 14. Effect of FFT digital filtering and interpolation on a noisy in-phase faradaic admittance polarogram. System: Hg-aqueous 1.0 M Na_2SO_4, 1.0 x 10^{-3} M Cd^{2+} interface, 25°C. Applied: Same as Figure 12, except 34 frequency components, staircase d.c. scan synchronized to a RDME of 0.25 s drop life, and d.c. potentials are vs. Ag/AgCl (saturated NaCl). Measured: As in Figure 12, except frequency is 195.3Hz, o = original sampled data, * = original data after digital filtering, and + = interpolated (x 2^3), filtered data.

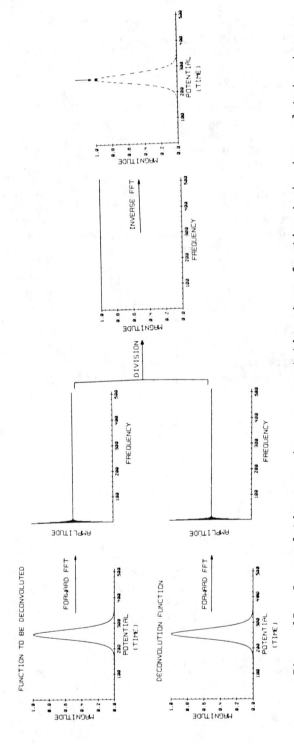

Figure 15. Deconvolution procedure representation where function to be deconvoluted and deconvolution function are identical (simulated data). $n = n_d = 2$.

Our work defined the deconvolution function as

$$f(x) = k/\cosh^2(n_d x) \quad [1]$$

where k is a proportionality factor and n_d controls the width of the deconvolution function, where the width decreases as n_d increases and vice versa. Integral values of n_d yield shapes identical to diffusion controlled (nernstian) FT-FAM peaks where the integer corresponds to the number of electrons transferred. In this work, n_d is not confined to integer values. Figure 16 is a simulated example of sharpening two overlapping FT-FAM peaks, separated by 30 mV.

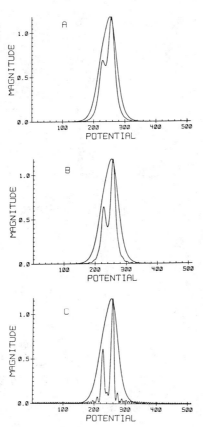

Figure 16. Different degrees of sharpening of simulated a.c. polarographic peaks. $\Delta E_{1/2}$ = 30 mV, C_1/C_2 = 2, n_1 = n_2 = 2: A) n_d = 5.5; B) n_d = 4.5; C) n_d = 3.2.

Sharpening increases from Figure 16A to 16C. One might characterize Figures 16A, B and C as insufficient sharpening, optimum sharpening (note faint indication of ringing onset at extremities of peak), and oversharpening, respectively. Figure 17 illustrates FT-deconvolution sharpening of actual experimental peaks involving the poorly resolved peaks of 2-nitroaniline (E_p = -810 mV.) and 4-nitrotoluene (E_p = -712 mV.) in an aprotic solvent-electrolyte system as a function of relative concentration values. It should be mentioned that FT-deconvolution of experimental admittance polarograms is preceded by application of FT-filtering and interpolation to optimize the data array for deconvolution. Despite the fact that resolution improvement in Figure 17 is far from ideal, it is sufficient to obtain linear

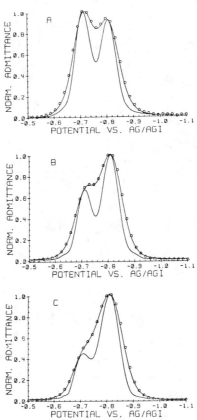

Figure 17. Sharpening of the overlapped a.c. admittance peaks of 2-nitroaniline (2NA), E_p = -810 mV, and 4-nitrotoluene (4NT), E_p = -712 mV in 0.1 TBAI/DMF solutions, n = 1 C_{4NT} = 2.19x10^{-4} M, n_d = 3.5
A. C_{2NA} = 2.05x10^{-4} M
B. C_{2NA} = 4.10x10^{-4} M
C. C_{2NA} = 9.23x10^{-4} M

calibration curves of the component with varying concentration to relative peak height units as small as 0.10 (peak height of concentration varying component = 0.10 x peak height of other component). Although it has not been studied further in our work, FT-deconvolution should be of considerable help in analytical applications involving overlapping peaks.

SOME APPLICATIONS OF FT-FAM MEASUREMENTS

Measurements of the Heterogeneous Charge Transfer Kinetic Parameters, k_s and α

Early in our investigations of the use of the Fourier transform in multiple frequency a.c. measurements, the $\cot\phi$-$\omega^{1/2}$ profile of a model system was measured using extensive frequency domain averaging (100 passes).[31] The system was the 1-electron reduction of the tris-oxalatoferric ion to the corresponding ferrous complex

$$Fe(C_2O_4)_3^{3-} + e \underset{Hg, H_2O}{\overset{\substack{1.0\text{ M K}_2C_2O_4 \\ 0.05\text{ M H}_2C_2O_4}}{\rightleftharpoons}} Fe(C_2O_4)_3^{4-}$$

Specialized measurements on this system in the microsecond range had yielded k_s values of 1.16±0.04 cm s^{-1} [48] and 1.44±0.05 cm s^{-1}.[49] A $\cot\phi$ spectrum obtained in our 100-pass Fourier transform is shown in Figure 18. A linear least-squares analysis of the $\cot\phi$-$\omega^{1/2}$ gave a slope of (1.06±0.04)x10^{-3} and an intercept of 1.003±0.05 cm s^1, where the uncertainties represent two standard deviations calculated by the linear regression method. Converting the slope magnitude to a k_s value produced 1.48±0.05 cm s^{-1}, using known values of the diffusion coefficients. These results are particularly impressive considering the facts that the charge transfer kinetic contribution in Figure 18 ranges from 2% to about 10% of the diffusional contribution (unity) and that the highest frequency used, 985 Hz, is approximately three orders-of-magnitude lower than those used previously. This provides a clear demonstration that extensive frequency domain averaging made possible by the speed of the Fourier transform can provide considerable benefits in heterogeneous charge transfer kinetics. With the more capable SYDAGES-based instrument described, k_s values of this order of magnitude can be measured rather routinely, with considerably less frequency averaging (15-20 averages).

Shortly after completion and check-out of the SYDAGES instrument, it was applied to the two electron reduction of Cd^{2+} at the DME-aqueous 1.0 M $ZnSO_4$, 0.18 M H_2SO_4 interface. Total admittance and $\cot\phi$ spectra at the peak potential are given in

Figure 19. A total admittance polarogram at 22.4 KHz is given in Figure 20. Using a calculated α of 0.35, the k_s value obtained from the slope of the peak $\cot\phi - \omega^{1/2}$ plot was $(8.5 \pm 0.5) \times 10^{-2}$ cm s^{-1}. The highest frequency in the faradaic admittance profile was 44.73 KHz and the corresponding value in the $\cot\phi$ plot was 20.64 KHz. This represents a new era in accessible faradaic frequencies with automated instrumentation. Only the infinitely slower bridge methods can compete.

The ease of making heterogeneous charge transfer kinetic measurements makes obvious the use of homologous series of molecules to study relationships between molecular structure and electrochemical behavior. One such series being investigated in our laboratory is aliphatic and aromatic nitro compounds in acetonitrile (ACN) or dimethylformamide (DMF). The supporting electrolye cation is either tetrabutylammonium or the tetraethylammonium, with the perchlorate

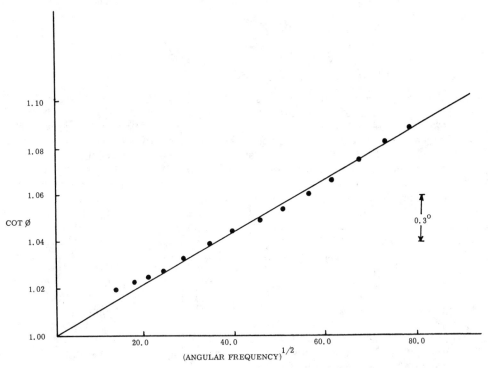

Figure 18. FT-FAM measurement of 5.0×10^{-3} M $Fe(C_2O_4)_3^{3-}$ in 1.0 M $K_2C_2O_4$ + 0.05 M $H_2C_2O_4$ at Hg, 25°C after 100 frequency domain replicates.

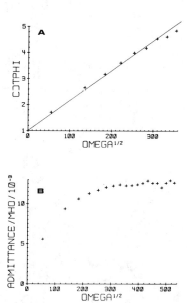

Figure 19. Peak faradaic admittance spectra for $Cd^{2+}/Cd(Hg)$ system after nonfaradaic compensation. System: DME-aqueous 1.0 M $ZnSO_4$, 0.18 M H_2SO_4 at 25°C. Applied: Pseudo-random odd-harmonic a.c. waveform with 1.5 mV per frequency component, 26 components superimposed on d.c. voltage of 0.523 volt vs. Ag/AgCl (saturated NaCl). Measured: Average of 10 replicates at 3.0 s in life of mechanically controlled drop life. A = phase angle cotangent. B = total admittance spectrum.

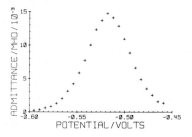

Figure 20. Total admittance polarogram of $Cd^{2+}/Cd(Hg)$ at 22.4 KHz. Conditions as in Figure 19 except as a function of d.c. potential using a staircase d.c. potential scan.

Table 5A

Compound	a_N(Gauss)	k_s(cm sec^{-1}) DMF	ACN
$(CH_3)_3CNO_2$	26	0.005±0.005	
$(CH_3)_2CHNO_2$	26	0.018±0.002	0.009±0.001
H₃C–⟨C₆H₂(CH₃)₂⟩–NO₂ (2,6-dimethyl)	16	0.22±0.01	0.30±0.01
H₂N–⟨C₆H₄⟩–NO₂	11.5	1.01±0.05	1.65±0.10
o-NH₂-C₆H₄-NO₂		1.30±0.05	2.30±0.10
o-CH₃-C₆H₄-NO₂	11	0.83±0.02	2.0±0.2
m-CH₃-C₆H₄-NO₂	10.4	1.40±0.05	1.7±0.1
p-CH₃-C₆H₄-NO₂	10	1.80±0.05	1.6±0.05

Table 5B

Compound	Solvent	Cation	k_s(cm sec^{-1})
NC-C$_6$H$_4$-NO$_2$ (para)	DMF	TBA$^+$ TEA$^+$	0.19±0.02 2.2±0.2
C$_6$H$_4$(CN)(NO$_2$) (ortho)	DMF	TBA$^+$ TEA$^+$	0.15±0.02 1.5±0.1
NC-C$_6$H$_4$-NO$_2$ (para)	ACN	TBA$^+$ TEA$^+$	0.63±0.05 4.0±0.5

iodide or tetrafluoroborate anion. All electrode reactions are one-electron reductions to the radical anion. Heterogeneous rate parameters are given in Table 5, together with odd-electron nitrogen coupling constants when they are available in the literature. It should be noted that a_N = 26 designates that all the odd-electron density is on the nitro group. This is found for the aliphatic compounds, as one would expect. The aromatic compounds show varying degrees of odd-electron delocalization. Some of the compounds listed in Table 5 (in DMF) had been studied by Peover.[50] We have confirmed his results and extended his work far beyond what could be done with a bridge. The data in Table 5 is not yet corrected for the Frumkin double-layer effect. This will be done subsequently. The information one obtains in Table 5 is that the rate determining step is solvent reorganization. That was clearly indicated by Peover, and our extensions of his work have changed nothing in this regard. The most dramatic effect found in our work has been the remarkable difference between the k_s values for the tetrabutyl- and tetraethylammonium salts. The tetraethylammonium salt yields apparent rates approximately an order of magnitude larger than for the tetrabutylammonium salt. This effect previously was reported for the t-nitrobutane molecule by Evans, et al. In our hands, it appears to be a general effect (Table 5B).

A.C. Cyclic Voltammetric Measurements

A technique which is becoming increasingly important in our laboratory is a.c. cyclic voltammetry. This experiment is run on a stationary electrode [Hanging Hg drop (HMDE), Pt, Au, graphite, etc.]. The d.c. potential staircase is swept first in one direction and then the other. The slopes for the forward and reverse scans usually are equal in magnitude, but opposite in sign. The ramp amplitude encompasses one or more admittance peaks. The FT-FAM measurement is performed in this context.

Some typical data obtained from the chromicyanide system are given in Figure 21.[51] These admittance magnitude polarograms show

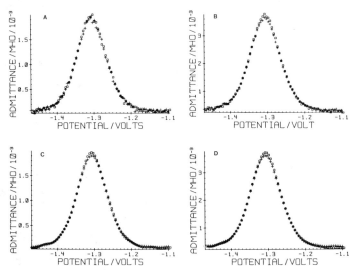

Figure 21. Typical FFT a.c. cyclic voltammograms of $Cr(CN)_6^{3-}/Cr(CN)_6^{4-}$ couple at HMDE in aqueous cyanide media. System: 1.0×10^{-3} M $Cr(CN)_6^{3-}$ at Hg-1.0 M KCN, water interface, 25°C. Applied: Pseudo-random, odd-harmonic a.c. waveform with 1.5 mV per frequency component, 32 total components; superimposed on staircase d.c. scan (5 mV per step) with triangular envelope whose scan rate = 50 mV s^{-1}. Measured: Faradaic admittance magnitude at 1840.8 rad s^{-1} (A,C) and 7877 rad s^{-1} (B,D) obtained on single measurement pass (30 other frequency components measured simultaneously, but not shown). (A,B) Raw data. (C,D) digitally filtered (smoothed) data. (+) Forward scan, (□) reverse scan. Abscissa = potential vs. Ag/AgCl.

Fourier Transform Faradaic Admittance Measurements (FT-FAM)

both forward and reverse scans. Curves A and B differ in the frequency applied (see Figure legend). Curves C and D depict the effect of FT-digital filtering on Curves A and B, respectively. The small difference in forward and reverse peak currents arises because of small differences in oxidized and reduced form diffusion coefficients. The cotϕ admittance polarograms gave $\alpha = 0.53 \pm 0.03$ and the cotϕ peak spectrum produced $k_s = 0.39 \pm 0.03$ cm s^{-1}, which is completely consistent with results obtained at a DME.

The a.c. cyclic voltammetric reduction of the TMPD$^+$ cation to the neutral radical is depicted once again in Figure 22. The working electrode was Pt, as before. This particular data set shows the original admittance polarograms at three frequencies: d.c., 488 Hz, and 1.466 KHz. The difficulty in locating the crossover point without interpolation is apparent. The fact that zero frequency is part of the Fourier transform also is revealed.

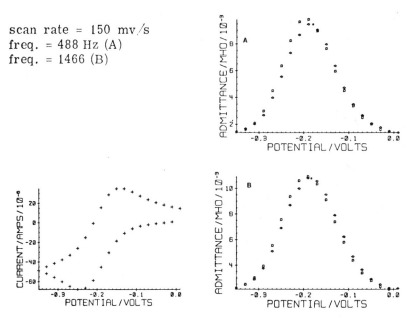

scan rate = 150 mv/s
freq. = 488 Hz (A)
freq. = 1466 (B)

Figure 22. FFT d.c. and a.c. cyclic voltammograms of TMPD$^+$/TMPD couple at platinum in acetonitrile--TBAP electrolyte. System: 8×10^{-4} M TMPD$^+$ at Pt-0.1 M TBAP,, CH$_3$CN interface, 25°C. Applied: Same as in Figure 21, except scan rate = 150 mV s^{-1}, and 40 frequencies applied. Measured: d.c. and a.c. components raw unsmoothed data shown, a.c. frequencies indicated in Figure.

We have used the FT-FAM approach to a.c. cyclic voltammetry to investigate systems with first-order preceding or with first-order following chemical reactions. Figure 23 gives an example of a process involving the one electron oxidation of cis-Cr(CO)$_2$(dpm)$_2$ [where dpm = bis(diphenylphosphino)methane], to the singly charged cation. The cation isomerizes to the trans form following charge transfer. The major effect seen is that the reverse scan peak is much smaller than the forward peak. This peak height ratio (reverse/forward) permits one to compute the homogeneous isomerization rate constant. At 20°C we obtain an isomerization rate constant of 0.50 ± 0.03 s^{-1}.[52]

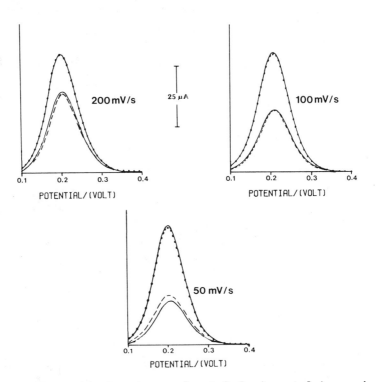

Figure 23. Theoretical and experimental fundamental harmonic a.c. cyclic voltammograms for cis-Cr(CO)$_2$(dpm)$_2$ system as described in Figure 8 but with T = 293 K, k_i = 0.50 s^{-1}, and k_s = 1.0 cm s^{-1}. Dependence on scan rate is shown.
———: theoretical
·····: experimental forward scan
-----: experimental reverse scan

It should be mentioned that the measurement which produces the time domain input-response data can be repeated as rapidly as once every 0.1 s, or at 10 Hz rate. Consequently, moderately rapid scanning is possible, although it is not possible to perform super-fast scans (>1 v/s).

Rapid Drop Time Measurements

In addition to having the ability to repeat measurement of time domain response arrays every 0.1 s, we have succeeded in making single FT-FAM measurements with the DME to time scales as short as 0.1 s.[53] Such short time measurements with the "RDME" electrode can be beneficial in the following instances: (a) minimization or complete suppression of polarographic maxima; (b) reduction or elimination of electrode reaction inhibition effects due to adsorption and precipitation phenomena; (c) enabling measurements in agitated solution; (d) permitting significant alteration of coupled chemical reaction effects, relative to normal drop life observations; (e) the production of smaller ohmic potential drop distortions; (f) minimization of spherical diffusion effects with amalgam forming systems; (g) enhancement of slow charge transfer kinetic effects; (h) enabling one to effect polarographic measurements in unstable situations, such as when a solution component is decomposing, or when electrode and/or cell materials are attacked by the solution (e.g., glass capillaries in acid F^- media). Obviously, if FT-FAM measurement can be repeated every 0.1 s, then one's ability to perform ensemble averaging is tremendous. With a 0.1 s drop life, one could complete a 100-pass FT-FAM measurements at a particular d.c. potential in 10 s, instead of requiring approximately 500 s which was done in the early 100-pass study discussed in the first part of this section. Some typical RDME data are given in Figure 24.

FT-FAM Measurements Using a D.C. Waveform of the Normal Pulse Polarographic Type

Another type of "d.c." waveform we are using is the so-called normal pulse polarographic waveform (NPPW). This waveform is characterized by an invariant initial potential and a pulse which occurs the last 30-70 ms of the mercury drop life. Each successive pulse increases linearly in amplitude. The a.c. input waveform is applied shortly before the pulse is applied. When measurement is completed (one drop life), the a.c. input and pulse are terminated and the d.c. potential returns to its initial potential. The net result is that the electrolysis occurs only for a short period during each drop life. This type of d.c. waveform carries many of the advantages one has in the RDME.[53] We were particularly interested in how well the FT-FAM procedure would withstand the large pulses characterizing the NPPW waveform. The results were a complete success. This method was applied successfully to the usual model systems[54] [$Cr(CN)_6^{3-}$/$Cr(CN)_6^{4-}$ in 1 $\underline{M}$ KCN; and Cd^{2+}/Cd(Hg) in 1.0 $\underline{M}$ Na_2SO_4]. It then

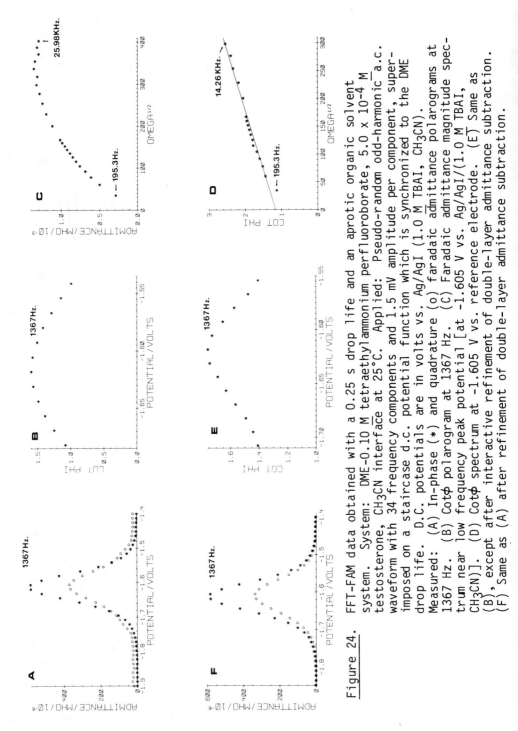

Figure 24. FFT-FAM data obtained with a 0.25 s drop life and an aprotic organic solvent system. System: DME-0.10 M tetraethylammonium perfluoroborate, 5.0 x 10⁻⁴ M testosterone, CH₃CN interface at 25°C. Applied: Pseudo-random odd-harmonic a.c. waveform with 34 frequency components and 1.5 mV amplitude per component, superimposed on a staircase d.c. potential function which is synchronized to the DME drop life. D.C. potentials are in volts vs. Ag/AgI (1.0 M TBAI, CH₃CN). Measured: (A) In-phase (*) and quadrature (o) faradaic admittance polarograms at 1367 Hz. (B) Cotϕ polarogram at 1367 Hz. (C) Faradaic admittance magnitude spectrum near low frequency peak potential [at -1.605 V vs. Ag/AgI/(1.0 M TBAI, CH₃CN)]. (D) Cotϕ spectrum at -1.605 V vs. reference electrode. (E) Same as (B), except after interactive refinement of double-layer admittance subtraction. (F) Same as (A) after refinement of double-layer admittance subtraction.

was used to examine how well the NPPW waveform would minimize effects of electroinactive surfactants.[55] What was learned was when the concentration of the surfactant was a factor of five or more smaller than the electroactive material, essentially complete removal of the surfactant influence occurred. If the surfactant concentration became larger than about one-fifth the electroactive material, then significant surfactant effects would appear. The rationale for these observations is that mass transport for both the surfactant and electroactive substance are controlled by diffusion and the diffusion rate is proportional to concentration (among other things). Ignoring differences in diffusion coefficients as a first approximation, one concludes that the electroactive substance must be present in significant excess (>5), relative to the surfactant amount for the NPPW waveform to be effective in eliminating surfactant influence. Some typical FT-FAM $\cot\phi$ peak spectra data are given in Figure 25 for $Cd^{+2}/Cd(Hg)$ showing the influence of the surfactant on using both the staircase (SCW) and NPPW waveforms. It is evident that the NPPW waveform nearly totally eliminates the surfactant influence under these conditions. FT-FAM admittance polarograms using the SCW d.c. scan with and without surfactant are shown in Figure 26, illustrating the major influence the surfactant has under these conditions.

Measurement of Homogeneous Redox Reaction Rates

Until 5-6 years ago, only the classical catalytic mechanism,

$$O + ne \rightleftharpoons R \underset{k_C}{\longrightarrow}$$

simple variations thereof, and the simple disproportionation case

$$2O + ne \rightleftharpoons 2R \rightarrow O + Y$$

have provided access to fast homogeneous redox reaction rates through electrochemical relaxation measurements, including FT-FAM. However, recent theoretical rate law derivations have made it clear that numerous other possibilities exist which have not been fully exploited. In our laboratory the mechanism:

$$A + e \rightleftharpoons B$$
$$C + e \rightleftharpoons D$$
$$A + D \rightleftharpoons B + C$$

has been investigated theoretically and scrutinized in detail experimentally[56-60] in the context where the A/B and C/D couples reduce in the order expected thermodynamically. The rate law

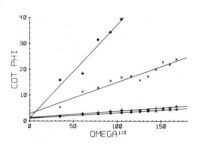

Figure 25. Peak $\cot\phi$-$\omega^{1/2}$ data for the Cd^{2+} system in the presence and absence of 1-octanol using the SCW (o,*) and NPPW (+,X) d.c. scans. System: 1.01×10^{-3} M Cd^{2+} at the Hg-aqueous 1.0 M Na_2SO_4-0.36 M H_2SO_4 interface, 25°C, where o = SCW with no added 1-octanol (+,X,*) = 2.0×10^{-4} M 1-octanol added. Applied: pseudo-random odd-harmonic a.c. waveform with 1.5 mV per frequency component, 35 total components, 195.3 Hz fundamental harmonic superimposed on SCW or NPPW d.c. scans. Pulse width = 61.5 ms NPPW E_i = +0.10 (+) and -0.45 (X) V. Measured: peak $\cot\phi$ at 5.0 s in the life of Hg drop, using five replicates and no FFT digital filtering.

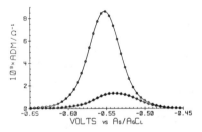

Figure 26. Admittance polarograms using SCW d.c. scan in the presence (*) and absence (o) of 2.0×10^{-4} M 1-octanol at 586 Hz. System: applied and measured as in Figure 25, except for FFT digital filtering (points) and FFT interpolation (solid curve).

indicates that if one observes only the more negative wave in the presence and absence of the more easily reduced material, one has the basis for calculating the second-order rate constant. One plots as a function of d.c. potential the ratio of the admittance for the more negative wave in the presence of the more easily reduced substance over the admittance in its absence. A sigmoidal plot initiating at unity (positive E_{dc}'s) and terminating at some larger plateau value is observed. While the theoretical rate law is complicated, the experimental procedure is quite simple. Table 6 presents some results we have obtained for the Eu^{2+} reduction of $Co(NH_3)_5X^{2+}$ complexes and the $Co(NH_3)_5PO_4$ neutral complex. Included with the data in Table 6 are some earlier results obtained by stopped flow[61,62] measurements. Earlier work in our lab[63] had yielded results in agreement with pulsed radiolysis measurements.[64] The agreement between these three widely divergent techniques is remarkable. One of the key advantages of the electrochemical approach is the fact that an optical absorption or emission spectrum is not required as it is in stopped flow and pulse radiolysis.

Kinetic Deconvolution

In the context of electrochemical relaxation technique analysis (in-phase a.c., differential pulse, etc.) one might question the utility of multiple frequency measurements when one simply needs, e.g., a single differential pulse, a.c. or admittance polarogram on which to base an assay. After all, considerable success has been realized using conventional electrochemical assay procedures which are based on measurement at a single frequency (fundamental and second harmonic a.c. polarography or voltammetry), a single point in time (differential and normal pulse polarographic magnitude, d.c. polarography at the end of drop life), some form of time average response (d.c. polarographic average currents at the DME, chronocoulometry) or some form of convolution or deconvolution response (semi-integration or semidifferentiation of a linear sweep or cyclic voltammogram). In addition, the measured detailed character of the analytical response usually is ignored and one bases an assay on the response at a single peak potential or in a potential region where the response is potential independent, as in the limiting currents characterizing d.c. or normal pulse polarography. Except for cursory qualitative inspection to insure that the wave seems "normal" and is properly located on the d.c. potential axis, an electrochemical assay procedure ignores the detailed d.c. potential-response profile. Further, most commerically available electroanalytical instruments do not automatically provide response-time or response-frequency profiles. Thus, a careful kinetic-mechanistic investigation of an electrode response is seldom a part of a routine electrochemical assay. It is our feeling that the FT-FAM procedure eventually will change that outlook.

Table 6. Second-order rate constants for the Eu(II) reduction of some Co(III)pentammine complexes.[d]

Co(III) Complex	k_{2nd} ($\underline{M}^{-1}$ s^{-1})[e]
RF^{2+}	2.6 x 10^4 [b]
	2.1 x 10^4 [a]
	1.2 x 10^4 [pH = 1][a]
	9.0 x 10^3 [pH = 2][a]
	7.1 x 10^3 [pH = 3][a]
	5.1 x 10^3 [pH = 4][a]
	1.5 x 10^4 [pH = 6][c]
	2.8 x 10^3 [pH = 6, μ = 0.06][c]
RCl^{2+}	3.9 x 10^2 [b]
	5.6 x 10^2 [a]
	4.7 x 10^2 [pH = 6][c]
	3.0 x 10^2 [17.2°C][b]
	4.7 x 10^2 [31.2°C][b]
RBr^{2+}	2.5 x 10^2 [b]
	3.0 x 10^2 [a]
	2.0 x 10^2 [17.8°C][b]
	3.0 x 10^2 [31.3°C][b]
RI^{+2}	1.2 x 10^2 [b]
	1.1 x 10^2 [a]
RN$_3$$^{2+}$	1.9 x 10^2 [b]
	1.5 x 10^2 [18°C][b]
	2.4 x 10^2 [31.7°C][b]
	3.5 x 10^2 [a]
	1.6 x 10^2 [pH = 1][a]
	1.3 x 10^2 [pH = 1.5][a]
RNO$_3$$^{2+}$	1.0 x 10^2 [b]
	1.2 x 10^2 [a]
RNCS^{2+}	10[c]
	30[a]
RPO$_4$	29[a]

Co(III) Complex	k_{2nd} ($\underline{M}^{-1}$ s^{-1})[e]
RNO_2^{2+}	43[a]
$RONO^{2+}$	58[a]

[a] References 59 and 60

[b] Reference 61.

[c] Reference 62.

[d] Unless noted otherwise, 25°C, pH = 0.3 and unity ionic strength mixture of $NaClO_4$ and $HClO_4$ represent measurement conditions.

[e] Rate constant uncertainties approximately 2-3 units at second significant figure for Reference 59 and 60 data.

Perhaps the most obvious use of a multiple frequency FT-FAM measurement is selection of an optimum frequency during <u>development</u> of an analytical procedure. A somewhat more sophisticated application of FT-FAM data would be assay methods based on direct observation of electrode reaction kinetics. For example, one can envision applications such as analyzing the water content of a nonaqueous solvent based on its rate of protonation of a proton-sensitive organic or organometallic electrode reaction product in a manner analogous to use of homogeneous reaction rate assays. One also can imagine the assay of chloride ion based on its profound kinetic influence on the $Bi^{3+}/Bi(Hg)$ heterogeneous charge transfer rate[10] or the assay of surfactants based on their major effect on the $Cd^{2+}/Cd(Hg)$ heterogeneous rate.[65] This type of assay procedure based on electrochemical kinetics is relatively unused, particularly where k_s measurements are concerned. Nevertheless, the potentialities are definitely nontrivial.

Possibly less obvious than the foregoing is a procedure we refer to as "kinetic deconvolution." We have examined this method carefully in the situation where heterogeneous charge transfer kinetics represent the kinetic problem. Other classes of "kinetic problems," such as coupled chemical reactions, also could be addressed by a more complicated procedure, but this has not yet been attempted. The basic concept involves use of the detailed kinetic-mechanistic information content of admittance spectra to effect a highly sensitive, computerized monitoring of the status of electrode reactions employed for assay work. Through such procedures the operator can be informed of deviations from expected response behavior, which might arise from unusual sample background constituents, improper cell preparation or a subtle instrument malfunction. Such response perturbations can invalidate an assay run, but may go unnoticed with the conventional single frequency or single time point readout modes. The possibility of using the observed response spectrum as a basis for automatically correcting undesirable kinetic induced fluctuations in the course of an assay procedure is viewed by this writer as highly important. There is no question that, in <u>real life</u> electrochemical assay procedures, one encounters problems with surfactants, and other types of excipients that are not present in the standard. Frequently, the problem will manifest itself as nonreproducible k_s and α values causing assay errors if not detected, as is often the case. If detected, various chemical procedures such as solvent variation, extraction, etc., are available, usually at significant cost in assay time. What we have proposed and demonstrated[66] is that use of the multiple frequency FT-FAM data allows complete correction ($\pm$ noise) for heterogeneous charge transfer kinetic influence for systems which one would consider good candidates for electrochemical analysis.

The procedure employed assumes that the heterogeneous charge transfer process is quasi-reversible on the a.c. time scale and reversible (nernstian) on the d.c. time scale (quasi-reversible systems on the d.c. time scale normally are not selected for assay work). Under these conditions, the faradaic rate law may be written as

$$| A(\omega t) | = A_{rev} G(\omega) \qquad [1]$$

$$G(\omega) = \left[\frac{2}{[1 + (\cot\phi)^2]} \right]^{1/2} \qquad [2]$$

A_{rev} is the reversible faradaic admittance magnitude and $G(\omega)$ is the kinetic correction which is related to the <u>observed</u> linear $\cot\phi$ spectrum as shown in Equation 2 ($\cot\phi$ contains the important thermodynamic ($E°$, E_{dc}, n, T, F, R) and kinetic parameters (k_s, α, and diffusion coefficients). All one does is rearrange Equation 1 to the form

$$A_{rev} = \frac{| A(\omega t) |}{G(\omega)} \qquad [3]$$

use the observed $\cot\phi$ peak spectrum to compute $G(\omega)$ and divide in accord with Equation 3 to obtain the heterogeneous kinetic deconvoluted reversible response, which is linear with $\omega^{1/2}$. Figure 27A shows an original FT-FAM peak response spectrum and the kinetic deconvoluted counterpart plotted versus $\omega^{1/2}$ for the $Cd^{2+}/Cd(Hg)$-1.0 M $ZnSO_4$ system. In this context our normal procedure is to divide the admittance by $\omega^{1/2}$ (Figure 27B) and plot $A(\omega t)/\omega^{1/2}$ vs. $\omega^{1/2}$. This presentation mode theoretically gives a horizontal line for the kinetic deconvoluted spectrum, which is matched quite well experimentally, although not perfectly. The average magnitude of the horizontal line is proportional to concentration, allowing one to use all the peak frequencies for a statistically more sound assay result than one would obtain with a single frequency measurement. Some before and after peak kinetic deconvolution data are given for four other systems in Figure 28. Figure 29 shows some results of kinetic

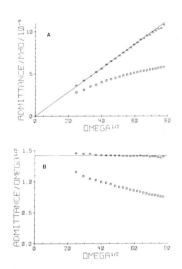

Figure 27. Peak faradaic admittance spectra for $Cd^{2+}/Cd(Hg)$ system in 1.0 M $ZnSO_4$, showing observed admittance and computed A_{rev} values. System: 1.0×10^{-3} M Cd^{+2} at DME-aqueous 1.0 M $ZnSO_4$ interface. Applied: Pseudo-random, odd-harmonic a.c. waveform with 1.5 mV per frequency component, 32 components; superimposed on staircase d.c. ramp synchronized to DME. Measured: (A) Peak $|A(\omega t)| - \omega^{1/2}$ profile (□) and computed A_{rev} counterpart (◇) at 3.0 s in life of DME, 10 replicates. (B) Frequency normalized presentation of data in Figure 3A.

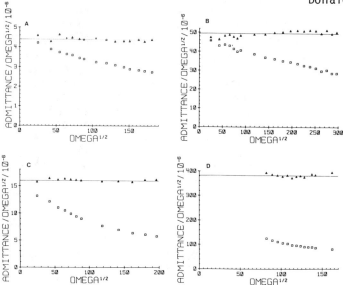

Figure 28. Frequency normalized peak faradaic admittance spectra for other systems investigated. Systems: (A) 2.0×10^{-3} M Cd^{2+} at DME-aqueous 1.0 M Na_2SO_4 interface, 25°C; peak admittance, 5 replicates. (B) 1.0×10^{-3} M $Cr(CN)_6^{3-}$ at HMDE-aqueous 1.0 M KCN interface, 25°C; peak admittance, one pass. (C) 1.0×10^{-3} M pyrene at DME-0.1 M TBAP, acetonitrile interface, 25°C; peak admittance, 2 replicates. (D) 1.0×10^{-3}, $TMPD^+$ at Pt-0.1 M TBAP, acetonitrile interface, 25°C; crossover point admittance, one pass. Applied: Same as Figure 27. Measured: Same as Figure 27B. □ = raw data, ▲ = kinetically deconvoluted data.

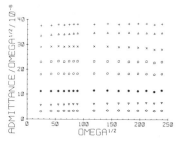

Figure 29. Frequency normalized A_{rev} spectra at various E_{dc}-values along faradaic admittance wave for $Cr(CN)_6^{3-}/Cr(CN)_6^{4-}$ system in 1.0 M KCN. System: As in Figure 21, except at DME. Applied: Same as in Figure 27. Measured: $A_{rev}/\omega^{1/2}$ spectra (5 replicates) at E_{dc} = -1.185 (◇), -1.220 (*), -1.245 (□), -1.280 (+), -1.300 (△), -1.310 (X), -1.330 (o), -1.365 (▽). A_{rev} spectra computed using least-squares line through $\cot\phi - \omega^{1/2}$ data, rather than raw data.

deconvolution along the faradaic admittance polarogram (varying E_{dc}) of the $Cr(CN)_6^{3-}/Cr(CN)_6^{4-}$ system in 1.0 M KCN at a Hg electrode. In our hands it appears that the heterogeneous charge transfer kinetic deconvolution is completely successful in both aqueous and aprotic solvents with both Hg and Pt working electrodes.

Drug Analysis Using Aprotic Organic Solvents

Various electrochemical methods of analysis have been applied to organic pharmaceuticals. Most of the work has been performed in protic solvents such as water and/or alcohols. Very little work[67,68] had been done in aprotic organic solvents, despite several major advantages of such solvent systems. Electrode adsorption of the reactant and/or product of reduction or oxidation is minimized or eliminated by the solvating ability of many aprotic organic solvents. Higher solubility means that the solvent has greater ability to extract an organic compound from a solid matrix, e.g., a drug from a tablet formulation. Reduction or oxidation of organic molecules with conjugated π-electron systems usually results in a simple, facile one electron transfer forming a radical anion or cation. In most cases the radical ions are stable, particularly the radical anions. Aprotic solvents also offer a greater d.c. potential range for measurement than protic solvents, allowing oxidation or reduction waves to be observed which are not observed in protic solvents.

Recent work in our laboratory has examined the reduction of the alkaloids reserpine and colchicine, the steroids testosterone, methyltestosterone, progesterone, prednisolone and hydrocortisone and the sulfa drugs, sulfadiazine and sulfamerizine.[69,70] Molecular structures and uses are given in Table 7. Solvents employed were acetonitrile (ACN) and dimethylformamide (DMF). The recommended supporting electrolyte is 1.0×10^{-2} M tetraethylammonium perchlorate (TEA$^+$BF$_4^-$), although the corresponding tetrabutylammonium salt (TBA$^+$BF$_4^-$) also was used. The unusually low supporting electrolyte concentration was justified on the basis of electrolyte cost savings, the fact that the electroactive drug species concentration always was lower than 10^{-3} M and usually between 5×10^{-5} and 1×10^{-4} M for assay conditions, and because the enhanced ohmic potential drop (≈ 3000 ohms) caused no special problems as far as analysis is concerned. Both FT-FAM and single frequency analog a.c. instruments were used in this work. It had been our intention to use the heterogeneous kinetic deconvolution procedure described above as part of the data refinement process. While certain drugs exhibited follow-up chemical reactions, the one electron reduction processes were so facile that heterogeneous charge transfer kinetic correction was judged to be unnecessary. Finally, the only purification step used was drying the solvent with a column of activated alumina.

Table 7. Structures and uses of various steroids

Testosterone

17β-Hydroxyandrost-4-ene-3-one

Testosterone is used for the eunuchoid male to stimulate and maintain the secondary sexual characteristics associated with the adult male.

Methyltestosterone

17-Methyl-17β-hydroxyandrost-4-ene-3-one

Methyltestosterone is used for the eunuchoid male as testosterone above.

Progesterone

Pregna-4-ene-3,20-dione

Progesterone is used in amenorrhea, and abnormal uterine bleeding due to hormonal imbalance in the absence of organic pathology.

Prednisolone

11β,17,21-Trihydroxypregna-1,4-diene-3,20-dione

Prednisolone is an adrenocortical steroid and is used as an anti-inflammatory agent.

Hydrocortisone

11β,17,21-Trihydroxypregna-4-ene-3,20-dione

Hydrocortisone is an adrenocortical steroid and is used as an anti-inflammatory agent.

Table 7. Structures and proper chemical names for reserpine and colchicine

Reserpine

11,17α-Dimethoxy-18β-[(3,4,5-trimethoxybenzoyl)oxy]-3β,20α-yohimban-16β-carboxylic acid methyl ester

Reserpine is an alkaloid used in relieving anxiety, tension, and headache in the hypertensive patient.

Colchicine

(S)-N-(5,6,7,9-Tetrahydro-1,2,3,10-tetramethoxy-9-oxobenzo[a]heptalen-7-yl)acetamide

Colchicine is an alkaloid used for the treatment of gout.

Table 7. Structures and uses of sulfa drugs

Sulfadiazine

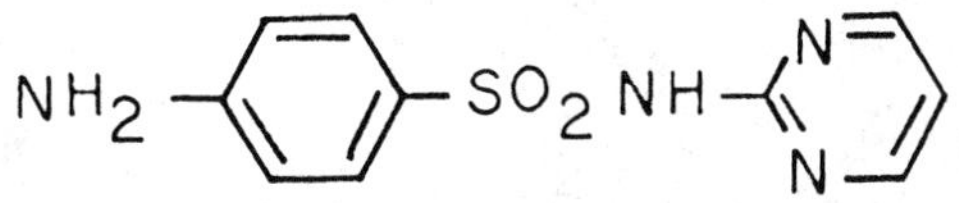

N^1-2-Pyrimidinylsulfanilamide

Sulfadiazine is an antibacterial drug used for the treatment of urinary tract infections.

Sulfamerazine

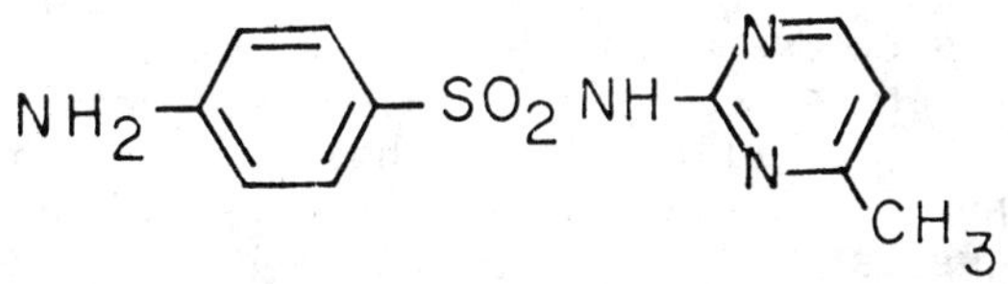

N^1-(4-methyl-2-pyrimidinyl)sulfanilamide

Sulfamerazine is an antibacterial drug used for the treatment of urinary tract infections.

Some typical results obtained for reserpine, methyltestosterone, and prednisolone with dry ACN and DMF will be given here, including composite and single tablet assays. Reserpine and methyltestosterone may be considered as representative of nearly ideal systems for analysis. Prednisolone in dry DMF and ACN shows clear evidence for a follow-up second order chemical reaction, making it a nonideal system.

Figure 30 shows calibration curves for methyltestosterone and reserpine in dry ACN and DMF, as well as in aqueous base which is the best protic solvent available. The response is somewhat larger in ACN than in DMF because ACN is less viscous leading to faster mass transfer (diffusion). FT-FAM admittance polarograms for standards and tablet extracts are compared for methyltestosterone (Figure 31) and reserpine (Figure 32). It is evident that in both cases standard and tablet extracts give essentially identical results except for expected magnitude differences. Composite and single tablet assays using analog and FT-FAM instrumentation are given in Tables 8-10 for the methyltestosterone and reserpine. At the bottom of each table is given the USP assay result, also performed in our laboratory. It is clear that the overall agreement between single-frequency analog and FT-FAM instrumentation is good to excellent for both composites and single tablet assays.

Figure 33 depicts a prednisolone calibration curve in dry ACN. The nonlinearity of this profile is obvious. Because there is absolutely no indication of adsorption effects, we hypothesize that there is a following second order chemical reaction. Hydrocortisone gives similar behavior in dry aprotic solvents. Both of these steroids are characterized by an aliphatic alcohol group which is not associated with the other three ideally behaving steroids. This circumstantial evidence suggests that the anion radical electrode reaction product is being protonated by reactant diffusing toward the electrode. The likelihood that this is the process occurring is further enhanced by the fact that addition of sufficient amounts of a weak proton donor (e.g., H_2O) leads to a linear calibration curve with lower sensitivity, as shown in Figure 34. The latter figure illustrates calibration curves for dry and wet ACN and DMF, as well as aqueous base. For correct evaluation of Figure 34, it is important to recognize that abscissa sensitivity is one order-of-magnitude lower than in Figure 33. One finds that only wet ACN and DMF give calibration curves which appear linear and pass through the origin. What we believe is occurring is that the added H_2O donates protons to the steroids via a pseudo-first order process, short-circuiting the second order reaction occurring in dry ACN and DMF. Further investigation is necessary to prove this hypothesis. It is well known that first order coupled chemical reactions lead to linear calibration curves. Despite these difficulties the use of wet aprotic solvents yields sensitivities which are still significantly better than the best protic media, even though they cannot compare to the so-called

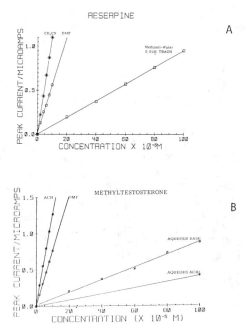

Figure 30. In-phase fundamental-harmonic a.c. polarographic peak current vs. concentration profiles for reserpine and methyltestosterone. System: (A) CH_3CN: 2,4,6,8,10 x 10^{-5} M reserpine at DME, 2s drop life, 0.01 M TEAPFB in CH_3CN, (A) DMF: 2,4,6,8,10 x 10^{-5} M reserpine at DME, 2s drop life, 0.01 M TEAPFB in DMF, (A) Methanol-water 0.01 M TBAOH: 2,4,6,8,10 x 10^{-4} M reserpine at DME, 2s drop life, 0.01 M TBAOH in 60% methanol-40% water, (B) CH_3CN: 2,4,6,8,10 x 10^{-5} M methyltestosterone at DME, 2s drop life, 0.01 M TEAPFB in CH_3CN, (B) DMF: 2,4,6,8,10 x 10^{-5} M methyltestosterone at DME, 2s drop life, 0.01 M TEAPFB in DME, (B) Aqueous Base: 2,4,6,8,10 x 10^{-4} M methyltestosterone at DME, 2s drop life, 0.01 M TBAOH in 60% methanol-40% water, (B) Aqueous Acid: 10 x 10^{-4} M methyltestosterone at DME, 2s drop life, 0.1 M TBAI in 50% ethanol-50% aqueous 0.1 M HCl, Hg-solution interface, room temperature, 25°C±2°C. Applied: 30 mV pp a.c. waveform, 100 Hz, superimposed on a linear 100 mV/min d.c. ramp, except (A) Methanol-Water 0.01 M TBAOH, 200 mV/min d.c. ramp. Measured: 100 Hz in-phase current.

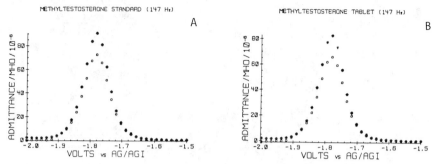

Figure 31. Fundamental-harmonic a.c. polarograms of a methyltestosterone standard and a tablet extract. System: (A) 1.3×10^{-4} M methyltestosterone standard at DME, 0.01 M TEAPFB in ACN interface, 2s drop life, room temperature, $25° \pm 2°C$. (B) same as (A) except methyltestosterone extracted from a tablet at about 1.3×10^{-4} M. Applied: pseudo-random, odd-harmonic a.c. waveform, 27 components, superimposed on a staircase d.c. scan. D.C. potentials are in V vs. Ag/AgI, saturated TEAI in ACN. Measured: 5 replicate average of faradaic admittance magnitude at 147 Hz. Notation: * = in-phase, o = quadrature.

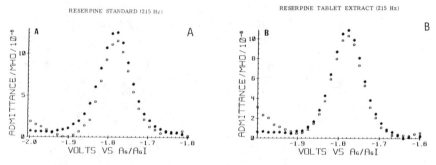

Figure 32. Fundamental-harmonic a.c. polarograms of a reserpine standard and a tablet extract. System: (A) 1.62×10^{-5} M reserpine standard. (B) 1.65×10^{-5} M reserpine extracted from a tablet, at DME, 0.01 M TEAPFB-ACN interface, 2s drop life, room temperature, $25° \pm 2°C$. Applied: pseudo-random, odd-harmonic a.c. waveform, 30 components superimposed on a staircase d.c. scan. Measured: 5 replicate average of faradaic admittance magnitude at 215 Hz. Notation: * = in-phase, o = quadrature.

Table 8. Assay results for methyltestosterone (25 mg tablet)

Analog System

	Composite % Label Claim	Single Tablet % Label Claim
	96.4	94.4
	98.4	94.5
	98.4	96.6
	99.0	94.8
	98.0	93.4
	96.4	89.8
	96.6	91.5
	98.8	92.5
	95.8	90.4
	96.2	93.2
Average:	97.4	93.1
Standard Deviation:	1.2	2.1
Relative Standard Deviation:	1.3%	2.3%

Computerized FFT System

	Composite % Label Claim	Single Tablet % Label Claim
	93.1	90.7
	94.0	87.6
	89.8	96.6
	94.2	93.6
	95.4	94.2
	91.8	90.2
	105.1*	91.6
	91.5	91.8
	92.1	87.9
	91.4	92.3
Average:	92.6	91.7
Standard Deviation:	1.7	2.8
Relative Standard Deviation:	1.9%	3.0%

USP Assay: 91.0% Label Claim

* = Value rejected by Q test, 90% confidence, not used for calculations.

Table 9. Assay results for reserpine (0.25 mg tablet)

Analog System

	Composite % Label Claim	Single Tablet % Label Claim
	94.2	97.3
	94.7	98.1
	95.1	99.0
	93.6	97.6
	92.3	96.1
	92.6	98.1
	88.0*	96.9
	94.2	96.9
	93.9	95.6
	93.4	95.9
Average:	93.9	97.2
Standard Deviation:	0.9	1.1
Relative Standard Deviation:	1.0%	1.2%

Computerized FFT System

	Composite % Label Claim	Single Tablet % Label Claim
	86.5*	94.1
	91.3	106.9
	92.5	106.5
	93.8	98.2
	95.0	101.0
	92.0	95.0
	94.7	95.0
	96.5	96.9
	97.9	90.6
	98.1	90.4
Average:	94.7	97.2
Standard Deviation:	2.5	6.1
Relative Standard Deviation:	2.6%	6.2%

* = Value rejected by Q test, 90% confidence, not used for calculations.

USP Assay: 97.7% Label Claim (average of two runs).

Table 10. Assay results for reserpine (0.1 mg tablet)

Analog System

	Composite % Label Claim	Single Tablet % Label Claim
	99.8	99.1
	97.2	108.3
	99.3	101.5
	98.8	85.8
	99.3	94.9
	95.6	88.0
	99.8	97.6
	96.7	91.1
	98.8	95.8
	97.0	93.5
Average:	98.2	95.6
Standard Deviation:	1.5	6.6
Relative Standard Deviation:	1.5%	6.9%

Computerized FFT System

	Composite % Label Claim	Single Tablet % Label Claim
	82.4	87.2
	86.1	88.9
	83.6	88.6
	86.9	86.8
	90.4	89.2
	88.7	84.3
	86.5	89.5
	91.5	82.2
	89.2	97.6
	92.8	86.6
Average:	87.8	88.1
Standard Deviation:	3.3	4.1
Relative Standard Deviation:	3.8%	4.6%

USP Assay not run.

Table 11. Assay results for prednisolone (5 mg tablet)

Analog System

	Composite % Label Claim	Single Tablet % Label Claim
	103.0	95.8
	100.6	99.0
	103.6	100.2
	103.6	99.2
	101.0	99.8
	104.4	99.8
	105.4	98.4
	103.4	101.6
	101.2	98.0
	98.8	93.8
Average:	102.4	98.4
Standard Deviation:	2.0	2.3
Relative Standard Deviation:	2.0%	2.4%

Computerized FFT System

	Composite % Label Claim	Single Tablet % Label Claim
	100.8	99.8
	101.4	99.8
	104.0	99.4
	100.0	99.2
	99.4	100.2
	106.8	100.4
	101.6	99.2
	100.0	102.0
	100.4	97.8
	99.4	100.2
Average:	101.4	99.8
Standard Deviation:	2.3	1.1
Relative Standard Deviation:	2.3%	1.1%

USP Assay: 98.2% Label Claim

"ideal" systems like reserpine and methyltestosterone. Figure 35 compares admittance polarograms for a prednisolone standard and tablet extract using wet ACN (3% H_2O), again demonstrating remarkable agreement and further enhancing the concept that appropriate organic solvents have superior extracting and dissolution ability for organic drugs. Composite and single tablet assay results are given in Table 11 for prednisolone. Results are again excellent.

Figure 36 provides an analog in-phase a.c. polarogram of 6.3 x 10^{-7} M methyltestostone indicating special sensitivity advantages of aprotic solvents.

Full details regarding assay solution preparation methods are given in the original publications.[69,70]

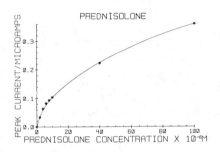

Figure 33. In-phase fundamental harmonic a.c. polarographic peak current vs. concentration profile for prednisolone in dry acetonitrile. System: 2,4,6,8,10,20,40,100 x 10^{-6} M prednisolone at DME, 2s drop life. 0.01 M TEAPFB in dry ACN interface, room temperature 25°+2°C. Applied: 30 mV p-p a.c. waveform, 100 Hz superimposed on a linear d.c ramp of either 200 mV/min (for 2,4,6,8,10 x 10^{-6} M) or 100 mV/min (for 20,40,100 x 10^{-6} M). Measured: 100 Hz in-phase current.

CONCLUSIONS

This monograph has attempted to bring the reader up to date on the current status of FT-FAM measurements. We have tried to point out advantages and disadvantages of the instrument system we are using. The advantages include high speed time domain waveform acquistion, which eventually leads to the Fourier domain results. The software contains a variety of Fourier transform based data

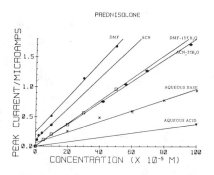

Figure 34. In-phase fundamental-harmonic a.c. polarographic peak current vs. concentration profiles for prednisolone. System: DMF: 1,3,5 x 10^{-4} $\underline{M}$ prednisolone at DME, 2 s drop life, 0.01 $\underline{M}$ TEAPFB in DMF, ACN: 1,2,4,10 x 10^{-5} $\underline{M}$ prednisolone at DME, 2s drop life, 0.01 $\underline{M}$ TEAPFB in ACN, DMF-15% H_2O: 5,10,20,30,50 x 10^{-5} $\underline{M}$ prednisolone at DME, 2s drop life, 0.01 $\underline{M}$ TEAPFB in 15% water-DMF, ACN-3% H_2O: 4.2,5.6,6.9,9.6 x 10^{-4} $\underline{M}$ prednisolone at DME, AQUEOUS BASE: 2,4,6,8,10 x 10^{-4} $\underline{M}$ prednisolone at DME, 2s drop life, 0.01 $\underline{M}$ TBAOH in 60% methanol-40% water, AQUEOUS ACID: 1 x 10^{-3} $\underline{M}$ prednisolone at DME, 2s drop life, 0.1 $\underline{M}$ TBAI in 50% aqueous 0.1 M HCl, Hg-solution interface, room temperature 25°C+2°C. Applied: ACN-3% H_2O: 20 mV p-p a.c. waveform, 250 Hz, superimposed on a linear 100 mV/min d.c. ramp, ALL OTHERS: 30 mV p-p a.c. waveform, 100 Hz, superimposed on a linear 100 mV/min d.c. ramp. Measured: ACN-3% H_2O: 250 Hz in-phase peak current, ALL OTHERS: 100 Hz in-phase peak current.

enhancement procedures. These capabilities are being applied to
thermodynamic and rate measurements of homogeneous and heterogeneous
rate processes, as well as drug analysis using a unique electroyte
media. Without question the instrumentation is as efficient as we
have ever used for small amplitude a.c. measurements.

The principal problem with the described instrumentation is FFT
computational speed (0.8 s for one 1024 word array) which forces us

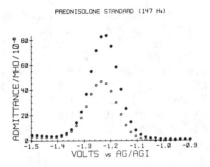

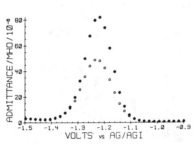

Figure 35. Fundamental-harmonic a.c. polarograms of a prednisolone
standard and a tablet extract. System: (A) 6.9×10^{-4}
$\underline{M}$ prednisolone standard at DME, 0.01 $\underline{M}$ TEAPFB in 3%
water in ACN interface, 1s drop life, room temperature
25°+2°C. (B) same as (A) except prednisolone extracted
from a single tablet at about 6.9×10^{-4} $\underline{M}$. Applied:
pseudo-random, odd-harmonic a.c. waveform, 27 components,
superimposed on a staircase d.c. scan. D.C. potentials
are in V vs. Ag/AgI, saturated TEAI in ACN. Measured:
5 replicate average of faradaic admittance at 147 Hz.
Notation: * = in-phase, o = quadrature.

to store two 1024 word (16 bits) time domain data arrays for each spectral measurement performed and process the data after the experimental phase is completed. The worst aspect of this procedure is the extremely inefficient use of disk memory. An obvious solution to this problem is a high speed array processor. Such a device has been funded and is on order. Perhaps a future monograph will outline the benefits of array processor technology in the context of electrochemical relaxation techniques.

ACKNOWLEDGEMENTS

The following figures have been reprinted with permission of the American Chemical Society: 1 (ref. 71); 2 (ref. 22); 3, 4, 19 and 20 (ref. 27); 9, 10 and 11 (ref. 72); 12 and 13 (ref. 73); 18 (ref. 31); 24 (ref. 53), 25 and 26 (ref. 55); 27, 28 and 29 (ref. 66). The following figures have been reprinted with permission of the Elsevier Scientific Publishing Company: 7 and 8 (ref. 38); 15, 16 and 17 (ref. 74); 21 and 22 (ref. 51); 23 (ref. 75); and 30 (ref. 69). The following tables have been reprinted with permission of the Elsevier Scientific Publishing Company: 2, 3 and 4 (ref. 38); and 9 (ref. 69).

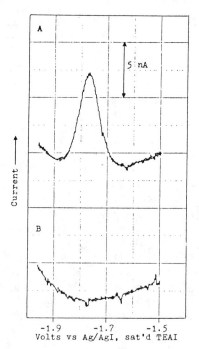

Figure 36. In-phase fundamental-harmonic a.c. polarogram of 6.3 x 10^{-7} M (190 ng/ml) methyltestosterone.

BIBLIOGRAPHY

1. Bard, A. J., and Faulkner, L. R., 1980, <u>Electrochemical Methods</u>, John Wiley and Sons, New York, N.Y.
2. Bard, A. J., Ed., <u>Electroanalytical Chemistry</u>, M. Dekker, Inc., New York, N.Y., ALL VOLUMES.
3. Bond, A. M., 1980, <u>Modern Polarographic Methods in Analytical Chemistry</u>, M. Dekker, Inc., New York, N.Y.
4. Delahay, P. and Tobias, C. W., <u>Advances in Electrochemistry and Electrochemical Engineering</u>, Wiley-Interscience, New York, N.Y., ALL VOLUMES.
5. Herovsky, J. and Kuta, J., 1966, <u>Principles of Polarography</u>, Academic Press, New York, N.Y.
6. MacDonald, D. D., 1977, <u>Transient Techniques in Electrochemistry</u>, Plenum Press, New York, N.Y.
7. Gerischer, H., 1951, Zeit. Physik. Chem., $\underline{198}$, 286.
8. Gerischer, H., 1951, Zeit. Electrochem., $\underline{55}$, 98.
9. Gerischer, H., 1952, Zeit. Physik. Chem., $\underline{201}$, 55.
10. Randles, J. E. B., 1947, Disc. Faraday Soc., $\underline{1}$, 11, 47.
11. Randles, J. E. B. and Somerton, K. W., 1952, Trans. Faraday Soc., $\underline{48}$, 937, 951.
12. Breyer, B. and Gutman, F, 1947, Disc. Faraday Soc., $\underline{1}$, 19.
13. Breyer, B. and Bauer, H. H., 1963, <u>Alternating Current Polargraphy and Tensammetry</u>, Wiley-Interscience, New York, N.Y.
14. Delahay, P., 1954, <u>New Instrumental Methods in Electrochemistry</u>, Interscience, New York, N.Y.
15. Smith, D. E. in Bard, A. J., Ed., 1966, <u>Electroanalytical Chemistry</u>, Vol. I, M. Dekker, Inc., New York, N.Y., Chapt. 1.
16. Sluyters-Rehbach, M. and Sluyters, J. H. in Bard, A. J., Ed., 1970, <u>Electroanalytical Chemistry</u>, Vol. IV, M. Dekker, Inc., New York, N.Y, Chapt. 1.
17. Delahay, P., 1966, J. Electrochem. Soc., $\underline{113}$, 967 and discussion.
18. Timmer, B., Sluyters-Rehbach, M. and Sluyters, J. H., 1970, J. Electroanal. Chem., $\underline{24}$, 287.
19. Brown, E. R., McCord, T. G., Smith, D. E. and DeFord, D. D., 1966, Anal. Chem., $\underline{38}$, 1119.
20. Brown, E. R., Smith, D. E. and Booman, G. L., 1968, Anal. Chem., $\underline{40}$, 1411.
21. Brown, E. R., Hung, H. L., McCord, T. G., Smith, D. E., and Booman, G. L., 1968, Anal. Chem., $\underline{40}$, 1424.
22. Huebert, B. J. and Smith, D. E., 1972, Anal. Chem., $\underline{44}$, 1179.
23. Pilla, A. A., 1970, J. Electrochem. Soc., $\underline{117}$, 467.
24. Pilla, A. A., 1971, J. Electrochem. Soc., $\underline{118}$, 702.
25. deLevie, R., Thomas, J. W. and Abbey, K. M., 1975, J. Electroanal. Chem., $\underline{62}$, 111.
26. Pilla, A. A., in Mattson, J. S., MacDonald, J. D., Jr. and Mark, H. B., Jr., 1972, <u>Computers in Chemistry and Instrumentation</u>, Vol. 2, M. Dekker, New York, N.Y., Chapt. 6.

27. Schwall, R. J., Bond, A. M., Loyd, R. J., Larsen, J. G. and Smith, D. E., 1977, Anal. Chem., 49, 1797.
28. K. F. Drake, 1979, Doctoral Dissertation, Northwestern University, Evanston, IL.
29. Smith, D. E., 1971, CRC Critical Rev. Anal. Chem., 2, 247.
30. Smith, D. E., Borchers, C. E. and Loyd, R. J., 1971, Physical Methods in Chemistry, Part 1B, Wiley and Sons, Inc., New York, N.Y., Chapt. 8.
31. Creason, S. C. and Smith, D. E., 1973, Anal. Chem., 45, 2401.
32. Birke, R. L., 1971, Anal. Chem.,43, 1253.
33. Gabrielli, C. and Keddam, M., 1974, Electrochem. Acta., 19, 355.
34. Bechet, B., Epelboin, I. and Keddam, M., 1977, J. Electroanal. Chem., 76, 129.
35. Brigham, E. O., 1974, The Fast Fourier Transform, Prentice-Hall, Inc., Englewood Cliffs, N.J.
36. Gold, B and Rader, C. M., 1969, Digital Processing of Signals, McGraw-Hill, New York, N.Y.
37. Otnes, R. K. and Enochson, L., 1971, Digital Time Series Analysis, Wiley-Interscience, New York, N.Y.
38. Creason, S. C., Hayes, J. W. and Smith, D. E., 1973, J. Electroanal. Chem., 47, 9.
39. Matsuda, H., 1958, Zeit. Elektrochem., 62, 977.
40. Lephardt, J. O., in Griffiths, P., Ed., 1978, Transform Techniques in Chemistry, Plenum Press, New York, N.Y., Chapt. 11.
41. Griffiths, P. R., 1975, Appl. Spectrosc., 29, 11.
42. Horlick, G. and Yuen, W. K., 1976, Anal. Chem., 48, 1643.
43. Bond, A. M., O'Halloran, R. J., Ruzic, I. and Smith, D. E., 1976, Anal. Chem., 48, 872.
44. Erdy-Gruz, T. and Volmer, M., 1930, Zeit. Physik. Chem., 150A, 203.
45. Kirmse, K. W. and Westerberger, A. W., 1971, Anal. Chem., 43, 1035.
46. Horlick, G., 1972, Anal. Chem., 44, 943.
47. Grabaric, B. S., O'Halloran, R. J. and Smith, D. E., Anal. Chim. Acta., (Section on Computer Techniques and Optimization).
48. deLeeuwe, R., Sluyters-Rehbach, M. and Sluyters, J. H., 1969, Electrochim. Acta., 14, 1183.
49. Rohko, T., Kogoma, M. and Aoyagui, S., 1972, J. Electroanal. Chem., 38, 45.
50. Peover, M. E. and Powell, J. S., 1969, J. Electroanal. Chem., 20, 427.
51. Bond, A. M., Schwall, R. J. and Smith, D. E., 1977, J. Electro-Chem., 85, 231.
52. Grabaric, B. S. and Smith, D. E., in preparation.
53. O'Halloran, R. J., Schaar, J. C. and Smith, D. E., 1978, Anal. Chem., 50, 1073.

54. Hayes, J. W. and Smith, D. E., 1980, J. Electroanal. Chem., 114, 283.
55. Hayes, J. W. and Smith, D. E., 1980, J. Electroanal. Chem., 114, 293.
56. Ruzic, I., Smith, D. E. and Feldberg, S. W., 1974, J. Electro-Anal. Chem., 52, 157.
57. Schwall, R. J., Ruzic, I. and Smith, D. E., 1975, J. Electroanal. Chem., 60, 117.
58. Schwall, R. J. and Smith, D. E., 1978, J. Electroanal. Chem., 94, 227.
59. Matusinovic, T. and Smith, D. E., 1979, J. Electroanal. Chem., 98, 133.
60. Matusinovic, T. and Smith, D. E., in press, Inorg. Chem.,
61. Candlin, J. P., Halperin, J. and Trimm, D. L., 1964, J. Amer. Chem. Soc., 86, 1019.
62. Faragi, M. and Feder, A., 1973, Inorg. Chem., 12, 236.
63. Schwall, R. J., Ruzic, I. and Smith, D. E., 1975, J. Electroanal. Chem., 60, 117.
64. Dulz, G. and Sutin, N., 1964, J. Amer. Chem. Soc., 86, 829.
65. Flato, J. B., 1972, Anal. Chem., 44 (11), 75A.
66. Schwall, R. J., Bond, A. M. and Smith, D. E., 1977, Anal. Chem., 49, 1805.
67. Woodson, A. L. and Smith, D. E., 1970, Anal. Chem., 42, 242.
68. Taira, A. and Smith, D.E., 1978, J. Assoc. Official Anal. Chem., 61, 941.
69. Schaar, J. C. and Smith, D. E., 1979, J. Electroanal. Chem., 100, 145.
70. Schaar, J. C., 1981, Doctoral Dissertation, Northwestern University, Evanston, IL.
71. Reprinted with permission of ACS, 1963, Anal. Chem., 35, 1811.
72. Reprinted with permission of ACS, 1973, Anal. Chem., 45, 277.
73. Reprinted with permission of ACS, 1978, Anal. Chem., 50, 1391.
74. Reprinted with permission of Elsevier Scientific Publishing Company, 1981, Anal. Chim. Acta., ACA Comp., 133, 349.
75. Reprinted with permission of Elsevier Scientific Publishing Company, J. Electroanal. Chem., in press.

OPTICAL DIFFRACTION BY ELECTRODES: USE OF FOURIER TRANSFORMS IN SPECTROELECTROCHEMISTRY

Richard L. McCreery and Paula Rossi

Department of Chemistry
The Ohio State University
Columbus, OH 43210

INTRODUCTION

The information content of an electrochemical experiment can be greatly enhanced by the addition of an optical probe to provide spectral information about material generated at the electrode surface. Methods combining UV-visible spectroscopy with electrochemistry were developed in the 1960's, mainly by Kuwana[1,2] and Murray[3] and have proved very valuable for the characterization and monitoring of electrogenerated species. The technique described in this chapter also involves UV-visible spectroscopy of electrochemical processes, but makes use of the diffraction of light by an electrode. The diffraction pattern is a spatial Fourier transform of the illuminated electrode, and contains information about chromophores generated by an electrode process. Advantages of the diffraction approach to spectroelectrochemistry include high sensitivity, fast time response, and the possibility of describing the spatial distribution of chromophore in solution. After a discussion of the objectives of the diffractive approach, its theory and experimental verification will be described.

It is an inherent property of electrochemistry that all the events of interest occur either at an electrode surface or in the solution very close to that electrode. The thickness of the relevant solution layer is of the order of 0.1 millimeter (100 micron) for an experiment lasting ten seconds, and much thinner for shorter experiments. The thinness of this solution layer puts severe constraints on the methods which can be used to monitor events occurring in it, such as mass transport and chemical reactions. The most common approach is to monitor the current, which is an indication of the surface concentration of electroactive species either present initially or generated by electrochemical processes. Measurement of current comprises the oldest and most familiar form of electrochemistry, and the resulting plot is usually one of current vs. potential or current vs. time.[4] There is an enormous amount of information to be gathered by monitoring the current, including the concentration of an electroactive species, something about its thermodynamic properties, and the presence of reactions which may be occurring after or preceding charge transfer. As an indication of the popularity of electrochemical techniques in which current is monitored, the paper laying the theoretical foundation of cyclic voltammetry[5] is the sixth most cited paper in chemistry. However, since any electrochemical method is based on a Nernstian process, methods monitoring current are fairly low-resolution techniques. For example, the width of a typical cyclic voltammetric peak shown in Figure 1 is of the order of 100 millivolt or more, depending upon the behavior of the system. Yet the total scale available for observation is about 2 volt for a given solvent, leading to a resolution of only about 1 part in 20. It is apparent that it would be difficult to resolve a large number of peaks such as is possible with NMR or mass spectrometry. This deficiency leads to two drawbacks of electrochemistry when only current is monitored. One is that it is usually difficult to monitor trace solution components when a trace constituent occurs in the presence of other electroactive materials, as might be the case in a biological or environmental sample. Pulse voltammetry[6] has been used successfully for such samples at the ca. 10^{-8} M level, but the sample is usually subjected to a clean-up procedure, such as chromatography, to remove interferences. A second problem occurs in the area of organic reaction mechanisms which are initiated electrochemically, where several species may have oxidation or reduction potentials near the species of interest. It is often difficult to sort out a complex reaction mechanism in which several species have similar redox poten-

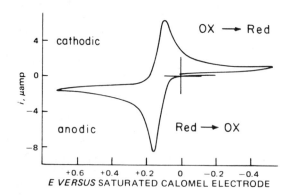

Figure 1. Typical cyclic voltammogram of a solution containing a redox couple in its reduced form. Potential (E) was scanned positive initially at 0.1 volt/sec.

tials which are not resolved by cyclic voltammetry. So although cyclic voltammetry and related methods have had wide applications, these methods have drawbacks when analyzing trace materials in mixtures or examining complex reactions.

When an optical probe is combined with electrochemistry, significant improvement in selectivity is acquired. A UV-visible spectrophotometer combined with an electrochemical apparatus can monitor both the optical properties of the materials being generated or consumed at the electrode as well as their electrochemical properties. A wide variety of applications has resulted from such combinations, including determining reaction mechanisms and monitoring kinetics.[2,7] The configurations of several such experiments are shown in Figure 2. The transparent electrode technique is the most

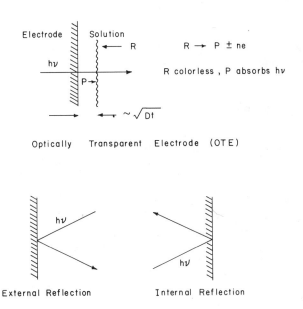

Figure 2. Three common configurations for spectroscopic monitoring of solution components generated at an electrode. With an optically transparent electrode, a beam passes through the electrode and absorbance is governed by an effective path length of approximately $\sqrt{Dt}$. The wavy line indicates an imaginary boundary layer of electrogenerated material (P) moving out into the solution. In the case of internal reflection, absorbance is determined by penetration of the evanescent wave into the solution. R indicates a reactant initially present in solution; P is an electrogenerated product.

common form of spectroelectrochemistry and has been developed quite extensively. The main advantage of such techniques is the increase in selectivity caused by the combination of two probes. If one is dealing with a complex reaction scheme in which several species may have similar redox potentials, one can selectively monitor a given species using its UV-visible absorption properties as well as redox properties. In addition, there may be species which are not electroactive which are involved as intermediates in the overall process, and can be monitored through their optical properties.

Although spectroelectrochemistry with transparent electrodes has been used extensively to obtain new information inaccessible to other methods, there remains room for significant improvement in the area of sensitivity. Consider reaction 1, in which R is a colorless reactant which is transformed electrochemically to some colored product P.

$$R \rightleftharpoons P \pm ne^- \qquad [1]$$

The absorbance of a beam passing through a transparent electrode is given by Equation 2

$$A(t) = \frac{2}{\pi^{1/2}} \epsilon_P \, C_R^b \, (D_R t)^{1/2} \qquad [2]$$

in which ϵ_P is the molar absorptivity of P, C_R^b is the bulk concentration of R, D_R its diffusion coefficient, and t the time from the beginning of electrolysis. Notice that by analogy to Beer's law, the effective path length for an optically transparent electrode is given by $(2/\pi^{1/2})(Dt)^{1/2}$. Typically, $D = 10^{-5}$ cm^2/sec and $t = 10$ sec, yielding an effective path length of 113 microns (0.0113 cm). Therefore, the absorbance is quite small for this technique, and decreases greatly at shorter times. Table 1 shows the absorbance for several times and concentrations for an electrogenerated species which has a fairly strong molar absorptivity of 10,000. Even for a strong chromophore, the absorbances are usually less than 0.01, and at low concentrations the absorbance is effectively unmeasurable. Because of this effective path length limitation, the optically transparent electrode technique has been used primarily for strong chromophores at moderate concentrations. It is not generally useful for very dilute solutions with concentrations less than 10^{-4} M, nor has it been used for weak chromophores. Although it is a very useful technique, improvements need to be made if it is to have the generality of other kinetic or analytical methods. If spectroelectrochemistry is to be a valuable method for analysis, its detection limit should be of the order of 10^{-6} or 10^{-8} M, rather than the present value of about 10^{-4} M.

Fourier Transforms in Spectroelectrochemistry

It is apparent from Figure 2 that techniques employing a beam passing through the entire diffusion layer will yield an absorbance which represents the integral of the concentration throughout the layer. This information is complementary to the measurement of current, since current can only reflect instantaneous concentrations at the electrode surface. Internal reflection spectroscopy also measures near-surface concentrations, since the light penetrates only about 1000 Å into the solution, while the diffusion layer is usually hundreds of times thicker. There have been several attempts to determine concentration as a function of distance from the electrode, rather than as an integrated concentration. Concentration vs. distance profiles, often called diffusion profiles, are fundamental to electrochemistry because they dictate the rate at which material reaches the electrode surface, and therefore determine the current. However, although they have been calculated extensively for different systems, different mass transport conditions, and different kinetic schemes, they have not been observed with a high degree of accuracy because of the thinness of the diffusion layer. The techniques that have been tried are based mainly on refractive index changes caused by the generation or removal of an absorbing chromophore.[8] In interferometry, for example, a beam of light passes parallel to a planar electrode surface, and the phase shifts induced by refractive index changes are measured. In a region of high concentration of electrogenerated species, there will be a relatively high change in refractive index, therefore a high phase shift and an apparent shift in the interferogram. Although techniques based on refractive index have been developed extensively, they do have some serious drawbacks when applied to many chemical systems. For example, interferometry requires a large enough change in refractive index to cause a perceptible change in the interferogram. Refractive index is not very sensitive to concentration, so fairly high concentrations of electroactive species are usually required. Interferometry has been applied successfully to high concentrations of copper sulfate, for example, where 0.1 M copper sulfate is reduced to copper metal on the surface

Table 1. Absorbance measured at an optically transparent electrode for chromophore generation from a colorless reactant.

Reactant concentration (M)	Generation Time (sec)		
	1	10^{-3}	10^{-5}
10^{-3}	0.036	1.1×10^{-3}	3.6×10^{-5}
10^{-5}	3.6×10^{-4}	1.1×10^{-5}	3.6×10^{-7}
10^{-7}	3.6×10^{-6}	1.1×10^{-7}	3.6×10^{-9}

$\epsilon_p = 10{,}000$ M^{-1} cm^{-1}, $D_R = 1 \times 10^{-5}$ cm^2/sec, chromophore generated by a potential step to a diffusion-limited potential. Absorbance calculated from Equation 2.

of the electrode. Concentration vs. distance profiles have been constructed from such experiments for certain systems, but the techniques cannot be applied to the dilute solutions of chromophores which are usually used in electroanalytical chemistry. If one has a complex reaction scheme in which three or more components are present, it would be very difficult to predict the changes in refractive index resulting from those reactions. With the case of copper sulfate, it is relatively straightforward to predict such changes because only one component is present. In addition, the concentrations used in copper sulfate experiments are much higher than those encountered in examining organic mechanisms. So the techniques based on refractive index changes are not generally applicable to the majority of electrochemical problems. The physical reasons for this lack of applicability are a lack of selectivity and sensitivity usually required for all but the simplest systems.

A method for monitoring diffusion profiles which is based on absorptivity rather than refractive index would have much greater generality, and one such method has been described.[9] Concentration vs. distance profiles were obtained for a dilute organic system, but the accuracy and time response of the method were limited by diffraction of the light near the electrode.[10] Despite difficulties with previous attempts to observe diffusion profiles, it remains an important goal to obtain accurate concentration vs. distance profiles for a variety of conditions and chemical systems. Such profiles are central to the area of mass transport, and contain information about reactions of electrogenerated species. Spatial resolution of the diffusion profile represents a new type of information, distinct from the integrated concentrations of transparent electrode experiments or from the surface concentrations derived from voltammetry or internal reflection spectroscopy.

Past efforts allow us to formulate three objectives for the present work. First we would like a technique that is roughly 2 to 4 (or more) orders of magnitude more sensitive than existing spectroelectrochemical methods. If this were achieved, the techniques could be applied to high-sensitivity analysis where one has a complex mixture and one makes use of the selectivity of spectroelectrochemistry. Second, it would be valuable to lower the usable time scale of spectroelectrochemistry down into the microsecond region for a variety of chemical systems. With an optically transparent electrode and virtually all spectroelectrochemical methods, the response is limited by an effective path length which decreases with the time scale. Therefore, it is very difficult to monitor species on a microsecond time scale simply due to the low sensitivity of the techniques. The third objective is spatial resolution of the diffusion layer. It would be very informative from both fundamental and practical standpoints to be able to accurately observe concentration vs. distance profiles.

Fourier Transforms in Spectroelectrochemistry

The approach described below uses diffraction of light by the electrode to deduce information about chromophores in the diffusion layer. The experiment is highly analogous to an x-ray diffraction experiment, with the only major difference being the wavelength domain of the light employed. Light passing close to an electrode surface is diffracted, and the diffracted light can be analyzed to provide information about the spectral properties, concentration, and spatial distribution of chromophore near the electrode. Figure 3 shows a simplified diagram of the diffraction approach. Clearly, the light passing distant from the electrode will not be diffracted and will not be affected by events occurring near the electrode. However, light passing close to the electrode (how close will be discussed later) will be diffracted, and some diffracted intensity will appear in the shadow of the electrode. Since the electrode is responsible for the diffracted light, any chromophore near the electrode will have a pronounced effect upon the diffraction pattern. The concentration, absorptivity, and spatial distribution of the chromophore will all affect the diffraction process.

As will be shown below, the diffraction pattern on the screen is a Fourier transform of the intensity distribution of the light passing the electrode after interacting with the chromophore. Therefore, the relationship between chromophore generation and diffracted intensity is embodied in the Fourier transform, and all that can be learned about events at the electrode can be learned by analyzing the diffraction pattern using Fourier transform techniques.

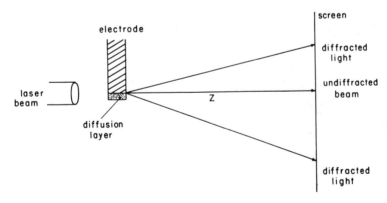

Figure 3. Simplified diagram of the diffraction experiment. The diffusion layer containing electrogenerated chromophore partially absorbs the beam, and affects the diffracted intensity.

THEORY

The diffraction experiment is shown in greater detail in Figure 4. The bottom edge of the electrode is the active region, and electrogenerated chromophore will diffuse downward in the Figure, as shown by the shaded area. The laser beam is partially occluded by the electrode, and the remaining part of the beam passes parallel to the active surface. During passage past the electrode, the beam is partially absorbed, depending upon the local concentration of chromophore near the electrode. Once past the electrode, the beam will impinge on the screen with the diffraction caused by the electrode edge being modified by the presence of chromophore. The objective of a theoretical analysis of this process is a description of the diffraction pattern of an electrode with and without a chromophore present.

A straightforward procedure for solving the diffraction problem is the use of Kirchoff-Huygens diffraction theory,[11] which has been used to solve a wide variety of similar problems. In this approach, a propagating wavefront (in this case, the laser beam) is assumed to

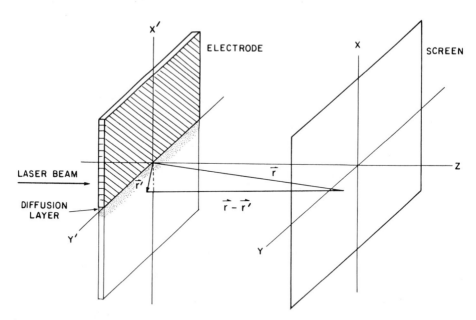

Figure 4. Coordinate system for determining the effect of electrochemical events on diffracted light. The shaded region represents electrogenerated chromophore.

consist of infinitesimally small spherical radiators, which interfere with each other as the wave propagates. At any point on the screen in Figure 4, the intensity is a summation of contributions from the radiators in the beam, taking interference into account. Thus each point in the beam, which has been attenuated by the electrode or chromophore, radiates spherically, and a given point on the screen will contain contributions from all points in the beam. The utility of this approach lies in the ability to assign each radiator in the beam a different amplitude, depending on how much it has been attenuated by the chromophore. In addition, any arbitrary electrode shape can be used, since an electrode is optically equivalent to a region in which the Huygens radiators have zero amplitude.

Using the coordinates indicated in Figure 4, the screen amplitude along one dimension (x) is given by

$$A(x) = \frac{ik}{2\pi} \int_{-\infty}^{\infty} \frac{e^{-ik(\vec{r}-\vec{r}')}}{|\vec{r}-\vec{r}'|} \phi(x') \, dx' \quad [3]$$

in which λ is the laser wavelength, k is $2\pi/\lambda$, and x' is the coordinate perpendicular to the electrode surface. A rigorous derivation would include the y-coordinate,[12] but diffraction in the y-direction is not important to the problem and results in an overall constant in the screen intensity. $\phi(x')$ is the amplitude function for the Huygens radiators and has the following form: $\phi(x')$ is zero for $x' > 0$, since the beam is blocked by the electrode; for $x' < 0$, $\phi(x')$ reflects the Gaussian shape of the beam, attenuated by any chromophore that may be present. The exponential term in Equation 3 represents a spherically radiating wave, with the function ϕ providing its source amplitude. Thus the whole integral is the summation of all radiators contributing to the point (x,0) on the screen.

The function ϕ can be calculated from Beer's law and the shape of the incident beam, using a path length equal to the thickness of the electrode along the beam axis. This calculation requires the assumption that the beam travels parallel to the electrode plane until it leaves the electrode region, an assumption which is strictly true only for very short electrodes. For the purpose of this derivation, it will be assumed that the light does travel parallel to the active surface, and experiences an attenuation according to Beer's law. The validity of this assumption will be examined later, but once it is made, an expression for $\phi(x')$ can be presented as Equation 4,

$$\phi(x') = I_0^{1/2}(x') \, 10^{-\epsilon b C(x')} \quad [4]$$

in which $I_0^{1/2}(x')$ is the incident amplitude as a function of x', ϵ the molar absorptivity of the chromophore, b the length of the electrode along the optical axis, and $C(x')$ the concentration of chromophore as a function of distance from the electrode.

For a given C(x'), Equation 3 can be evaluated to produce a diffraction pattern appearing at the screen. The integral may be solved for any set of experimental conditions, but the problem is simplified significantly if the screen is placed at some large distance away from the electrode. This is a standard approximation in solving diffraction problems known as the Fraunhofer approximation,[13] and can easily be realized experimentally. When this approximation is made, the expression for $|\vec{r}-\vec{r}'|$ is simplified, and Equation 3 becomes

$$A(x) = B \int_{-\infty}^{\infty} \phi(x') \, e^{i(\frac{kx}{z})x'} \, dx' \qquad [5]$$

in which B is a constant related to the overall beam amplitude. The integral in Equation 5 has the form of a complex Fourier transform of $\phi(x')$ to a function of x (or rather (kx/z)). Thus the Fraunhofer diffraction pattern of the electrode region is given by the Fourier transform of the amplitude distribution $\phi(x')$ created when the beam is truncated by the electrode and attenuated by the electrogenerated chromophore. It is a general rule of Fourier optics that a diffraction pattern can be calculated for any arbitrary object by taking the Fourier transform of the corresponding amplitude distribution $\phi(x)$.[14] For example, a single slit has an amplitude function shaped like a square pulse, and a single slit diffraction pattern is the Fourier transform of such a pulse. In the present case, diffraction from any arbitrary electrode shape or chromophore distribution can be calculated, provided $\phi(x')$ can be determined. Conversely, if the diffraction pattern can be measured, the shape of the function ϕ, and therefore the chromophore distribution, can be calculated. In the present discussion, it will be assumed that $\phi(x')$ is known, so that the diffraction pattern will be calculated and verified experimentally.

An informative physical analogy[15] to Equation 5 is presented in Figure 5. Consider a transmission diffraction grating with a sinusoidally varying optical density. The periodicity of the variation is assigned a "spatial frequency" ν_S. When a beam of monochromatic light falls on the grating, it is split into three components, the undiffracted beam, and diffracted beams above and below the main beam. The angle of the diffracted beam, and therefore the distance x on the screen, are directly related to ν_S, the spatial frequency of periodicity in the optical density of the grating. There is a one-to-one correspondence between x and ν_S, provided z and the wavelength of light are constant. Note that high diffraction angles correspond to a large screen coordinate x, the two being related by the screen distance.

Consider an arbitrary distribution of optical density, which in the present case may be an electrode with a chromophore in solution nearby. Any arbitrary distribution of optical density can be imagined to be constructed of sinusoidal components, similar to the diffraction grating above. Each component will generate a diffracted intensity at a particular x on the screen, yielding a complete diffraction pattern, with the higher diffraction angles (higher x) corresponding to higher frequency sinusoidal components. Thus, the intensity at each value of x on the screen represents the contribution of a particular sinusoidal component to the original optical density distribution, and the diffraction pattern is the spatial Fourier transform of the optical density distribution represented by the electrode and associated chromophore. So if one knows the optical density profile (and therefore $\phi(x')$), one can calculate the diffraction pattern, and vice versa. The diffraction pattern results from $\phi(x')$ which is a combination of optical density and incident beam intensity. The spatial frequency approach applies to the arbitrary amplitude profile $\phi(x')$ in a fashion similar to that for arbitrary optical density profiles.

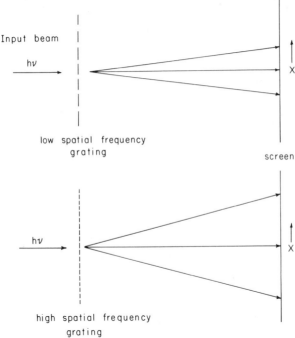

Figure 5. Effect of a transmission diffraction grating on the propagation of a monochromatic beam of light. The beam is much larger than the spacing of the gratings, but only the direction of propagation is shown. The lines above and below the main beam represent diffracted light, with the angle of the beams corresponding to the first angle of constructive interference. Closer spacing of the grating yields a higher "spatial frequency" and a correspondingly higher angle of diffraction. An identical effect is observed if the optical density of the grating varies sinusoidally.

Returning to the electrochemical problem, the Fourier transform approach is summarized in Figure 6. Without an electrochemical diffusion layer present, the beam profile will be a Gaussian beam truncated by an electrode. The Fourier transform of such a pattern is symmetrically disposed about the origin of the screen when intensity is plotted, and has a rather featureless monotonically decreasing shape. When an absorbing electrochemical diffusion profile is created at the electrode, the beam will lose intensity near the electrode surface where the concentration of chromophore is highest, and the beam distant from the electrode will be unchanged. The resulting diffraction pattern falls off faster than for a bare electrode, as shown in the Figure. Diffraction at higher angles (higher screen coordinate) is attenuated more than at lower angles because the large angle diffraction represents higher spatial frequencies, which were important to describing the sharpness of the edge. Since the chromophore has its greatest effect at the edge (where its concentration is highest), those frequencies related to the edge are most affected by

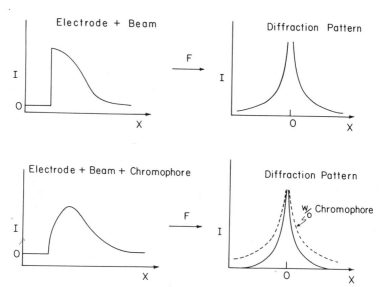

Figure 6. Beam profiles and diffraction patterns for the experiment shown in Figure 3. Intensity vs. x' for Electrode + Beam represents the cross-sectional intensity of the laser beam, truncated by the electrode: its diffraction pattern was calculated using a Fourier transform algorithm. The lower pair of plots shows the same case with a chromophore near the electrode. The sharpness of the electrode edge is decreased, and the wings of the diffraction pattern are affected the most. The dashed line shows the diffraction from a bare electrode for comparison.

Fourier Transforms in Spectroelectrochemistry

the presence of chromophore. Viewed differently, a synthesis of the more rounded profile represented by the electrode in the presence of chromophore requires lower Fourier frequencies than that of a sharp, bare edge.

Detailed theoretical profiles for ϕ are shown in Figure 7, with the dashed line being an electrode without chromophore placed in the middle of a Gaussian laser beam. The solid line is the same electrode with a stable, diffusing chromophore present, after 10 sec of electrolysis. These profiles are the shape of the beam which would be observed immediately at the electrode, before diffraction occurred. The chromophore was generated from a non-absorbing precursor, and both species were assumed to obey Fick's laws of diffusion. Note that the beam is attenuated most at the electrode surface, as predicted from the fact that the chromophore concentration is highest at the electrode surface. Figure 8 shows diffraction patterns calculated from the profiles of Figure 7; again, the dashed line corresponds to the bare edge. As expected, the higher x-coordinates (higher frequency components) are attenuated more than the low frequencies. Figure 9 is a plot of absorbance, defined as the logarithm of the ratio of intensity without chromophore to that with chromophore, observed at a particular electrolysis time. It is obvious from this plot that absorbance (and therefore attenuation of diffrac-

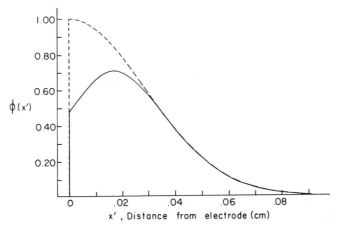

Figure 7. Beam profile for a Gaussian beam truncated at its center by an electrode. $\phi(x')$ is the beam amplitude relative to its center, and x' is the distance from the electrode (downward in Figure 4). The dashed line shows the profile for the beam before any chromophore is generated, and the solid line is the beam cross-section after 10 sec of chromophore generation.

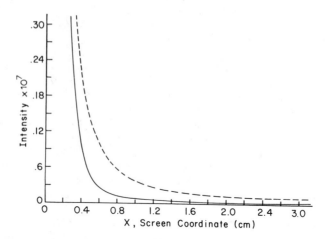

Figure 8. Half of the theoretical diffraction pattern for the profiles shown in Figure 7 (the pattern is symmetric about x = 0). Intensity is relative to that of the initial beam center; screen is 36.5 cm away from the electrode, so 1.0 cm corresponds to a 1.6° diffraction angle. Dashed line is bare electrode; solid line shows effect of chromophore.

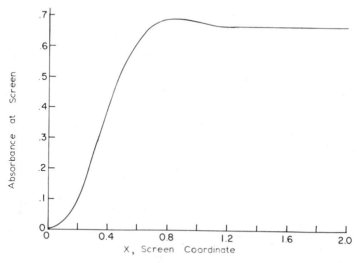

Figure 9. Absorbance as a function of screen coordinate, calculated from curves similar to those in Figure 8. Absorbance is defined as the common log of the ratio of intensity without chromophore to that with chromophore.

ted light) increases rapidly with angle, then reaches a constant value independent of angle. This value is equal to the Beer's law absorbance corresponding to a path length equal to the dimension of the electrode along the optical axis and a chromophore concentration equal to the bulk concentration of its precursor. This absorbance value thus corresponds to complete conversion of precursor to chromophore, followed by attenuation along the electrode length according to Beer's law. Complete conversion of precursor to chromophore occurs only at the electrode surface, so the region of constant absorbance must largely represent distances very close to the surface. The fact that the absorbance is constant with angle at large angles indicates that all the higher frequencies have been attenuated by the Beer's law value, and one is monitoring the essentially constant concentration of chromophore very near the surface. The limiting case of this process occurs at infinite electrolysis times, when the entire solution is filled with chromophore, and the whole diffraction pattern is attenuated by the same (Beer's law) absorbance. It should be noted that there is not a one-to-one correspondence between diffraction angle and distance from the electrode, but it is clear that the higher diffraction angles represent higher spatial frequencies, which relate to the electrode edge, and are most sensitive to chromophore generation. To a first approximation, high angles are sensitive to events near the surface, while lower angles respond to more gradual changes in the entire beam profile.

As chromophore is generated at the electrode surface, it will diffuse into the solution, so both $\phi(x')$ and the diffraction pattern will depend upon the time from the beginning of electrolysis. Figure 10 is a plot of several successive diffraction patterns, again expressed as absorbance relative to the diffracted intensity in the absence of chromophore. As expected, the higher angles reach the limiting absorbance more rapidly than lower angles, and in theory all angles will eventually reach the constant, Beer's law absorbance (0.65 in this case). Figure 11 shows similar data re-plotted as absorbance vs. time at particular angles, a format that is simpler to duplicate experimentally. At long times, all angles approach the Beer's law absorbance. Thus, the entire diffraction pattern approaches an attenuation dictated by Beer's law, but the high angles do so at a faster rate.

Three important features of the diffractive technique are apparent from the theoretical results shown in Figures 9-11. First, the absorbance values are much larger than similar experiments using an optically transparent electrode. Enhancement of at least a factor of 1000 over transparent electrode experiments is possible, simply because the effective path length is determined by the electrode dimension rather than the diffusion layer thickness, and path lengths of 1 cm or more are possible. Second, the absorbance measured with diffracted light changes very rapidly as a function of time after electrolysis, when measured at angles greater than 1.0°. For exam-

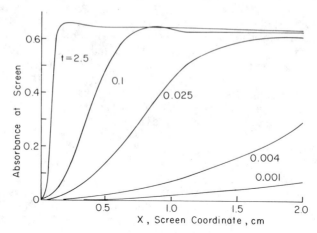

Figure 10. Absorbance vs. screen position (as in Figure 9) as a function of time after electrolysis begins. Number next to each curve is the time in sec after the beginning of chromophore generation.

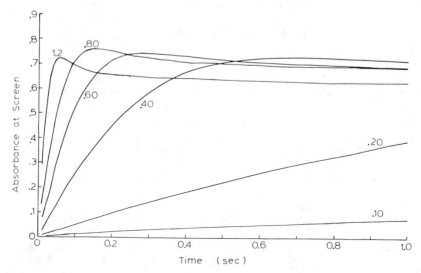

Figure 11. Theoretical absorbance vs. time profiles at various diffraction angles. Chromophore ($\epsilon = 10{,}750$) was generated from 1.2×10^{-3} M of a colorless precursor, at an electrode with 0.5 mm path length along the optical axis (Beer's law absorbance = 0.65). The diffusion coefficient was 1.5×10^{-5} cm^2/sec and the laser wavelength was 632.8 nm.

Fourier Transforms in Spectroelectrochemistry

ple, the absorbance at 1.2° reaches 0.7 units in 25 msec, for typical conditions. This absorbance is 100 times the value expected for a transparent electrode experiment with the same parameters. Both the high sensitivity and fast time response of the diffractive method stem from the ability to monitor light which has passed close to the electrode surface along a path that is long compared to the diffusion layer thickness. The relevant effect being exploited is merely Beer's law, but the diffractive process allows one to define a sampled region very close to the electrode surface.

The third important observation derived from the theory is that the diffracted light contains spatial information about the distribution of electrogenerated chromophore. This feature is useful in two complementary ways. First, it is possible to select, at least approximately, which region of the diffusion layer is monitored. High angles monitor events occurring primarily at short distances away from the electrode, while smaller angles reflect events at greater distances. As noted earlier, the correspondence between diffraction angle and electrode distance is not direct, but it is still useful for approximate experiments. The second aspect of spatial properties lies in the inverse Fourier transform of the diffraction pattern to yield the chromophore distribution. This process should generate a snapshot of the diffusion layer with a resolution comparable to the wavelength of the light. This procedure is completely analogous to x-ray diffraction experiments, except optical diffraction is used to reconstruct a diffusion layer with features having dimensions greater than the optical wavelength.

EXPERIMENTAL VERIFICATION

The experimental apparatus[12] used for diffraction measurements is shown schematically in Figure 12. The electrode dimension along the optical axis, and therefore the Beer's law path length used to calculate ϕ, is 0.5 mm. The glass shields adjacent to the electrode assure linear diffusion to and from the walls of the cell. The lens following the cell is present to ensure that the Fraunhofer approximation is valid. Since the screen is placed at the focal length of the lens, light focussed at the screen must have been parallel when it entered the lens. Thus the lens samples parallel light and the screen is effectively at infinity, validating the approximation. The electrode surface was polished with conventional procedures, and care was taken to orient the surface parallel to the laser beam. Diffracted light was sampled by a small aperture placed before a photomultiplier tube, and the detector assembly was moved vertically to sample different portions of the diffraction pattern. Once the detector was positioned at a particular angle, a computer monitored the intensity before and during a potential step which generated chromophore. The absorbance vs. time transient was calculated by taking the log of the ratio of intensity before a potential was applied to that after chromophore generation began. Before any electrochemistry was per-

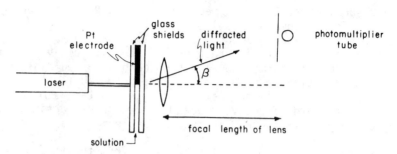

Figure 12. Experimental apparatus used to verify the theory for optical diffraction. Bottoms of shields were immersed in solution, and the electrode region was filled by capillary action. Auxiliary and reference electrodes were placed in the solution container below the shields. For more details, see the original work.[12]

formed, it was verified that the observed diffraction pattern had the shape predicted in Figure 8.

For the present work, the test reaction was the oxidation of tri-(p-methoxyphenyl)amine, which is colorless, to its cation radical, which has a molar absorptivity of 10,750 at the He-Ne laser wavelength (632.8 nm). The oxidation was carried out at a platinum electrode in acetonitrile, with tetraethylammonium perchlorate as supporting electrolyte. This system is known to be well-behaved, with both precursor and cation radical obeying Fick's laws.[16]

Figure 13 shows experimental absorbance vs. time transients measured as described above. All conditions and parameters (ϵ, λ diffusion coefficient, etc.) were identical to those used to calculate Figure 11, so direct comparison of theory and experiment can be made. Qualitative agreement is good at all angles, and theory and experiment agree quantitatively at angles above 0.4°. The important theoretical predictions made earlier are all verified by Figure 13. First, the sensitivity of the diffractive approach is high, with the absorbance reaching a limiting value proportional to the Beer's law path length of the electrode. This limiting value is reached at shorter times at larger angles, and is two orders of magnitude larger than a comparable experiment with a transparent electrode. Second, the absorbance at large angles rises rapidly at short times, since the sampled region is being restricted to the solution very close to the electrode. The rapid rise to a constant absorbance at 1.2° corresponds to "filling" the sampled region with chromophore at a concentration equal to the initial bulk concentration of precursor. Third, the response is very dependent on angle, indicating that the

Fourier Transforms in Spectroelectrochemistry

diffraction pattern shape changes as the diffusion layer changes. Thus the diffraction pattern contains spatial information that can be used to deduce information about the distribution of chromophore in the diffusion layer. Although the full potential of the spatial properties of the diffraction pattern remains to be realized, it is apparent that regions very close to the electrode surface can be monitored, or conversely, that the whole diffusion layer may be monitored by examining a large portion of the diffracted beam.

Although the agreement between theory and experiment supports the basic principles of diffractive spectroelectrochemistry, there are two limitations that will inevitably creep into experimental results. First, the electrode edge is not infinitely thin, as was assumed for the Huygens-Kirchoff approach. For very long electrodes, the diffractive process occurring near the surface is hard to imagine. The light waves will be perturbed by the surface, but since the surface is much longer than the wavelength of the light, it is difficult to predict the resulting diffraction. Although the

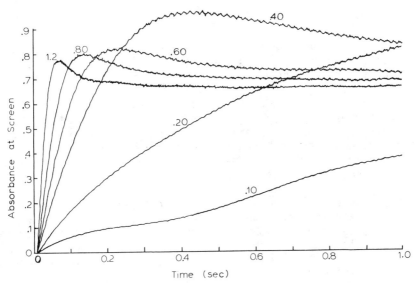

Figure 13. Experimental absorbance vs. time transients for generation of the tri-(p-methoxyphenyl)amine cation radical from its colorless reduced form. All conditions and parameters are identical to those used to calculate Figure 11, so direct comparison of theory and experiment can be made. Numbers next to each curve give diffraction angle in degrees (0.80° = 0.51 cm screen coordinate).

agreement between theory and experiment indicates that this problem is not severe for the present experiment, it is likely that it will decrease the spatial resolution of diffraction experiments using long paths. However, even with poorer resolution, the absorbance will reach a limiting value proportional to the longer path length. This conclusion is supported by the easy observation of Beer's law value for absorbance from a 10 micromolar solution with a 1 cm electrode. A second limitation of the technique stems from the small changes in refractive index accompanying chromophore generation. Such changes would introduce phase shifts in the light as it propagates along the electrode, and would result in changes in the diffraction pattern. Based on interferometric data, the refractive index changes are too small for the dilute solutions used in most electrochemistry to have a significant effect on the diffractive results. Neither of these potential problems is apparent in the experimental results of Figure 13, but they may appear under more extreme conditions of concentration or electrode length. The disagreement between theory and experiment at small angles results from the position of the electrode relative to the beam. Small-angle diffraction is dependent upon overall beam shape, while the high angles are derived primarily from the edge; thus small angles are more sensitive to positioning of the electrode relative to the center of the beam. This difficulty will be corrected with improved alignment.

SUMMARY

Diffractive spectroelectrochemistry is a technique which is very much in its infancy, but it is clear that it has the potential to improve on the performance of previous methods. The low sensitivity of previous spectroelectrochemical methods limited their applicability to strong chromophores with millimolar concentrations. In the diffractive approach, the path length is not constrained by the diffusion process, so the sensitivity is much higher, and dilute solutions of weak chromophores can be examined. A wide variety of both analytical and kinetic experiments are possible with diffractive spectroelectrochemistry with its potentially increased generality over previous methods. In addition, a spatially sensitive probe of the thin layer of solution near an electrode permits many new questions about mass transport and chemical reactions to be addressed. Diffractive spectroelectrochemistry should evolve into an excellent probe of the diffusion layer, a region of utmost importance to electrochemistry, but one which is exceedingly difficult to probe with conventional methods.

ACKNOWLEDGMENTS

This work was supported in part by the National Science Foundation (CHE-7828068) and the National Institute of Mental Health (NIMH 28412).

REFERENCES

1. Kuwana, T., Darlington, R. K., and Leedy, D. W. 1964, Anal. Chem. 36, 2023.
2. For reviews, see: Kuwana, T., and Winograd, N. 1978, in Electroanalytical Chemistry, Vol. 7, ed. A. J. Bard, Dekker, N.Y.; Heineman, W. 1978, Anal. Chem. 50, 390A; and Kuwana, T., and Heineman, W. R. 1976, Acct. Chem. Res. 9, 241.
3. Murray, R. W., Heineman, W. R., and O'Dom, G. W. 1967, Anal. Chem. 39, 1666.
4. See, for example, Bard, A. J., and Faulkner, L. R. 1980, Electrochemical Methods, John Wiley & Sons, N.Y.
5. Nicholson, R. S., and Shain, I. 1964, Anal. Chem. 36, 706.
6. Flato, J. 1972, Anal. Chem. 44, 75A.
7. Kuwana, T. 1973, Ber. Bunsenges Phys. Chem. 77, 858.
8. Muller, R. H. 1973, Adv. Electrochem. Eng. 9, 281.
9. Pruiksma, R., and McCreery, R. L. 1979, Anal. Chem. 51, 2253.
10. Pruiksma, R., and McCreery, R. L. 1981, Anal. Chem. 53, 202.
11. The Kirchoff-Huygens approach is discussed in most physics textbooks. Clear treatments are presented in: Longhurst, R. S. 1957, Geometrical and Physical Optics, Lungmans, Green & Co., N.Y., p. 193; and Marcuse, D. 1972, Light Transmission Optics, Van Nostrand Reinhold, N.Y., p. 31.
12. Rossi, P., McCurdy, C. W., and McCreery, R. L. 1981, J. Amer. Chem. Soc. 103, 2524.
13. See ref. 11a, p. 195; ref. 11b, p. 39.
14. Gaskill, J. P. 1978, Linear Systems, Fourier Transforms, and Optics, John Wiley & Sons, N.Y., p. 376.
15. Vest, C. M. 1979, Holographic Interferometry, John Wiley & Sons, N.Y., p. 21.
16. See Pruiksma, R. 1980, Ph.D. thesis, The Ohio State University.

CONTRIBUTORS

Jess H. Brewer, Department of Physics, University of British Columbia, Vancouver, B.C., Canada   V6T 2A6

Robert H. Cole, Department of Chemistry, Brown University, Providence, Rhode Island   02912

Melvin B. Comisarow, Department of Chemistry, University of British Columbia, Vancouver, B.C., Canada   V6T 1Y6

N. S. Dalal, Department of Chemistry, West Virginia University, Morgantown, West Virginia   26506

James A. de Haseth, Department of Chemistry, University of Alabama, University, Alabama   35486

C. L. Dumoulin, Department of Chemistry, The Florida State University, Tallahassee, Florida   32306

Donald G. Fleming, Department of Chemistry, University of British Columbia, Vancouver, B.C.   V6T 1Y6

Willis H. Flygare[*], Noyes Chemical Laboratory, University of Illinois, Urbana, Illinois   61801

Tomas B. Hirschfeld, Lawrence Livermore National Laboratory, L-325, Livermore, California   94550

Stanley M. Klainer, Lawrence Berkeley Laboratory, 90/1140, Berkeley, California   94720

George C. Levy, Department of Chemistry, The Florida State University, Tallahassee, Florida   32306

Robert A. Marino, Hunter College of CUNY, 695 Park Avenue, New York, New York   10021

[*] (deceased)

CONTRIBUTORS

Alan G. Marshall, Departments of Chemistry and Biochemistry, The Ohio State University, Columbus, Ohio   43210

Richard L. McCreery, Department of Chemistry, The Ohio State University, Columbus, Ohio   43210

W. B. Mims, Bell Laboratories, Murray Hill, New Jersey   07974

Gareth A. Morris, Physical Chemistry Laboratory, South Parks Road, Oxford OX1 3QZ, England

Robert J. Nordstrom, Principal Research Scientist, Physico-Chemical Systems, Battelle Columbus Laboratories, Columbus, Ohio   43201

Paul W. Percival, Department of Chemistry, Simon Fraser University, Burnaby, B.C., Canada   V5A 1S6

Paula Rossi, Department of Chemistry, The Ohio State University, Columbus, Ohio   43210

Neil J. A. Sloane, Mathematics and Statistics Research Center, Bell Laboratories, Murray Hill, New Jersey   07974

Donald E. Smith, Department of Chemistry, Northwestern University, Evanston, Illinois   60201

Paul Winsor, IV, Department of Chemistry, Brown University, Providence, Rhode Island   02912

INDEX

Absolute-value spectrum, see Magnitude spectrum.
Absorption spectrum, 1-6, 99-102, 109, 220, 281, 289, 358, 443
A.C. polarography, see Polarography, a.c.
A.C. steady-state dielectric response, 105
Acoustic NQR, 180
Activation energy, 336, 340
Accuracy,
  frequency, 132, 396, 438-439
  in peak area determination, 75
  mass determination, 132-134
  timing, 196
Address space; addressing, see Memory
Admittance, 183-197, 456-457, 463, 466
  faradaic, 457, 476
Adsorption, 117, 122, 455, 507
Advantage,
  etendue, see Throughput
  Fellgett, see Fellgett
  Jacquinot, see Throughput
  multichannel, see Fellgett
  throughput, see Throughput
Algorithm,
  FFT, see Cooley-Tukey
  FHT, see H.T.
  peak-picking, see Peak-picking
Aliasing, 40-41, 199, 372, 445-446, 466-467
Alternating current, 126-129
  polarograhy, see Polarography, a.c.
Analog-to-digital converter, 130-131, 222, 224, 266, 349, 404, 439, 457-459
Antibiotic, 297-298
Aperture, 398-399, 427-434, 543

Apodization, 20-26, 288, 316, 355, 393, 426-427, (see also Weight functions)
Array processor, 85, 522
Array, virtual, 75-89
Asymmetry parameter
  NMR, 151-153
Atomic spectroscopy, 446
Auto correlation, 274, 463, 466

Balance analogy, see Weighing designs
Bank switching, see Memory
Baseline correction, 22-25, 71-73, 109
  drift, 107
Background subtraction, 353-354, 439
Beam, molecular, 228
  supersonic, 243
Beam splitter, 388, 395, 399-400, 404, 423-424
Benzene, on silica gel, ^{1}H NMR, 117
Bessel filter, 461
  function, 158, 221, 253
Bipolar RAM, see Memory
Bit inversion, 80
  reversal, 81-84
Block, see Memory block
Block averaging 489
  F.T. of, 80
Bloch equations, 208, 212, 333, 335
Bode relation, see Hilbert transform
Boxcar function, see Square wave
Broad-band excitation
  from frequency-sweep, 13-15, 131
  from pulse, 10-12
Bytes, see Memory

^{13}C NMR, 272-273, 275, 278, 283, 294, 295, 299-300
Capacitance, 186
Capacitor detector for FT-IR, 126-129
Carbohydrates, 279, 280, 284, 292-293, 295
Carr-Purcell, see Spin-echo
Cat's eye retroreflector, see Retroreflector
Causal function,
  F.T. of, 16-17
  Hilbert transform of, 108
Cavity,
  Fabry-Perot, 207-208
  relaxation time, 208
Central Procesing Unit, see CPU
Charge transfer, 462, 471, 484, 489-493, 496
Chemical exchange, NMR, 106, 116
Chemical shift, 115, 122, 274-275, 278, 282, 286, 297-298
Contact shift, 326, 328, 331, 333, 338
Chromatography,
  2-dimensional, 271-272
  gas, see GC
  liquid, see LC
Chromium carbonyls, nitrosyls, 136-144
Circle, DISPA, 100-102
Circularly polarized radiation, 18-19, 311
Coal, 122
Coalesence of peaks, 116
Coaxial line cell, 189-190
  connector, 240
  sample line with matched termination, 190-191
Codes,
  A-optimal, 54
  conference matrix, 63
  cyclic, 35, 59-60
  D-optimal, 55
  Fourier, 36-42
  Hadamard, 34-36, 53
  multiplex, 34-42
  S-matrix, 51, 58, 62-63
  single-channel, 33-34
  well-conditioned, 34
Coherent ion motion, 129
  radiation source, 129, 207, 229
  rotational motion, 229
Cole-Cole plot, 101

Complex F.T., 9, 36-42, 79, 167
  quantities, 6
Conductance, 190-194, 252
Conference-matrix, 63
Continuous-wave excitation, 1-2, 153, 234, 330, 406
  response, 1-6, 153, 234, 330-332, 335, 406
Convolution (see also Deconvolution), 26-32, 70, 197, 317, 392
  difference method, 288, 297
  graphical, 28-30
Convolution theorem, 29-30, 197, 392, 427, 485
Cooley-Tukey algorithm, 39, 69, 80, 85, 108, 157, 198, 266, 421, 456, 461
  speed of, 88-89
Coordination complexes, 136-141, 311
Correlation, heteronuclear 2D-NMR, 276, 282-285, 297-300
  homonuclear 2D-NMR, 276, 281-282, 296-297
  NMR, 70 (see also frequency-sweep)
Correlation function, 185
  time, 16, 329-330, 340
Cosine F.T., see F.T., cosine
Cosine bell function, 316
CPU (Central Processing Unit), 76, 88, 459
Cross correlation, 274, 463, 466
Cross relaxation, 285, 312
Cross section through 2D FT-NMR plots, 272, 291, 293-294, 301
Cyclic voltammetry, see Voltammetry, cyclic
Cyclically permuted codes, 35, 59
Cyclotron frequency, 125

Data table, see Memory
D.C. conductance, 183
D.C. polarography, see Polarography, D.C.
D.C. spectral component, 316 (see also Baseline correction)
Debye relaxation, 186, 198
Deconvolution, 31-32, 197, 461, 463, 485-489, 501-507

INDEX

Decoupling in NMR,
  gated, 279-280, 296
  heteronuclear, 279-285, 295-296, 297-300
  homonuclear, 275-279, 293-294, 296-197
Density matrix, 209
  interaction representation, 209-210
Derivative, 70 (see also Semi-derivative)
  by F.T. methods, 26, 187
Derivative spectrum, 4, 109, 330-332
Detector, linear
  capacitor, 126-129
  cooled bolometer, 402-404
  dielectric, 188-191
  FAM, 457-463
  Fabry-Perot, see Fabry-Perot
  Golay, thermocouple, 402-404
  heterodyne (see detector, phase)
  IR, 390, 402-404
  marginal oscillator, 127
  muon, 348-350
  phase, 131-132, 155-156, 220, 265
  photoconductive, 402-404
  photomultiplier, 434, 455, 543-544
  pyroelectric, 402-404
  remote, 153, 157
  silicon photodiode, 434
  superheterodyne, 207, 229
  superregenerative NQR, 153-154
  tank circuit (NMR), 154
  trapped-ion cell, 129
  waveguide, 207, 212, 215-234, 257
    Ku-band, 234
    X-band, 240
Deuterium NMR, 337-340
Dielectric constant, 101
  loss, 198
  relaxation, 101
    experimental, 204-205
    theory, 185-203
  sample cells, 188-191
Difference spectra, 205, 439
Differentiation using transforms (see also Semi-derivative), 26, 187
Diffraction, 237, 533-543

Diffraction grating, see Grating, diffraction
Diffusion coefficient, 529, 542, 544
  layer, 530-533, 538, 541
  limited current, 531, 471
Diffuse reflectance, 414
Digital-to-analog converter, 457-459, 461
Digitization, see Sampling
Digital filtering, see Filtering, digital
Dipole-dipole coupling, 328-329, 340-341
  correlation function, 185
Dipole moment,
  electric, 214, 231, 248
Direct-current, see D.C.
Direct memory access, see Memory
Discrete sampling, 33-42
  (see Foldover; Sampling; F.T., discrete)
Disk memory, see Memory, disk
  based F.T., 78-84
DISPA, definition, 102
  method for line shape analysis, 99-123
Dispersion-mode spectrum, 1-6, 101-102, 220, 358
  from absorption-mode spectrum, 109
  origin from causal response, 16-17
Dispersion relation, 210
  vs. absorption, see DISPA
Doppler distribution, 230
  broadening, 232, 252-253, 257, 260
Double-beam optics, 401, 409-410
Double F.T., see F.T., 2-dimensional
Double-layer admittance subtraction, 498
Double-layer capacitance, 454-455, 457
Double resonance
  electron-electron, see ELDOR
  electron-nuclear, see ENDOR
  ion-ion, 137-144
  NQR, 153
  nuclear-nuclear, 272-286
DPPH, 324, 325, 330-340
Dropping mercury electrode, see Electrode, dropping mercury

Drug assay, electrochemical, 504, 507-522
 analysis, 297-298
 shelf life (NQR) 180
Dynamic range, 404

Electrochemistry, spectro-, 529-533, 546
Electric field gradient, 150
Electrolysis, 539-541
Electrostatic potential at nucleus, 148-150
Electrode, dropping mercury, 461, 471, 497
 stationary, 494
 transparent, 529-532, 541, 543
ELDOR, 323-326
Electron Spin Resonance, 109, 119-121, 307, 323-326 (see also Spin labels)
ENDOR, 307-326
Emission spectra, 402, 431, 437, 440, 445-446, 448
Encoding masks, see Masks
End capacitance cell, 188-189
Energy levels,
 electron-nuclear magnetic, 309
 μSR, 367-372, 376
 NQR, 151-153
 rotational, 208
Enhancement, resolution, see Resolution enhancement
 S/N, see Signal-to-noise enhancement
ESR, see Electron-Spin Resonance
Etendue, see Throughput
Even function, 16-17
Exchange broadening in ESR, 107, 120
 chemical, 106-107, 116
 electron, 330
Excitation (see also double-resonance), coherent, 207
 step, 186
Excited ion cyclotron motion, 126
Exponential
 apodization, see Weight function, exponential
 F.T. of, 13-15
 relaxation, 7-8, 199, 260
 truncated, 199-200

Fabry-Perot cavity, 207-208, 212, 234-241, 244-269

Far-infrared, 205, 402, 423
Faradaic admittance, see Admittance
Fast Fourier Transform, see Cooley-Tukey
Fellgett advantage, 34, 50, 165, 406-407, 415, 421-422, 434-438, 449, 477
FFT, see Cooley-Tukey
FID, see Free Induction Decay
Filling, see Zero-filling
Filtering, analog, 41, 196, 226-227, 396, 461
 digital, 22-25, 356, 461, 477-481 494, 500
First derivative EPR spectrum, 109
Floating point data, 76
 F.T. of, see F.T., floating point
Fluorouracil, 115
Folding, see Convolution; Aliasing
Foldover, 40-41, 226
Fourier Transform, see F.T.
Fraunhofer lineshape, see SINC function
Free Induction Decay, 154, 158, 163, 168, 170, 172, 174, 176-177, 271-272, 275, 287, 311, 349, 378
Free radicals, 330, 338, 341, 345, 495, (see also Muonic radicals)
Frequency, carrier, 224, 254
 dielectric excitation, 195
 domain, 10-16, 167
 driving, 1-6, 101
 ion cyclotron, 125
 Larmor, 349, 370, 382
 μSR, 367-373, 376, 380
 natural, 7, 101, 125
 negative, 18-19
 NQR, 151-153, 380
 Nyquist (foldover), 39-41
 sampling, 40
 shifts, dynamic, 122
  from solid angle effect, 431
 spatial, 536-537, 541
 suppression in F.T. ENDOR, 314-315
 sweep excitation, 13-15, 32, 130
Friction coefficient, from steady-state response, 1-4
 from transient response, 7-8
Fringing effects in dielectric measurements, 109

INDEX

FT, complex, 9, 37, 79, 167, 295, 357, 361, 372, 375, 377, 379, 381, 390, 392, 463, 536
  computations, 69-98
    speed of, 88
  continuous, 9-16, 390, 425, 536
  cosine (see FT, real)
  definition, 79-80, 167, 198, 357
    continuous, 9
    discrete, 36-39
  dielectric spectroscopy, 183-206
  discrete, 36-42, 78-90, 198, 225, 266, 357
  disk-based, 78
  double, see Two-dimensional FT
  ENDOR (2D), 307-322
  FAM, 453-526
  fast, see Cooley-Tukey
  floating point, 78
  FFT, see Cooley-Tukey
  Hilbert transform from, 109
  ICR, 125-146
  imaginary, 9-16, 79, 167-178, 220, 315, 318, 358, 362, 372, 390, 463
  inverse, 9, 37-38, 79-80, 109, 194, 426, 543
  IR, 387-420
  magnitude, 9, 30, 32, 167-178, 317
  μSR, 345-386
  multiplex code, 36-42
  negative frequencies, 18-19
  NMR (2D), 271-306
  NMR, paramagnetic species, 323-344
  NQR, 147-182
  of circularly polarized signal, 18-19
  of data blocks, 80
  of frequency sweep, see convolution theorem
  of linearly polarized signal, 18-19
  of noise, 14-16, 359
  of product or quotient, see convolution theorem
  pairs (pictorial library), 8-16, 393, 428
  power spectrum, 9-16, 167-178, 262, 352, 358, 361, 371, 463

FT (continued)
  real, 9-16, 79, 167-178, 220, 315, 318, 358, 362, 372, 390, 463
  relation between spatial and frequency domain, 536-537
  relation between time and frequency domains, 7, 10-16, 154, 167, 263, 351, 363-367, 520
  relation to derivative, 26, 187
  rotational spectroscopy, 207-270
  sine (see FT, imaginary)
  spatial, 536-537
  spectroelectrochemistry, 527-548
  spectroscopy, 1-44
  two-dimensional, see Two-dimensional FT
  two-dimensional NMR, 271-305
  UV-VIS spectroscopy, 421-452
g-factor, see Magnetogyric ratio
Gas chromatography, see GC
Gated decoupling, see Decoupling, gated
Gaussian beam profile, 535, 538-539
  distribution in peak position, 104, 107
  F.T. of, 13-15, 256
  standing wave, 211, 236
  to Lorentzian transformation, 22-23, 288
  weight function, 13-15, 288
GC/FT-IR, 408-411
Gel, benzene adsorption on silica, 117
Grating, diffraction, 537

H-code, see Hadamard codes
^{1}H NMR, 115-117, 279-280, 292, 299-300
  vs. Mu, 347
Hadamard codes, 34-36
  criteria for existence of, 57
  fast transform, 39
  IR spectroscopy, 36, 64
  inverse of, 35-36, 57
  matrix, 35-36, 53, 57
  transform, 36, 63
Hamiltonian, nuclear quadrupole, 150
  rotational, 208
  spin, 321, 328, 367, 370, 376, 380

Heme-a, 319
Heterodyne detection, see
  Detection, phase
Heteronuclear
  2D correlation spectroscopy,
    276, 282-285, 297-300
  2D J-spectroscopy, 276, 279-281,
    295-296
  multiquantum 2D FT-NMR, 277
Hilbert transform, 7, 108
Homogeneous line-broadening, 107,
  120
Homogeneous redox reaction rates,
  499-507
Homonuclear 2D correlation
    spectroscopy, 276, 281-282,
    296-297
  2D J-spectroscopy, 275-279, 293-
    294
  multiquantum 2D FT NMR, 277
Hyperfine coupling, EPR, 121,
    323-342, 493
  interactions, see ENDOR
μSR, 367, 370, 376, 378, 380

ICR, see Ion cyclotron resonance
Image reconstruction, Hadamard 1-
    and 2-dimensional, 65
  NMR 2-dimensional, 274
  spectrometric, 65
Imaginary component (see also
    Dispersion-mode), 6
Impedance (see also Admittance),
    183, 188
  matching, 208
Impulse excitation, time-domain
    response to, 7
Impurity, FT-NQR detection, 179-180
Inductance, 188
Inhomogeneity,
  in electric field strength, 255
  in line position, 99, 107
  in line width, 99, 107
  in magnetic field strength, 115,
    232, 286, 297
  in rf field, 286
Inorganic complexes, see
    Coordination compounds
Instrumental lineshape,
  deconvolution of, see Decon-
    volution
Integration by F.T., 26
  by Simpson rule, 72, 74-75
  of ESR absorption derivative, 109

Intensity of spectral peak, 72, 289
Interaction representation, 209-
    210
Interferogram, 288, 301, 389, 391,
    422
  two-sided vs. one-sided, 17
Interferometer, 387-388, 395-397,
    423, 441, 444, 531, 546
  mirror and drive, 400-401, 423-
    424, 441, 444
Interpolation (see also Zero-
    filling),
  linear, 72
  parabolic, 72
Intramolecular coupling, 129, 329
Intermolecular coupling, 129
Inverse F.T., see FT, inverse
  H.T., see Hadamard transform,
    inverse
  Moore-Penrose generalized, 54
Inversion symmetry, 346
  of matrix, see Code

I/O (input-output) devices, 76
  disk, 82
Ion cyclotron resonance, 125-146
Ion-molecule reactions, 136-144
Irradiation, see Double-resonance
Isomers, 414
Isotope effect, ^{2}D vs. ^{1}H, 338
  Mu vs. ^{1}H, 347
Isotopes, 226, 267

$J^{13}C^{-13}C$, 278, 301
$J^{13}C^{-1}H$, 273, 278-279, 283, 296
$J^{19}F^{-1}H$, 115
J-coupling, 115, 268, 275, 279, 281
J-spectra, 274, 278, 285-286
Jacquinot avantage, see
    Throughput
Jeener pulse sequence, 276, 281,
    283, 286, 296
Jitter, timing, 196

Kinetic studies (see also
    Chemical exchange),
  electroactive species, 529, 489-
    493, 499-507
  ion-molecule reactions, 136-144
  radical reactions, 374
Kinetic deconvolution, 501-507
Kramers-Kronig relation, see
    Hilbert transform

INDEX

Larmor frequency, see Frequency, Larmor
Laser F.T.-interferometry, 404-405, 441-442, 444
   F.T.-spectroelectrochemistry, 542-544
LC/FT-IR, 411-414
Leakage, 200-204, 467, 470, 478
Linear response theory, 185
   F.T. relation for, 108
Linearized DISPA plot, 110-112
Linearly independent code rows, 34
Linearly polarized radiation, 18-19, 311
Liquid chromatography, see LC
Liquid crystal, 279
Log-Gaussian distribution
   in correlation time, 107
   in relaxation time, 105, 107, 111
Log-scale, effect on FFT, 198-199
Look-up-table, sine and cosine, 85
Lorentz lineshape, 1-6, 13-15, 101, 153, 216, 239, 331, 351
   circular DISPA plot, 100-102
   integrated area, 74-75
   superposition of two or more, 99 ff.
Lorentzian-to-Gaussian transformation, 22-23, 288
Lumped capacitance cell, 188

M-sequence, see Pseudo-random sequence
Macromolecules, see protein, RNA, polymers
Magnetic resonance, 115-121, 349
   field, inhomogeneous, 115
   relaxation mechanisms, 107
   tape, 79
Magnetogyric ratio, 301, 311, 347, 367
Magnitude spectrum, 9-16, 32, 167-178, 289, 292
Manganese carbonyls, 139
Mapping, see Memory mapping
Marginal oscillator, 127
Mask
   coding, 49
   Hadamard, 64
   S-matrix mask, 51, 58, 60
Mass calibration, 132-135
Mass range, FT-ICR, 125-127, 134-136
Mass resolution, 130

Mass spectroscopy, FT-ICR, 125-146
Matrix, see Code
Memory address space, 75, 88
   bank-switching, 76
   bipolar, 225
   bit, see Bit
   block, 79
   buffer, 457-459
   bytes, 75
   core, 75, 224, 287
   data table, 76
   disk, 76, 79, 82 88, 287, 462
   extended, 76
   magnetic tape, 459
   mapping, 77-84
      explicit, 77
      transparent, 77
   pages, 76
   random access, 225
   shift register, 131-132, 459
   virtual, 75-77
   window, 80-84
Metal-metal bonds, 142-143
Michelson, see Interferometer
Microprocessor, 75
Microwave spectrometry, 205, 207-270
   EPR excitation, 307
   FT spectrometer, 222, 263
Minerals, FT-NQR of, 180
Minicomputer, 196
   FFT efficiency and speed, 88-89
Mixer detector, see Detector, phase
Mixture analysis, see FT-ICR, GC-IR, and LC-IR
Modulation (see also ENDOR, Phase modulation, and Spin-Echo)
   broadening, 107, 119
   J-, see J-spectroscopy
   magnetic field, 330
   Stark, 229-230
   zero-field in μSR, 382
   (see also ENDOR, Phase modulation, and Spin-echo)
Modulus transform, see Magnitude-mode
Molecular beam, see Beam, molecular
Mu, 345
Muon, 345, 347
Muon spin rotation, 345-385
Muonic atom, 346

Muonic radicals, 367, 369-374
Muonium, 345
Muonium atom, 346
Multichannel detector, 49
Multiple quantum coherences, 274, 285
    reflection expressions (dielectric relaxation), 192-193
Multiplex advantage spectrometer, 34, 50, 455

^{14}N
    NQR, 151, 159, 161, 163-165, 176-178, 323-326
    NMR, 323-344
    ENDOR, 308, 315, 317, 319-321, 323, 325, 330-337
Negative
    frequencies in F.T., 18-19, 296, 372
    ion cyclotron resonance, 133
    muon, 346, 382
Nernstian process, 487, 528
Network theory, 187
Noise
    detection of, 435-436
    effect on lineshape analysis, 114, 355-357
    flicker (1/f), 468
    fluctuation (modulation; scintillation), 438
    level detection, 71
    photon, 335-360, 436-437
    power as a radiation source, 13-15
    pseudorandom, 468, 472-474
    white, 229, 405, 468, 472-474
NQR, see FT-NQR
Nuclear quadrupole moment, see Quadrupole
Numerical F.T., see F.T., discrete
Nyquist frequency, 39-41, 131, 229, 395-396, 456, 466-467

Odd function, 16-17
Odd-harmonic waveforms, 472-477
Offset, see Baseline
Optical throughput, see Throughput
Overlap of peaks, 99, 103-104, 107, 115, 117, 121
Overmodulation, see Modulation broadening

Oxidation potential, see Potential, oxidation

^{31}P, see Phosphorus-31
Paramagnetic species, (see also ESR),
    FT-NMR of, 323-344
Peak analysis (see also DISPA), 70-72
    distributions, see Gaussian; Log-Gaussian; Chemical exchange; Inhomogeneity
    identification, 71
    -picking algorithm, 71-72
    position, 71, 99
    -to-peak separation, dispersion vs. derivative, 4
    width, 72, 99, 394-395
Permittivity,
    complex, 185-197
Pharmaceutical assay, see Drug assay
    characterization, 180, 297-298
Phase angle cotangent, 461, 466, 471, 478
    correction, 166, 226, 287, 317, 358, 447
    cycling, 281, 282, 293, 296
    detector, 222, (see also Detector, phase)
    lock, 222
    memory decay, see T_2
    modulation, 275, 296
    misadjustment, 107
    shift,
        from filter, 226
        from time-delay, 166-175, 196, 285, 546
Phosphorus-31 NMR, 118, 283, 284
Photoacoustic spectroscopy, 414-415, 440, 450
Photomultiplier, see Detector, photomultiplier
Pions, 346
Plane-polarized radiation, 209
Poisson statistics, 355, 358
Polarization
    complex, 219
    electric rotational, 214, 244-263
    muon, 349-350, 368, 376
Polarized radiation, see Circular-, Linear-, and Plane-polarized

INDEX

Polarography, a.c.
    fundamental, 453-454, 501
    second harmonic, 501
  d.c., 453, 501
  pulse, 453, 485, 497-499
  square-wave, 453, 485
Polymers (see also Protein, RNA),
  DISPA analysis, 118
Polynomial
  fit-subtract-transform, 481
  fit to baseline, 73
  fit to transient, 316
  primitive, 60-61
Porphyrins, 341
Positive ion mass spectroscopy, 133-144
  muon, 347
Potential, electrostatic, at nucleus, 148-150
  oxidation, 528
  redox, 530
  reduction, 528
Powders, 180, 341
Power broadening, 107, 216, 218, 323
  noise, 14-16, 360
Power spectrum, 9-16, 167-178, 262, 268, 352, 358, 361, 371, 375, 377, 379, 381, 463
Precession, muon, see Mu
  nuclear, see Larmor
Precision of spectral line shape analysis, 114
  of spectral measurements, 45-59, 353
Pressure broadening (see also Doppler broadening), 129, 257
Protein, stellacyanin, 314-315
  viral coat, 118
Proton NMR, see ^{1}H NMR
Pseudorandom noise, see Noise, pseudorandom
  as radiation source, 16
  sequence, 62
Pulse sequence,
  rotational, 265
  spin-echo ENDOR, 307, 312
  two-dimensional NMR, 273-286
Pulse width, 221, 273, 275
Pulsed decoupling, 273-286
  excitation, 158
  molecular nozzle, 235, 241-244, 265-269

Q-value, 237-240, 248, 254-255
Quadrature phase detection, 88, 361-363, 454
Quadrupole coupling constant, 151-153, 268, 309, 329, 340
  moment, 150, 378
  nuclei
    experiments, 330-342
    NMR of, 326-330
    theory of, 326-330

Radar, 181
Radiation damage, 180
Radiofrequency, see RF
Raman spectroscopy
  baseline-flattening in, 22-25
Random noise
  as radiation source, 13-15
  effect on lineshape analysis, 114
Rapid-scan excitation, see Frequency-sweep
Rate constants, see Chemical exchange
  electroactive reaction, see Kinetics
  ion-molecule reaction, 136-144
Reaction kinetics, see Kinetics; Chemical exchange
Real spectrum, see Absorption-mode
Real-time averaging, 225
Reconstruction of waveforms in FT-ENDOR, 318-320
Redfield relaxation theory, 285
Redox reaction, 528-529, 545, 499-507
Reduction potential, see Potential, reduction
Reflection spectroscopy, 531-532
Reflectometry, time-domain, 193-195
Relaxation, Debye, 186
  dielectric, see Dielectric relaxation
  distribution, 105, 107, 118
  longitudinal, see T_1
  mechanical, 5
  mechanisms, see DISPA
  Mu, 350
  relation to line width (see also T_1, T_2), 2-3
  spin-lattice, see T_1
  spin-spin, see T_2
  time, 101, 128
  transverse, see T_2
  wall, 257

Remote sensing, 153, 157
Reproducibility, see Precision
Resistance
 Faradaic, 455
 non-Faradaic ohmic, 454, 457
Resolution
 elements, 33-42, 227-228
 enhancement, 20-21, 69, 274
 vs. signal-to-noise ratio, 12, 20, 230-234, 327
Resolution, frequency, 12
 FT/IR, 398, 427-430, 442
 Rayleigh criterion, 394
 time, 12
Retardation, 391, 398, 404, 424-425, 427, 439
Retroreflector, 400, 441-442, 445
Reverse transform, see FT, inverse
RF detection, 125-129
 dielectric relaxation, 204-205
 spectroscopy, 148
Ribosomes, P-31 NMR of, 118
Ringdown after pulse, 257
RNA, see t-RNA; Ribosomes
Ribonucleic acid, see t-RNA
Rotating frame
 microwave, 213
 NMR, 284
Rotational correlation time, 107, 329-330, 340
Rotational spectroscopy, see FT rotational spectroscopy

S-matrix
 code, 49, 58
 construction of, 62
 cyclic, 59
 -transform, 63
Sample spinning, NMR, 115
Sampling, discrete, 37-42, 108-109
 frequency, 40, 225, 395-397
 line, 184, 187
 optics, FT/IR, 401-402, 404-405
 theorem, 40, 131, 225, 395
Saturation, DISPA detection of, 107
 microwave, 216, 218, 232
 recovery, see T_1
Scalar coupling, see J-coupling
Scanning spectrometer, see Continuous-wave
Selection rules
 NQR, 151-153
Semiconductor impurities, 374, 415-418

Semi-derivative, 501
Sensitivity, see Signal-to-noise ratio
Shift register sequences, 60-61
 buffer memory, 131-132
Short-circuited sample (dielectric measurements), 191
Shuffling in FT data handling, 88
Signal averaging, 34, 130, 196, 226-227, 266
Signal-to-noise ratio (see also Weighing designs), 221, 230-234
 enhancement, 20-21, 69
 in DISPA analysis, 114
 in FT/IR, 406-407, 434-438
 in magnetic relaxation, 327, 332, 341
 in μSR, 355-360
 in time domain vs. frequency domain, 363-367
 vs. resolution, 12, 20, 132
Silica gel, adsorption on, 117
Silicon impurities, 416-418
SINC function, 10-12, 198, 392, 430
Sine, see Look-up table
Sine FT, see FT, sine
Single-channel detectors, 33-34
Smoothing, 203-204 (see also Baseline flattening, Digital filtering, Signal-to noise enhancement, and Window smoothing)
S/N, see Signal-to-noise ratio
Solid angle, FT/IR, 398, 427-432
Solid-state samples, EPR, 380
 IR, 415
 μSR, 374-380
 NMR, 117, 274, 341
 NQR, 148, 179-180
Source optics, FT-IR, 397-399, 423
Spatial imaging
 by Hadamard IR, 65
 by NMR, 274
Spectral analysis, DISPA, 99-123
 segment extraction, 131-132
Spectrum, absolute-value, see Spectrum, magnitude
 absorption, 1-6, 99-102, 109
 complex, 6
 dispersion, 1-6, 101-102
 imaginary, 6
 noise, 13-15
 power, 9
 real, 1-6

INDEX

Spin-coupling, scalar (see also J-coupling), 268
Spin-density, 333-334, 373
Spin echoes
  ENDOR, 307, 310
  μSR, 347
  NMR, 274-275, 282, 297
  NQR, 153, 158-165, 169, 171, 173, 175, 178
Spin Hamiltonian, see Hamiltonian, spin
Spin-labels, EPR, 121
Spin-lattice relaxation, see T_1
Spin-lock spin echo, 159-160
Spin-spin relaxation, see T_2
Square wave, see Weight function, rectangular
Standing wave, see Fabry-Perot cavity
Stark modulation, see Modulation, Stark
Steady state response (see also Continuous-wave), 1-6
Step apodization, see Weight function, rectangular
Step voltage pulse, 186, 468, 543
Steroids, 279, 280, 507-522
Stochastic excitation, 16, 468
Strain determination by NQR, 180-181
Strong off-resonance comb, see Spin-echo, NQR
Supercon magnets, 287
Superheterodyne detector, see Detector, superheterodyne
Superhyperfine splittings, see ENDOR
Superregenerative NQR detector, 153-154
Supersonic nozzle, 241-244, 265-269
Surface studies,
  electrode, 499, 527-533, 539
  solid, 117
SYDAGES, 457-463

T_1, ESR, 323
  ICR, 329, 341
  NMR, 329, 341
  NQR, 154
  rotational, 215, 257
T_2, ESR, 325-326
  μSR, 350-351
  NMR, 297, 329, 340
  NQR, 154, 161
  rotational, 208, 215, 257

Table, memory, see Memory
  sine-cosine, see Look-up table
Tailored excitation, 468
Tank circuit, NQR detector, 154
Tape, magnetic, see Memory, magnetic tape
Temperature measurement by NQR, 180
Throughput advantage, 42, 407, 415, 417, 422, 432-434, 437, 449
  optical, 407
Time-delay, relation to phase-shift, see phase-shift
Time-domain reflectometry, 183-206
  and frequency-domain response, see F.T., relation between time and frequency-domain
  -resolved admittance measurements, 462
  response, 7-8 (see also Transient, time-domain)
Transfer-RNA, see t-RNA
t-RNA, 23
Transient, time-domain signal, 7-8, 153, 219, 227, 232
  relation to frequency-domain FT, 38-39, 154, 167, 263, 267, 462, 497, 545
Transition frequency,
  μSR, 368, 370-371, 380
  NQR, 151-153, 309
Transition metal compounds, 137-144
Transition probability, 219, 310
Transmission line theory, 185-197
Trapped-ion cell, 129
Traveling wave, see Waveguide
Truncation of data set (see also Filtering, digital) 30, 199-200, 392
Tunnel diode generator, 184
Two-dimensional FT-ENDOR, 307-322
  F.T. NMR, 271-305
  Hadamard IR imaging, 65
  NMR imaging, see Zeugmatography

Ultra-high resolution
  FT-ICR, 132-133
  FT-rotational spectroscopy, 233-234, 268
  FT-UV-VIS, 442, 445-449
Uncertainty principle (see also Leakage), time/frequency, 12
Undersampling, 395-397

Van der Waals molecules, 268
Vector,
  electric dipole, 209, 214
  electric field, 185, 209, 247-251
  Maxwell displacement, 185
Virus, coat protein, 118
Virtual memory, see Memory, virtual
Voltammetry
  cyclic, 461, 483, 485, 494-497, 501, 528-529, 543
  linear sweep, 501, 543

Wave, standing, see Fabry-Perot
  traveling, see Waveguide
Waveforms for FT-FAM, 468, 472-474
Waveguide, 207
Weighing designs (see also Fourier and Hadamard codes), 47
  A-optional, 54
  D-optional, 55
  spectrometers, 48-49

Weight function, 10-11
  cosine bell, 316, 356
  exponential, 13-15, 20-21, 256, 288, 296, 299, 355-357, 363, 372
  rectangular, 30, 356, 392, 426-427, 478-480
  triangle, 393, 428
Weight-on-a-spring
  massless, 5
  steady-state, 1-6, 101
  transient, 7-8
Window, see Memory
Window smoothing, 316-318

Zeeman effect, 234 (see also Hamiltonian, spin
Zero-filling, 40
  EPR, 108
  FAM, 481-485, 500
  μSR, 356, 364
  NMR, 118, 301
Zeugmatography, 274